Colloids and Interface Science Series
Volume 3

Colloid Stability and Application in Pharmacy

Edited by
Tharwat F. Tadros

1807–2007 Knowledge for Generations

Each generation has its unique needs and aspirations. When Charles Wiley first opened his small printing shop in lower Manhattan in 1807, it was a generation of boundless potential searching for an identity. And we were there, helping to define a new American literary tradition. Over half a century later, in the midst of the Second Industrial Revolution, it was a generation focused on building the future. Once again, we were there, supplying the critical scientific, technical, and engineering knowledge that helped frame the world. Throughout the 20th Century, and into the new millennium, nations began to reach out beyond their own borders and a new international community was born. Wiley was there, expanding its operations around the world to enable a global exchange of ideas, opinions, and know-how.

For 200 years, Wiley has been an integral part of each generation's journey, enabling the flow of information and understanding necessary to meet their needs and fulfill their aspirations. Today, bold new technologies are changing the way we live and learn. Wiley will be there, providing you the must-have knowledge you need to imagine new worlds, new possibilities, and new opportunities.

Generations come and go, but you can always count on Wiley to provide you the knowledge you need, when and where you need it!

William J. Pesce
President and Chief Executive Officer

Peter Booth Wiley
Chairman of the Board

Colloids and Interface Science Series
Volume 3

Colloid Stability and Application in Pharmacy

Edited by
Tharwat F. Tadros

WILEY-VCH Verlag GmbH & Co. KGaA

The Editor

Prof. Dr. Tharwat F. Tadros
89 Nash Grove Lane
Wokingham, Berkshire RG40 4HE
Great Britain

Library of Congress Card No.: applied for

British Library Cataloguing-in-Publication Data
A catalogue record for this book is available from the British Library

Bibliographic information published by the Deutsche Nationalbibliothek
The Deutsche Nationalbibliothek lists this publication in the Deutsche Nationalbibliografie; detailed bibliographic data are available in the Internet at http://dnb.d-nb.de

Printed in the Federal Republic of Germany
Printed on acid-free paper

Cover illustration Adam Design, Weinheim
Composition K+V Fotosatz GmbH, Beerfelden
Printing betz-druck GmbH, Darmstadt
Bookbinding Litges & Dopf Buchbinderei GmbH, Heppenheim
Wiley Bicentennial Logo Richard J. Pacifico

ISBN 978-3-527-31463-8

Contents

Colloids and Interface Science Series, Vol. 3
Colloid Stability and Application in Pharmacy. Edited by Tharwat F. Tadros

ISBN: 978-3-527-31463-8

Preface

Volume 3 of this series is dedicated to the subject of colloid stability and some of the colloid and interface applications in Pharmacy. The volume consists of thirteen chapters that cover diverse topics. Only three chapters are directly relevant to applications in Pharmacy: Chapter 2 describes the solubilization of poorly insoluble drugs by micellar solutions, whereas Chapter 13 gives an interesting application of nano-particles for drug delivery to the brain. These two chapters were dedicated to Professor A.T. Florence on the occasion of his 65th birthday. The third chapter that is relevant to application to Pharmacy is that on controlling the stability of liposomal colloids. This chapter was dedicated to Professor D. Exerowa and Professor D. Platikanov on the occasion of their 70th birthday. The remaining ten chapters cover wide range of topics: Chapter 1 describes the preparation of iron oxide nano-particles. This gives a comprehensive review of 55 pages on the subject. Chapter 4 describes a mathematical model of coagulation and flocculation. Chapter 5 deals with emulsion stability and interfacial properties. It gives a good account on the correlation between emulsion stability and physicochemical parameters. Chapter 6 describes wetting film dynamics and stability. The mechanism of film rupture is well described. Measurement of surface forces using the atomic force microscope is reviewed in Chapter 7, whereas Chapter 8 deals with a specific topic of ion specificity in colloidal systems. The stabilization of thin films, emulsions and bifluid gels with surface-active solid particles is reviewed in Chapter 9 with particular reference to the use on nano-particles. Chapter 10 deals with the DLVO and structural forces in nanofilms containing polyelectrolytes, whereas Chapter 11 describes the association between polyelectrolytes and oppositely charged surfactants both in bulk solution and at the solid/liquid interface. The use of nonionic micelles for stabilization of foams, emulsions and suspensions is described in Chapter 12.

I should mention that ten of the thirteen chapters in this volume are dedicated to Professor Exerowa and Professor Platikanov. Their biography was given in Volumes 1 and 2 of this series and hence this is not reproduced in this volume. As mentioned above two chapters are dedicated to Professor Florence. The only independent chapter is that by Ignác Capek (Chapter 1) which I decided to include in this volume due to its important application in nano-technology.

Colloids and Interface Science Series, Vol. 3
Colloid Stability and Application in Pharmacy. Edited by Tharwat F. Tadros

ISBN: 978-3-527-31463-8

I would like to thank all the authors for their dedication in producing these excellent reviews, which made my editing task fairly easy. I would like to thank the staff of Wiley-VCH for producing this volume quickly.

Wokingham, April 2007

Tharwat Tadros
Editor of the Series

List of Contributors

Karine Andrieux
Université Paris-Sud
Faculty of Pharmacy
Laboratory of Biopharmacy
and Pharmaceutical Technology
UMR CNRS 8612
5, rue J. B. Clément
92296 Châtenay-Malabry
France

David Attwood
University of Manchester
School of Pharmacy and
Pharmaceutical Sciences
Oxford Road
Manchester M13 9PT
United Kingdom

Colin Booth
University of Manchester
School of Chemistry
Manchester M13 9PT
United Kingdom

Ignác Capek
Slovak Academy of Sciences
Polymer Institute
Dúbravská cesta 9
84236 Bratislava
Slovakia
and
Trenčín University of A. Dubček
Faculty of Industrial Technologies
Ul. I. Krasku 30
02001 Púchov
Slovakia

Per M. Claesson
Royal Institute of Technology
Department of Chemistry
Surface Chemistry
10044 Stockholm
Sweden
and
Institute for Surface Chemistry
Box 5607
11486 Stockholm
Sweden

Danièle Clausse
Université de Technologie
de Compiègne
Département Génie des Procédés
Industriels
BP 20529
60205 Compiègne Cedex
France

Colloids and Interface Science Series, Vol. 3
Colloid Stability and Application in Pharmacy. Edited by Tharwat F. Tadros

ISBN: 978-3-527-31463-8

Martien A. Cohen Stuart
Wageningen University
Laboratory of Physical Chemistry
and Colloid Science
P.O. Box 8038
6700 EK Wageningen
The Netherlands

Patrick Couvreur
Université Paris-Sud
Faculty of Pharmacy
Laboratory of Biopharmacy
and Pharmaceutical Technology
UMR CNRS 8612
5, rue J. B. Clément
92296 Châtenay-Malabry
France

Ferenc Csempesz
Eötvös Lorand University
Laboratory of Colloid and
Supramolecular Systems
Institute of Chemistry
P.O. Box 32
1518 Budapest
Hungary

Christine Dalmazzone
Institut Français du Pétrole
1 et 4 Avenue Bois-Préau
92852 Rueil-Malmaison
France

Delphine Daniel-David
Université de Technologie
de Compiègne
Département Génie des Procédés
Industriels
BP 20529
60205 Compiègne Cedex
France

Andra Dedinaite
Royal Institute of Technology
Department of Chemistry
Surface Chemistry
10044 Stockholm
Sweden
and
Institute for Surface Chemistry
Box 5607
11486 Stockholm
Sweden

Marcel Giesbers
Wageningen University
Laboratory of Organic Chemistry
P.O. Box 9101
6700 HB Wageningen
The Netherlands

François Gomez
Université de Technologie
de Compiègne
Département Génie des Procédés
Industriels
BP 20529
60205 Compiègne Cedex
France

Katarzyna Hänni-Ciunel
Technische Universität Berlin
Stranski-Laboratorium für
Physikalische und Technische Chemie
Straße des 17. Juni 112
10623 Berlin
Germany

Hyun Ryoung Kim
Université Paris-Sud
Faculty of Pharmacy
Laboratory of Biopharmacy
and Pharmaceutical Technology
UMR CNRS 8612
5, rue J. B. Clément
92296 Châtenay-Malabry
France

J. Mieke Kleijn
Wageningen University
Laboratory of Physical Chemistry
and Colloid Science
P.O. Box 8038
6700 EK Wageningen
The Netherlands

Regine v. Klitzing
Technische Universität Berlin
Stranski-Laboratorium für
Physikalische und Theoretische
Chemie
Straße des 17. Juni 124
10623 Berlin
Germany

Ljepsa Komunjer
Université de Technologie
de Compiègne
Département Génie des Procédés
Industriels
BP 20529
60205 Compiègne Cedex
France

Frans A. M. Leermakers
Wageningen University
Laboratory of Physical Chemistry
and Colloid Science
Dreijenplein 6
6703 HB Wageningen
The Netherlands

Philippe Letocart
Saint-Gobain-Sekurit
Central Research Laboratories (ZAF)
Thin-Films Department
Glasstrasse 1
52134 Herzogenrath
Germany

Johannes Lyklema
Wageningen University
Laboratory of Physical Chemistry
and Colloid Science
Dreijenplein 6
6703 HB Wageningen
The Netherlands

Róbert Mészáros
Eötvös Loránd University
Department of Colloid Chemistry
P.O. Box 32
1518 Budapest 112
Hungary

Alex D. Nikolov
Illinois Institute of Technology
Department of Chemical
and Environmental Engineering
Chicago, IL 60616
USA

Christine Noïk
Institut Français du Pétrole
1 et 4 Avenue Bois-Préau
92852 Rueil-Malmaison
France

Isabelle Pezron
Université de Technologie
de Compiègne
Département Génie des Procédés
Industriels
BP 20529
60205 Compiègne Cedex
France

István Puskás
Eötvös Loránd University
Department of Colloid Chemistry
P.O. Box 32
1518 Budapest 112
Hungary

Dan Qu
Max-Planck-Institut für Kolloid- und Grenzflächenforschung
14424 Potsdam
Germany

Boryan Radoev
University of Sofia
Department of Physical Chemistry
1 James Bourchier Boulevard
1164 Sofia
Bulgaria

Daniel Rapoport
Max-Planck-Institut für Kolloid- und Grenzflächenforschung
14424 Potsdam
Germany

Venkataramana Runkana
Columbia University
NSF Industry/University Cooperative Research Center for Advanced Studies in Novel Surfactants
School of Engineering and Applied Science
New York, NY 10027
USA
and
Tata Research Development and Design Centre
54-B Hadapsar Industrial Estate
Pune 411013
India

Ponisseril Somasundaran
Columbia University
NSF Industry/University Cooperative Research Center for Advanced Studies in Novel Surfactants
School of Engineering and Applied Science
New York, NY 10027
USA

Klaus Werner Stöckelhuber
Leibniz Institute of Polymer Research Dresden e.V.
Hohe Straße 6
01069 Dresden
Germany

Roumen Tsekov
University of Karlsruhe
Institute of Physical Chemistry
76131 Karlsruhe
Germany

Imre Varga
Eötvös Loránd University
Department of Colloid Chemistry
P.O. Box 32
1518 Budapest 112
Hungary

Orlin D. Velev
North Carolina State University
Department of Chemical and Biomolecular Engineering
Raleigh, NC 27695
USA

Krassimir P. Velikov
Unilever R & D Vlaardingen
FSD, UFHRI
Olivier van Noortlaan 120
3133 AT Vlaardingen
The Netherlands

Darsh T. Wasan
Illinois Institute of Technology
Department of Chemical and Environmental Engineering
Chicago, IL 60616
USA

1
Iron Oxide Nanoparticle Dispersions*

Ignác Capek

Abstract

The various approaches to the synthesis of magnetite nanoparticles, such as microemulsion, miniemulsion, precipitation, homogeneous and heterogeneous particle nucleation, sonochemical synthesis and sol–gel reactions, are discussed. The microemulsion approach prepares metallic nanoparticles by simple mixing of inverse microemulsions containing a reducing agent and a metal precursor. The precipitation (particle nucleation) of the reduced metal precursor is based on the supersaturation of solution by precursors (metal salts). A typical homogeneous nucleation synthesis consists of a one-step process in which precursor(s), stabilizer(s) and other additives are stirred and heated in an oil or water continuous phase. The heterogeneous nucleation of metal particles consists of a several-step process in which the seeded particles are generated in the first step and growth occurs in the subsequent steps. The homogeneous and heterogeneous nucleation methods differ greatly in their ability to synthesize iron nanoparticles of different sizes. In seed synthesis, larger particles are formed from small seeded iron nanoparticles (the seeds) mixed with precursors and heated in the presence of stabilizers, costabilizers and reducing agents. This approach can be used to prepare magnetic metal–iron oxide nanoparticles and iron–metal (alloys) in various media. The cubic spinel-structured MFe_2O_4 or $MOFe_2O_3$ nanoparticles, for example, represent a well-known and important class of iron oxide–metal materials. By adjusting the chemical identity of M^{2+}, the magnetic configurations of MFe_2O_4 can be molecularly engineered to provide a wide range of magnetic properties. Chemically prepared FePt (alloy) nanoparticles have attracted great interest recently because of their ease of synthesis, chemical stability and potential applications in high-density data storage and high-performance permanent magnets. The preparation of highly charged, monodisperse superparamagnetic colloidal spheres is also summarized. These superparamagnetic, monodisperse and charged polymer particles, containing nanoscale iron

* A list of symbols and abbreviations can be found at the end of this chapter.

Colloids and Interface Science Series, Vol. 3
Colloid Stability and Application in Pharmacy. Edited by Tharwat F. Tadros

ISBN: 978-3-527-31463-8

oxide nanoparticles, can be synthesized through (mini)emulsion polymerization. The nanoparticle materials exhibit very interesting electrical, optical, magnetic and chemical properties, which cannot be achieved by their bulk counterparts. Nanoparticles can have unique properties owing to their small size. These effects are most pronounced for very small particles, smaller than about 10–20 nm and in most cases, essentially vanish by the time 40–50 nm is reached. Especially iron oxide and ferrite are important as ferrofluids. Due to the large surface area-to-volume ratio and magnetic dipole–dipole attractions, magnetic nanoparticles tend to aggregate. Therefore, the stabilization of a dispersion of magnetic nanoparticles is one of the critical steps in producing stable magnetic fluids. The reaction mechanism of hydrophobically or hydrophilically modified iron oxide particles is somewhat complex. The stabilizer formed by the imide linkage between surface-active additives allows adequate particle stabilization in dilute solution. The stabilization of small nanoparticles can also be achieved using different stabilizers including the organic ions. Organic ions are suggested to affect the formation of metal oxides or oxyhydroxides through the chelation and the adsorption of these ions on the nuclei.

1.1 Introduction

The synthesis of magnetic materials on the ultra-nanoscale is a field of intense study. Much of the research has been focused on preparing particles or particle systems and measuring the macroscopic properties of the particles, often magnetic properties such as magnetization and coercivity. Various methods have been reported for the synthesis of iron oxide nanoparticles carried out at low and high temperatures, in the presence and absence of surface-active additives, under ultrasound irradiation, etc. The development of uniform nanometer-sized particles has been intensively pursued because of their technological and fundamental importance. Nanoparticle materials often exhibit very interesting properties, which cannot be achieved by their bulk counterparts [1–3]. Especially iron oxide and ferrite have attracted a great deal of interest as ferrofluids [4]. In 1980, Massart obtained stable aqueous alkaline and acidic magnetite colloids by free precipitation and surface treatment [5]; the procedures were improved on preparing maghemite, manganese and cobalt ferrite magnetic fluids [6]. Maghemite (γ-Fe_2O_3) nanoparticles are especially attractive due to their technological and fundamental importance, such as for information storage, magnetic resonance imaging contrast agents and macroscopic quantum tunneling associated with size quantification and electronic quantum confinement effects [4, 7, 8].

Iron-based nanoparticles have been of interest in fundamental studies of monodomain magnets and have been studied or used for a wide range of applications including magnetic recording media [9], ferrofluids [10], magnetic cell separation [11], magnetic resonance imaging (MRI), contrast agents [12] and environmental remediation [13]. Historically, there have been three main challenges: controlling the particle size, achieving monodispersity and minimizing the amount of oxidation. These have been scrupulously controlled at the same time only under high-vacuum conditions using a cluster beam source [14], an approach suitable for fundamental studies requiring only minute amounts of material. The particle properties are extremely history dependent and the process is complicated and requires great efforts to optimize it. Smaller nanoparticles tend to have properties that represent wide departures from bulk properties, but as we move to sizes of several tens of nanometers in diameter, these properties gradually revert to the bulk properties. The size-dependent properties are not wholly proportional or inversely proportional to the particle size. The coercivity of small particles is zero but increases with increasing size and reaches a maximum in the size range of tens on nanometers, then declines and approaches bulk coercivity [15, 16].

Magnetic fluids are usually colloidal suspension of stabilized ferri- or ferromagnetic single-domain nanoparticles. The unique and novel size-dependent properties displayed by some nanocrystals initiated the current worldwide intense research on nanomaterials [17, 18]. Recently, keen interest has extended into controlling the shape of nanomaterials and also into understanding the correlations between a material's properties and its shape. The shape of nanomaterials can be as vital as the size in determining the uniqueness and novelty of material properties [19, 20]. Assembly of nano components is a key process in building devices and the shape of nano building blocks and related properties is crucial for assembly and device designs. Successes in shape control of nanocrystals have been reported in the syntheses of metals [21] and semiconductors [22]. Clearly, it is desirable to develop strategies for shape-controlled syntheses of metal particles possessing valuable properties, especially magnetic characteristics. Monodisperse metal oxide nanocrystals have been synthesized by several nonhydrolytic methods using precursors of organometallic and coordination compounds [23, 24]. It has been demonstrated that seed-mediated growth is an effective route for size-controlled synthesis of magnetite nanocrystals [24].

There are a number of reasons for the great interest in iron nanoparticles. Recently, novel uses of magnetic fluids in biomedical area increase significantly due to their novel physical properties. Generally, the average diameter of these magnetic particles is in the range 7–20 nm. Concerning medical applications, it is important to synthesize smaller magnetic particles ($d<5$ nm) dispersed in a physiological medium (pH=7). Dispersed in or encapsulated by appropriate polymeric matrixes, magnetic nanoparticles can be used in cell separation [25, 26], contrast agent techniques [27, 28] and advanced clinical applications such as cancer diagnosis and treatment. However, due to their large surface area-to-volume ratio and magnetic dipole–dipole attractions, magnetic nanoparticles tend to aggregate. This will limit the effect of magnetic fluids in these applications. The stabilization of a dispersion of magnetic nanoparticles is one of the critical steps in producing a stable magnetic fluid. Furthermore, magnetic nanoparticles are of great interest today due to their useful magnetic materials. Iron has the highest room-temperature saturation magnetization (M_{sat}) of any magnetic element and has a high Curie temperature (T_C): M_{sat} (A m^2 kg^{-1})/metal 222/Fe, 162/Co, 57/Ni and T_C (K)/metal 1043/Fe, 1388/Co, 627/Ni [29, 30].

Iron is a very soft magnetic material and also has a low magnetocrystalline anisotropy. Sufficiently small magnetic nanoparticles show superparamagnetic behavior and the maximum volume of a particle that can be superparamagnetic at a given temperature varies directly with the magnetocrystalline anisotropy. Iron nanoparticles are very reactive, specifically with water and oxygen. This general weakness causes the rapid and complete oxidation of iron nanoparticles. Maintaining iron nanoparticles in its zerovalent state generally limits it to applications where water and oxygen are largely excluded or it is maintained in a reducing atmosphere. The oxidation of iron particles can be decreased by a coated thin film of some metals. This condition prevents fusion of the nanoparticles during high-temperature reduction and then reduction in a hydrogen atmosphere. The iron particles can be exposed to ethanol vapor carried by nitrogen gas to passivate the surface of the particles. Without this step, the iron would rapidly oxidize upon exposure to air, despite the previously applied, for example, alumina coating. This mild treatment was apparently sufficient to provide a passivating coating on the surface of the particles that prevented or greatly reduced the rate of oxidation. However, the extreme reactivity of iron nanoparticles can be beneficial in a nonoxidizing environment. Nanoparticles tend to be more reactive than larger particles in chemical reactions. The particles possess an enormous amount of energy with their high surface-to-volume ratios, which changes their reactivity [30]. These nanoparticles, however, tend to agglomerate in order to reduce their surface energy. Therefore, one of the main problems in producing stable magnetic fluid is to prevent such agglomeration during the synthesis and coating process. The presence of surface-active compounds, the organic anions, the ligands, bipolar compounds, etc. can prevent the agglomeration of particles. For example, double surfactant layers were usually used for the stabilization of magnetic particles [31]. The first layer of surfactant attached to magnetic particles by hydrophilic groups and the second layer helped dispersion in an aqueous carrier liquid via its hydrophilic groups. A basic weakness in this design is its dependence on van der Waals forces between the two layers and the secondary layer is very easily removed by polar organic solvents such as acetone.

Organic anions such as carboxylate and hydroxycarboxylate ions were observed to be very efficient stabilizers during particle formation of ferric oxides or oxyhydroxides [32, 33]. In particular, citrate ions are known to interfere with the formation and growth of these oxides [32, 34, 35]. For example, the size of the maghemite particles can grow to up to ca. 8 nm without citrate. Under the same reaction conditions but with citrate ions present, the size of the particles decreases and can be as small as 2 nm. Based on these studies, citrate ions have been used for controlling particle size in many experiments. This approach is convenient method for preparing well-dispersed magnetic colloids with an average size smaller than 5 nm. Unlike Massart's method [36], magnetic washing and acidity of the initial mixture are not required and magnetic colloids can be directly produced and exhibit dissimilar results [37]. These fluids are used for studies of phase transition phenomena and also allow ionic magnetic fluids stable in the presence of large amounts of salt to be obtained. They can also be used for incorporation in complex colloidal systems for which the interlamellar distance is small and which may accept only very small grains [38].

The synthesis of magnetic iron nanoparticles is designed in response to a particular type of the study. Generally, typical surfactants are too bulky to allow close interaction of iron nanoparticles. For example, broadly used surfactants contain ca. 18 carbon units, providing spacing between the particles on the order of 3 nm. At this distance, for example, magnetic interactions are so weak that particles tend to form hexagonally packed, two-dimensional arrays, which are structures that are not favored by strongly interacting dipoles. An additional problem with traditionally used surfactant systems is the deleterious effect that they have on the magnetic properties of the iron nanoparticles. This effect is particularly pronounced in the case of very small nanoparticles. The main goal of the synthesis is to find a small surfactant that produces as little degradation of the physical properties as possible [30].

1.2 Preparation of Iron and Iron Oxide Nanoparticles

1.2.1 General

Various methods have been reported for the synthesis of iron oxide nanoparticles, such as microemulsion, miniemulsion, precipitation and coprecipitation (homogeneous and heterogeneous particle nucleation) in the absence or presence of surface-active additives (stabilizers), sonochemical synthesis, mercury-based methods, sol–gel reactions, etc. Mercury-based methods were reported to produce many of the particles and confirmed the properties of single-domain magnetic particles [39]. However, mercury-based methods have been replaced with water- and organic-solvent-based methodologies. This change occurred long before concerns about the toxicity of mercury vapor and probably had more to do with the confounding effects of the substantial solubility of iron in mercury. Today, iron oxide nanoparticles are commonly produced via coprecipitation of ferrous (Fe^{2+}) and ferric (Fe^{3+}) ions by a base, usually NaOH or $NH_3 \cdot H_2O$, in an aqueous solution [28–30], or they may be made by thermal decomposition of an alkaline solution of Fe^{3+} chelate in the presence of hydrazine [40] and by sonochemical decomposition of a hydrolyzed Fe(II) salt followed by thermal treatment [41]. The disadvantage of these aqueous solution syntheses is that the pH of the reaction mixture has to be adjusted in both the synthesis and purification steps and the process towards smaller (<20 nm) monodisperse nanoparticles has had only very limited success. There are many different possible iron oxide structures, such as magnetite (Fe_3O_4), hematite (α-Fe_2O_3), maghemite (γ-Fe_2O_3) and ferric hydroxide (β-FeOOH), which can be prepared by similar or different methods [42].

Iron pentacarbonyl, [$Fe(CO)_5$], is a metastable organometallic compound whose facile decomposition has made it an extremely useful reagent. The fact that the standard enthalpy of formation of iron pentacarbonyl is only ~ 185 kcal mol^{-1} and that the five carbon monoxide subunits each have an enthalpy of formation of ~ 110.5 kcal mol^{-1} account for its tendency to dissociate [29]. The molecule is easy to decompose, but the decomposition reaction is extraordinarily complicated. The decomposition pathway of intermediate iron carbonyls and iron clusters is complex (Scheme 1.1). It is supposed that the intermediate iron compounds can catalyze the decomposition. A great many surfactants and other species can also catalyze the decomposition. The result is that the decomposition frequently changes the growth rate and even order during the course of a reaction [43].

The sonochemical decomposition of iron carbonyl in the synthesis of iron nanoparticles was reported by Suslick et al. [44]. As in the case of thermal decomposition, the reaction is mechanistically very complicated. Although the mechanism appears to be different from the thermal decomposition of iron carbonyl [45], synthetic design of a reaction is equally interesting. In particular, size control in the sonochemical production of iron nanoparticles is considered difficult [46], although some control over size has been demonstrated by varying the concentration of iron carbonyl [47]. Iron pentacarbonyl can also be thermally decomposed by the application of heat in the form of microwave irradiation to form polymer-stabilized iron nanoparticles [48] or carbon-supported iron nanoparticles [49]. Although there are only a few examples of this type of synthesis, the heating of reactions through microwave irradiation is a rapidly growing area in particle synthesis.

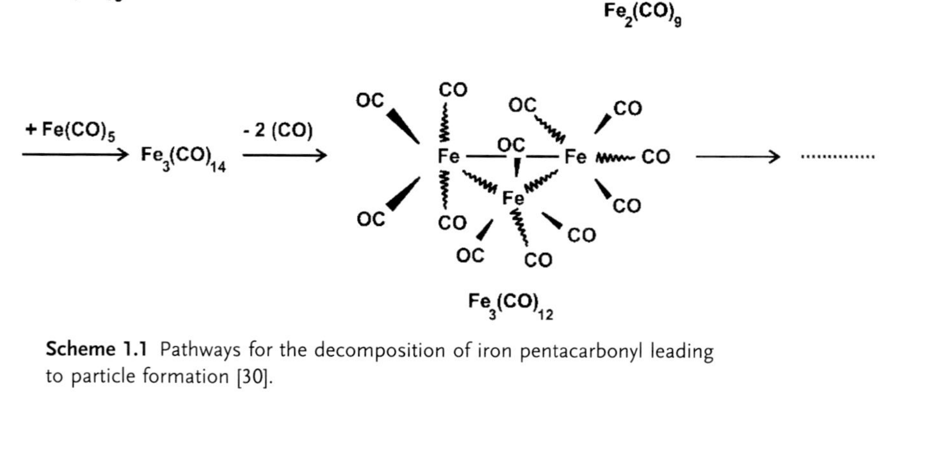

Scheme 1.1 Pathways for the decomposition of iron pentacarbonyl leading to particle formation [30].

One of the most suitable techniques for preparing ultrafine nanoparticles in a controlled way is the chemical coprecipitation method [8, 50]. Especially, Fendler et al. [51] and Stroeve et al. [52] reported a procedure for synthesizing Fe_3O_4 nanoparticles by bulk aqueous solution and chemical solution without using surfactant vesicles, respectively. These methods have the advantage that no surfactant needs to be removed from the nanoparticles before application in the precision industry and biomedical fields. Advances in the synthesis have demonstrated that direct decomposition of the Fe–ligand complex [53] or decomposition of $Fe(CO)_5$ followed by oxidation [4] can lead to high-quality monodisperse γ-Fe_2O_3 nanoparticles. Organic solution-phase decomposition of the iron precursor at high temperatures has been widely used in iron oxide nanoparticle synthesis.

Iron particles have also been prepared by mechanical milling. This method has been used to prepare ferrofluids containing iron nanoparticles in kerosene or similar organic liquids [10]. The particles were dispersed by the addition of a surfactant such as oleic acid to coat the particle surfaces. This approach yields a significant distribution of sizes and shapes, as do gas-phase methods [54], and the magnetizations were fairly large. Iron nanoparticles can be produced through high-energy milling [55] and although this is not the most common method of forming iron nanoparticles, it does have advantages. The primary advantage is the ease with which this method can be scaled up. The drawback, however, is that the product tends to be very polydisperse in size and irregular in shape. In addition to standard milling procedures, composites can be produced directly by co-milling iron and an additive material, such as alumina [56]. Additionally, reactive milling can produce iron *in situ* by co-milling magnetite and aluminum.

Another example of thermally unstable organometallics is $[Fe\{N(SiMe_3)_2\}_2]$. This molecule has been thermally decomposed in the presence of oleic acid and oleylamine to yield iron cubes that are about 7 nm on each side [57]. The reaction is conducted under hydrogen and produces nanocubes that are fully reduced. This hydrogen atmosphere likely prevents the oxidation problems that may otherwise be caused by the production of water by the acid and amine. The thermally decomposition of perovskite ceramic in a reducing atmosphere yields single-crystal iron nanorods embedded in an $LaSrFeO_4$ matrix [58]. The iron nanorods were regularly spaced in the matrix.

Gas-phase-grown Fe particles can be used to prepare larger particles with an average diameter of 20 nm [54]. There are many aqueous solution methods to prepare relatively monodisperse iron oxide (Fe_3O_4 or γ-Fe_2O_3) nanoparticles, although the particle surfaces may still be irregular [5]. Although the magnetizations of these phases are lower [59], by fully oxidizing the particles their magnetic properties are stable over time – an important requirement for applications of magnetic particles.

1.2.2 Microemulsion

The syntheses of iron nanoparticles by micellar or microemulsion approaches are governed by the choice of surfactant and cosurfactants, which determines the range of solvents and possible cosolvents to be used. Ionic (e.g. quaternary ammonium-based surfactants) [60, 61] and non-ionic (e.g. polyether-based surfactants) [62] long-chain surfactants are commonly used in micellar syntheses. Iron salts can be reduced using any of a wide range of common reducing agents, including hydrazine [61], sodium borohydride [60, 63] and lithium borohydride [62], although there have been reports of boron being incorporated into the reduced iron with some of the borohydride-based reducing agents [62].

The microemulsion method (water-swollen inverse micelles) is commonly used for metal nanoparticle syntheses but is not usually applied to the synthesis of iron nanoparticles. Although zerovalent iron can be produced in the presence of water, even oxygen-free water can oxidize iron over time. Iron nanoparticles are frequently produced and handled in dry boxes free of water. However, some microemulsion syntheses of iron nanoparticles have been reported [60, 64]. The microemulsion approach combined with a subsequent reduction step was successfully used for the preparation of iron nanoparticles. The iron oxides prepared by the microemulsion are reduced to iron nanoparticles in the next step. During reduction, the iron oxide particles pass through intermediate states such as Fe_3O_4 (magnetite) and FeO (wustite) before becoming fully reduced iron. The industrial process begins with the precipitation of iron hydroxide needles, which are subsequently dehydrated to form iron oxide (hematite, α-Fe_2O_3) and then reduced by hydrogen gas [65].

The microemulsion method is a simple and powerful approach for obtaining ultrafine iron and iron oxide particles by means of chemical reactions [66]. As an example, in Fig. 1.1 some particular results for iron particles obtained in sodium bis(2-ethylhexyl)sulfosuccinate (AOT) microemulsions with a stopped-flow technique. Both the nucleation and growth processes can be clearly seen. Nucleation implies an increase in the number of scattering centers which is associated with the increase in the scattering intensity. In contrast, the growth of the particles is associated with a decrease in the scattered intensity.

The particles are kept apart because of the surfactant molecules attached to their surfaces. The presence of surfactant was confirmed by the X-ray spectrum. This analysis showed, apart from the iron peak, the S and Na peaks corresponding to the surfactant AOT. The size of Fe particles estimated by TEM is $\sim$4 nm, which is similar to the droplet size of the microemulsion employed ($\sim$3.4 nm). Some of results for the radius of particles (r_p) obtained by dynamic light scattering (DLS) and also by small-angle X-ray scattering (SAXS) for different microemulsions and different concentrations of reactants are shown in Table 1.1.

All these results show that the size of the particles obtained is in the range of the size of the droplets of the microemulsions used. Several parameters, such as concentration of reactants (especially surfactant), flexibility of the surfactant

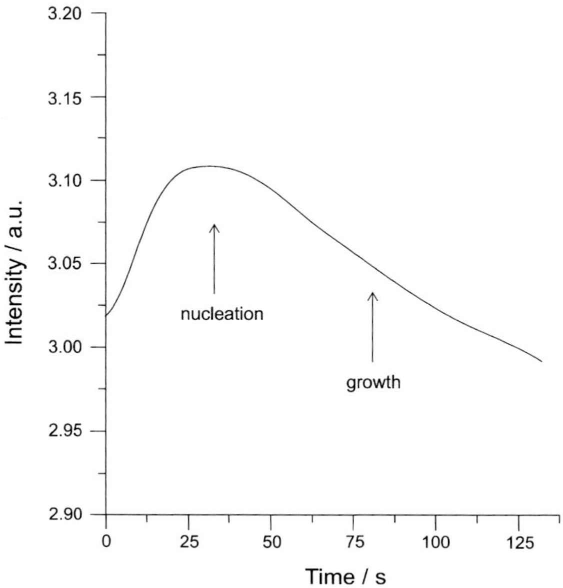

Fig. 1.1 Time course of the intensity obtained by SAXS during the formation of particles in microemulsions. Microemulsion: AOT–water–heptane, $W = 22$, [AOT]=0.05 M, $[FeCl_2] = 1.9 \times 10^{-4}$ M (reactant A), $[NaBH_4] = 8.8 \times 10^{-4}$ M (reactant B) [66, 67].

Table 1.1 Colloidal parameters of AOT microemulsions for different concentrations of reactants [66, 67].

Run No.$^{a)}$	[AOT] (mol dm^{-3})	$W^{b)}$	$[FeCl_2] \times 10^4$ (mol dm^{-3})	$[NaBH_4] \times 10^3$ (mol dm^{-3})	r_p (nm)
1	0.1	22	7.5	3.5	3.9
2	0.1	22	3.8	1.75	3.9
3	0.1	22	1.6	0.68	4.1
4	0.1	22	0.38	0.35	3.9
5	0.05	10	0.93	0.35	2.5
6	0.025	4.44	48	24	1.9
7	0.025	8.88	48	24	2.5
8	0.025	13.3	48	24	3.3
9	0.025	17.8	48	24	4.3
10	0.025	22.2	48	24	4.7
11	0.025	26.6	48	24	5.4
12	0.025	31.1	48	24	5.8

a) SAXS, runs 1–5; DLS, runs 6–12.
b) W=[water]/[surfactant].

film, and [water]:[emulsifier] ratio, may influence the final size of the particles [68]. The size of Fe magnetic particles has a large influence on their magnetic properties. For example, Fe particles with average size $r \approx 3.4$ nm show typical supermagnetic behavior due to their very small size. When the size is increased, a transition to the classical ferromagnetic behavior is observed. As another magnetic example is the production of a typical iron-rich spinel: magnetite. An aqueous solution–*n*-heptane–AOT ($W = 10$) microemulsion containing $FeCl_2$ and $FeCl_3$ was mixed with another similar microemulsion containing NH_4OH. The X-ray diffraction pattern of the separated (washed and dried) product indicates the presence of Fe_3O_4. The hysteresis loop at room temperature of the powders obtained shows that the particles are supermagnetic due to their small size ($r_p \approx 4$ nm).

Iron boride nanoparticles were prepared by the reduction of Fe(III) salt but the dependence of the average particle diameter on the water and salt concentrations was somewhat complex [69]. An aqueous solution–ethyl oleate–(labrasol–isostearic plurol) microemulsion containing $FeCl_2$ and $NiCl_2$ was mixed with another similar microemulsion containing $NaBH_4$ to obtain FeNi alloys. The magnetic properties of these samples measured in a magnetic field of 100 Oe at room temperature (grain size $\approx$ 30 nm $\approx$ droplet size) are magnetization $M_{mag} = 4.2$ emu g^{-1} and coercive field $H_c = 10$ Oe.

The preparation of metallic iron–gold nanoparticles takes advantages of two self-organizing processes [61]. First, the reverse micelle method is to synthesize metallic nanoparticles in the water pools of the reverse micelles. Metallic nanoparticles are synthesized in reverse cetyltrimethylammonium bromide (CTAB)–butanol–octane–water microemulsion solution using hydrazine as a reducing agent. CTAB act as a surfactant and butanol as cosurfactant. The metallic iron particles grow to fill the centers of the micelle and minimal aggregation of iron occurs since the organic portions of the micelle keep particles separate. In a second self-organizing process, since gold and sulfur form spontaneous bonds, the gold coating will direct the ferromagnetic nanoparticles into an ordered array on the surface of a thiol-functionalized substrate to produce a thin film (Fig. 1.2). Since the iron particles have formed inside the micelle, an aqueous solution of $HAuCl_4$ is added to the iron–CTAB mixtures. Addition of the aqueous gold solution increases the size of the reverse micelle and the Au(II) is reduced to Au(0) via excess hydrazine. Because gold and iron grow with complementary crystal structures, the metallic gold forms a coating on the outer surface of the iron particles. The gold shells on the iron particles provide functionality and thin films of the gold-coated particles have been made by self-assembly reactions between the gold surface of the particle and thiol-functionalized substrates. Multilayers of gold-coated iron nanoparticles can be formed by additional self-assembly reactions. An X-ray diffraction pattern obtained on a powder sample of gold-coated iron shows that only gold and iron exist in the sample and the 2.5-nm thick gold coating effectively prevents any oxidation of the metallic iron core. No aggregation is observed in a sample containing a thiol-functionalized grid.

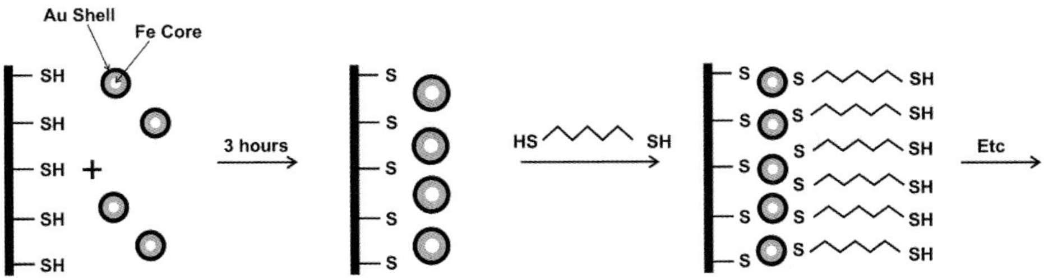

Fig. 1.2 Formation of mono- and multilayers of Au-coated Fe nanoparticles on a thiolated substrate based on chemical self-assembly [61].

1.2.3
Precipitation

The (co)precipitation (particle nucleation) of particles is based on the supersaturation of solution by reactants such as precursors (metal salts), reducing agent, stabilizers and additives. Increased solubility of a component in the continuous phase can be reached by the rising in temperature. The supersaturation state is then achieved by a reduction in temperature. Generation of supersaturation through *in situ* chemical reactions by converting highly soluble into less soluble chemicals is a good example of this approach.

A typical homogeneous nucleation synthesis consists of a one-step process in which precursor(s), stabilizer(s) and other additives are stirred in the oil or water continuous phase and then heated [4]. For example, Fe particles were prepared by simple stirring of oleic acid (OA) in dioctyl ether, then the solution was heated and a solution of $Fe(CO)_5$ precursor was added [70].

The heterogeneous nucleation of metal particles consists of several-step processes [71]. In a typical heterogeneous nucleation, the first step is the formation of primary or small (seed) metal particles and then their growth is achieved by the addition of precursor(s), stabilizer(s) and additives before the reaction mixture is heated. In the synthesis of 7-nm diameter particles, for example, platinum seed clusters were formed by the polyol reduction of platinum acetylacetonate, $Pt(acac)_2$, by 1,2-hexadecanediol in a dioctyl ether solution containing surfactant [1:1 molar ratio of oleic acid (OA):oleylamine (OAm)] [70]. The solution was heated while stirring and $Fe(CO)_5$ was added. The color change indicated the growth of Fe on the Pt seed clusters. An additional stage of growth of particle size can again follow. This can be obtained by the addition of precursors [$Fe(CO)_5$, $M_{met}(II)(acac)_2$, $M_{met}(III)(acac)_3$, etc.] while mixing and heating to higher temperature. To make smaller particles, less precursor (iron pentacarbonyl, metal acetylacetonate, etc.) was used. To increase the particle size, larger amounts of precursor were used and the heating times were longer during both stages. Larger particles were also prepared using reduced amounts of the metal seeding agent. Additional OA–OAm can be used to improve the dispersion stability. The surfactants used can be OA, OAm, OA–OAm, 1:1 molar mixtures of shorter-chain analogues, hexanoic acid and hexylamine, propanoic acid and propylamine and tributylphosphine and 1,2-hexadecanediol. When OA and OAm are mixed to make OA–OAm, heat is evolved. This is attributed to imide formation when the carboxylic acid group of OA reacts with the amine group of OAm, as in the formation of a peptide. Although both OA and OAm are liquid at room temperature, a mixture of the two heated to improve homogeneity solidifies when cooled. To exchange surfactants during the solvent exchange, the concentration of the new surfactant was chosen to match that of the original OA–OAm when the particles were redispersed in hexane.

1.2.3.1 **Hydrophobically Modified Nanoparticles**

The synthesis of iron oxide (Fe_3O_4) nanoparticles with sizes below 20 nm was reported by Sun and Zeng [24]. The high-temperature reaction of iron(III) acetylacetonate, $Fe(acac)_3$, in diphenyl ether in the presence of hydrophobic additives was applied to make monodisperse magnetite nanoparticles. According to this procedure, the small Fe_3O_4 nanoparticles can be prepared. 1,2-Hydrocarbon diols, including 1,2-hexadecanediol and 1,2-dodecanediol, were reported to react well with $Fe(acac)_3$ to yield Fe_3O_4 nanoparticles [72]. The process involves high-temperature reaction of metal acetylacetonate with 1,2-hexadecanediol, oleic acid (OA) and oleylamine (OAm) (Scheme 1.2). The size of the oxide nanoparticles can be controlled by varying the reaction temperature or changing the metal precursor. As illustrated in Scheme 1.2, reaction of $Fe(acac)_3$ with surfactants at high temperature leads to monodisperse Fe_3O_4 nanoparticles, which can be easily isolated from the reaction byproducts and the high-boiling ether solvent. When diphenyl ether was used as solvent, 4-nm Fe_3O_4 nanoparticles were separated, whereas the use of dibenzyl ether led to 6-nm Fe_3O_4. As the boiling point of diben-

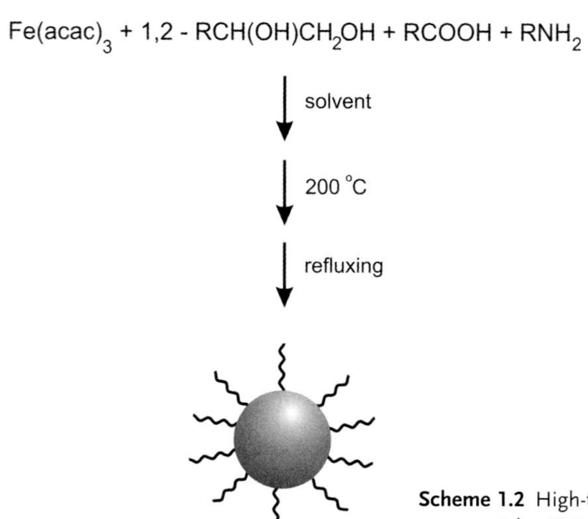

Scheme 1.2 High-temperature preparation of iron nanoparticles [72].

zyl ether (298 °C) is higher than that of diphenyl ether (259 °C), the larger-sized Fe_3O_4 particles obtained from benzyl ether solution seem to indicate that a higher reaction temperature will yield larger particles. However, regardless of the size of the particles, the key to the success of making monodisperse nanoparticles is to heat the mixture to 200 °C first and maintain that temperature for some time before heating to reflux at 265 °C in diphenyl ether or at ~300 °C in dibenzyl ether. Directly heating the mixture to reflux from room temperature would result in Fe_3O_4 nanoparticles with a wide size distribution from 4 to 15 nm, indicating that the nucleation of Fe_3O_4 and the growth of the nuclei under these reaction conditions are not a fast process.

The mechanism leading to Fe_3O_4 in the reactions presented is very complex. However, evidence suggests that reduction of the Fe(III) salt to an Fe(II) intermediate occurs, followed by the decomposition of the intermediate at high temperature. The formation of an Fe(II) intermediate was indicated by the fact that product separated after a short refluxing time (5 min) instead of 30 min showed no magnetic response and contained FeO, as evidenced by X-ray diffraction (XRD). Furthermore, in the presence of a slight excess of 1-hexadecanethiol, a black powder corresponding to FeS could be separated. If Fe(II) D-gluconate or Fe(II) acetylacetonate was used, the same product was obtained. No metallic Fe was detected in the final product. The imide linkage between OA and OAm was reported to allow adequate particle stabilization in dilute solution [70].

The larger Fe_3O_4 nanoparticles can be made by seed-mediated growth. This method has recently been applied to larger metallic nanoparticle and nanocomposite synthesis [73, 74] and is an alternative way of making monodisperse nanoparticles along with LaMer's method through fast supersaturated-burst nucleation [75] and Finke's method via slow, continuous nucleation and fast, autocatalytic surface growth [76]. In the seed synthesis, the small Fe_3O_4 nanoparticles (the seeds) are mixed with more materials and heated and particle diameters can be increased, allowing the diameter to be tuned to up to about 20 nm [72]. Fe_3O_4 nanoparticles prepared according to Scheme 1.2 or the seed-mediated growth method are monodisperse. By controlling the quantity of nanoparticle seeds, Fe_3O_4 nanoparticles with various sizes can be formed. For example, mixing and heating of 8-nm Fe_3O_4 nanoparticles with $Fe(acac)_3$, stearyl alcohol, oleic acid and oleylamine led to 12-nm Fe_3O_4 nanoparticles, whereas changing the mass of seeds into larger amount of nanoparticles led to 16-nm Fe_3O_4 nanoparticles. HRTEM images showed that these nanoparticles are single crystalline. The position and relative intensity of all diffraction peaks match well those of the commercial magnetite powder. The average particle diameter estimated using Scherrer's equation [77] was ca. 15 nm, close to the 16-nm average size determined by statistical analysis of the TEM images, indicating that each individual particle is a single crystal. Figure 1.3 shows typical TEM images from representative 6- and 12-nm Fe_3O_4 nanoparticles deposited from their hexane (or octane) dispersions and dried under ambient conditions. It can be seen that the particles have a narrow size distribution.

$Fe(acac)_2$ or Fe(II) acetate can also be used but yields no better results than $Fe(acac)_3$. Fe(II) D-gluconate is another good precursor for Fe_3O_4 synthesis. In

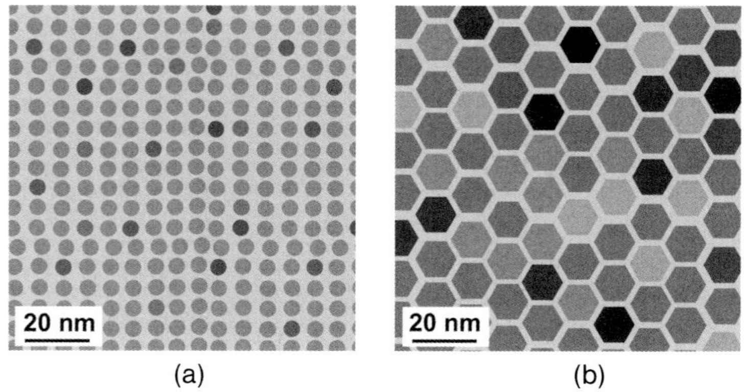

Fig. 1.3 TEM bright field images of (a) 6- and (b) 12-nm Fe_3O_4 nanoparticles deposited from their hexane dispersion on an amorphous carbon-coated copper grid and dried at room temperature [72].

dibenzyl ether, the reaction of Fe(II) D-gluconate with a 3-fold excess of each of oleic acid and oleylamine and a 5-fold excess of 1,2-hexadecanediol led to nearly monodisperse 8-nm Fe_3O_4 nanoparticles. Several different alcohols and polyalcohols have been tested for their reactions with $Fe(acac)_3$. Long-chain monoalcohols, such as stearyl alcohol and oleyl alcohol, were used to prepare Fe_3O_4 nanoparticles, but the particle quality was poor. However, in the seed-mediated growth process, these monoalcohols can be used to form larger Fe_3O_4 nanoparticles. Oleic acid and oleylamine are necessary for the formation of particles. Sole use of oleic acid during the reaction resulted in a viscous red–brown product that was difficult to purify and characterize. On the other hand, the use of oleylamine alone produced iron oxide nanoparticles in a much lower yield than the reaction in the presence of both oleic acid and oleylamine.

In the synthesis, nanoparticles were prepared via decomposition of $Fe(CO)_5$ stabilized by OA and OAm with average diameters ranging from 5 to 20 nm (Table 1.2, Figs. 1.4 and 1.5) [70]. The particle size was most sensitive to the surfactant:iron molar ratio and to the reaction temperature. Evidence of Fe nanoparticle formation was not seen in the homogeneous nucleation preparation until the temperature reached the boiling point of dioctyl ether. This was attributed to the formation of an iron–oleate complex when the $Fe(CO)_5$ is added to the oleic acid solution [4]. The black color associated with Fe(0) does not appear until the solution has aged. In the heterogeneous nucleation method, the Fe(0) forms at lower temperature because there is less oleate to bind to the Fe, requiring less thermal energy to be driven off.

The homogeneous and heterogeneous nucleation methods differed greatly in their ability to synthesize large Fe particles (>9 nm). Using the homogeneous nucleation approach, 11.2-nm particles were prepared via decomposition of $Fe(CO)_5$ in the presence of excess OA alone and 9-nm particles in an excess of OAm. Particles as large as 19 nm could be made using $Fe(CO)_5$ and OA–OAm

Table 1.2 Sample types and particle concentrations [70].

Sample name	Nucleation method [a)]	Fe (mol)	Particle diameter, d_{TEM} (nm)	No. of particles, N_p (dm^3)
Fe112 [b)]	Het	0.051	5.8	3.6×10^{18}
Fe07 [c)]	Het	0.0957	7.2	3.5×10^{18}
Fe01 [d)]	Het	0.0614	8.6	1.3×10^{18}
Fe021 [d)]	Het	0.080	8.4	1.9×10^{18}
Fe022	Hom	0.0284	11.2	
Fe04	Hom		9.1	
Fe052	Hom		19	
Fe090	Het		9.2	
Fe10	Het		7.0	

a) Het, heterogeneous; Hom, homogeneous.
b) Stabilized by hexadecanediol.
c) Stabilized using OA–OAm.
d) Synthesized in decalin instead of dioctyl ether.

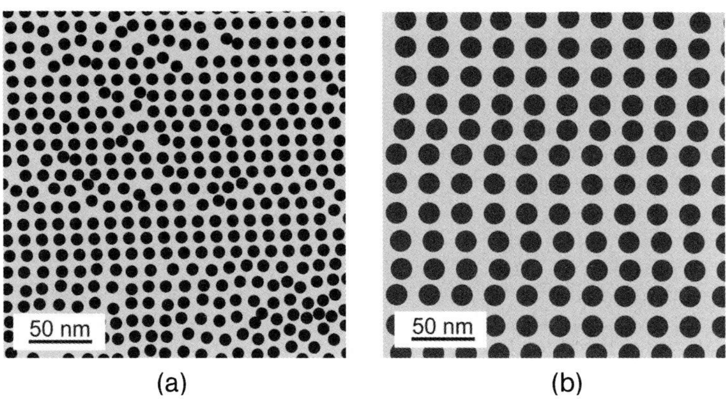

Fig. 1.4 TEM images of heterogeneously nucleated particles: (a) ~7.0 nm (sample Fe10); (b) ~9.2 nm (sample Fe090) [70].

mixture during each of two growth stages (Fig. 1.5 c). By seeding the reaction with Pt, a large number of Fe particles could be prepared during short reactions, but the size of the particles is limited. Particles 4.8 nm in diameter were synthesized using $Fe(CO)_5$ in a two-stage reaction with no surfactant present. For the heterogeneous method, the maximum size for monodisperse particles was ~9 nm. The time at reflux was the determining factor for uniformity of particle shape and size monodispersity. Heating for 1 h or more during particle growth was needed for maximum size and an Ostwald ripening period of >1 h was also required. Heating times >70 min, however, had no discernible effect on size but compromised the particle shape for the homogeneously nucleated particles. Particles with size <3 nm, arising from secondary nucleation, were

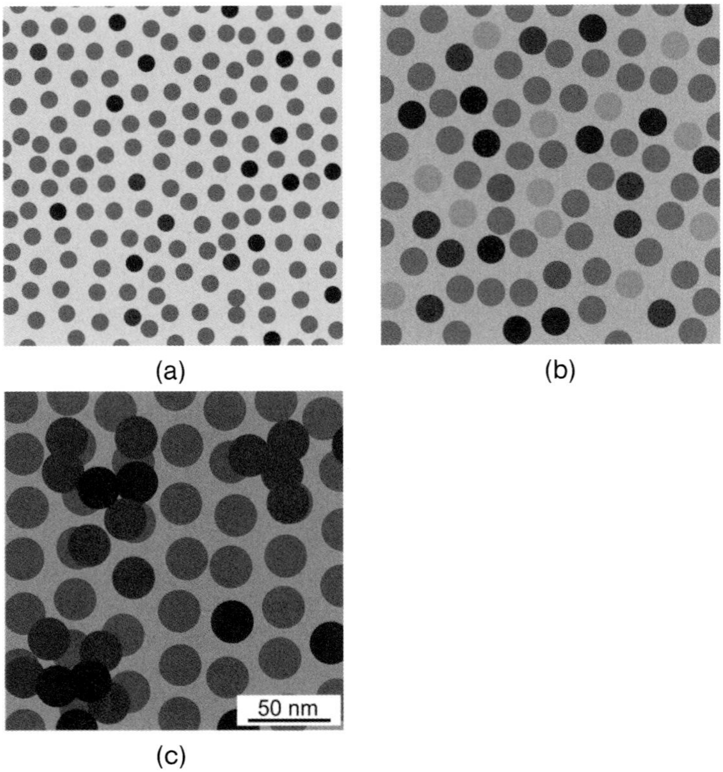

Fig. 1.5 TEM images of homogeneously nucleated nanoparticles: (a) ∼9.1 nm (sample Fe04); (b) ∼11.2 nm (sample Fe022); (c) ∼19 nm (sample Fe052) [70].

dissolved with sufficient time and temperature through Ostwald ripening. The particle concentration of samples is on the order of 10^{17} particles dm^{-3}, an order of magnitude lower than that for the heterogeneously nucleated samples [70].

The hydrodynamic diameter, d_h, found from size-exclusion chromatography (SEC) and light scattering is larger than the inorganic Fe core size as measured by TEM because of the presence of the organic ligand on the surface [70]. This depends on whether the organic ligand binds irreversibly to the Fe clusters during the measurements or whether dynamic exchange between bound and unbound surfactant occurs. In high-performance liquid chromatographic (HPLC) measurements, for example, a polystyrene column separates the Fe clusters from the stabilizer, effectively preventing their complete elution from the column. The presence of a small amount of hexadecylamine in the toluene mobile phase helps to eliminate this effect. This effect is similar to that observed in transition metal clusters such as Ag, Au or Pt in which alkanethiols improve cluster elution and peak shape [78]. Addition of oleic acid, hexadecylamine or both to the cluster solution itself was ineffective at stabilizing the Fe clusters because little or no elution occurred, demonstrating that the observed improvement in elution was due to a dynamic equilibrium between free and bound surfactant. The formation of a link between the two surfactants (OA, OAm) added for Fe nanoparticles was confirmed by the observation of an elution peak [70]. The elution time observed for OA–OAm did not match that of either pure OA or pure OAm. SEC can separate by both size and shape, so the size might not be expected to agree with simple TEM measurements. The average TEM size for Fe samples is smaller than the hydrodynamic SEC size by 1.4 nm.

The results of X-ray fluorescence studies on heterogeneously nucleated samples were compared with the Pt and Fe concentrations expected from the abundance of the starting materials in the syntheses [70]. The high measured masses of the Fe samples indicate that some surfactant or diol remains on the particle surface despite repeated washing. The results of these particle concentration studies indicate that half to two-thirds of the surfactant mass remains on the particles after thorough washing. Homogeneously nucleated particles coated with OA displayed greater contrast variation between particles, indicating differing crystallographic orientations. The absence of this variation in heterogeneously nucleated and OAm-stabilized homogeneously nucleated particles indicated that these particles have a more disordered structure. The higher temperatures at which the Fe formed in the OA-stabilized particles led to this improved crystallinity. Electron diffraction patterns showed enhanced crystallinity in the homogeneously nucleated particles. Figure 1.6 shows the electron diffraction patterns of the particles in Fig. 1.5. The typical pattern of a homogeneously nucleated sample consisted of three or four diffuse rings corresponding to lattice spacings of 0.255, 0.211, 0.150 and 0.115 nm. The diffuseness indicates that the particles are not highly crystalline. The observed lattice spacings were compared with the (*hkl*) spacings for body-centered cubic (bcc) Fe and face-centered cubic

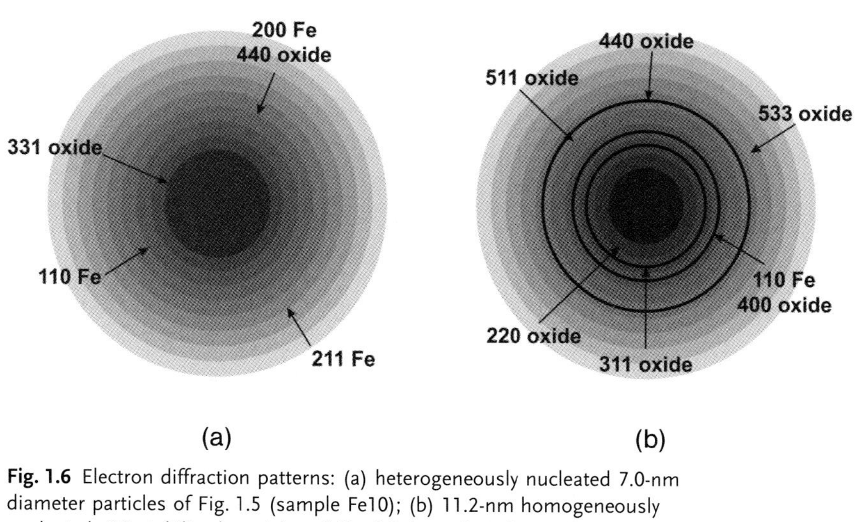

Fig. 1.6 Electron diffraction patterns: (a) heterogeneously nucleated 7.0-nm diameter particles of Fig. 1.5 (sample Fe10); (b) 11.2-nm homogeneously nucleated, OA-stabilized particles of Fig. 1.5 (sample Fe022) [70].

(fcc) Fe_3O_4. The strongest reflections for bcc Fe are 0.203 (110), 0.117 (211), 0.143 (200) and 0.0906 nm (310) and the strongest oxide reflections are 0.253 (311), 0.148 (440), 0.297 (220) and 0.162 nm (511). The observed patterns contain the two strongest oxide reflections and one or two of the strongest Fe reflections. γ-Fe_2O_3 and Fe_3O_4 have the same lattice spacings for the first five rings; indexing to the specific phase requires more precise angle and intensity information than is available from an electron diffraction pattern.

Unless the oxides have been treated at very high temperatures (>1000 °C) to achieve the equilibrium structure, their vacancy locations are most likely disordered. Particles that were synthesized by the heterogeneous nucleation method but using extended heating periods had electron diffraction patterns with additional rings at lattice spacings of 0.295 and 0.160 nm, corresponding to strong oxide reflections. Heating also sharpened the contrast in these diffraction patterns, indicating enhanced crystallinity. The patterns for these particles matched well with the patterns from particles of roughly the same size synthesized by homogeneous nucleation in the presence of OAm and a longer heating period.

Many groups have used metal carbonyl precursors. Fe nanoparticles can be prepared from $Fe(CO)_5$ in organic solvents [79, 80]. Furthermore, a platinum catalyst can improve the particle formation process [70]. Detailed analysis shows that particles can be synthesized with magnetic properties equal to (and in some cases somewhat better than) those achieved using polymeric surfactants. The reaction begins by the *in situ* formation of platinum(0) cores modeled after a reaction designed to synthesize iron–platinum alloy particles [71], then using these cores to catalyze the decomposition of iron carbonyl in the presence of a small-molecule surfactant mixture (a 1:1 molar ratio of oleic acid and oleylamine) at 287 °C. This was followed by a second addition of iron carbonyl and more moderate heating at 260 °C. The 7-nm particles synthesized by this method had an M_{sat} value of 175 A m^2 kg^{-1}, and increasing the size to about 9 nm increased the M_{sat} value to 200 A m^2 kg^{-1}. In both cases the authors noted the existence of an oxide layer, which was estimated to be 0.5 nm thick. There are several possible explanations for this oxide, e.g. the existence of an oxidizing impurity in the reagents, oleic acid oxidizing the iron or the reaction of the acid and amine leading to the formation of water. The room-temperature combination of oleic acid and oleylamine leads to the formation of an ammonium carboxylate salt, but heating forms the amide and liberates water, which would be expected to contribute to oxide formation [30].

The decomposition of pentacarbonyl was observed to decrease in the presence of some polymers, that is, it proceeded at 140–160 °C under an inert atmosphere to yield magnetic iron nanoparticles [29, 30]. Polymers and copolymers such as the polybutadiene (polyBd), poly(styrene-*co*-butadiene) (polyS-Bd) and poly(styrene-*co*-4-vinylpyridine) (polyS-VP) acted as surfactant and catalyst. All of the polymers showed an increased reaction rate when compared with the uncatalyzed reaction. The more reactive polymers gave greater enhancements in reaction rate, but yielded nanoparticles of lower magnetism, with M_{sat} values ranging from 82 A m^2 kg^{-1} for roughly 8-nm particles made with polyS-VP to

172 A m^2 kg^{-1} for similarly sized particles made with polyBd. If particles are too weakly magnetic or the surfactants are too long and bulky, then the particles will not magnetically attract each other and we often see hexagonally packed spheres. If particles are too magnetic or the surfactant too small, the particles very quickly agglomerate into three-dimensional structures. The polymers used in this procedure provided just enough of a restorative force repelling neighboring particles to allow them only to attract and agglomerate based upon their dipolar interactions.

Stable magnetic fluid consisting of amorphous iron oxide nanoparticles was synthesized by ultrasound irradiation of a decalin solution of $Fe(CO)_5$ at 30 °C in the presence of oleic acid as solutions of $Co[CH_3(CH_2)_3CH(C_2H_5)CO_2]_2$ or $Co[CH_3(CH_2)_3CH(C_2H_5)CO_2]_2$ and $Fe(CO)_5$ in decalin, which were irradiated sonochemically at –60 °C [81]. The authors discussed the preparation of ferrofluids containing iron oxide and Co- and Cu-based ferrite nanoparticles. In this method, the role of a stabilizing agent is played by the organic ligand $[CH_3(CH_2)_3CH(C_2H_5)CO_2]^-$. Varying the ratio between components in initial solution yields cobalt or copper ferrite fluids of different stoichiometry. TEM images of Cu-based ferrofluids synthesized without the surfactant additives demonstrate the presence of initially rounded particles of mean size <8 nm. Aged for 1 week, the magnetic liquids begin to arrange in larger clusters with the lower magnetization. The sample aged for 6 weeks showed much larger round aggregates of the mean size ~50 nm. The decrease in the magnetic signal may be explained considering disorientation of the elementary magnetic moments in aggregates, so the net magnetic moment is small. The assemblies formed in Co-based ferrofluids during aging form network-like structures. The difference in the geometry of clusters formed during aging can be explained by the difference in the magnetic properties of the components: the magnetic moment is largest in the case of Fe–Co nanoparticles, intermediate for Fe_2O_3 and smallest for Fe–Cu particles. The competition between surface and magnetic dipole energies is a driving force determining the cluster shape; weak magnetic interactions in Cu-based ferrofluids cannot prevent surface area minimization and result in round-shaped clusters. Intermediate-strength Fe_2O_3 results in separated, but elongated particles. The magnetically strongest Fe–Co-based particles aggregate in the form of chains closing a magnetic field inside the formation. Apparently, this behavior is a self-organizing process. Once it has appeared, the magnetic anisotropy assists the growth of the clusters in particular directions. The increase in a cluster size and the corresponding change of its shape amplify the anisotropy of magnetic interactions.

Suslick et al. introduced sonochemical methodology by taking neat iron carbonyl to produce nanoscopic amorphous iron powders [82]. These powders crystallized and fused when heated above 300 °C. In later work, sonicating iron carbonyl in the presence of a polyvinylpyrrolidone (PVP) stabilizer at 0 °C produced well-dispersed iron nanoparticles [44]. The nanoparticles formed were shown to be of amorphous iron in the 3–8-nm size range and were superparamagnetic. While initially amorphous, the particles could be annealed *in situ* to produce α-

iron without fusing. The same study demonstrated the utility of small-molecule surfactants by stabilizing sonochemically produced particles with oleic acid. The oleic acid-stabilized particles had an average size of about 8 nm and a smaller distribution of sizes. Ultrasound radiation was used to generate polymer-stabilized iron nanoparticles beginning with a solution of iron carbonyl and monomer [83]. Monomers such as methyl methacrylate and styrene were successfully polymerized while sonochemically generating iron nanoparticles [84, 85].

Gedanken and coworkers have studied a number of small-molecule surfactants, including alkylchlorosilane [86], sodium dodecyl sulfate [86], a series of alkanethiols [87], a series of alcohols [88], a series of carboxylic acids [89], a series of alkanesulfonic acids [90], octadecanephosphonic acid [90] and the bifunctional molecule 11-mercaptoundecanoic acid [91], to produce iron nanoparticles. These techniques were applied to functionalize the iron particles with alcohols, carboxylic acids, octadecylphosphonic acid and octadecylsulfonic acid and then determine the effect of the surfactant on the magnetic properties [92]. Hydrophobic agents (octanoic, decanoic and dodecanoic acid) were used in the recipe to stabilize the hydrophobic magnetic iron fluids [93]. Spherical γ-Fe_2O_3 nanocrystals having the same average size (10 nm) and with different coatings, such as octanoic acid (C_8), decanoic acid (C_{10}) and dodecanoic acid (C_{12}), were prepared [93]. The deposition of iron oxide nanoparticles was observed to vary with the shell structure of nanoparticles and the reaction conditions (in the presence or absence of a magnetic field). When a magnetic field (0.59 T) is applied during the deposition of C_8 spheres, the nanocrystals are organized in fairly long chain-like structures, oriented along the direction of this field. The maximum length, average diameter and average spacing between the chains are 40 μm, 400 nm and 3 μm, respectively. On increasing the nanocrystal concentration, spherical and highly compact aggregates having an average diameter of 1.5 μm appear. In an applied magnetic field (0.59 T), long superimposed cylinders with an average diameter of 3 μm are observed. With a dilute C_{12} nanocrystals ferrofluid, in contrast to the findings with C_8, the deposited nanocrystals in the absence and presence of an applied magnetic field did not show any influence of the magnetic field on the organization. On increasing the nanocrystal concentration, a very flat surface with high compactness of the film is observed. In an applied magnetic field, stripes are observed on a large scale with a highly dense and undulating structure.

C_{10} nanocrystal aggregates show an intermediate behavior between those of C_8 and C_{12}. With no magnetic field, a dilute solution of ferrofluid forms small aggregates whereas lines are observed with a magnetic field. On increasing the nanocrystal concentration, the SEM images show a flat surface in the absence and poorly defined cylinders in the presence of a magnetic field. C_{10} nanocrystals coated with short alkyl chain acids self-organize in large clusters in dilute solution, whereas a rough film of highly compact spherical aggregates is observed in a concentrated solution. Similarly, when these nanocrystals are subjected to a magnetic field, long chain-like structures are obtained in dilute solutions whereas superimposed cylinders are formed in concentrated systems. It was concluded that the magnetic properties of isolated nanocrystals with different coatings are the same, which means that the changes in mesostructures are due mainly to the coating.

Amorphous Fe nanoparticles can also be made provided that the temperature is kept low and the decomposition of the carbonyl is stimulated sonochemically [94]. Particles of 30 nm made in this way contained less than 12 atom% carbon and had a magnetization of 173 emu g^{-1} [94]. For elevated temperature preparations, the carbon content can be as high as 30 atom%, lower temperatures favoring a higher proportion of carbon [95]. With a large carbon fraction, the annealed particles crystallize at temperatures as low as 250 °C to yield mixtures of C–Fe and various iron carbide phases [96]. The presence of ammonia during the synthesis can lead to the less oxidation-sensitive ε-Fe_3N phase, if it is abundant [97]. The specific magnetization of particles made with $Fe(CO)_5$ and NH_3 was reported to be as high as 140 emu g^{-1} of Fe for 30-nm particles, based on iron masses estimated from the precursor feed ratios [98]. It is possible to narrow the size distribution of $Fe(CO)_5$-based nanoparticles sufficiently that, when coated with a surfactant, they spontaneously assemble into arrays when the solvent is evaporated [99].

1.2.3.2 Hydrophilically Modified Nanoparticles

Bee et al. [36] investigated the effect of citrate ions on the size of γ-Fe_2O_3 particles. The particle size of maghemite particles decreases with increasing concentration of citrate ions {represented by the molar ratio of citrate to metallic species [Fe(II) + Fe(III)] (R_{mr})} (Fig. 1.7). Thus, for $R_{mr}=0$ (without citrate), the mean diameter is ~8 nm, and for $R_{mr}=3\%$, the mean particle size is much smaller, ~2 nm. The adsorption of citrate ions at the oxide surface is based on a ligand exchange where the surface hydroxyl groups are replaced by the adsorbed anion. The formation of surface complexes requires both deprotonated carboxy and/or hydroxy groups [100]. An unusual ferrofluid ($R_{mr}=0$) is stable at pH < 5 (acidic ferrofluid) and at pH > 8 (alkaline ferrofluid). For $R_{mr} > 3\%$, the ferrofluid obtained by peptization of the small particles is stable from pH 3.5 to 11. In an acidic medium, the two free carboxylate functions of the adsorbed ligand are protonated (pK=4.30, 5.65); the particles therefore bear no charge and flocculate. Nevertheless, the ferrofluid is stable in a physiological medium (pH 7) in the anionic state and so can be used for biomedical applications. The peaks in X-ray diffractograms become broader with increasing R_{mr}. Assuming that this broadening of the diffraction peaks reflects only the decrease in crystallite size, d_{RX} can be evaluated from the peak due to the (311) plane reflection; d_{RX} is plotted versus R_{mr} in Fig. 1.7.

In the range where the saturation magnetization (M_{sat}) can be determined, it is found to be independent of the particle size except for the smallest particles ($d_M < 5$ nm). The experimental value of M_{sat} is about 65% of the saturation magnetization of bulk γ-Fe_2O_3 material. This means that the magnetic order is not perfect in the whole particle. For $R_{mr} > 2\%$ the saturation magnetization is not

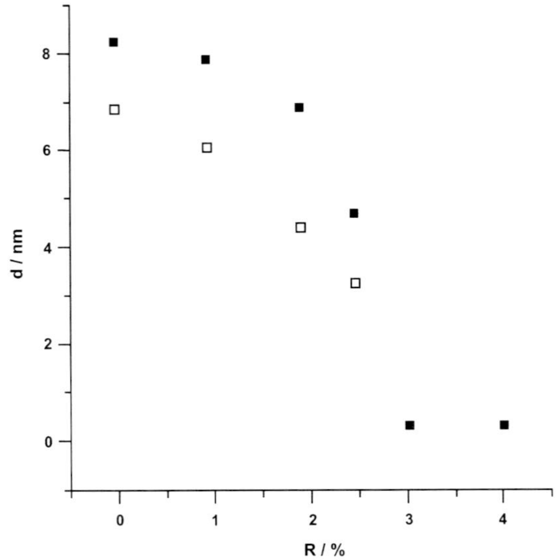

Fig. 1.7 Crystallite size [■ $d(\mathrm{RX})$] and magnetic size [□ $d(M_{\mathrm{mag}})$] versus R.

reached. The size values obtained from the magnetization curves are in good agreement with the TEM data (Fig. 1.7). Colloidally stable magnetic iron fluids were obtained in the presence of hydrophilic (citrate ions) coating agents [93].

Liu et al. [101] described a convenient method for preparing well-dispersed magnetic colloids with size smaller than 5 nm. These colloids have a narrow particle size distribution and are stable in a wide pH range ($4 < \mathrm{pH} < 12$). The size of particles was also controlled through changing the amount of citrate, that is, it decreased with increasing amount of citrate. The morphology of these particles can be seen on the high-resolution TEM image (Fig. 1.8).

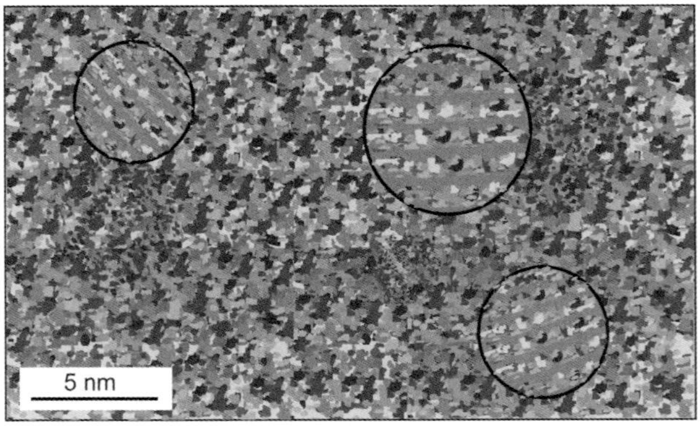

Fig. 1.8 High-resolution TEM image of maghemite particles [101].

The nanoparticles are almost spherical and well dispersed with an average diameter of 5.1 nm (Fig. 1.9). When R_{mr} increases from 25 to 125%, the sizes of particles decrease. For $R_{mr} < 25\%$, the magnetic colloid is not very stable and turns to aggregate in 2 weeks. For $R_{mr} > 50\%$, the prepared particles are too small to be detected by TEM. When R_{mr} exceeds 125%, the synthesis is greatly influenced and black magnetic colloid could not be successfully obtained. The XRD patterns demonstrate that the product is maghemite, which has the inverse spinel structure although the amount of citrate ions varied. Debye–Scherrer calculations predict an average diameter of 3.6 nm for particles, which is smaller than the radius measured by TEM.

The synthesis of nanosized superparamagnetic hematite particles by dissolving ferric salts ($FeCl_3$) in hydrochloric acid and heating at 100 °C was described by Raming et al. [102]. The hydrolysis reaction causes the formation of hematite particles. The moment at which precipitation was first visible was different for the three varieties of precipitation of the 0.02 M ferric solution used. In the second method, with a medium heating rate, precipitation occurred during addition of the concentrated ferric solution (0.5 M). For the third method, with the slowest heating rate, precipitation occurred on heating the 0.02 M ferric chloride solution from room temperature to 100 °C. If the ferric salt was added directly (the first method), an orange–red suspension resulted after 1 week, whereas in the other two cases a purple–brown suspension was produced. Regularly shaped hematite crystals of 30–60 nm had formed after heating at 100 °C for 1 day and for 1 week (method 1). The form was hexagonal to sub-rounded. XRD revealed the presence of hematite only. The average particle size determined from TEM images was 41 nm, whereas X-ray line broadening (XRLB) gave 46 nm as the average diameter. The size polydispersity of these particles was 15%. The morphology of the precipitated particles was identical for each precipitation, but the average crystalline size, as determined by XRLB, varied from 47 to 71 nm. Methods 2 and 3 led to larger hematite crystals than method 1. The majority of

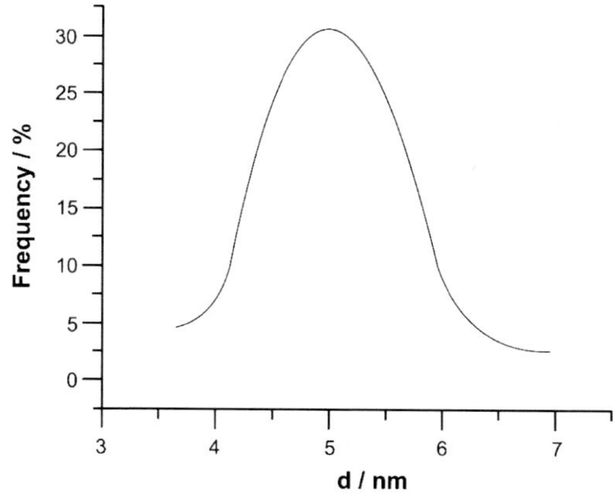

Fig. 1.9 The corresponding particle size histogram from TEM data [101].

these hematite crystals had the same shape and size as after 1 day, but a minority of crystals had grown to larger particles of up to 500 nm diameter.

The results of the precipitation experiments with $FeCl_3$ show that very small changes in the precipitation conditions may cause very large differences in the resulting crystal structure, size and morphology. Most investigations obtained the same morphology after heating at 100 °C for 1 day and reported average crystallite sizes varying from 46 [103] to 80 nm [104], just as was found by Raming et al. by repeating the same precipitation reaction several times. According to Matijevic and Scheiner, the hydrolysis of 0.02 M $FeCl_3$ solution by heating at 100 °C for 1 week resulted in sub-rounded particles with an average size of 46 nm [105]. This hydrolysis was performed by addition of concentrated (3.0 M) $FeCl_3$ solution to an HCl solution. In Raming et al.'s study [102], the aqueous $FeCl_3$ concentration could not be increased above 0.5 M without the occurrence of precipitation reactions. Much larger hematite particles were formed in Matijevic and Scheiner's recipe. This difference can probably be explained by the difference in pH between the two stock solutions. In the 0.5 M Fe^{3+} solution, hydrolysis of the ferric ions may have occurred that did not occur in the more acidic 3.9 M Fe^{3+} solution. The lower concentration of the $FeCl_3$ solution, the higher the pH of the solution will be and the faster hydrolysis will occur. These deviations from the results of Matijevic and Scheiner are probably caused by small differences in the hydrolysis rate. Bailey et al. [106] reported for the same hydrolysis reaction that only after 7 days were spherical particles with an average size of 250 nm formed.

Jiang et al. [107] improved the coprecipitation procedure, through which the average size and size distribution of Fe_3O_4 magnetite nanoparticles can be easily controlled via the use of NH_4OH and urea to manipulate the pH of ferrite solution homogeneously. In addition, because of its non-toxicity, hydrophilicity and high bioaffinity, dextran was selected as a surfactant on nanoparticles here to obtain a water-based magnetic fluid. Many bio-probes can be further coated on to dextran through suitable chemical reactions to make the magnetic nanoparticles biocompatible. The XRD pattern confirms the synthesized Fe_3O_4 particles (Fig. 1.10). It is clear that only the phase of Fe_3O_4 is detectable. There is no other phase such as $Fe(OH)_3$ or Fe_2O_3, which are the usual coproducts in a chemical coprecipitation procedure. The results shown in Fig. 1.10 reveal a high purity for the synthesized Fe_3O_4 magnetic particles.

With the XRD pattern, the average core size of the particles from the Scherrer equation [108] was evaluated as 12.24 nm. According to DLS results, the ratio of the standard deviation to the average value of the diameter distribution of the Fe_3O_4 particles ranges from 0.2 to 0.25 (the value for the particle diameter obtained from the XRD pattern means the particle core size, whereas the size detected using DLS refers to the hydrodynamic diameter of particles). The standard deviation is 5.9 nm and the average is 25.3 nm for this diameter distribution. It is worth noting that the value of 0.2–0.25 for the ratio is much lower than values in other reports obtained through a conventional chemical coprecipitation procedure, in which the ratio is generally higher than 40% [109, 110].

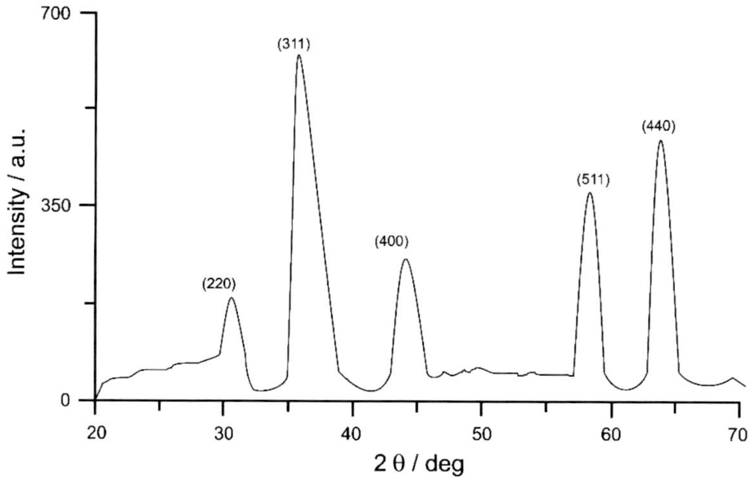

Fig. 1.10 θ–2θ X-ray powder diffraction pattern of the synthesized magnetic particles. The indices are for the Fe_3O_4 phase [107].

This gives evidence that the size distribution of particles can be significantly narrowed with the addition of urea to the ferrite solution. The possible mechanism is as follows. As is known, urea is decomposed at around 80–90 °C:

$$CO(NH)_2 + H_2O \rightarrow 2NH_3 + CO_2 \tag{1}$$

In water, the NH_3 can generate hydroxyl ions to raise the pH of the solution for particle formation:

$$NH_3H_2O \rightarrow NH_4^+ + OH^- \tag{2}$$

Thus, OH^- was generated throughout the solution when heated to 80–90 °C. This led to a uniform increase in the pH of the solution and a homogeneous environment was achieved for the formation of Fe_3O_4 particles as a concentrated NH_4OH solution was titrated. Therefore, a more uniform distribution was obtained for the Fe_3O_4 particles. In addition to the reduction in the size distribution, the control of the decomposition of urea can vary the average sizes of Fe_3O_4 particles. In the synthesis process with larger amounts of urea at 80–90 °C, larger particles (52.6 nm) were observed. Through the preparation procedure, the average diameter of the synthesized Fe_3O_4 particles can be adjusted from 8 to 50 nm, depending on the decomposition of urea in the ferrite solution.

Jeong et al. [111] synthesized γ-Fe_2O_3 nanoparticles by a chemical coprecipitation technique through the typical pipette drop and the novel piezoelectric nozzle method. Figure 1.11 clearly shows that the particle size obtained by the piezoelectric method is smaller than that obtained by the pipette method.

It is worth noting that the size distribution of the maghemite nanoparticles prepared by the typical pipette drop method is 5–8 nm, as shown in Fig. 1.12 a. How-

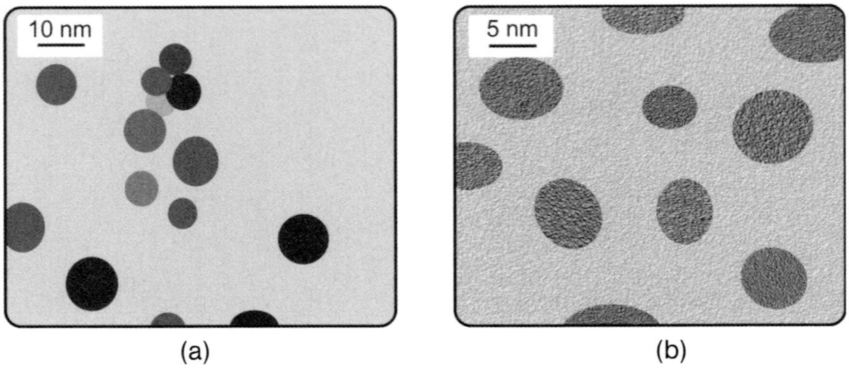

Fig. 1.11 TEM images of γ-Fe_2O_3 nanoparticles for (a) the typical pipette drop method and (b) the piezoelectric nozzle method [111].

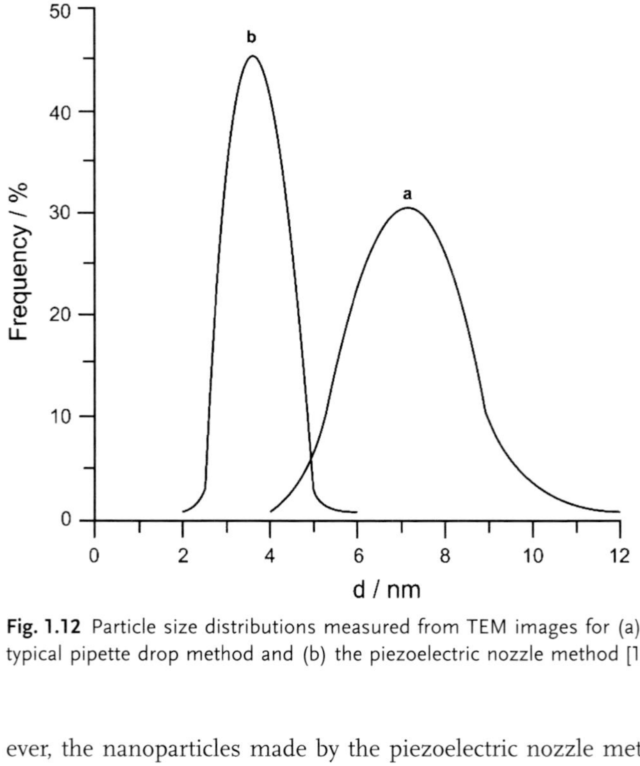

Fig. 1.12 Particle size distributions measured from TEM images for (a) the typical pipette drop method and (b) the piezoelectric nozzle method [111].

ever, the nanoparticles made by the piezoelectric nozzle method show a smaller size and very narrow size distribution from 3 to 5 nm, as shown in Fig. 1.12 b.

A mean diameter of $d \approx 3.8$ nm with a standard deviation $\sigma_d = 0.15$ was obtained for the γ-Fe_2O_3 nanoparticles prepared by the piezoelectric nozzle method and $d = 7.0$ nm with $\sigma_d = 0.17$ for the those prepared by the pipette method [112]. The mean diameter of the γ-Fe_2O_3 nanoparticles was also verified by fitting the magnetization curve using the theoretical Langevin function [113]. All magnetization curves fit well with the Langevin function. From analysis of these curves, the average size of magnetic nanoparticles is found to be 4.0 ± 0.3 and 7.1 ± 0.3 nm for the piezoelectric method and the pipette drop method, respectively [111]. These values show good agreement with those obtained from TEM measurements.

Usually, stable ionic ferrofluids can be obtained by surface modification of metal particles with citric acid [114, 115]. The surface of metal nanoparticles can be also modified with appropriate bipolar additives. This can be performed by shaking particles with bipolar molecules such as tetramethylammonium 11-aminoundecanoate [72]. Shaking a hexane dispersion of 6-nm Fe_3O_4 nanoparticles with a suspension of tetramethylammonium 11-aminoundecanoate in dichloromethane gave Fe_3O_4 nanoparticles that were hydrophilic and water soluble. The absorption bands at 1565 and 1478 cm^{-1} from the hydrophilic nanoparticles match those from free tetramethylammonium 11-aminoundecanoate (1566 and 1487 cm^{-1}), indicating the existence of free –COO– groups in the hydrophilic nanoparticles. The TEM image of 6-nm Fe_3O_4 nanoparticles indicated that nanoparticles in water are well dispersed without any noticeable agglomeration. Magnetic nanoparticles dispersed in water are superparamagnetic and under a tickling field can yield good magnetic signals that are readily sensed by a spin valve sensor [116–118]. Several experiments were performed on a monolayer of 16-nm Fe_3O_4 nanoparticles deposited on 0.3-μm wide spin valve sensors via polyethylenimine-mediated self-assembly [119]. These submicron spin valve sensors produced signals on the order of 10 μV due to the presence of a layer of Fe_3O_4 nanoparticles. These magnetic nanoparticles, if functionalized with single-stranded DNA and immobilized on a similarly functionalized spin valve surface via DNA hybridization, could be used as labels for highly sensitive and quantitative DNA detection.

Kundig et al. [120] reported a critical size of 13.5 nm for hematite particles made by impregnation of silica with ferric salt solution and subsequent calcinations at 500 °C. For spherical hematite particles the critical size was reported to be 8 nm [121]. The largest critical particle size for the transition from ferromagnetic to superparamagnetic behavior for hematite particles reported so far is 32 nm [122], which is slightly lower than that reported by Raming et al. of ~40 nm [102].

Sun et al. [123] improved the method of coating magnetite nanoparticles with silica. First, stable ferrofluids were obtained by modifying the magnetite surface with citric acid. Then the ferrofluids were treated with an ethanol solution containing a critical volume of silicate component. Finally, silica coating was performed by the Stober process [124]. XRD and electron diffraction (ED) were used to determine the crystal structure of the magnetic nanoparticles obtained by the reduction coprecipitation method (Table 1.3). The *d* spacings calculated from the patterns are close to those of magnetite and maghemite (Table 1.3), both of which are inverse spinel structures. The magnetite particles are approximately spherical and the mean size was about 7.6 nm according to the TEM im-

Table 1.3 ED and XRD data for the magnetic nanoparticles prepared by reduction coprecipitation method [123].

	1	**2**	**3**	**4**	**5**	**6**	**7**
$d_{s,ED}$ (Å)	4.84	3.00	2.55	2.12	1.74	1.63	1.49
$d_{s,XRD}$ (Å)	4.78	2.95	2.53	2.09	1.71	1.61	1.48
$d_{s1,calc}$ (Å)	4.85	2.97	2.53	2.10	1.71	1.62	1.48
$d_{s2,calc}$ (Å)	4.82	2.95	2.51	2.09	1.70	1.61	1.47
Cpl (*hlk*)	(111)	(220)	(311)	(400)	(422)	(511)	(440)

Cpl = crystalline plane, d_s = spacing parameter determined by ED ($d_{s,ED}$) and XRD ($d_{s,XRD}$) and calculated d_s for Fe_3O_4 ($d_{s1,calc}$) and γ-Fe_3O_3 ($d_{s2,calc}$).

age. This agrees well with the mean diameter (7.1 nm) calculated from the XRD pattern using the Scherrer equation, which is usually smaller than that of TEM.

Magnetite–silica core–shell structures can be developed by tetraethyl orthosilicate (TEOS) hydrolysis on magnetite nanoparticles. The magnetite nanoparticles were pretreated with an ethanolic sodium silicate solution. Due to the low solubility in ethanol (5×10^{-4} wt% SiO_2) [125], the silicate solution tended to precipitate and a sudden turbidity appeared on adding excess sodium silicate. The TEOS precursor condensed on the small nuclei and the silica coating on magnetite nanoparticles was inhibited. Without sodium silicate, irregular clusters of magnetite nanoparticles were coated with silica. Hence the appropriate amount of sodium silicate solution added to the ethanol was a prerequisite. Silica coating stabilizes the magnetite nanoparticles in two different ways. One is by sheltering the magnetic dipole interaction through the silica shell. The magnetic dipole attraction in a simplified head-to-tail configuration is proportional to the $1/a_d^3$ (where a_d is the center distance of two dipoles) [126]. The attraction decreases quickly with increase in thickness of the silica shell. The other is by bringing charges to the magnetic particles. The points of zero charge (PZC) of silica and magnetite are about 2.0 and 7.3 in water, respectively. In neutral biological media, the pristine magnetite nanoparticles are nearly free of charges. On the other hand, the silica nanoparticles are negatively charged. The z-potential of magnetite nanoparticles changed from nearly zero (10 mV) to negative (–42.2 mV) through silica modification. Also, the ζ-potential of final core–shell with TEOS precursor agrees well with that of pure silica (–40.5 mV). So the silica coating enhanced the coulomb repulsion of the magnetic nanoparticles. Hence the magnetic dipole interactions between the magnetic nanoparticles were greatly screened.

Typical magnetic core–shell structures prepared with different volume of TEOS precursor, in which the sizes were 10.5, 30.2, 44.9 and 135 nm, respectively. The unmodified and those modified with little TEOS precursor existed in general aggregation. When the thickness of the silica shell increased, the magnetic nanoparticles changed from aggregates to discrete. In contrast to the results obtained by the TEM, the hydrodynamic diameter of the core–shell structures did not increase steadily with the amount of TEOS precursor. The mean hydrodynamic diameters (pH 7) and the polydispersity measured by photon correlation spectroscopy (PCS) decreased compared with those of bare magnetite nanoparticles and those modified with citric acid (CA) or with TEOS (from 5 to 250 μL) as follows [123]: *D* (nm)/additive: 250/none, 185/CA, 240/5 μL TEOS, 200/20 μL TEOS, 150/50 μL TEOS, 145/250 μL TEOS. The larger hydrodynamic diameters are due to the particle aggregations caused by the incompletely screened magnetic dipole interactions without or with less TEOS precursor. In contrast, those of relative smaller hydrodynamic sizes are the results of improved sheltering of magnetic interactions by the silica coating.

1.3 Metal–Iron Oxide Nanoparticles

Magnetic metal–iron oxide nanoparticles and their dispersions in various media have long been of scientific and technological interest. The cubic spinel structured $M_{met}Fe_2O_4$ and $M_{met}OFe_2O_3$ represent a well-known and important class of iron oxide materials where oxygen forms an fcc close packing and M_{met}^{2+} and Fe^{3+} occupy either tetrahedral or octahedral interstitial sites [59]. By adjusting the chemical identity of M_{met}^{2+}, the magnetic configurations of $M_{met}Fe_2O_4$ can be molecularly engineered to provide a wide range of magnetic properties. Depending on the chemical identity of M_{met}^{2+}, the densely packed solid-state form of nanocrystalline $M_{met}Fe_2O_4$-based materials, on the other hand, can have either high magnetic permeability and electrical resistivity (for M_{met} representing one or the mixed components from Co, Li, Ni, Zn, etc.) or half-metallicity (for M_{met} = Fe) and may be a potential candidate for future high-performance electromagnetic [127–129] and spintronic devices [130–133]. To use $M_{met}Fe_2O_4$ nanoparticles for highly sensitive magnetic nanodevices, a practical route to monodisperse $M_{met}Fe_2O_4$ nanoparticles with diameters smaller than 20 nm and a tight size distribution is needed. A commonly used solution-phase procedure for making small particles has been the coprecipitation of M_{met}^{2+} and Fe^{3+} ions by a base, usually NaOH or NH_4OH in an aqueous solution [134–136] or in a reverse micelle template [137, 138]. Although this precipitation method is suitable for mass production of magnetic $M_{met}Fe_2O_4$ ferrofluids, it does require careful adjustment of the pH of the solution for particle formation and stabilization and it is difficult to control sizes and size distributions, particularly for particles smaller than 20 nm. An alternative approach to monodisperse iron oxide nanoparticles is via high-temperature organic phase decomposition of an iron precursor, for example, decomposition of $FeCup_3$ [Cup = *N*-nitrosophenylhydroxylamine, $C_6H_5N(NO)O^-$] [53] or decomposition of $Fe(CO)_5$ followed by oxidation to Fe_2O_3 [139, 140]. The latter process has recently been extended to the synthesis of monodisperse cobalt ferrite ($CoFe_2O_4$) nanoparticles [23].

Sun and Zeng [24] reported a convenient organic-phase process for making monodisperse $M_{met}Fe_2O_4$ nanoparticles through the reaction of $Fe(acac)_3$ and a long-chain alcohol. These experiments were extended to the synthesis of different $M_{met}Fe_2O_4$ nanoparticles (with M_{met}=Co, Ni, Mn, Mg, etc.) by simply adding a different metal acetylacetonate precursor to the mixture of $Fe(acac)_3$ and 1,2-hexadecanediol [24, 72]. For example, when $Co(acac)_2$ was partially substituted for $Fe(acac)_3$ in a 1:2 ratio under the same reaction conditions as in the synthesis of $CoFe_2O_4$, nanoparticles were formed. When $Mn(acac)_2$ was used, $MnFe_2O_4$ nanoparticles were made. The Co:Fe and Mn:Fe ratios in both cobalt ferrite and manganese ferrite were retained from the ratios of the initial metal precursors and the final Co–Fe and Mn–Fe compositions could be readily controlled. $CoFe_2O_4$ and $MnFe_2O_4$ nanoparticles of 14 nm were made from seed-mediated growth. The XRD results for both samples are very similar to those for Fe_3O_4, indicating a cubic spinel structure of the particles. At temperatures up to 300 K, 16-nm $CoFe_2O_4$ nanoparticles are ferromagnetic. Figure 1.8 shows the hysteresis loops of 16-nm $CoFe_2O_4$ nanoparticles measured at both 10 and 300 K. The coercivity of the assembly is about 400 Oe at 300 K, but reaches 20 kOe at 10 K, much larger than that of the 16-nm Fe_3O_4 nanoparticles (450 Oe at 10 K), indicating that the incorporation of the Co cation in the Fe–O matrix greatly increases the magnetic anisotropy of the materials. Such anisotropy enhancement of $CoFe_2O_4$ vs. Fe_3O_4 has also been observed in films deposited from aqueous solution [141]. In contrast, the incorporation of Mn cation in the Fe–O matrix reduces the magnetic anisotropy of the materials [142], as the 14-nm $MnFe_2O_4$ nanoparticles show a coercivity of only 140 Oe at 10 K.

Teng and Yang [143] presented a synthetic route to $PtFe_2O_3$ core–shell nanoparticles and face-centered tetragonal (fct) FePt nanoparticles. $PtFe_2O_3$ core–shell nanoparticles were made in dioctyl ether with a reactant mixture of platinum acetylacetonate $[Pt(acac)_2]$, iron carbonyl $[Fe(CO)_5]$, oleic acid, oleylamine and 1,2-hexadecanediol [71, 143, 144]. Using these core–shell nanoparticle precursors, they could make FePt magnetic nanoparticles not only with the preferred fct phase but also in a size range that was not achievable using wet chemistry synthetic approaches. Figure 1.14 is an illustration of the synthetic

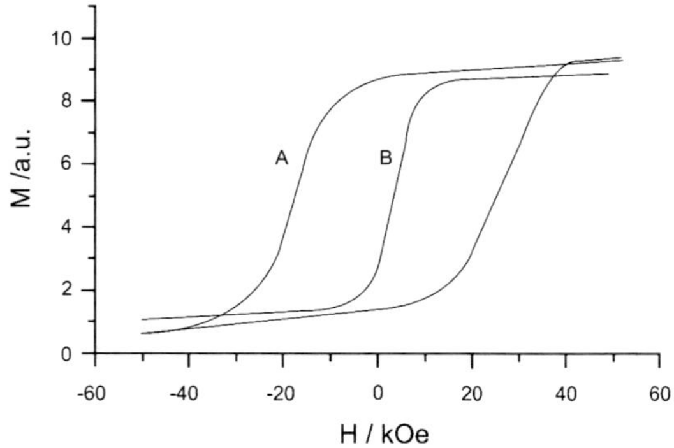

Fig. 1.13 Hysteresis loops of 16-nm $CoFe_2O_4$ nanoparticle assembly measured at (A) 10 and (B) 300 K [24].

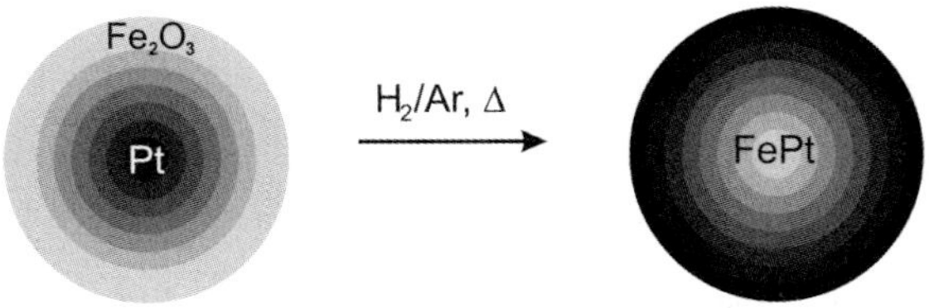

Fig. 1.14 Synthesis of FePt nanoparticles from $PtFe_2O_3$ core–shell nanoparticles [143].

approach for making FePt nanoparticles from $PtFe_2O_3$ core–shell nanoparticles [143]. The reduction of iron oxide and alloy formation of FePt can be accomplished at temperatures above ~450 °C in the presence of reducing gas (hydrogen) and an inert atmosphere [145].

The $PtFe_2O_3$ core–shell particles consist of Pt cores with an average diameter of ~10 nm and iron oxide shells of ~3.5 nm (Fig. 1.15 a). The cores and shells were clearly distinguished because of the sharp contrast between iron and platinum due to the large difference in their electron penetration efficiencies. FePt nanoparticles were obtained after conversion at 550 °C on a thin amorphous carbon support (Fig. 1.15 b). Core–shell nanoparticles that spread using drop casting could form a multilayered assembly and coalesce during the solid-state conversion to FePt alloy at enhanced temperature. The coalescence led to the broadening of the size distribution of FePt nanoparticles. The nanoparticles obtained using Langmuir–Blodgett monolayers were uniform in size in large areas (Fig. 1.15 c). The average diameter of the FePt nanoparticles was estimated to be ca. 17 nm.

Figure 1.16 shows a representative selected area electron diffraction (SAED) pattern of a single layer of 17-nm FePt nanoparticles converted from Pt–Fe_2O_3 core–shell nanoparticles. The observed SAED patterns was assigned to the fct FePt alloy [145]. The electron diffraction rings of {002} and {200} planes in fct FePt nanoparticles are too close to be differentiated from each other, as are the rings for {220} and {202} planes. Relatively high crystallinity can be obtained for these nanoparticles, judging by the spotted electron diffraction patterns. The results of EDX analysis on randomly selected FePt nanoparticles led to the conclusion that the atomic ratios of these nanoparticles varied around $Fe_{50}Pt_{50}$.

The diffraction at 39.8°, 46.3°, 67.5° and 81.3° 2θ obtained in the XRD spectra were indexed to (111), (200), (220) and (311) planes of platinum in a cubic phase, respectively. The absence of iron oxide diffractions was ascribed to the low crystallinity of the iron oxide and the heavy atom effect from platinum [144]. After conversion, these XRD peaks shifted to 41.1°, 47.1°, 68.9° and 83.4° 2θ and additional peaks could also be observed. These diffractions match very well those from (002), (202) and other crystal planes of fct FePt alloy. The XRD peaks of the film samples were relatively narrow compared with those of $PtFe_2O_3$ core–shell nanoparticles [144], most likely because of the increased crystallinity and the coalescence of the nanoparticles during the reaction/sintering at 550 °C. The converging evidence from SAED, EDX and powder X-ray dif-

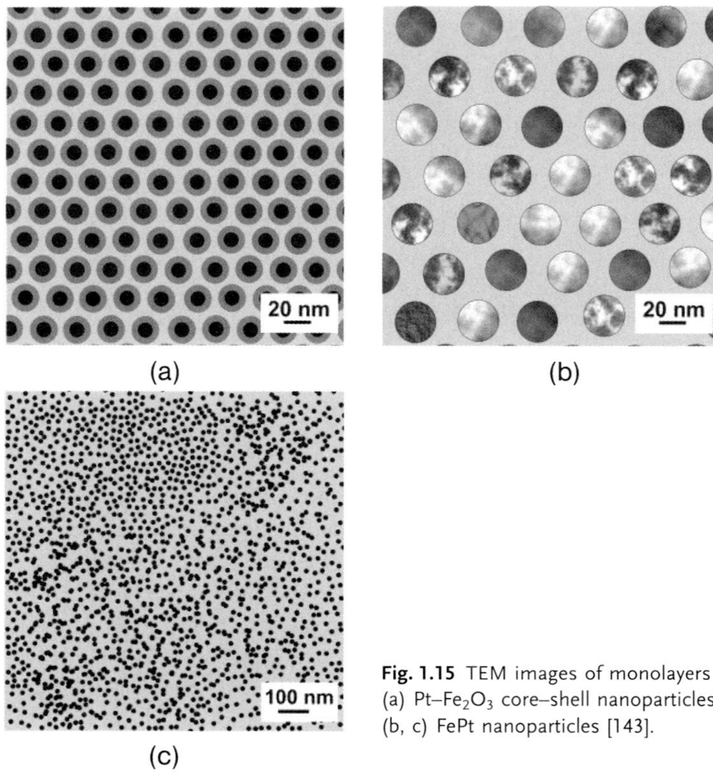

Fig. 1.15 TEM images of monolayers of (a) Pt–Fe_2O_3 core–shell nanoparticles and (b, c) FePt nanoparticles [143].

fraction (PXRD) analyses indicates that the final products were made of fct FePt alloys. The fct FePt alloys have been shown to possess high coercivity after annealing at enhanced temperatures [146, 147].

The hysteresis loop of an FePt film shows the characteristics of a ferromagnetic material (Fig. 1.17). The coercivity of this film was 8.0 kOe measured at 5 K. It is known that FePt alloys can have various coercivities, depending on the synthetic conditions, annealing temperatures and compositions [148–151], because coercivity is not an intrinsic property [152]. The coercivity of the FePt granular films made from Pt–Fe_2O_3 core–shell nanoparticles at 650 °C was 9.1 kOe at 5 K and 7.0 kOe at 300 K. The coercivity increased when the reaction/annealing temperature increased in this range [71].

Chen et al. [153] presented a one-step synthesis of FePt nanoparticles with controlled composition and size tunable up to 9 nm in diameter. By mixing oleic acid and $Fe(CO)_5$ with a dibenzyl ether solution of $Pt(acac)_2$ first and heating the mixture at 130 °C before oleylamine was added, a portion of highly faceted FePt nanoparticles was obtained. If the mixture contained excess of $Fe(CO)_5$, by refluxing in a shorter period of time and air oxidation, core–shell-structured FePt–Fe_3O_4 nanoparticles were separated [4, 154]. Figure 1.18 shows the TEM image of the 7-nm–1.2-nm FePt–Fe_3O_4 nanoparticles prepared with

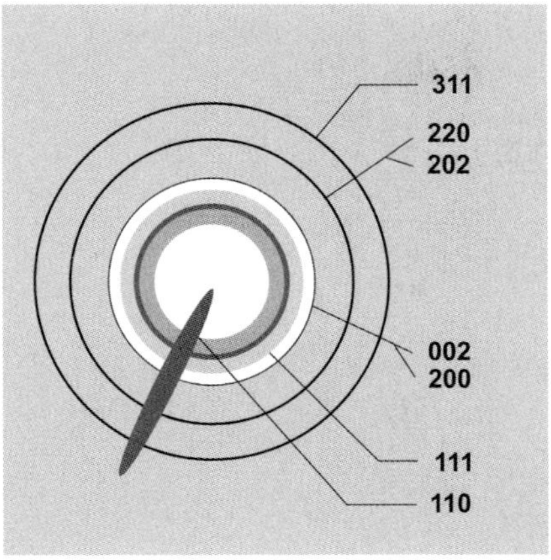

Fig. 1.16 SAED of 17-nm FePt nanoparticles [143].

$Fe(CO)_5 : Pt(acac)_2 = 3$. Further experiments indicated that various $FePt–Fe_3O_4$ nanoparticles with tunable core and shell can be readily prepared by this one-step synthesis. The synthetic results seem to support the following particle formation mechanism illustrated in Fig. 1.19. The Pt-rich nuclei (a) are formed from the reduction of $Pt(acac)_2$ either at temperatures >200 °C or by Fe atoms from the decomposed $Fe(CO)_5$ or both. More Fe atoms will then coat over the existing Pt-rich nuclei, forming larger clusters (b). Exposing these clusters to air leads to the formation of Pt-rich $FePt–Fe_3O_4$ as evidenced by XRD analysis. Heating the clusters (b) to reflux at 300 °C leads to atomic diffusion and formation of fcc-structured FePt nanoparticles (c), as shown in Figs. 1.18 and 1.20. In the presence of excess of $Fe(CO)_5$, the extra Fe will continue to coat over (c), leading to core–shell-structured FePt–Fe that is further oxidized to $FePt–Fe_3O_4$ (d), as shown in Fig. 1.19. The as-synthesized larger FePt nanoparticles with controlled

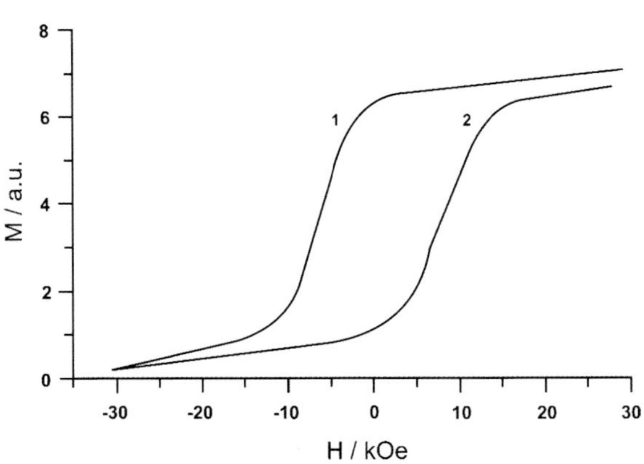

Fig. 1.17 Hysteresis loop of a granular film of fct FePt alloy converted from a multilayered $Pt–Fe_2O_3$ nanoparticle assembly at 550 °C. The measurements were conducted at 5 K [143].

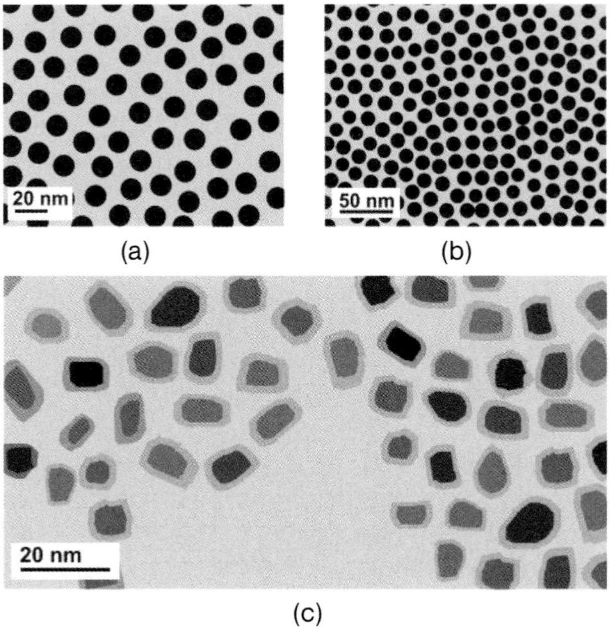

Fig. 1.18 TEM bright field images of (a) 6-nm $Fe_{53}Pt_{47}$, (b) 9-nm $Fe_{44}Pt_{56}$ and (c) 7-nm–1.2-nm FePt–Fe_3O_4 nanoparticles deposited from their hexane dispersions on an amorphous carbon surface and dried at room temperature [153].

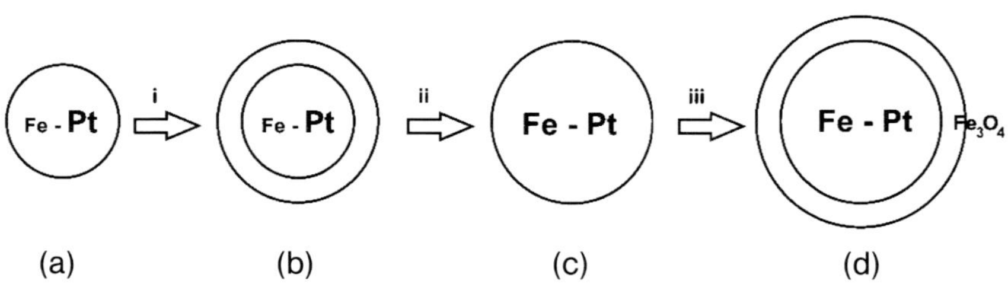

Fig. 1.19 (i) Fe coating, (ii) Fe,Pt interface diffusion and (iii) Fe coating, then oxidation [153].

composition are suitable for size-dependent structural and magnetic studies. When annealed at 560 °C for 30 min, 6-nm FePt nanoparticles show a coercivity of 1.2 T, but 3–4-nm particles have a coercivity of only 0.5 T. Annealing results also indicate that the particles can withstand higher temperatures than their 4 nm counterparts before they start to sinter. The coating of Fe_3O_4 around the FePt core will further stabilize the core. For an assembly containing core–shell 7-nm–1.2-nm FePt–Fe_3O_4 nanoparticles, annealing at 650 °C did not lead to noticeable aggregation of the FePt core, as evidenced by both TEM and XRD studies, and the coercivity of the assembly reached 1.3 T [72]. Considering the fact

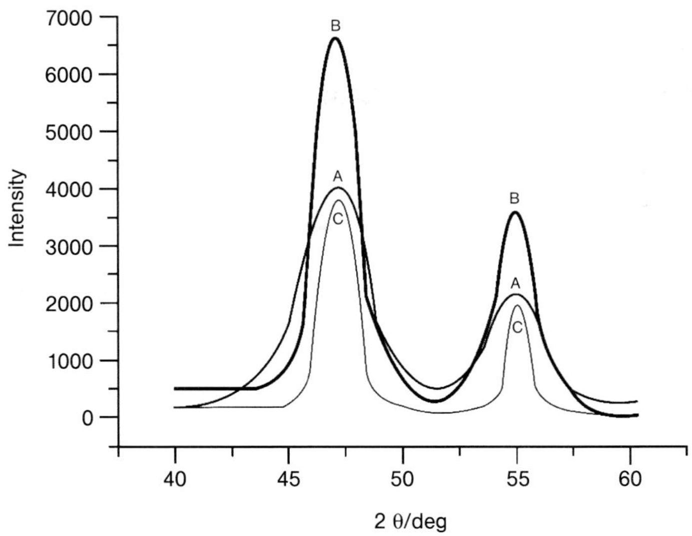

Fig. 1.20 X-ray diffraction patterns of the as-synthesized FePt nanoparticles with different sizes at (A) 3 mm, (B) 6 nm and (C) 9 nm [153].

that larger magnetic particles are thermally more stable than smaller particles [155, 156], the FePt nanoparticles prepared by the current one-step synthesis will have great potential for single particle magnetism studies and as useful building blocks for the construction of ultra-high density information and energy storage devices.

It was also found that, at a fixed stabilizer: $Pt(acac)_2$ ratio, both heating rate and interim heating temperature were important in making FePt particles with a tunable size. A heating rate of $\sim 15\,^{\circ}C\,min^{-1}$ and an interim heating temperature of 240 °C finally yielded 6-nm FePt, whereas a rate of $\sim 5\,^{\circ}C\,min^{-1}$ and a heating temperature of 225 °C led to 9-nm FePt [72]. The composition of the particles is controlled by varying the molar ratio of $Fe(CO)_5$ and $Pt(acac)_2$. Under the current reaction conditions, an $Fe(CO)_5$: $Pt(acac)_2$ ratio of 2 gave 6-nm $Fe_{53}Pt_{47}$ and 9-nm $Fe_{44}Pt_{56}$ nanoparticles. Figure 1.18 shows TEM images of (a) 6-nm $Fe_{53}Pt_{47}$ and (b) 9-nm $Fe_{44}Pt_{56}$ nanoparticles obtained from the one-step synthesis. In both cases, the particles have a narrow size distribution with standard deviation <10%. The FePt particles have a well known chemically disordered fcc structure [71] and the diffraction peaks are narrower when the size of the particles is larger. Estimation using the Scherrer's equation [77] on the (111) peak gives an average particle size that is consistent with that observed from TEM images, indicating that the as-synthesized FePt nanoparticles are of good crystallinity (Fig. 1.20).

A seed-mediated growth approach to monodisperse spinel cobalt ferrite, $CoFe_2O_4$, was applied to grow nanocrystals into a nearly spherical shape or an almost perfect cube in a highly controllable manner [157]. From the nonhydrolytic process as the first step of the synthesis, spherical $CoFe_2O_4$ nanocrystals with a diameter of 5 nm were obtained (Fig. 1.21a) [157]. X-ray and electron diffraction studies confirmed that such nanocrystals and all other subsequently obtained nanocrystals possessed a spinel structure. Chemical analysis using inductively coupled plasma atomic emission spectrometry (ICP-AES) showed the Co:Fe molar ratio to be 1:2 in all these nanocrystals. The 5-nm nanocrystals served as seeds for the seed-mediated growth as the second step of the synthesis for the nanocrystals with a larger size. The growth procedures generated a sample consisting of 8-nm spherical nanocrystals together with a tiny fraction of nanocrystals having a diameter of ~5 nm. Monodisperse 8-nm nanocrystals (Fig. 1.21) were obtained after a small amount of acetone had been added to a hexane suspension of as-grown nanoparticulate sample. Monodisperse nanocrystals with a diameter of 9 nm can be obtained using the same procedures but with a higher concentration of precursors in the particle growth solution. Therefore, monodisperse nanocrystals with larger diameters were typically produced by incrementally increasing the size of the nanoparticulate seeds. For instance, 8-nm nanocrystals were used in seed-mediated growth to synthesize monodisperse nanocrystals with a diameter of 10 or 12 nm.

The shape of $CoFe_2O_4$ nanocrystals can be tuned to cubic during the seed-mediated growth process. Using the same 5-nm spherical seeds and the same growth solution as for the preparation of 8- or 9-nm spherical nanocrystals, cubic $CoFe_2O_4$ nanocrystals were produced with an edge length of 8 or 9 nm when the temperature was raised to only 210 °C at a rate of 2–3 °C min^{-1}. The product was mainly monodisperse nanocubes (Fig. 1.21). A tiny fraction of the product consisted of ~5-nm spherical nanocrystals, which can be easily separated from the cubes through addition of acetone to the hexane suspension. By using 8-nm spherical seeds, nanocubes with an edge length of 10 or 11 nm were produced. Due to the decisive effect of the heating rate and growth temperature on shape control, Fig. 1.21e shows almost spherical monodisperse $CoFe_2O_4$ nanocrystals of 12-nm diameter, which were prepared by using nanocube seeds with an 8-nm edge length. Such reversible shape formation clearly demonstrated that the shape of $CoFe_2O_4$ nanocrystals can be precisely controlled through the temperature and heating rate. Figure 1.21 f displays a histogram from TEM studies showing a narrow size distribution of nanocubes with an almost uniform shape. The inset shows more than 80% of the nanocubes having an aspect ratio between 1 and 1.05. Both spherical and cubic $CoFe_2O_4$ nanocrystals are single crystals, as high-resolution TEM studies have shown. Figure 21 b and c show an interesting short-range order in the nanocube assembly.

The nanocrystal growth rate is the key in shape control. Oleic acid stabilized the nanocrystals and oleylamine was used to provide basic conditions essential for the formation of spinel oxides. The long-chain alcohol also was essential in nucleation and growth processes since it probably made ferrous cations available, which peculiarly always facilitated the formation of $CoFe_2O_4$ nanocrystals [158, 159]. A slow heating rate kept a low concentration of available metal cations from the decomposition of precursors, which, combined with the low

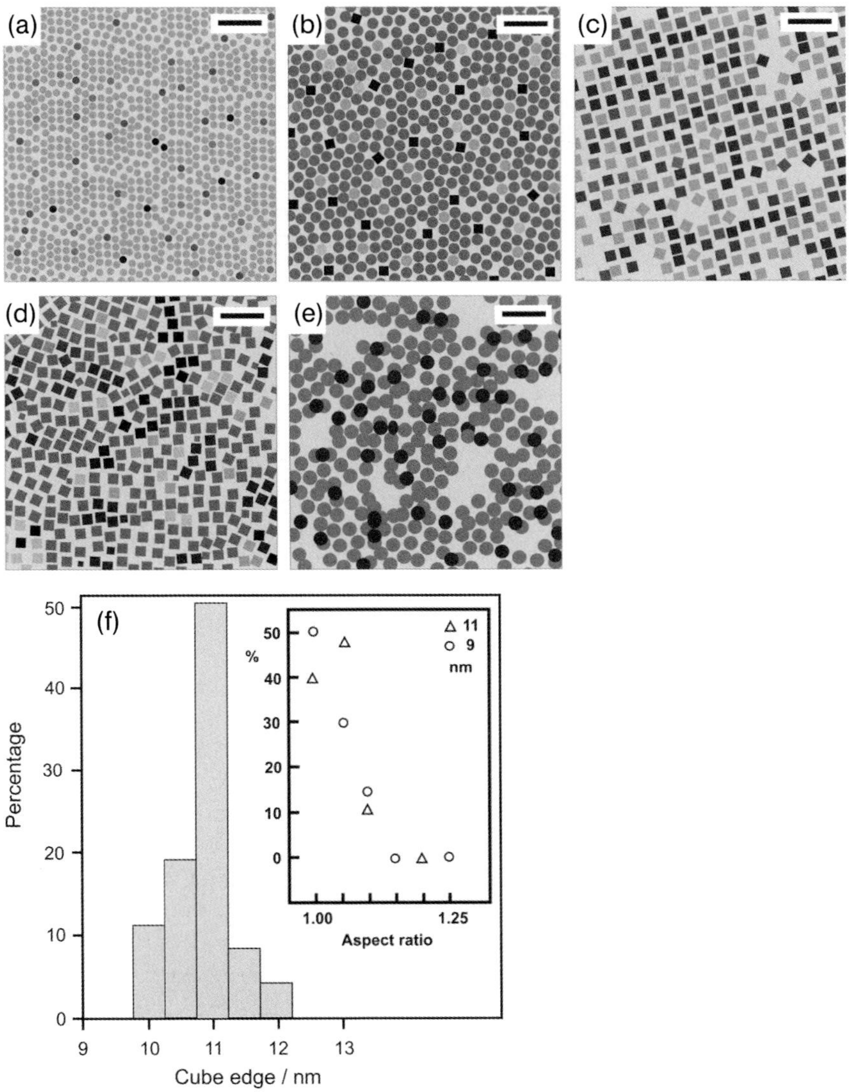

Fig. 1.21 TEM micrographs with the scale bar = 50 nm. The samples in (a) and (b) are 5.2-nm and 7.9-nm spherical $CoFe_2O_4$ nanocrystals, respectively; (c) and (d) are 9.1-nm and 10.9-nm cubic $CoFe_2O_4$ nanocrystals, respectively; (e) is 11.8-nm spherical $CoFe_2O_4$ nanocrystals. The representative histogram in (f) shows the size distribution of cubic nanocrystals in (d) after measuring over 400 nanocrystals. The inset in (f) displays the aspect ratios of the cubic nanocrystals in (c) and (d) [157].

growth temperature, gave a slow rate of crystal growth. The growth of cubic $CoFe_2O_4$ nanocrystals was terminated at {100} planes (Fig. 1.22), which were predicted to have the lowest surface energy [160, 161]. When a much faster heating rate was employed, a high concentration of metal cations was present. A

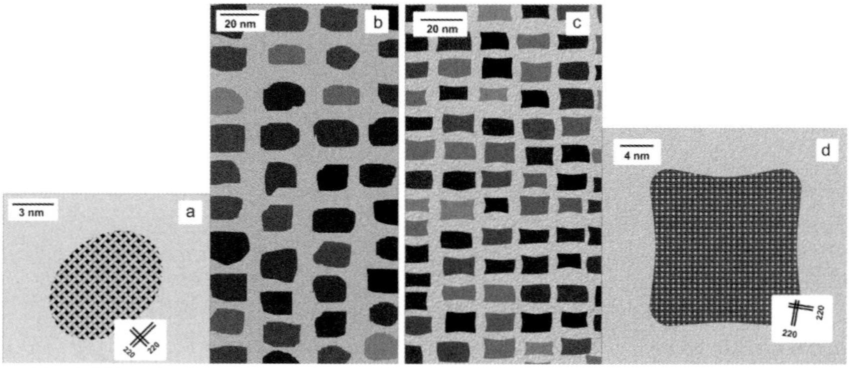

Fig. 1.22 HRTEM images: (a) is an ~8-nm spherical $CoFe_2O_4$ nanocrystal, (b) and (c) display the short-range order in ~9-mm and ~11-nm nanocube assemblies and (d) shows an ~12-nm cubic $CoFe_2O_4$ nanocrystal. The spacing of the atomic lattice plane (220) in (a) and (d) is 2.994 Å and the zone axis is [001] [157].

faster growth rate at a higher temperature with plenty of metal cations available resulted in the crystal growth being much less selective in direction and hence produced spherical $CoFe_2O_4$ nanocrystals.

1.4 Iron–Metal Alloys

Nanoparticles of iron–platinum alloys have been under intensive study in recent years because of their magnetic properties [119, 145, 162, 163]. Chemically prepared FePt nanoparticles have generated great interest recently because of their ease of synthesis, chemical stability and potential applications in high-density data storage [145, 164] and high performance permanent magnets [145, 164]. The surfactants surrounding each FePt nanoparticle can be replaced by other bifunctional stabilizers, rendering the particles water soluble and suitable for site-specific biomolecule attachment and magnetic field-assisted bioseparation [165, 166]. The particles are commonly synthesized via decomposition of iron pentacarbonyl, $Fe(CO)_5$, and reduction of platinum acetylacetonate, $Pt(acac)_2$ [167, 168], co-reduction of iron salt and $Pt(acac)_2$ [169–171] or seed-mediated growth [71], where smaller FePt nanoparticles are used as seeds and more FePt is coated over the seeds. Although the size distribution of the particles prepared from all these methods can be well controlled with standard deviations usually within 10%, tuning the size in the desired range along with tight control of composition has been difficult to achieve thus far. The decomposition and reduction or co-reduction methods can yield FePt nanoparticles with controlled composition but fail to produce particles larger than 4 nm, whereas the seed-mediated growth method gives larger FePt particles without accurate control on

Fe:Pt ratio at different sizes. As both the structural and magnetic properties of FePt nanoparticles depend not only on the size, but also on the composition of the particles [147, 172, 173], this size limitation and composition variation at different sizes prevent one from studying property evolution with size and composition of the particles, an understanding that is essential for mapping the scaling limits of future high-density magnetic storage technology.

There are three major different types of ordered FePt alloys, namely Fe_3Pt, FePt and $FePt_3$ [174]. The fct (also known as $L1_0$ phase) FePt alloy is particularly desirable because of its high magnetic anisotropy, high coercivity, small domain wall width (2.8–3.3 nm), small minimal stable grain sizes (2.9–3.5 nm) and chemical stability [175]. These properties make fct FePt nanoparticles excellent candidates and widely studied systems for applications in ultra-high density magnetic storage media and for making advanced magnetic materials [71, 140, 145].

Currently, monodisperse magnetic nanoparticles of FePt alloy are synthesized through the simultaneous reduction of metal salts and thermal decomposition of organometallic compounds in the presence of mixed surfactants [46, 119]. The stabilizing agent is important for the formation of monodisperse FePt nanoparticles in the disordered fcc phase [119]. The assembled nanoparticles can be converted into fct FePt alloys after annealing at enhanced temperatures. It is important to note that the monodisperse FePt nanoparticles made using wet chemistry methods typically have diameters in the size range 4–6 nm [119]. For many practical applications, magnetic nanoparticles larger than 6 nm are preferred because the coercivity, H_c, and remanence to saturation magnetization ratio, M_r/M_{sat}, of the nanoparticles are closely related to the volume or size of magnetic nanoparticles [59, 176]. These two properties have maximum values when the nanoparticles reach the critical sizes, which are around a few tens of nanometers in diameter depending on the chemical composition and crystalline structure of the particles. For instance, 20–40-nm nanoparticles have the highest coercivities for the CoNi alloy system [176]. Nanoparticles gradually become superparamagnetic due to random anisotropy when they are smaller than the critical size. Domain structure exists in particles larger than the critical size and is responsible for the decrease in coercivity and remanent magnetization. Unfortunately, Ostwald ripening, which has been successfully applied to the production of semiconductor quantum dots [177] and silver nanowires [178] of various sizes, does not seem to work on FePt nanoparticles.

1.5 Polymer–Iron Oxide Nanoparticles

Xu et al. [179] demonstrated the fabrication of highly charged, monodisperse superparamagnetic colloidal spheres. These superparamagnetic, monodisperse and charged polystyrene particles, containing nanoscale iron oxide nanoparticles, were synthesized by emulsion polymerization [180, 181]. The authors estimated from the magnetization measurements and the reaction stoichiometry that ~3.5% of the polystyrene particles contained iron oxide particles. The XRD spectra of iron oxide nanoparticles and polystyrene iron oxide composite particles are identical, indicating that the iron oxide structure was not changed during the emulsion polymerization (Fig. 1.23). Although these particles show XRD *d* spacings close to those of both magnetite and maghemite, which have inverse spinel structures, the relative diffraction intensities indicate that they are closer to maghemite than magnetite. The increased oxygen content may result from the oxidation of particles in air [182–185].

The iron oxide nanoparticles have a broad size distribution (2–15 nm) with an average diameter of ~10 nm (Fig. 1.24 a). The polystyrene–iron oxide composite particles have a number-average particle diameter of 134 nm (Fig. 1.24 b). These particles show a polydispersity of 7.5%. TEM shows that the iron oxide particles occur as clumps within the polystyrene particles. The aggregates of iron oxide nanoparticles appear as black dots in the center of the larger polystyrene–iron oxide composite particles.

Energy-dispersive X-ray fluorescence spectrometry (Fig. 1.25) shows the expected iron peaks. The peak due to Fe is clearly evident, while the strong Cu peak is derived from the copper grid used to support the sample.

A one-step inverse miniemulsion polymerization was adopted to synthesize magnetic (iron oxide, Fe_3O_4) polyacrylamide particles using a small molecular emulsifier (Span 80) and these magnetic polymeric particles can be attracted easily by a magnet [186]. A suitable amount of monomers [acrylamide (AAm) and *N,N*-methylenebisacrylamide (MBA)] and magnetic fluid (iron oxide), initiator, emulsi-

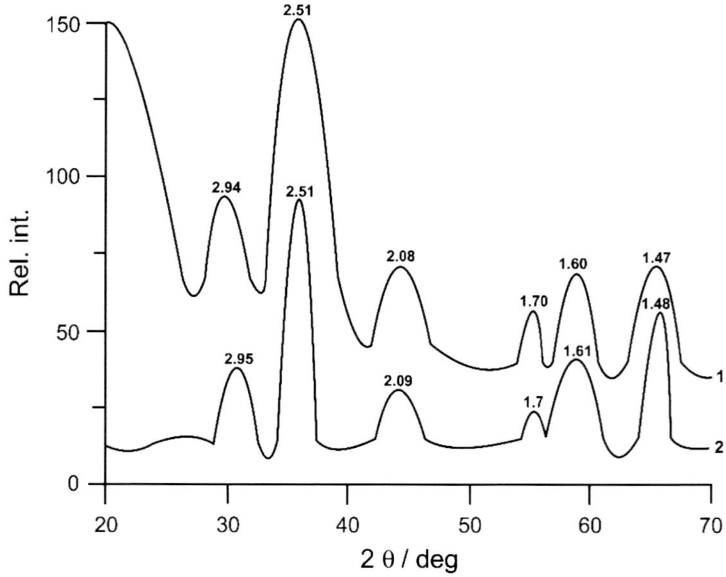

Fig. 1.23 XRD patterns (Cu Kα radiation) of nanoscale iron oxide (1) and polystyrene–iron oxide composite particles (2) [179].

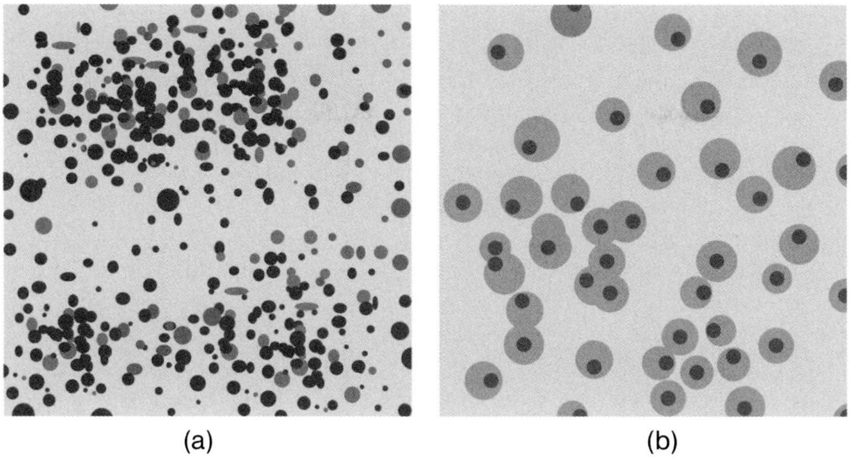

Fig. 1.24 TEM measurement of nanoscale iron oxide (a) and polystyrene–iron oxide composite particles (b) [179].

fier and cyclohexane were first mixed homogeneously, then ultrasonicated and finally polymerized. This reaction was very rapid and was completed in a few minutes. After the polymerization, the magnetic polymeric particles were washed three times with cyclohexane under the help of a magnet. The Fe_3O_4 particle size was determined as ~ 10 nm, which is a common value for monodomains of superparamagnetic iron oxide nanoparticles, as reported by Shinkai et al. [187].

A series of magnetic polymeric particles were prepared by miniemulsion polymerization (Table 1.4). On increasing the amount of Span 80 from 0.25 to 0.75 g,

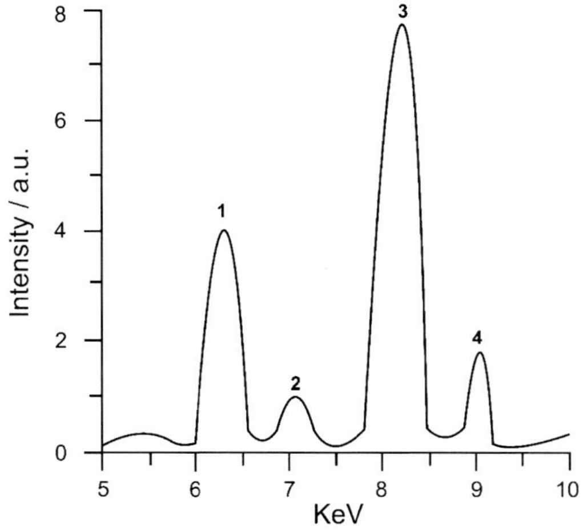

Fig. 1.25 Energy-dispersive X-ray fluorescence spectrum of polystyrene–iron oxide composite particles [179].

the particle size of the magnetic microspheres decreased from 109 to 81 nm. At the same time, the residue in the system decreased sharply with increase in emulsifier, similarly to that reported by Landfester et al. [188]. This observation indicates that the amount of Span 80 in this polymerization system has a significant effect on the particle size and formation of stable latex. Table 1.4 indicates that a stable miniemulsion can be achieved with 0.75 g of Span 80.

On increasing the ultrasonication time from 4 to 16 min, the particle size of the magnetic polymer microspheres decreases from 136 nm (ML4) to 67 nm (ML7). Table 1.4 also shows the effect of the ferrofluid content on the particle diameter. An increase in the ferrofluid content in the polymerization system reduced the particle size of the magnetic polymer microspheres, which means that the amount of Fe_3O_4 also has a great influence on the particle size. This is because during the fusion/fission process in ultrasonication, the droplets favored fission rather than fusion with increase in the amount of Fe_3O_4. Thus, Fe_3O_4 seems to act as a hydrophile, that is, it decreases the amount of water in the organic continuous phase and the transfer of water between the droplets. DLS analysis showed that the particle size of the magnetic polymeric latex (ML6, ML11, ML12 and ML13) fluctuated around 80 nm, which means that the amount of MBA (a cross-linking agent) in this polymerization system did not have much effect on the particle size. For example, the z-average particle size of the polymeric magnetic particles was 81 nm (ML11). DLS studies showed that the microspheres obtained had values ranging from 40 to 170 nm. The phase inversion method was applied to change sample ML11 from a water-in-oil latex

Table 1.4 Influence of varying synthesis parameters on the particle size of magnetic polymeric latex [186] [a].

Code	Ultrasonication time (min)	AAm (g)	MBA (g)	d_h [b]
ML1	10	2.25	0	109
ML2	10	2.25	0	92
ML3	10	2.25	0	81
ML4	4	2.25	0	136
ML5	8	2.25	0	104
ML6	10	2.25	0	84
ML7	16	2.25	0	67
ML8	10	2.25	0	160
ML9	10	2.25	0	116
ML10	10	2.25	0	67
ML11	10	2.14	0.11	81
ML12	10	2.04	0.21	82
ML13	10	1.95	0.3	81

a) Cyclohexane, 24.0 g; Fe_3O_4 dispersion (20 wt%) + H_2O, 3.0 g; AIBN, 75 mg; Span 80, 0.75 g (ML1 = 0.25, ML2 = 0.5 g).

b) The z-average particle size of magnetic polymeric latex without further treatment.

to an oil-in-water dispersion. The z-average particle size (200 nm) is much larger than that before phase inversion, because the particles in the aqueous phase had been swollen. A representative TEM image of the magnetic polymeric particles obtained from ML9 (Table 1.4) is shown in Fig. 1.26.

In Fig. 1.26, a large number of nanosized iron oxide particles can be observed in one magnetic polymeric particle and the iron oxide particles are mainly located inside the polymer particles. An SEM image of ML9 (Fig. 1.27) clearly shows that the surface of the magnetic polymeric particles is very smooth and this hints that the iron oxide nanoparticles were mainly located inside of the polymeric particles instead of on the surface of the particles. The size of these magnetic polymeric particles was about 50–200 nm, which was similar to the DLS result (116 nm) in Table 1.4.

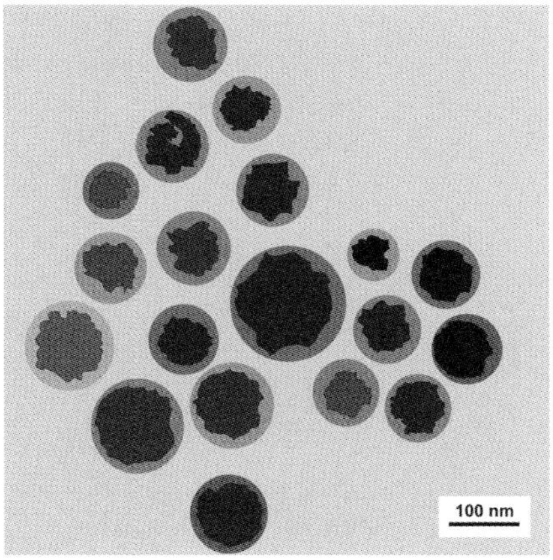

Fig. 1.26 TEM image of magnetic Fe_3O_4–PAAm particles (ML9) [186].

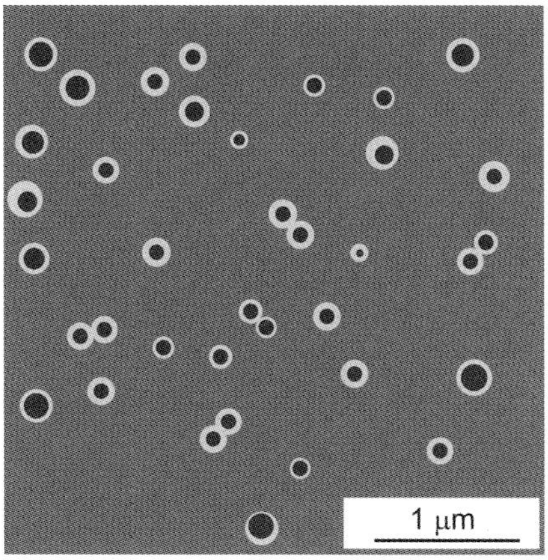

Fig. 1.27 SEM image of magnetic Fe_3O_4–PAAm particles (ML9) [186].

In an inverse system, hydrophilic monomer was dissolved in the above magnetic fluid and inverse miniemulsion polymerization was conducted successfully. The experimental results demonstrate that inverse miniemulsion polymerization is an effective way to synthesize magnetic polymeric particles. The magnetic polymeric particles are spherical and their size ranges from 60 to 160 nm depending on the reaction parameters. The nanosized iron oxide particles can be well encapsulated in polyacrylamide particles and the magnetic polymeric particles are superparamagnetic.

Hydrophilic magnetic nanoparticles were prepared according to the procedure reported by Shen et al. [189]. In addition to Shen et al.'s recipe, a dispersant agent, poly(methacrylic acid), was used to stabilize the iron oxide nanoparticles during the coprecipitation of $FeCl_2$ and $FeCl_3$ by ammonia under an N_2 atmosphere at 60 °C [186]. The particle dispersion was pretreated and then particles were redispersed in dilute ammonia for further use (the solid content was about 20 wt%). The PMAA-coated magnetic iron oxide nanoparticles dispersed in monomers were dispersed into cyclohexane and polymerized. The final polymer dispersion contained both the latex particles that do not contain magnetic iron oxide nanoparticles and those which contain magnetic iron domains. The polymer particles can be easily separated from the magnetic polymeric particles by a magnetic field.

1.6 Oxidation of Iron Oxide Particles

When the diluted aqueous solution is not deoxygenated and the synthesis proceeds under air without the protection of N_2 gas, the iron colloids obtained are reddish brown, which indicates that there is a contamination of other iron oxides in the colloids as a result of strong oxidation. Fe_3O_4 might be oxidized as follows:

$$Fe_3O_4 + 0.25O_2 + 4.5H_2O \rightarrow 3Fe(OH)_3 \quad (3)$$

This would critically affect the magnetic properties of nanosized particles and M_{sat} is decreased compared with previous former reactions.

It was reported that Fe_3O_4 can be oxidized to γ-Fe_2O_3, which can be further transformed into α-Fe_2O_3 at higher temperature [190]. Figure 1.28 A shows the XRD pattern of the as-synthesized, black 16-nm Fe_3O_4 nanoparticle assembly. After oxidation under O_2 at 250 °C, the black assembly is transformed into a red–brown form. Figure 1.28 B shows that all XRD peak positions and relative intensities of this red–brown material match well those of commercial γ-Fe_2O_3 powder materials, indicating that the oxidation of Fe_3O_4 under O_2 leads to γ-Fe_2O_3. Compared with Fig. 1.28 A, the large-angle peaks in Fig. 1.28 B shift slightly to larger angles, whereas at smaller angles there exist additional weak diffraction peaks of (110), (113), (210) and (213) that are characteristic of γ-Fe_2O_3 [191]. Figure 1.28 C shows the XRD pattern of the dark red–brown ma-

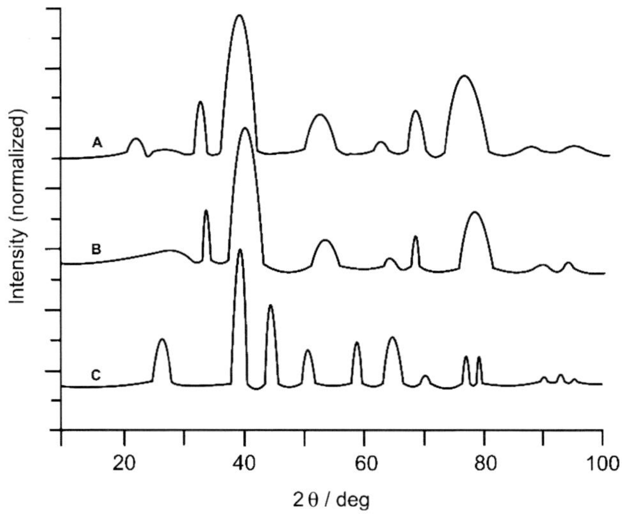

Fig. 1.28 XRD patterns of (A) a 16-nm Fe_3O_4 nanoparticle assembly, (B) a γ-Fe_2O_3 nanoparticle assembly obtained from the oxidation of (A) under oxygen at 250 °C for 6 h and (C) an γ-Fe_2O_3 nanoparticle assembly obtained from the further annealing of (B) under Ar at 500 °C for 1 h [72].

terial obtained after annealing of γ-Fe_2O_3 under Ar at 500 °C. The diffraction pattern matches that from known α-Fe_2O_3 materials [191], indicating the transformation of γ-Fe_2O_3 to α-Fe_2O_3 at high temperature. However, the as-synthesized Fe_3O_4 nanoparticles do not go through such a change if annealed under an inert atmosphere. Even at 650 °C, the Fe_3O_4 structure is still retained, as evidenced by both XRD and HRTEM. This confirms that the valence state of the iron cations in the as-synthesized sample closely matches that of Fe_3O_4 rather than similarly structured γ-Fe_2O_3. The transformation of Fe_3O_4 to Fe_2O_3 was further confirmed by near-edge X-ray absorption fine structure (NEXAFS) spectrometry in total electron yield mode.

When the Fe_3O_4 nanoparticles were oxidized by bubbling oxygen through the dispersion at room temperature, they precipitated from hexane as a red–brown powder [72]. Adding more oleic acid did not cause redispersion of this powder into hexane. However, adding oleylamine did so, leading to an orange–brown hexane dispersion. This is consistent with the observation that γ-Fe_2O_3 nanoparticles can be stabilized by alkylamine surfactants [192, 193], suggesting that $-NH_2$ coordinates with Fe(III) on the surface of the particles.

In the colloids, the very fine magnetite nanoparticles are oxidized to maghemite at room temperature under air; this phenomenon agrees with studies on the diffusion mechanism for Fe_3O_4 [101, 194]. The oxidation of magnetite to maghemite is known to occur through the outward diffusion of Fe^{2+} cations once it reaches the surface. The fractional conversion can be described by the following equation [195]:

$$M_t/M_\infty = 6\pi^{-\frac{1}{2}}(Dt/r_p^2)^{\frac{1}{2}} - 3(Dt/r_p^2) \quad (4)$$

where M_t is the quantity of Fe that has diffused out after time t, M_∞ is the corresponding quantity after infinite time, D is the diffusion constant and is a function of temperature T_a and r_p the average radius of the particles. The oxidation is strongly dependent on particle size and colloid dispersibility. A short oxidation time implies the good dispersibility of the magnetic particles.

The additional oxidation in the heterogeneously nucleated Fe oxide sample suggests that the presence of an oxygen-containing ligand (OA) encourages oxidation of surface Fe atoms [70]. The electron diffraction patterns of OA-stabilized particles made by extended heating cycles of homogeneously nucleated dispersions differed from those without OA. The greatest extent of oxidation was seen in particles synthesized with extended heating in the presence of OA, particularly the homogeneously nucleated particles. The oxidation evident in the electron diffraction patterns could arise either during the synthesis or during a later stage of TEM sample preparation. The least amount of oxidation was seen for heterogeneously nucleated particles coated with tributylphosphine ligands. These particles were studied through a series of annealing experiments performed on heterogeneously nucleated particles synthesized using small amounts of OA–OAm and on particles synthesized via homogeneous nucleation using a large amount of OA. The OA–OAm coating on heterogeneously nucleated 6.7-nm Fe particles was exchanged with tributylphosphine to limit the available oxygen at the particle surface and the particles were dried and annealed under a partial argon pressure for 30 min at 400 °C.

A TEM image of the post-annealed particles is shown in Fig. 1.29 a. Holes have appeared at the core of the particles and the particle diameter is larger than that prior to annealing. Particles coated with OA–OAm and annealed under the same conditions did not form holes at their centers. Because the annealing temperature is above the decomposition temperature for oleic acid [196], these particles sintered together, forming large iron oxide crystals. The morphology of the annealed tributylphosphine-coated particles suggests that a rapid reaction has occurred during annealing. This is consistent with Fe particles upon exposure to oxygen, which would occur after decomposition and removal of the tributylphosphine during annealing. The absence of this feature in OA–OAm-coated particles suggests that the oxygen from the carboxylic acid groups has begun to oxidize the particle surface prior to decomposition of the surfactant, preventing rapid reaction of the surface with ambient oxygen upon surfactant removal. The results of these annealing experiments also suggest that the amount of OA present during heterogeneous nucleation experiments is not sufficient to oxidize the particle surface significantly unless thermal energy is added to the particles.

Annealing experiments were repeated for a sample of 9-nm particles synthesized via homogeneous nucleation in the presence of excess OA to test for surface oxidation in these particles. The "donut" morphology is not observed for an-

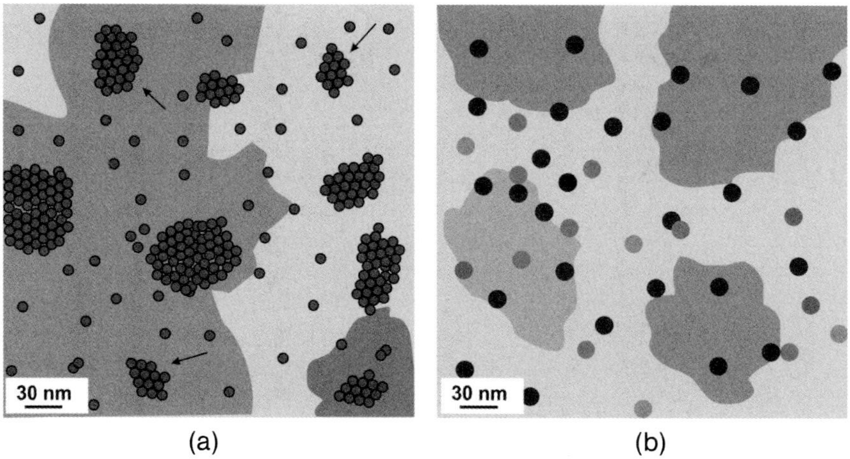

Fig. 1.29 TEM images of annealed tributylphosphine-coated Fe: (a) annealed 6.7-nm Fe particles made by the heterogeneous nucleation method (arrows point to particles with hollow centers); (b) annealed 9-nm Fe–Fe oxide particles made by the homogeneous nucleation method, showing no hollow centers [70].

nealed tributylphosphine-coated particles, shown in Fig. 1.29b, indicating that rapid oxidation of these particles does not take place and that an oxide shell is in place on these particles prior to drying.

1.7 Applications

Liquids containing small magnetic particles (ferrofluids) are of immense technological importance [197, 198]. Novel uses of magnetic fluids in the biomedical area increase significantly due to their novel physical properties. Dispersed in or encapsulated by appropriate polymeric matrixes, magnetic nanoparticles can be used in cell separation [26, 199], drug delivery [3], MRI techniques [27, 28] and advanced clinical applications such as cancer diagnosis and treatment. Ferrofluids made with the small magnetic particles thus obtained are suitable for biomedical applications. In particular, *in vivo*, the small size of these particles enables them to cross the endothelial barrier and so reach various target cells [200].

To use $M_{met}Fe_2O_4$ (M = metal ion) nanoparticles for future highly sensitive magnetic nanodevice and biomedical applications, a practical route to monodisperse $M_{met}Fe_2O_4$ nanoparticles with diameters smaller than 20 nm and a tight size distribution is needed. The $M_{met}Fe_2O_4$ nanoparticles may have numerous applications in magnetic nanodevices and biomedicine, but additional requirements may arise from particular applications. For example, in biomagnetic applications, the superparamagnetic nanoparticles often need to be water soluble [109, 201–204]. Superparamagnetic Fe_3O_4 nanoparticles can be made water soluble and yield a good magnetic signal that is suitable for spin valve sensor detection.

Due in part to this versatility, nanometer-scale $M_{met}Fe_2O_4$ materials have been among the most frequently chosen systems for studies of nanomagnetism and have shown great potential for many important technological applications, ranging from information storage and electronic devices to medical diagnostics and drug delivery. Dispersions of magnetic $M_{met}Fe_2O_4$ nanoparticles have been used widely not only as ferrofluids in sealing, oscillation damping and position sensing [197] but also as promising candidates for biomolecule tagging, imaging, sensing and separation [202, 204].

$PtFe_2O_3$ core–shell nanoparticles could be very useful in the study of magnetic properties in the mesoscopic range [205] and have broad ramifications in developing practical systems for data storage and in making solid-state magnetic nanodevices. This synthetic method can potentially be used to make FePt nanoparticles with different diameters and other nanoparticles and thin films, such as CoPt and AgCo alloys from their corresponding core–shell particles [206–209].

Based on their unique mesoscopic, physical, tribological, thermal and mechanical properties, superparamagnetic nanoparticles offer high potential for several applications in different areas. High electrical resistance and lower saturation magnetization make them suitable for magnetic and magneto-optical applications [112]. For the purpose of fabricating next-generation high-density disks, self-assembled ferromagnetic FePt nanoparticles were first produced by the IBM group [71]. The morphology of these nanoparticles has been observed by TEM and the crystallographic structure of each nanoparticle was studied by wide-angle XRD. However, the coalescence and long-range packing structure of the whole self-assembled layers have not been thoroughly studied. Further, in order to enhance the magnetic properties, a thermal annealing procedure is necessary to convert the structure of each particle from a chemically disordered structure (fcc) to the chemically ordered, L10, structure (fct) [210].

1.8 Particle Size Estimations

In order to analyze the size distribution quantitatively, it can be fitted using a log-normal function [201]:

$$P\ln(d/d_0) = 1/A\sigma_d(2\pi)^{0.5}\exp\{(-1/2\sigma_d^2)[\ln(d/d_0)^2]\} \tag{5}$$

where σ_d is the standard deviation of the diameter and d_0 is the mean diameter. For example, a mean diameter of d_0, as determined from Eq. (5), is about 3.8 nm with a standard deviation σ_d=0.15 for nanoparticles prepared by the piezoelectric nozzle method and d_0=7.0 nm; σ_d=0.17 for nanoparticles prepared by the pipette method.

With the XRD pattern, the average core size of the particles can be evaluated from the Scherrer equation [77, 108]:

$$L = 0.94\lambda / B(2\theta) \cos\theta \tag{6}$$

where L is equivalent to the average core diameter of the particles, λ is the wavelength of the incident X-ray, $B(2\theta)$ denotes the full width in radians subtended by the half-maximum intensity width of the powder peak, for instance (3 1 1) and θ corresponds to the angle of the (3 1 1) peak. For the (3 1 1) peak in the XRD pattern shown in Fig. 1.10, 2θ is observed as 35.52° and $B(2\theta)$ is 0.701. With λ being 0.15148 nm, L is obtained as 12.24 nm via Eq. (6).

The mean diameter of the nanoparticles is also verified by fitting the magnetization curve using the theoretical Langevin function [113]:

$$M_{mag}/M_{sat} = \coth(\mu H/k_B T_a) - (k_B T_a/\mu H) \tag{7}$$

where μ is the true magnetic moment of each particle, k_B is the Boltzmann constant, T_a is the absolute temperature, H is the applied field and M_{sat} is the saturation magnetization. Furthermore, the diameter of the magnetic cluster under different reaction conditions can be estimated using the Langevin function.

1.9 Conclusions

Various hydrolytic and nonhydrolytic methods have been reported for the synthesis of iron oxide nanoparticles with different structures [magnetite (Fe_3O_4), hematite (α-Fe_2O_3), maghemite (γ-Fe_2O_3) and ferric hydroxide (β-FeOOH)]. Iron oxide nanoparticles are commonly produced via coprecipitation of ferrous (Fe^{2+}) and ferric (Fe^{3+}) ions by a base, usually NaOH or NH_4OH, in an aqueous solution or they may be made by thermal decomposition of alkaline solution of Fe^{3+} chelate in the presence of reducing agent (e.g. hydrazine) and by sonochemical decomposition of hydrolyzed Fe(II) salt followed by thermal treatment. The formation of iron oxide particles can proceed in both the aqueous solution and micellar systems (especially in reverse microemulsions). The advantage of this method is the simple reaction conditions under which the reaction proceeds. The disadvantage of these aqueous solution syntheses is that the pH of the reaction mixture has to be adjusted in both the synthesis and purification steps and the process towards smaller monodisperse nanoparticles has had only very limited success. Monodisperse metal oxide nanoparticles can be synthesized by several nonhydrolytic methods using precursors of organometallic and coordination compounds. The nonhydrolytic synthesis has been developed to produce high-quality and monodisperse small iron oxide and metal–iron oxide nanocrystals. The shape of the nanocrystals is remarkably controllable through adjusting the particle growth rate. This synthesis method is possibly applicable to the synthesis of different spinel ferrite and other metal oxide nanoparticles. It has been demonstrated that seed-mediated growth is an effective route for the size-controlled synthesis of magnetite.

The nucleation of iron particles is mostly based on the supersaturation of the reaction solution by precursors (metal salts). In the absence of surface-active compounds, the formed particle nuclei agglomerate and form the primary particles. In the presence of stabilizer, the growth and agglomeration of particle nuclei are restricted. The particle dispersion consists of unstable metal nuclei, the stable primary particles and stable premature (large) particles. The particle growth may proceed by several simultaneous processes: (1) the agglomeration of colloidal unstable particle nuclei themselves leads to the formation of the stable primary particles; (2) the adsorption of metal nuclei by the primary particles leads to the particle growth and the formation of premature particles; (3) the reduction of precursor adsorbed on the surface of large particles leads to the particle growth; and (4) Ostwald ripening dissolves the primary particles due to which the mean particle size increases (Fig. 1.30). Metal particles nucleate and grow until the reagent is consumed.

The seeded synthesis is used to prepare fine metal particles, metal alloy particles and core–shell particles. The seed particles are used to control the particle growth rate and the heterogeneous particle nucleation mechanism. The small seeded iron nanoparticles (the seeds) can be mixed with one or several precursors to obtain a simple or multiplet core–shell structure of particles. This approach can be also used to prepare magnetic metal–iron oxide nanoparticles and iron–metal alloys in various media. For example, the cubic spinel structured $M_{met}Fe_2O_4$ or $M_{met}O^{\bullet}Fe_2O_3$ nanoparticles represent a well-known and important class of iron oxide–metal materials. By adjusting the chemical identity of M_{met}^{2+}, the magnetic configurations of $M_{met}Fe_2O_4$ can be molecularly engineered to provide a wide range of magnetic properties.

The high-temperature reaction of iron(III) acetylacetonate, $Fe(acac)_3$, in high-boiling solvents in the presence of a reducing agent and surface-active compounds was used to prepare monodisperse magnetite nanoparticles. Reduction of the Fe(III) salt to an Fe(II) intermediate occurs, followed by decomposition of the intermediate at high temperature. In the presence of a slight excess of 1-hexadecanethiol, a black powder corresponding to FeS could be separated. No metallic Fe was detected in the final product. When oleic acid (OA) and oleylamine (OAm) are mixed to make OA–OAm, heat is evolved. This is attributed to imide formation when the carboxylic acid group of OA reacts with the amine group of OAm, as in the formation of a peptide. Although both OA and OAm are liquid at room temperature, a mixture of the two heated to improve homogeneity solidifies when cooled. This is attributed to imide formation when the carboxylic acid group of OA reacts with the amine group of OAm, as in the formation of a peptide. The imide linkage between OA and OAm was reported to allow adequate particle stabilization in dilute solution. OA and OAm are necessary for the formation of particles. The sole use of OA during the reaction resulted in a viscous red–brown product that was difficult to purify and character-

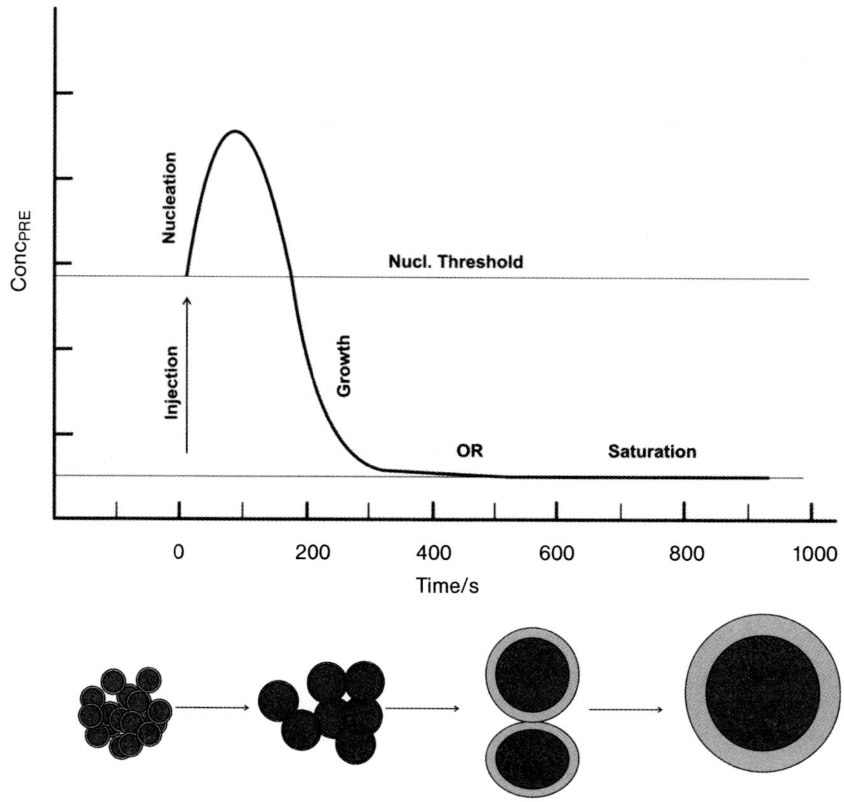

Fig. 1.30 Scheme depicting the stages of nucleation and growth for the preparation of monodisperse particles in the framework of the LaMer model [75].

ize. On the other hand, the use of OAm alone produced iron oxide nanoparticles in a much lower yield than the reaction in the presence of mixed OA–OAm. Long-chain monoalcohols, such as stearyl alcohol and oleyl alcohol, were used to prepare Fe_3O_4 nanoparticles, but the particle quality was poor. However, in the seed-mediated growth process, these monoalcohols can be used to form larger Fe_3O_4 nanoparticles.

Numerous groups have used metal carbonyl precursors. Fe nanoparticles can be prepared by the thermal decomposition of $Fe(CO)_5$ in organic solvents in the presence of stabilizers. Amorphous Fe nanoparticles can also be made provided that the temperature is kept low and the decomposition of the carbonyl is stimulated sonochemically. It is possible to narrow the size distribution of $Fe(CO)_5$-based nanoparticles sufficiently that, when coated with a surfactant, they spontaneously assemble into arrays when the solvent is evaporated.

Stabilizers or surface-active compounds such as organic ions are suggested to affect the formation of metal oxides or oxyhydroxides through two processes:

(1) the chelation of these ions with metal ions preventing nucleation and (2) the adsorption of these ions on the nuclei produced by hydrolysis inhibiting the growth of the nuclei. The presence of some additives can inhibit the nucleation. This leads to the formation of larger particles because the number of nuclei formed decreases and after nucleation only particle growth takes place. Organic (citrate) irons, for example, are known to affect the formation of iron oxides through superficial complexation. When citrate ions are adsorbed on the nuclei of iron oxides, the further growth of particles is inhibited. In these experiments, Fe_3O_4 nanoparticles are synthesized in the presence of excess citrate, which ensures that enough citrate ligands are adsorbed on the surface of Fe_3O_4 once the coprecipitation reaction occurs. The citrate ligands not only influence the growth of Fe_3O_4 nanoparticles, but also bring negative electric charges on particle surface to prevent their aggregation.

In the coprecipitation procedure, the average size and size distribution of Fe_3O_4 magnetite nanoparticles can be easily controllable via the use of NH_4OH or urea to manipulate the pH of a ferrite solution homogeneously. In addition, because of its non-toxicity, hydrophilicity and high bioaffinity, an oligosaccharide or polysaccharide can be selected as a surfactant on nanoparticles here to obtain a water-based magnetic fluid. Many bioprobes can be further coated on to polysaccharide through suitable chemical reactions to make the magnetic nanoparticles biocompatible. The control of the decomposition of urea can vary the average sizes of Fe_3O_4 particles. In the synthesis process with urea, larger particles were observed. Through the preparation procedure, the average diameter of the synthesized Fe_3O_4 particles can be adjusted from 8 to 50 nm, depending on the decomposition of urea in the ferrite solution.

Stable ionic ferrofluids can be obtained by surface modification of metal particles. The surface of metal nanoparticles can be also modified with appropriate bipolar additives. This can be performed by shaking of particles with bipolar molecules such as tetramethylammonium 11-aminoundecanoate. The Fe_3O_4 nanoparticles obtained are well dispersed in water without any noticeable agglomeration. Magnetic nanoparticles dispersed in water are superparamagnetic and under a tickling field can yield good magnetic signals that are readily sensed by a spin valve sensor.

The preparation of highly charged, monodisperse superparamagnetic metal–polymer spheres is also very important. These superparamagnetic, monodisperse and often charged metal–polymer particles, containing nanoscale iron oxide nanoparticles, can be synthesized through micellar [(mini)emulsion] polymerization.

The nanoparticle materials exhibit very interesting electrical, optical, magnetic and chemical properties, which cannot be achieved by their bulk counterparts. Especially iron oxide and ferrite have attracted a great deal of interest as ferrofluids. Due to the large surface area-to-volume ratio and magnetic dipole–dipole attractions, magnetic nanoparticles can also be very reactive.

Acknowledgments

This research was supported by the Science and Technology Assistance Agency through grants APVT-20-017304, APW-0173-06 and APVT-51-021702. The author is also indebted to the Alexander von Humboldt Stiftung for support.

List of Symbols and Abbreviations

a_d	center distance of two dipoles
AAm	acrylamide
AIBN	α,α'-azoisobutyronitrile pure
AOT	sodium bis(2-ethylhexyl)sulfosuccinate
bcc	body-centered cubic
CA	citric acid
CCAs	crystalline colloidal arrays
$Co(acac)_2$	cobalt acetyl acetonate
Cpl	crystalline plane
CTAB	cetyltrimethylammonium bromide
D	diffusion constant
d	average particle diameter
d_0	mean diameter
DLS	dynamic light scattering
d_M	magnetic size
DNA	deoxyribonucleic acid
d_{RX}	crystallite size
d_s	spacing parameter
d_{TEM}	particle diameter
E_A	magnetocrystalline anisotropy of a nanocrystal
ED	electron diffraction
EDS	energy-dispersive spectrometry
EDX	energy-dispersive X-ray spectrometry
EMR	electron magnetic resonance
EXAFS	extended X-ray absorption fine structure
fcc	face-centered cubic
fct	face-centered tetragonal
$Fe(acac)_3$	iron acetylacetonate
$Fe(CO)_5$	iron carbonyl
FeO	wustite
β-FeOOH	ferric hydroxide
α-Fe_2O_3	hematite
γ-Fe_2O_3	maghemite
Fe_3O_4	magnetite
FePt	ironplatinum
FHYD	ferrihydrite

H	external magnetic field
$HAuCl_4$	tetrachloroauric
H_c	coercivity
HPLC	high performance liquid chromatography
HRTEM	high-resolution TEM
ICP-AES	inductively coupled plasma atomic emission spectrometry
K	anisotropy constant
k_B	Boltzmann's constant
L	average core size
LB	Langmuir-Blodgett
M	magnetization
M	mol
M_∞	corresponding quantity of metal that has diffused out after infinite time
MBA	N,N-methylenebisacrylamide
MFe_2O_4	metal–iron oxide
M_{mag}	magnetisation
M_{met}	metal
M_r	remanent magnetization
M_r/M_{sat}	remanence to saturation magnetization ratio
MRI	magnetic resonance imaging
M_{sat}	saturation magnetization
M_t	quantity of metal that has diffused out after time t
N_p	number of particles
$NaBH_4$	sodium borohydride
NEXAFS	near-edge X-ray absorption fine structure
NM	nucleation method
OA	oleic acid
OA–OAm	oleic acid–oleyl amine
OAm	oleyl amine
PAAm	polyacrylamide
PCS	photon correlation spectroscopy
PMAA	poly(methacrylic acid)
polyBd	polybutadiene
polyS-Bd	poly(styrene-co-butadiene)
polyS-VP	poly(styrene-co-4-vinylpyridine)
$Pt(acac)_2$	platinum acetyl acetonate
PVP	polyvinylpyrrolidone
PXRD	powder X-ray diffraction
PZC	points of zero charge
R	radius
R_{mr}	the molar ratio of organic ions to metallic species
r_p	average radius of the particle
s	edge-to-edge interparticle spacing
SAED	selected area electron diffraction

SAXS	small-angle X-ray scattering
SD	single domain
SEC	size-exclusion chromatography
t	time
T	Tesla
T_a	absolute temperature
T_B	blocking temperature
T_C	Curie temperature
TEM	transmission electron microscopy
TEOS	tetraethyl orthosilicate
T_M	Morin transition temperature
T_N	Néel temperature
TOPO	trioctylphosphine oxide
V	volume of nanoparticle (nanocrystal)
XANES	X-ray absorption near-edge structure
XAS	X-ray absorption spectrometry
XRD	X-ray diffraction
XRLB	X-ray line broadening
ZFC	zero-field cooled curve
μ	dipole moment
μ_0	universal constant of permeability in free space
μ_{pt}	average saturation magnetic moment per polystyrene–iron oxide composite particle
ϕ_{mag}	interaction energy
σ_d	standard deviation of the diameter
θ	angle between the easy axis and magnetization direction
θ	temperature constant
τ_0	microscopic attempt time
χ	initial particle susceptibility

References

1 R.C. Ashoori, *Nature 379* (**1996**) 413.

2 B.H. Sohn, R.E. Cohen, G.C. Papaefthymiou, *J. Magn. Magn. Mater. 182* (**1998**) 216.

3 G.C. Hadjipanayis, R.W. Siegel (Eds.), *Nanophase Materials: Synthesis, Properties, Application*, Kluwer, Dordrecht, **1994**.

4 T. Hyeon, S.S. Lee, J. Park, Y. Chung, H.B. Na, *J. Am. Chem. Soc. 123* (**2001**) 12798.

5 R. Massart, *IEEE Trans. Magn. 17* (**1981**) 1247.

6 F.A. Tourinho, R. Franck, R. Massart, *J. Mater. Sci. 25* (**1990**) 3249.

7 J. Tejada, X.X. Zhang, Li Balcells, *J. Appl. Phys. 73* (**1993**) 6709.

8 L. Zhang, G.C. Papaefthymiou, J.Y. Ying, *J. Phys. Chem. B 105* (**2001**) 7414.

9 F. Jorgensen, *The Complete Handbook of Magnetic Recording*, McGraw-Hill, New York, **1995**.

10 R.E. Rosensweig, *AIAA J. 4* (**1966**) 1751.

11 R.S. Molday, D.J. Mackenzie, *Immunol. Methods 52* (**1982**) 353–367.

12 L.X. Tiefenauer, A. Tscgirky, G. Kuhne, R.Y. Andres, *Magn. Reson. Imaging 14* (**1996**) 391–402.
13 D.W. Elliott, W.X. Zhang, *Environ. Sci. Technol. 35* (**2001**) 4922–4926.
14 I.M.L. Billas, A. Chatelain, W.A. de Heer, *Science 265* (**1994**) 1682.
15 F.E. Luborsky, *J. Appl. Phys. 32* (**1961**) S171–S183.
16 B. Cullity, *Introduction to Magnetic Materials*, Addison-Wesley, Reading, MA, **1972**.
17 A.P. Alivisatos, *Science 271* (**1996**) 933.
18 J.R. Heath, *Acc. Chem. Res. 32* (**1999**) 388.
19 V.F. Puntes, K.M. Krishnan, A.P. Alivisatos, *Science 291* (**2001**) 2115.
20 Z.W. Pan, Z.R. Dai, Z.L. Wang, *Science 291* (**2001**) 1947.
21 Y.C. Rongchao Jin, C.A. Mirkin, K.L. Kelly, G.C. Schatz, J.G. Zheng, *Science 294* (**2001**) 1901.
22 X. Peng, U. Manna, W. Yang, J. Wickham, E. Scher, A. Kadavanich, A.P. Alivisatos, *Nature 404* (**2000**) 59.
23 T. Hyeon, Y. Chung, J. Park, S.S. Lee, Y.W. Kim, B.H. Park, *J. Phys. Chem. B 106* (**2002**) 6831.
24 S. Sun, H. Zeng, *J. Am. Chem. Soc. 124* (**2002**) 8204.
25 N.M. Pope, Y.A. Chang, A.K.J. Sonith, *Biomed. Mater. Res. 28* (**1994**) 499.
26 Z.L. Liu, Z.H. Ding, K.L. Yao, J. Tao, *J. Magn. Magn. Mater. 265* (**2003**) 98.
27 M. Abo, Z. Chen, L.J. Lai, T. Reese, B. Bjelke, *Neuroreport 12* (**2001**) 1543.
28 R.W. Briggs, Z. Chen, et al., *Magn. Reson. Imaging 15* (**1997**) 559.
29 D.R. Lide, *CRC Handbook of Chemistry and Physics*, 82nd edn, CRC Press, Boca Raton, FL, **2001**.
30 D.L. Huber, *Small 1*, No. 5 (**2005**) 482–501.
31 A. Wooding, M. Kilner, D.B. Lambrick, *J. Colloid Interface Sci. 144* (**1991**) 236.
32 T. Ishikawa, S. Kataoka, K. Kandori, *J. Mater. Sci. 28* (**1993**) 2693.
33 H. Kodama, M. Schnitzer, *Geoderma 19* (**1977**) 279.
34 K. Kandori, Y. Kawashima, T. Ishikawa, *J. Colloid Interface Sci. 152* (**1992**) 284.
35 P.S. Sidhu, R.J. Gilkes, A.M. Posner, *J. Inorg. Nucl. Chem. 40* (**1978**) 429, 10.
36 A. Bee, R. Massart, S. Neveu, *J. Magn. Magn. Mater. 149* (**1995**) 6.
37 R. Massart, *C. R. Acad. Sci. Paris, Ser. C 291* (**1980**) 1.
38 P. Fabre, R. Ober, M. Veyssie, V. Cabuil, *J. Magn. Magn. Mater. 85* (**1990**) 77.
39 F.E. Luborsky, *J. Phys. Chem. 61* (**1957**) 1336–1340.
40 R.S. Sapieszko, E. Matijevic, *J. Colloid Interface Sci. 74* (**1980**) 405.
41 R. Vijayakumar, Y. Koltypin, I. Felner, A. Gedanken, *Mater. Sci. Eng. A286* (**2000**) 101.
42 R. Zboril, M. Mashlan, D. Petridis, *Chem. Mater. 14* (**2002**) 969.
43 T. Smith, D. Wychick, *J. Phys. Chem. 84* (**1980**) 1621–1629.
44 K. Suslick, M. Fang, T. Hyeon, *J. Am. Chem. Soc. 118* (**1996**) 11960–11961.
45 T. J. Mason, *Ultrasonics 24* (**1986**) 245–253.
46 T. Hyeon, *Chem. Commun.* (**2003**) 927–934.
47 X. Cao, Y. Koltypin, G. Kataby, R. Prozorov, A. Gedanken, *J. Mater. Res. 10* (**1995**) 2952–2957.
48 H. Srikanth, R. Hajndl, C. Chirinos, J. Sanders, A. Sampath, T.S. Sudarshan, *Appl. Phys. Lett. 79* (**2001**) 3503–3505.
49 J. Brenner, J. Harkness, M. Knickelbein, G. Krumdick, C. Marshall, *Nanostruct. Mater. 8* (**1997**) 1–17.
50 J.L. Dormann, D. Fiorani (Eds.), *Magnetic Properties of Fine Particles*, North-Holland, Amsterdam, **1992**.
51 J.H. Fendler, et al., *J. Phys. Chem. 94* (**1990**) 2573.
52 P. Stroeve, et al., *Chem. Mater. 3* (**1996**) 2209.
53 J. Rockenberger, E.C. Scher, P.A. Alivisatos, *J. Am. Chem. Soc. 121* (**1999**) 11595.
54 S. Gangopadhyay, G.C. Hadjipanayis, C.M. Sorensen, K.J. Klabunde, V. Papaefthymiou, A. Kostikas, *Phys. Rev. B 45* (**1992**) 9778.
55 L. Kerekes, J. Hakl, S. Meszaros, K. Vad, P. Gurin, I. Kis-Varga, I. Uzonyi, S. Szabo, D. Beke, *Czech. J. Phys. 52* (**2002**) A89–A92.
56 M. Pardavi Horvath, L. Takacs, *Scr. Metall. Mater. 33* (**1995**) 1731–1740.
57 F. Dumestre, B. Chaudret, C. Amiens, P. Renaud, P. Fejes, *Science 303* (**2004**) 821–823.

58 L. Mohaddes-Ardabili, H. Zheng, S.B. Ogale, B. Hannoyer, W. Tian, J. Wang, S.E. Lofland, S.R. Shinde, T. Zhao, Y. Jia, L. Salamanca-Riba, D.G. Schlom, M. Wuttig, R. Ramesh, *Nat. Mater. 3* (**2004**) 533–538.

59 R.C. O'Handley, *Modern Magnetic Materials, Principles and Applications*, Wiley, New York (**2000**) 125–132, 434–437.

60 F. Li, C. Vipulanandan, K. Mohanty, *Colloids Surf. A 223* (**2003**) 103–112.

61 C. Sei, C. O'Connor, *Nanostruct. Mater. 12* (**1999**) 183–186.

62 A. Martino, M. Stoker, M. Hicks, C. Bartholomew, A. Sault, J. Kawola, *Appl. Catal. A 161* (**1997**) 235–248.

63 G.N. Glavee, K.J. Klabunde, C.M. Sorensen, G.C. Hadjipanayis, *Inorg. Chem. 34* (**1995**) 28–35.

64 L. Guo, Q. Huang, X. Li, S. Yang, *Phys. Chem. Chem. Phys. 3* (**2001**) 1661–1665.

65 S. Hisano, K. Saito, *J. Magn. Magn. Mater. 190* (**1998**) 371–381.

66 M.A. Lopez-Quintela, J. Rivas, *J. Colloid Interface Sci. 158* (**1993**) 446.

67 I. Capek, *Adv. Colloid Interface Sci. 110* (**2004**) 49.

68 A. Khan-Lodhi, B.H. Robinson, T. Towey, C. Hermann, W. Knoche, U. Thesing, in *The Structure, Dynamics and Equilibrium Properties of Colloidal Systems*, D.M. Bloor, E. Wyn-Jones (Eds.), NATO ASI Series C, Vol. 324, Kluwer, Dordrecht, (**1990**) 373.

69 N. Lufimpadio, J.B. Nagy, E.G. Derouane, in *Surfactants in Solution*, Vol. 3, K.L. Mittal, B.L. Lindman (Eds.), Plenum Press, New York (**1984**) 1483.

70 D. Farrell, S.A. Majetich, J.P. Wilcoxon, *J. Phys. Chem. B 107* (**2003**) 11022–11030.

71 S. Sun, C.B. Murray, D. Weller, L. Folks, A. Moser, *Science 287* (**2000**) 1989.

72 S. Sun, H. Zeng, D.B. Robinson, S. Raoux, P.M. Rice, S.X. Wang, G. Li, *J. Am. Chem. Soc. 126* (**2004**) 273–279.

73 N.R. Jana, L. Gearheart, C.J. Murphy, *Chem. Mater. 13* (**2001**) 2313.

74 H. Yu, P.C. Gibbons, K.F. Kelton, W.E. Buhro, *J. Am. Chem. Soc. 123* (**2001**) 9198.

75 V.K. LaMer, R.H. Dinegar, *J. Am. Chem. Soc. 72* (**1950**) 4847.

76 M.A. Watzky, R.G. Finke, *J. Am. Chem. Soc. 119* (**1997**) 10382.

77 H.P. Klug, L.E. Alexander, *X-Ray Diffraction Procedures for Polycrystalline and Amorphous Materials*, Wiley, New York (**1962**) 491–538.

78 J.P. Wilcoxon, S.A. Craft, *Nanostruct. Mater. 9* (**1997**) 85–88.

79 C.H. Griffiths, M.P. O'Horo, T.W. Smith, *J. Appl. Phys. 50* (**1979**) 7108.

80 P.H. Hess, P.H. Parker Jr., *J. Appl. Polym. Sci. 10* (**1966**) 1915–1927.

81 T. Prozorov, R. Prozorov, K.V.P.M. Shafi, A. Gedanken, *Nanostruct. Mater. 12* (**1999**) 669–672.

82 K.S. Suslick, S.B. Choe, A.A. Cichowlas, M.W. Grinstaff, *Nature 353* (**1991**) 414–416.

83 P. Kruus, *Ultrasonics 25* (**1987**) 20–22.

84 S. Wizel, R. Prozorov, Y. Cohen, D. Aurbach, S. Margel, A. Gedanken, *J. Mater. Res. 13* (**1998**) 211–216.

85 S. Wizel, S. Margel, A. Gedanken, *Polym. Int. 49* (**2000**) 445–448.

86 O. Rozenfeld, Y. Koltypin, H. Bamnolker, S. Margel, A. Gedanken, *Langmuir 10* (**1994**) 3919–3921.

87 G. Kataby, Y. Koltypin, X. Cao, A. Gedanken, *J. Cryst. Growth 166* (**1996**) 760–762.

88 G. Kataby, R. Prozorov, A. Gedanken, *Nanostruct. Mater. 12* (**1999**) 421–424.

89 G. Kataby, M. Cojocaru, R. Prozorov, A. Gedanken, *Langmuir 15* (**1999**) 1703–1708.

90 C. Yee, G. Kataby, A. Ulman, T. Prozorov, H. White, A. King, M. Rafailovich, J. Sokolov, A. Gedanken, *Langmuir 15* (**1999**) 7111–7115.

91 G. Kataby, A. Ulman, M. Cojocaru, A. Gedanken, *J. Mater. Chem. 9* (**1999**) 1501–1506.

92 G. Kataby, Y. Koltypin, A. Ulman, I. Felner, A. Gedanken, *Appl. Surf. Sci. 201* (**2002**) 191–195.

93 M.P. Pileni, A.T. Ngo, *Chem. Phys. Chem. 6* (**2005**) 1027–1034.

94 M.W. Grinstaff, M.B. Salamon, K.S. Suslick, *Phys. Rev. B 48* (**1993**) 269.

95 P.E. Jönsson, S. Felton, P. Svedlindh, P. Norblad, M.F. Hansen, *Phys. Rev. B 64* (**2001**) 212402.

96 J. van Wonterghem, S. Mørup, S.W. Charles, S. Wells, J. Villadsen, *Phys. Rev. Lett. 55* (**1985**) 410–413.

97 I. Nakatani, M. Hijikata, K.J. Ozawa, *Magn. Magn. Mater. 122* (**1993**) 10.

98 N.A.D. Burke, H.D.H. Stöver, F.P. Dawson, *Chem. Mater. 14* (**2002**) 4752–4761.

99 M.D. Bentzon, A.R. Thølen, *Ultramicroscopy 38* (**1991**) 105.

100 R.M. Cornell, P.W. Schindler, *Colloid Polym. Sci. 258* (**1980**) 1171.

101 Z.L. Liu, H.B. Wang, Q.H. Lu, G.H. Du, L. Penga, Y.Q. Du, S.M. Zhang, K.L. Yao, *J. Magn. Magn. Mater. 283* (**2004**) 258–262.

102 T.P. Raming, A.J.A. Winnubst, C.M. van Kats, A.P. Philipse, *J. Colloid Interface Sci. 249* (**2002**) 346.

103 J. Zhang, J. Buffle, *J. Colloid Interface Sci. 174* (**1995**) 500.

104 N. Kallay, I. Fischer, E. Matijevic, *Colloids Surf. 13* (**1985**) 145.

105 E. Matijevic, P. Scheiner, *J. Colloid Interface Sci. 63* (**1978**) 509.

106 J.K. Bailey, C.J. Brinker, M.L. Mecartney, *J. Colloid Interface Sci. 157* (**1993**) 1.

107 W. Jiang, H.C. Yang, S.Y. Yang, H.E. Hong, J.C. Hung, Y.C. Chen, C.-Y. Hong, *J. Magn. Magn. Mater. 283* (**2004**) 210–214.

108 B.E. Warren, *X-Ray Diffraction*, Dover Publications, Inc., New York (**1990**).

109 L. Fu, V. Dravid, D. Johnson, *Appl. Surf. Sci. 181* (**2001**) 173.

110 M. Lacava, R. Azevedo, L. Silva, Z. Lacava, K. Neto, N. Buske, A. Bakuzis, P. Morais, *Appl. Phys. Lett. 77* (**2000**) 1876.

111 J.R. Jeong, S.C. Shin, S.J. Lee, J.D. Kim, *J. Magn. Magn. Mater. 286* (**2005**) 5–9.

112 D.K. Kim, Y. Zhang, W. Voit, K.V. Rao, M. Muhammed, *J. Magn. Magn. Mater. 225* (**2001**) 30.

113 B.D. Cullity, *Introduction to Magnetic Materials*, Addison-Wesley, Reading, MA, **1972**, pp. 410–418.

114 T. Goetze, C. Gansau, N. Buske, M. Roeder, P. Gornert, M. Bahr, *J. Magn. Magn. Mater. 252* (**2002**) 399.

115 J.C. Bacri, R. Perzynski, D. Salin, V. Cabuil, R. Massart, *J. Magn. Magn. Mater. 85* (**1990**) 27.

116 G. Li, V. Joshi, R.L. White, S.X. Wang, J.T. Kemp, C. Webb, R.W. Davis, S. Sun, *J. Appl. Phys. 93* (**2003**) 7557.

117 R.L. Edelstein, C.R. Tamanaha, P.E. Sheehan, M.M. Miller, D.R. Baselt, L.J. Whitman, R.J. Colton, *Biosens. Bioelectron. 14* (**2000**) 805.

118 M.M. Miller, P.E. Sheehan, R.L. Edelstein, C.R. Tamanaha, L. Zhong, S. Bounnak, L.J. Whitman, R.J. Colton, *J. Magn. Magn. Mater. 225* (**2001**) 138.

119 S. Sun, S. Anders, T. Thomson, J.E.E. Baglin, M.F. Toney, H.F. Hamann, C.B. Murray, B.D. Terris, *J. Phys. Chem. B 107* (**2003**) 5419.

120 W. Kundig, H. Bommel, G. Constabaris, R.H. Lindquist, *Phys. Rev. 142* (**1966**) 327.

121 R. Zysler, D. Fiorani, J.L. Dorman, A.M. Testa, *J. Magn. Magn. Mater. 133* (**1994**) 71.

122 T. Takada, N. Kawai, *J. Phys. Soc. Jpn. 17* (**1962**) 691.

123 Y. Sun, L. Duan, Z. Guo, Y. DuanMu, M. Ma, L. Xu, Y. Zhang, N. Gu, *J. Magn. Magn. Mater. 285* (**2005**) 65–70.

124 W. Stober, A. Fink, E. Bohn, *J. Colloid Interface Sci. 26* (**1968**) 62.

125 P.A. Buining, L.M. Liz-Marzan, A.P. Philipse, *J. Colloid Interface Sci. 179* (**1996**) 318.

126 C. Rocchiccioli-Deltcheff, R. Franck, V. Cabuil, R. Massart, *J. Chem. Res. (S)* (**1987**) 126.

127 P.C. Fannin, S.W. Charles, D. Vincent, A.T. Giannitsis, *J. Magn. Magn. Mater. 252* (**2002**) 80.

128 Matsuchita, C.P. Chong, T. Mizutani, M. Abe, *J. Appl. Phys. 91* (**2002**) 7376.

129 T. Nakamura, T. Miyamoto, Y. Yamada, *J. Magn. Magn. Mater. 256* (**2003**) 340.

130 R.G.C. Moore, S.D. Evans, T. Shen, C.E.C. Hodson, *Physica E 9* (**2001**) 253.

131 J.J. Versluijs, M.A. Bari, J.M.D. Coey, *Phys. Rev. Lett. 87* (**2001**) 26601.

132 S. Soeya, J. Hayakawa, H. Takahashi, K. Ito, C. Yamamoto, A. Kida, H. Asano, M. Matsui, *Appl. Phys. Lett. 80* (**2002**) 823.

133 T.A. Sorenson, S.A. Morton, G. Dan Waddill, J.A. Switzer, *J. Am. Chem. Soc. 124* (**2002**) 7604.

134 T. Fried, G. Shemer, G. Markovich, *Adv. Mater. 13* (**2001**) 1158.

135 Z. J. Zhang, Z. L. Wang, B. C. Chakoumakos, J. S. Yin, *J. Am. Chem. Soc. 120* (**1998**) 1800.

136 S. Neveu, A. Bec, M. Robineau, D. Talbol, *J. Colloid Interface Sci. 255* (**2002**) 293.

137 M. P. Pileni, N. Moumen, *J. Phys. Chem. B 100* (**1996**) 1867.

138 C. Liu, B. Zou, A. J. Rondinone, Z. J. Zhang, *J. Phys. Chem. B 104* (**2000**) 1141.

139 Q. Guo, X. Teng, S. Rahman, H. Yang, *J. Am. Chem. Soc. 125* (**2003**) 630.

140 F. X. Redl, K. S. Cho, C. B. Murray, S. O'Brien, *Nature 423* (**2003**) 968.

141 T. Y. Kim, M. S. Lee, Y. I. Kim, C. S. Lee, J. C. Park, D. Kim, *J. Phys. D: Appl. Phys. 36* (**2003**) 1451.

142 A. R. West, *Basic Solid State Chemistry*, Wiley, New York, **1988**, pp. 356–359.

143 X. Teng, H. Yang, *J. Am. Chem. Soc. 125* (**2003**) 14559–14563.

144 X. Teng, D. Black, N. J. Watkins, Y. Gao, H. Yang, *Nano Lett. 3* (**2003**) 261–264.

145 H. Zeng, J. Li, J. P. Liu, Z. L. Wang, S. H. Sun, *Nature 420* (**2002**) 395–398.

146 D. Weller, S. H. Sun, C. Murray, L. Folks, A. Moser, *IEEE Trans. Magn. 37* (**2001**) 2185–2187.

147 S. H. Sun, E. E. Fullerton, D. Weller, C. B. Murray, *IEEE Trans. Magn. 37* (**2001**) 1239–1243.

148 T. S. Vedantam, J. P. Liu, H. Zeng, S. Sun, *J. Appl. Phys. 93* (**2003**) 7184–7186.

149 Y. Ding, S. Yamamuro, D. Farrell, S. A. Majetich, *J. Appl. Phys. 93* (**2003**) 7411–7413.

150 S. S. Kang, D. E. Nikles, J. W. Harrell, *J. Appl. Phys. 93* (**2003**) 7178–7180.

151 B. Bo, D. E. Laughlin, K. Sato, Y. Hirotsu, *IEEE Trans. Magn. 36* (**2000**) 3021–3023.

152 D. Givord, M. Rossignol, V. M. T. S. Bartherm, *J. Magn. Magn. Mater. 258–259* (**2003**) 1–5.

153 M. Chen, J. P. Liu, S. Sun, *J. Am. Chem. Soc. 126* (**2004**) 8394–8395.

154 J. Cheon, N. J. Kang, S. M. Lee, J. H. Lee, J. H. Yoon, S. J. Oh, *J. Am. Chem. Soc. 126* (**2004**) 1950–1951.

155 D. Weller, A. Moser, *IEEE Trans. Magn. 35* (**1999**) 4423–4439.

156 S. Sun, D. Weller, *J. Magn. Soc. Jpn. 25* (**2001**) 1434–1440.

157 Q. Song, Z. J. Zhang, *J. Am. Chem. Soc. 126* (**2004**) 6164–6168.

158 A. J. Rondinone, A. C. S. Samia, Z. J. Zhang, *J. Phys. Chem. B 103* (**1999**) 6876.

159 A. J. Rondinone, A. C. S. Samia, Z. J. Zhang, *Appl. Phys. Lett. 76* (**2000**) 3624.

160 M. J. Davies, S. C. Parker, G. W. Watson, *J. Mater. Chem. 4* (**1994**) 813.

161 C. M. Fang, S. C. Parker, G. De With, *J. Am. Ceram. Soc. 83* (**2000**) 2082.

162 R. Skomski, *J. Phys. Condens. Matter 15* (**2003**) R841–R896.

163 D. J. Sellmyer, C. P. Luo, M. L. Yan, Y. Liu, *IEEE Trans. Magn. 37* (**2001**) 1286–1291.

164 H. Zeng, J. Li, Z. L. Wang, J. P. Liu, S. Sun, *Nano Lett. 4* (**2004**) 187–190.

165 H. Gu, P. L. Ho, K. W. T. Tsang, L. Wang, B. Xu, *J. Am. Chem. Soc. 125* (**2003**) 15702–15703.

166 C. Xu, K. Xu, H. Gu, X. Zhong, Z. Guo, R. Zheng, X. Zhang, B. Xu, *J. Am. Chem. Soc. 126* (**2004**) 3392–3393.

167 E. Shevchenko, D. Talapin, A. Kornowski, J. Kotzler, M. Haase, A. Rogach, H. Weller, *Adv. Mater.* 12 (**2002**) 287–290.

168 B. Stahl, J. Ellrich, R. Theissmann, M. Ghafari, S. Bhattacharya, H. Hahn, N. S. Gajbhiye, D. Kramer, R. N. Viswanath, J. Weissmüller, H. Gleiter, *Phys. Rev. B 67* (14422) (**2003**) 1–12.

169 B. Jeyadevan, K. Urakawa, A. Hobo, N. Chinnasamy, K. Shinoda, K. Tohji, D. D. J. Djayaprawira, M. Tsunoda, M. Takahashi, *Jpn. J. Appl. Phys. 42* (**2003**) L350–L352.

170 T. Iwaki, Y. Kakihara, T. Toda, M. Abdullah, K. Okuyama, *J. Appl. Phys. 94* (**2003**) 6807–6811.

171 M. Nakaya, Y. Tsuchiya, K. Ito, Y. Oumi, T. Sano, T. Teranishi, *Chem. Lett. 33* (**2004**) 130–131.

172 M. Ulmeanu, C. Antoniak, U. Wiedwald, M. Farle, Z. Frait, S. Sun, *Phys. Rev. B 69* (54417) (**2004**) 1–5.

173 Y.K. Takahashi, T. Koyama, M. Ohnuma, T. Ohkybo, K. Hono, *J. Appl. Phys. 95* (**2004**) 2690–2696.

174 H.P.J. Wijin (Ed.), *Magnetic Properties of Metals: d-Elements, Alloys and Compounds*, Springer, Berlin, **1991**.

175 D. Weller, A. Moser, L. Folks, M.E. Best, W. Lee, M.F. Toney, M. Schwickert, J.U. Thiele, M.F. Doerner, *IEEE Trans. Magn. 36* (**2000**) 10–15.

176 D. Mercier, J.C.S. Levy, G. Viau, F. Fievet-Vincent, F. Fievet, P. Toneguzzo, O. Acher, *Phys. Rev. B 62* (**2000**) 532–544.

177 C.B. Murray, C.R. Kagan, M.G. Bawendi, *Annu. Rev. Mater. Sci. 30* (**2000**) 545–610.

178 Y. Sun, B. Mayers, T. Herricks, Y. Xia, *Nano Lett. 3* (**2003**) 955–960.

179 X. Xu, G. Friedman, K.D. Humfeld, S.A. Majetich, S.A. Asher, *Chem. Mater. 14* (**2002**) 1249–1256.

180 H. Noguchi, N. Yanase, Y. Uchida, T. Suzuta, *J. Appl. Polym. Sci. 48* (**1993**) 1539–1547.

181 N. Yanase, H. Noguchi, H. Asakura, T. Suzuta, *J. Appl. Polym. Sci. 50* (**1993**) 765–776.

182 F. Sauzedde, A. Elaissari, C. Pichot, *Colloid Polym. Sci. 277* (**1999**) 846–855.

183 L. Liz, A. Quintela, J. Mira, J. Rivas, *J. Mater. Sci. 29* (**1994**) 3797–3801.

184 J.A. Perez, M.A. Quintela, J. Mira, J. Rivas, S.W. Charles, *J. Phys. Chem. B 101* (**1997**) 8045–8047.

185 J. Lee, T. Isobe, N. Senna, *J. Colloid Interface Sci. 177* (**1996**) 490–494.

186 Z.Z. Xu, C.C. Wang, W.L. Yang, Y.H. Deng, S.K. Fu, *J. Magn. Magn. Mater. 277* (**2004**) 136–143.

187 M. Shinkai, H. Honda, T. Kobayashi, *Biocatalysis 5* (**1991**) 61.

188 K. Landfester, M. Willert, M. Antonietti, *Macromolecules 33* (**2000**) 2370.

189 L. Shen, P.E. Laibinis, T.A. Hatton, *Langmuir 15* (**1999**) 447.

190 G. Bate, in *Magnetic Oxides, Part 2*, D.J. Craik (Ed.), Wiley, New York, **1975**, pp. 705–707.

191 R.M. Cornell, U. Schwertmann, *The Iron Oxides: Structure, Properties, Reactions, Occurrence and Uses*, VCH, New York, **1996**, pp. 167–168.

192 M. Rajamathi, M. Ghosh, R. Seshadri, *Chem. Commun.* (**2002**) 1152.

193 A.K. Boal, K. Das, M. Gray, V.M. Rotello, *Chem. Mater. 14* (**2002**) 2628.

194 P.S. Sidhu, R.J. Gilkes, A.M. Posner, *Soil Sci. Soc. Am. J. 44* (**1980**) 135.

195 J. Tang, M. Myers, K.A. Bosnick, L.E. Brus, *J. Phys. Chem. 107* (**2003**) 7501.

196 Y. Sahoo, H. Pizem, T. Fried, D. Golodnitsky, L. Burstein, C.N. Sukenik, G. Markovich, *Langmuir 17* (**2001**) 7907–7911.

197 K. Raj, R. Moskowitz, *J. Magn. Magn. Mater. 85* (**1990**) 233.

198 B.M. Berkovsky, V.F. Medvedev, M.S. Krokov, *Magnetic Fluids: Engineering Applications*, Oxford University Press, Oxford, **1993**, p. 218.

199 N.M. Pope, Y.A. Chang, A.K.J. Sonith, *Biomed. Mater. Res. 28* (**1994**) 499.

200 R. Weissleder et al., *Radiology 181* (**1991**) 245.

201 D.K. Kim, Y. Zhang, J. Kehr, T. Klason, B. Bjelke, M. Muhammed. *J. Magn. Magn. Mater. 225* (**2001**) 256.

202 Q.A. Pankhurst, J. Connolly, J. Dobson, *J. Phys. D: Appl. Phys. 36* (**2003**) R167.

203 P. Tartaj, M.P. Morales, S. Veintemillas-Verdaguer, T. Gonzalez-Carreno, C.J. Serna, *J. Phys. D: Appl. Phys. 36* (**2003**) R182.

204 C.C. Berry, A.S.G. Curtis, *J. Phys. D: Appl. Phys. 36* (**2003**) R198.

205 S.A. Majetich, Y. Jin, *Science 284* (**1999**) 470–473.

206 J.I. Park, J. Cheon, *J. Am. Chem. Soc. 123* (**2001**) 5743–5746.

207 N.S. Sobal, U. Ebels, H. Möhwald, M. Giersig, *J. Phys. Chem. B 107* (**2003**) 7351–7354.

208 N.S. Sobal, M. Hilgendorff, H. Mohwald, M. Giersig, M. Spasova, T. Radetic, M. Farle, *Nano Lett. 2* (**2002**) 621–624.,

209 C.J. O'Connor, J.A. Sims, A. Kumbhar, V.L. Kolesnichenko, W.L. Zhou, J.A. Wiemann, *J. Magn. Magn. Mater. 226* (**2001**) 1915–1917.

210 H. Zeng, S. Sun, T.S. Vedantam, J.P. Liu, Z.R. Dai, Z.L. Wang, *Appl. Phys. Lett. 80* (**2002**) 2583.

2
Solubilization of a Poorly Soluble Aromatic Drug by Micellar Solutions of Amphiphilic Block Copoly(oxyalkylene)s*

David Attwood and Colin Booth

Abstract

This chapter is a summary of our work on the design of block copolymer micellar systems with improved solubilization capacity for poorly soluble aromatic drugs. The copolymers of interest are block copoly(oxyalkylene)s with linear di- or triblock architecture which combine hydrophilic poly(ethylene oxide) with hydrophobic blocks formed from poly(propylene oxide), poly(1,2-butylene oxide), poly(styrene oxide) or poly(phenyl glycidyl ether). A comparison of the solubilization capacity per gram of hydrophobic block for the model poorly soluble drug griseofulvin shows an increase in solubilization capacity with increase in the hydrophobicity of the core-forming block and the volume of the core of spherical micelles. As a consequence of their larger core diameter, diblock copolymers are more effective solubilizers than triblock copolymers of similar composition and chain length. Copolymers with short hydrophilic blocks relative to the length of the hydrophobic block may form cylindrical or worm-like micelles of high aggregation number resulting in an increased solubilizing capacity. The formulation of mixed block copolymer systems of high solubilization capacity and favorable gelation characteristics for use in controlled drug delivery is discussed.

2.1 Introduction

It is estimated that almost half of the potentially useful drug candidates identified by high-throughput methods of screening have pronounced solubility problems. Although many of these are highly potent, their very low aqueous solubility usually gives rise to poor or erratic absorption characteristics and consequently these compounds are frequently not taken forward to formulation development. A variety of colloidal delivery systems, including, for example, liposomes, nanoparticles, micro-

* An explanation of the notation used can be found at the end of this chapter.

Colloids and Interface Science Series, Vol. 3
Colloid Stability and Application in Pharmacy. Edited by Tharwat F. Tadros

ISBN: 978-3-527-31463-8

spheres, dendrimers and cyclodextrins, have been investigated for their potential as drug carriers in attempts to overcome this problem. In this chapter, we focus on the application of block copolymer micelles as alternative carriers for poorly water-soluble drugs. In these systems, the hydrophobic micelle core is stabilized by the hydrophilic chains of the micelle corona and provides a suitable microenvironment for solubilization of the drug. A poly(oxyethylene) corona allows the micelles to evade scavenging by the mononuclear phagocyte system, resulting in lengthy circulation times in the bloodstream [1, 2]. Within limits, the composition of the core can be varied independently of the E-block corona, and copolymers with a wide range of hydrophobic blocks have been examined for their potential for solubilizing poorly soluble drugs, as described in a number of reviews (e.g. [2–5]), recently by Torchilin [6], who has reviewed the literature (including the numerous patents) relating to the use of amphiphilic block copolymers in the solubilization of poorly soluble drugs and strategies to target these micellar carriers to tumors. Allen et al. [3] have summarized the characteristics to be sought in a micellar system for drug delivery: small micelle size, narrow size distribution, a low critical micelle concentration (c.m.c.) giving a high extent of micellization and stability on injection, and high solubilization capacity. This last can be achieved by ensuring compatibility of the micelle core with the target drug, a point which is emphasized in [3] by the comparison of partition coefficients of pyrene between water and micelle, i.e. a variation in order from 10^2 for micelles with a poly(propylene oxide) core to 10^5 for micelles with a polystyrene core. We can add the requirement of mobility of the chains comprising the core, a feature which controls the rate of uptake and release of the drug. Allen et al. [3] also note that micelle morphology is an important variable, particularly when micelles are formed by predissolution in a water-compatible organic solvent followed by dialysis against water to remove the solvent.

Within the family of copoly(oxyalkylene)s, many studies of the solubilization of drugs and drug models have been carried out with the commercially available linear triblock copolymers of ethylene oxide and propylene oxide, type $E_mP_nE_m$. Relevant references can be found in published reviews and it is not our intention to revisit that work. However, we mention early work in our laboratory [7] and, particularly, the extensive work of Kabanov and coworkers (e.g. [4, 8]). Useful guidelines emerge from related work. Hatton and coworkers studied, both experimentally [9] and theoretically [10], the solubilization of polycyclic aromatics in solutions of $E_mP_nE_m$ copolymers. In addition to showing the beneficial effect of a high P content, they showed that linear copolymers are preferable to branched. Theory showed an increase in micelle association number with concentration of aromatic solute [10], an effect supported by the Monte Carlo simulations of Li and Mattice [11]. This effect implies a lowering of the c.m.c. in the presence of hydrophobic solutes, as predicted theoretically by Nagarajan and Ganesh [12] for diblock P_nE_m copolymers in water. Nagarajan and coworkers had used a diblock P_nE_m copolymer in an investigation of the solubilization of aliphatic and aromatic solutes [13], and corresponding theory was developed through the years, initially for P_nE_m copolymers [12, 14] and later for $E_mP_nE_m$

copolymers [15, 16]. Detailed information was obtained on the extent of solubilization in spherical micelles and the effect of an increase in extent of solubilization in inducing sphere-to-cylinder and cylinder-to-lamella transitions [16].

In this chapter, we draw together our work over several years aimed at improving the solubilization capacity of block copoly(oxyalkylene)s for poorly soluble aromatic drugs. We have used copolymers prepared and characterized under controlled conditions to explore effects of block architecture and composition on micellization in dilute solution [17] and on the gelation of concentrated micellar solutions [18] and additionally, as discussed below, to explore pharmaceutical application in drug delivery. The emphasis has been on block copoly(oxyalkylene)s which form spherical or near spherical micelles on dissolution in water. However, spherical micelles of copolymers with short E-block lengths relative to the length of the hydrophobic (core-forming) block may undergo a sphere-to-cylinder transition on heating the solution, an effect which has been recognized for certain $E_mP_nE_m$ copolymers for many years, e.g. for copolymers $E_{27}P_{39}E_{27}$ (denoted P85) and $E_{21}P_{47}E_{21}$ (denoted P94) (see, e.g., [19–24]). As indicated by Nagarajan [16], there is an advantage so far as solubilization capacity is concerned in using copolymers which form cylindrical micelles and we have synthesized copolymers which form cylindrical, even worm-like, micelles at room temperature or just above [25–27].

In our work, all solutions were prepared by dissolving the copolymers directly in water at room temperature and, since values of the glass transition temperature (T_g) of the hydrophobic blocks are low, it can be safely assumed that all results are for systems at equilibrium. Of the components of the copolymers considered in this chapter, only poly(styrene oxide) and poly(phenyl glycidyl ether) have values of T_g above 0 °C, namely *ca.* 40 and 18 °C, respectively [28–30], but these values apply to lengthy chains and T_g is much lower for the short blocks of the copolymers, as demonstrated directly for S blocks by investigation of chain dynamics in micellar cores [31].

Bearing in mind earlier studies by Elworthy and coworkers on poly(oxyethylene) alkyl ethers [32, 33], we have used griseofulvin (**1**) as the standard aromatic drug for comparing the solubilization capacity of micellar solutions of our different block copolymers. Griseofulvin is a crystalline solid with a very low water solubility, approximately 10 mg dm^{-3} at 25 °C and 12 mg dm^{-3} at 37 °C [34].

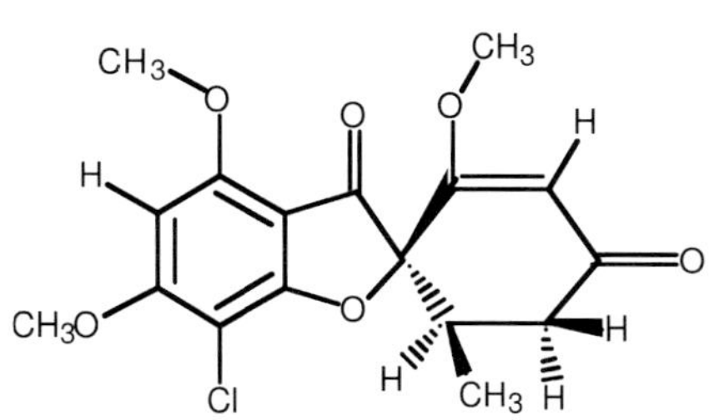

1

2.2 Copolymers

The preparation of the copoly(oxyalkylene)s by sequential anionic polymerization of two monomers from an appropriately chosen monofunctional or difunctional initiator is a well-established technique [35]. The underlying reaction can be illustrated by the polymerization of ethylene oxide (EO) initiated by a monofunctional diethylene glycol monomethyl ether, part of which is in the form of its alkali metal salt (e.g. $-O^-K^+$). Initiation is instantaneous and the reaction scheme is written as follows:

$$\text{Slow propagation:}\quad \mathrm{MeE}_m\mathrm{O}^-\mathrm{K}^+ + \mathrm{EO} \rightarrow \mathrm{MeE}_{m+1}\mathrm{O}^-\mathrm{K}^+$$

$$\text{Fast equilibration:}\quad \mathrm{MeE}_m\mathrm{O}^-\mathrm{K}^+ + \mathrm{MeE}_n\mathrm{OH} \rightleftharpoons \mathrm{MeE}_m\mathrm{OH} + \mathrm{MeE}_n\mathrm{O}^-\mathrm{K}^+$$

where E_m and E_n represent growing chains. The active center is an ion pair and the rapid equilibration ensures that all chains are equally likely to grow. This ideal reaction produces a polymer with a number-average molar mass (M_n) given by the mass of EO polymerized relative to the moles of initiator used and with a Poisson distribution of chain lengths [36]. In the absence of deliberate or accidental termination, the system is living and is ideal for preparation of block copolymers.

The same scheme applies to propylene oxide (PO), but with the addition of a hydrogen-abstraction transfer reaction [37]:

$$\text{Transfer:}\quad \mathrm{MeP}_n\mathrm{O}^-\mathrm{K}^+ + \mathrm{PO} \rightarrow \mathrm{MeP}_n\mathrm{OH} + \mathrm{CH}_2 = \mathrm{CHCH}_2\mathrm{O}^-\mathrm{K}^+$$

This reaction reduces M_n and broadens the chain length distribution, since the unsaturated alcohol initiates new chains. It has consequences for the compositions of E/P block copolymers, not only broadening the P-block-length distribution, but also giving rise to a proportion of diblock PE copolymers in purportedly $E_mP_nE_m$ triblocks and of homopoly(oxypropylene) if P is added in the second stage of sequential copolymerization. A similar reaction has been recognized in the polymerization of phenyl glycidyl ether [38], but it is absent in the anionic polymerization of 1,2-butylene oxide (BO) and styrene oxide (SO) under laboratory conditions [39, 40], although it may occur in polymerization under the more extreme conditions (high temperatures) used in industry [35]. Impurities arising from this effect, and from initiation by moisture accidentally introduced with the monomers, are either accounted for in interpretation of results or, preferably, are removed [41].

A second effect arises because monomers other than EO add to a growing chain to give secondary oxyanions and slow initiation of the polymerization of EO by a first-formed block terminated in this way may result, leading to a widened E-block distribution. This effect has been recognized in the copolymerization of EO on to a preformed B chain [42, 43]. If the E-block is short, uncapped ends may be present at the completion of the reaction. In forming a triblock copolymer with E end blocks, uncapped ends imply a proportion of diblock copolymer. This problem is avoided in the preparation of diblock copolymers by polymerizing the E-block first. For triblocks and diblocks polymerized in reverse order, the effect is minimized if the E blocks are lengthy.

We stress the importance of the thorough molecular characterization of our copolymers. Complete information is obtained by monitoring the copolymerization at each stage using analytical gel permeation chromatography (GPC) (appropriately calibrated) to check the chain length distribution and ^{13}C nuclear magnetic resonance spectroscopy (carried out under conditions which give complete relaxation between pulses and with due allowance, where appropriate, for differential nuclear Overhauser enhancements) for absolute number-average chain length and confirmation of the block structure. Valuable supplementary

Table 2.1 Poly(oxyalkylene)s investigated.

Copolymer	w_E (wt%)	M_n (kg mol^{-1})	M_w/M_n	Ref.
$E_{130}P_{58}$	0.63	9.1	1.05	44
$E_{107}P_{69}$	0.54	8.7	1.05	44
$E_{62}P_{39}E_{62}$ (F87)	0.70	5.0	–	–
$E_{21}P_{47}E_{21}$ (P94)	0.40	4.6	–	–
$E_{21}P_{67}E_{21}$ (P123)	0.30	5.8	–	–
$E_{98}P_{67}E_{98}$ (F127)	0.70	12.5	–	–
$E_{11}B_8$	0.46	1.1	1.03	25
$E_{49}B_9$	0.77	2.8	1.03	45
$E_{96}B_{18}$	0.76	5.5	1.03	46
$E_{134}B_{19}$	0.81	7.3	1.03	47
$E_{43}B_{14}E_{43}$	0.69	4.8	1.08	48
$E_{50}S_{3.5}$	0.84	2.6	1.02	49
$E_{50}S_5$	0.78	2.8	1.04	49
$E_{51}S_{6.5}$	0.74	3.0	1.03	49
$E_{17}S_8$	0.44	1.7	1.05	26
$E_{45}S_8$	0.67	2.9	1.06	50
$E_{45}S_{10}$	0.62	3.2	1.04	51
$S_{10}E_{135}$	0.83	7.1	1.04	50
$S_{13}E_{60}$	0.63	4.2	1.03	31
$S_{15}E_{63}$	0.61	4.6	1.04	51
$S_{17}E_{65}$	0.58	4.9	1.04	51
$S_{20}E_{67}$	0.55	5.3	1.05	51
$E_{82}S_8E_{82}$	0.88	8.2	1.07	52
$E_{20}S_{10}E_{20}$	0.59	3.0	1.03	50
$E_{66}S_{13}E_{66}$	0.79	7.4	1.04	52
$E_{67}S_{15}E_{67}$	0.77	7.7	1.04	52
G_5E_{67}	0.79	3.7	1.03	53
$E_{71}G_7E_{71}$	0.85	7.3	1.05	54
$E_{62}G_8E_{62}$	0.82	6.7	1.05	54
$E_{38}G_{12}E_{38}$	0.65	5.2	1.05	54

information can be obtained from matrix-assisted laser desorption/ionization time-of-flight (MALDI-TOF) mass spectrometry. These procedures are possible because the chain lengths involved seldom exceed 500 chain units ($M_n \leq$ 20 000 g mol^{-1}). Other methods of molecular characterization (e.g. static light scattering, membrane osmometry), which are indispensable at higher molar masses, may also provide useful supplementary information.

The copolymers used for solubilization in our work are listed in Table 2.1. The characteristics listed are the weight fraction of the E blocks (w_E) and the number-average molar mass (M_n) from NMR spectroscopy, and the ratio of weight- to number-average molar mass (M_w/M_n) from GPC. The references give additional information about their preparation and characterization. The $E_mP_nE_m$ copolymers were commercial samples (as indicated) and were used as received.

2.3 Micellization and Micelle Properties

Ideally copolymers for drug solubilization should be completely micellized in dilute solution in water at 25 °C. This would allow drugs to be solubilized and the solutions stored at room temperature prior to injection at body temperature. This restriction effectively rules out $E_mP_nE_m$ copolymers. For selected copolymers, Table 2.2 shows values of the critical micelle concentration at 25 °C and of the critical micelle temperature (c.m.t.) of 1 wt% solutions. These copolymers have values of the 1 wt% c.m.t. lower than those reported for other $E_mP_nE_m$ copolymers. The micellization range is *ca.* 20 °C [55–57], implying that P123, the copolymer with the lowest c.m.c. in wt% units, is completely micellized only at *ca.* 36 °C. It is interesting that much early experimental work on the solubilization of aromatics was carried out at 25 °C, i.e. under conditions where the extent of micellization could be very low [9, 13, 58].

It is characteristic of $E_mP_nE_m$ copolymers that the c.m.c. is very temperature dependent, e.g. values listed by Alexandridis et al. [59] for copolymer F127 vary

Table 2.2 Micellization of $E_mP_nE_m$ copolymers.

Copolymer	Nominal formula	c.m.c. (25 °C) (wt%)	c.m.t. (1 wt%) (°C)	Ref.
F87	$E_{62}P_{39}E_{62}$	2.6	30	61
P94	$E_{21}P_{47}E_{21}$	0.3	23	55
P103	$E_{20}P_{54}E_{20}$	0.07	20	59
P104	$E_{31}P_{54}E_{31}$	0.1, 0.3	19, 22	59, 61
P105	$E_{38}P_{54}E_{38}$	0.2	22	59
L122	$E_{13}P_{67}E_{13}$	0.01	18	62
P123	$E_{21}P_{67}E_{21}$	0.03	16	59
F127	$E_{98}P_{67}E_{98}$	0.7, 1.0	24, 25	57, 59, 63

from *ca.* 1 wt% at 25 °C to ca. 0.02 wt% at 37 °C. In this respect, we note that the values of the c.m.c. in Table 2.1 are considerably higher than those listed by Kabanov and coworkers in their review [4], reflecting the difference in solution temperature, 25 versus 37 °C [60].

Based on a comparison of values of c.m.c.s for diblock copolymers in molar units [17, 51], at 25 °C the relative hydrophobicities per chain unit are as follows:

$$P:C:B:S:G = 1:5:6:12:14 \tag{1}$$

with similar values for triblock copolymers [17, 52]. We include the value for the methylene unit of an alkyl ethoxylate (denoted C) to keep contact with work in that area. With the exception of copolymers with P blocks, the effect of temperature on these ratios can be ignored over the temperature range 25–37 °C. Using the standard (van't Hoff) enthalpy of micellization as a measure of the temperature dependence of the c.m.c., the values reported for copolymers with P blocks are high, 200–400 kJ (mol copolymer)$^{-1}$, whereas those for the other copoly(oxyalkylene)s are relatively low, even approaching zero for lengthy blocks [17, 51, 52], an effect attributed to collapse of the hydrophobic block prior to micellar association [64, 65].

In polymer solution theory (Flory–Huggins theory), the interaction parameter used to characterize the interaction between components is based on a common segment volume. Using known specific volumes of related homopolymers, the ranking of hydrophobicity for segments of equal volume becomes

$$P:B:S:G:C = 1:5:7:7:18 \tag{2}$$

The determining factor for micellization, i.e. the product of the interaction parameter and the chain length measured in chain segments, is unchanged by choice of segment volume. The ranking based on a common segment volume gives a more realistic view of the variation of hydrophobicity with chemical composition than that based on chain units, particularly so for alkyl chains, and more importantly because it relates directly to the hydrophobicity of micelle cores of equal volume. However, block lengths in chain units, which relate directly to chemical composition, are more usual and are used in this chapter.

The c.m.c.s at 25 °C and c.m.t.s of 1 wt% solutions are listed in Table 2.3. When values were not measured directly or interpolated from measurements at nearby temperatures, values of the c.m.c. were estimated making use of the plots of log(c.m.c./molar units) against hydrophobic block length, while values of the c.m.t. were estimated making use of van't Hoff plots of log c against reciprocal temperature: examples of these plots can be found elsewhere (e.g. [17, 51, 52]). The c.m.t.s of 1 wt% solutions are below 0 °C for all except the copolymers with P blocks. Values of the micelle hydrodynamic radius (r_h) determined by dynamic light scattering are listed as an indicator of micelle size, albeit one which is dominated by the width of the E-block corona. Values of the

Table 2.3 Micellization and micelle properties [a)].

Copolymer	c.m.c. (25 °C) (wt%)	c.m.t. (1 wt%) (°C)	r_h (25 °C) (nm)	Ref.
$E_{130}P_{58}$	*0.1*	*16*	–	66
$E_{107}P_{69}$	*0.02*	*14*	–	66
$E_{62}P_{39}E_{62}$ (F87)	2.6	30	9 (50 °C)	61
$E_{21}P_{47}E_{21}$ (P94)	0.3	23	8	23, 55
$E_{21}P_{67}E_{21}$ (P123)	0.03	16	10 (30 °C)	59, 67
$E_{98}P_{67}E_{98}$ (F127)	0.7	25	12 (35 °C)	57, 59, 67, 68
$E_{11}B_8$	0.07	b)	5	25
$E_{49}B_9$	*0.01*	b)	–	69
$E_{96}B_{18}$	0.001	b)	15	46
$E_{134}B_{19}$	*0.004*	b)	–	46
$E_{43}B_{14}E_{43}$	0.04	b)	6	48
$E_{50}S_{3.5}$	0.03	b)	7	49
$E_{50}S_5$	0.007	b)	7	49
$E_{51}S_{6.5}$	0.005	b)	7	49
$E_{17}S_8$	*0.003*	b)	8	26
$E_{45}S_8$	*0.003*	b)	–	51
$E_{45}S_{10}$	0.002	b)	10	51
$S_{10}E_{135}$	*0.002*	b)	–	51
$S_{13}E_{60}$	0.0005	b)	10	65, 70
$S_{15}E_{63}$	*0.0004*	b)	12	51
$S_{17}E_{65}$	*0.0002*	b)	13	51
$S_{20}E_{67}$	*0.0001*	b)	16	51
$E_{82}S_8E_{82}$	0.04	b)	7	52
$E_{20}S_{10}E_{20}$	*0.01*	b)	–	52
$E_{66}S_{13}E_{66}$	0.005	b)	8	52
$E_{67}S_{15}E_{67}$	0.003	b)	8	52
G_5E_{67}	0.006	b)	8	53
$E_{71}G_7E_{71}$	0.01	b)	6	54
$E_{62}G_8E_{62}$	<0.01	b)	6	54
$E_{38}G_{12}E_{38}$	<0.001	b)	7	54

a) Italics indicate estimated values based on results in the reference given.
b) Indicates a c.m.t. below 0 °C.

weight-average micelle association number, which relate to the size of the micelle core, can be found in Table 2.4.

2.4 Drug Solubilization

The extent of drug solubilization was determined absolutely by ^{1}H NMR spectroscopy [31, 50], after calibration by UV spectroscopy [25, 50, 53, 54], and by gas–liquid chromatography [71]. Copolymer concentrations were in the range

Table 2.4 Solubilization capacity (method 1).

Copolymer	T (°C)	N_w [a]	s_{cp} (mg g^{-1}) [b]	s_h (mg g^{-1})	Ref.
$E_{130}P_{58}$	25	*50*	1.5	4	31
$E_{107}P_{69}$	25	*150*	3.9	8	31
$E_{62}P_{39}E_{62}$ (F87)	25	[c]	0	0	74
	37	[c]	0.8	3	
$E_{21}P_{47}E_{21}$ (P94)	25	[c]	1.2	2	31
$E_{21}P_{67}E_{21}$ (P123)	25	[c]	3.0	4	67, 71
	37	90	3.8	5	
$E_{98}P_{67}E_{98}$ (F127)	25	[c]	2.2	6	71
$E_{11}B_8$	25	63	3.9	7	25
	37	200	9.0	17	
	40	339	21	38	
$E_{49}B_9$	25	*21*	2.9	12	31
$E_{96}B_{18}$	25	163	3.3	13	25
	35	170	3.1	13	
	40	174	3.9	12	
$E_{134}B_{19}$	25	*130*	3.9	19	31
$E_{43}B_{14}E_{43}$	25	9	2.3	7	31
$E_{50}S_{3.5}$	25	20	2.8	16	31
$E_{50}S_5$	25	35	4.0	17	31
$E_{51}S_{6.5}$	25	50	4.9	18	31
$E_{17}S_8$	15	160	—	—	26, 50
	25	250	29.5	52	
	30	420	—	—	
	37	*800*	35.0	62	
$E_{45}S_8$	25	*80*	7.5	22	50
$E_{45}S_{10}$	25	103	11.2	29	50
	37	108	19.1	50	
$S_{10}E_{135}$	25	*50*	6.0	35	50
$S_{13}E_{60}$	25	105	14.1	37	31
$S_{15}E_{63}$	25	140	11.4	28	50
	37	145	15.0	38	
$S_{17}E_{65}$	25	150	11.7	28	50
	37	153	17.2	41	
$S_{20}E_{67}$	25	189	12.4	27	50
	37	191	17.5	39	
$E_{82}S_8E_{82}$	25	11	2.7	21	50
$E_{20}S_{10}E_{20}$	25	*40*	11.8	28	50, 52
$E_{66}S_{13}E_{66}$	25	21	4.0	18	50
	37	24	4.3	19	
$E_{67}S_{15}E_{67}$	25	25	5.6	22	50
	37	28	7.1	29	
G_5E_{67}	25	50	12.4	54	53
$E_{71}G_7E_{71}$	25	16	6.0	40	54
$E_{62}G_8E_{62}$	25	20	8.0	44	54
$E_{38}G_{12}E_{38}$	25	77	17.8	51	54

a) Italics indicate estimated values based on results in the reference given.
b) Estimated uncertainty ±1 mg g^{-1}.
c) Partially or not micellized.

1–2.5 wt%. We were aware of the possibility of impurities in the griseofulvin and instability in aqueous solution, so background measurements were made after dissolution in pure water. Also, in order to assess the solubilization of griseofulvin in the E-block corona of the micelles, measurements were made using solutions of poly(ethylene glycol) of $M_n = 6000\ \mathrm{g\ mol^{-1}}$ (PEG6000, E_{136}, 5–30 wt% in water).

Two methods of solubilization were explored. In method 1, a portion of stock copolymer solution was added to finely ground griseofulvin powder and the mixture stirred at constant temperature for 3–5 days before being filtered at the same temperature to remove unsolubilized drug. This method is equivalent to the so-called shake-flask method (ASTM E1148-02). Because of the low water solubility of griseofulvin, we were concerned that solubilization should not be limited by slow diffusion of the drug into the micelle core. Accordingly, in certain experiments (method 2), ground griseofulvin was added to the copolymer melt at 65 °C before dissolution in water at the same temperature, cooling to 25 °C and proceeding as in method 1.

The solubilization capacity of the copolymer in solution (s_{cp}) is reported in milligrams of drug solubilized per gram of copolymer in solution after correction for the solubility of griseofulvin in water. No systematic differences in values of s_{cp} were observed for different copolymer concentrations and by the different methods of analysis and the values listed are averages of several determinations. Also calculated are values of the solubilization capacity per gram of hydrophobic block, calculated from average values of s_{cp} as $s_h = s_{cp}/w_h$, where w_h is the weight fraction of the hydrophobic block (i.e. $w_h = 1 - w_E$, values of w_E taken from Table 2.1) after correction for the small amount of drug solubilized in the micelle corona. This quantity gives a direct measure of the efficiency of solubilization of griseofulvin in the micelle core.

2.4.1 Method 1

Values of the solubilization capacity (s_{cp} and s_h) obtained using method 1 at temperatures in the range 25–40 °C are listed in Table 2.4. Values of the weight-average micelle association number (N_w) are also listed. These values were either determined directly by static light scattering, interpolated or extrapolated where necessary, or were estimated from related values using the scaling law established for copolymers of the type under discussion to describe the dependence of N_w for spherical micelles on hydrophobic and hydrophilic block length [72]. These estimated values are shown in italics.

2.4.1.1 Effect of Composition and Micelle Core Size: 25 °C

As would be expected from the discussion in Section 2.3, low values of the solubilization capacity were determined for solutions of copolymers with hydrophobic P blocks, reflecting high values of c.m.t., hence low extents of micellization.

Although the size of the micelles may be large (as high as N_w=50–150; see Table 2.4), they are few in number at 25 °C [55, 73]. The values of the solubilization capacity for griseofulvin determined for solutions at 37 °C are no higher, which indicates poor solubilization of this drug even under conditions when $E_mP_nE_m$ copolymers are relatively well micellized.

The other copolymers are essentially completely micellized in solution at 25 °C. With the exception of copolymers $E_{17}B_8$ and $E_{38}G_{12}E_{38}$, at this temperature all the micelles are spherical or near spherical. Apart from a change in core composition, the results may also be affected by changes in core size. It is convenient to use core volume as an approximate indicator of core size, calculated as

$$v_c = N_w n M_0 v_{sp} / N_A$$

where M_0 (g mol^{-1}) is the molar mass of a B, S or G chain unit, v_{sp} (cm^3 g^{-1}) is the specific volume of the corresponding high polymer (discussed in [25, 26, 51–54]) and N_A (mol^{-1}) is Avogadro's constant.

Plots of s_h against micelle core volume are shown in Fig. 2.1 for solutions of copolymers forming spherical micelles in solution at 25 °C. It is clear that the solubilization capacity increases with increase in core size, reaching a limiting value when v_c exceeds *ca.* 100 nm^3, corresponding to a core radius $r_c \approx 3$ nm. The dashed lines indicate limiting values of the solubilization capacity at 25 °C of s_h=16, 31 and 54 mg g^{-1} for copolymers with B, S and G cores, respectively. Thus, through Fig. 2.1, our results provide clear evidence of the benefit of a large core volume and an appropriate choice of core composition if solubilization capacity is to be maximized.

Results for solutions of copolymer $E_{38}G_{12}E_{38}$ are not included in Fig. 2.1. As seen in Table 2.4, the value of N_w determined for this copolymer is high compared with those of the other $E_mG_nE_m$ copolymers, which, together with other evidence [54], indicates the formation of nonspherical micelles. The relatively

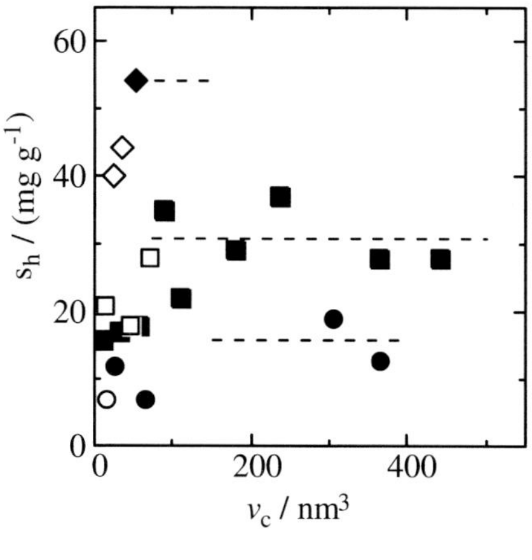

Fig. 2.1 Dependence on micelle core volume of solubilization capacity (s_h) for griseofulvin. Aqueous micellar solutions at 25 °C of (●) E_mB_n, (○) $E_mB_nE_m$, (■) E_mS_n and S_nE_m, (□) $E_mS_nE_m$, (◆) G_nE_m and (◇) $E_mG_nE_m$ copolymers. The dashed lines indicate the plateau values at high values of the core volume.

high value of s_h (see Table 2.4) is consistent with this assignment, as there is other evidence (see Section 2.4.1.2) [3, 16] that the extent of solubilization is enhanced if micelles are cylindrical rather than spherical.

2.4.1.2 Effect of Solubilization Temperature

Plots of s_h against micelle core volume are shown in Fig. 2.2 for solutions of copolymers $E_{11}B_8$ and $E_{96}B_{18}$ in the range 25–40 °C. It is well established that raising the temperature of micellar solutions of block copoly(oxyalkylene)s results in an increase in micelle association number [17, 75]. However, for spherical micelles a limit is reached when the hydrophobic blocks of the core are highly stretched and, as discussed in Section 2.1, any further increase in N_w can only be at the expense of a change from spherical to cylindrical geometry. The micelles of copolymer $E_{96}B_{18}$ are seen to approach, but not exceed, that limit over the range 25–40 °C (see Table 2.4). Consequently, the values of v_c plotted in Fig. 2.2 are almost invariant and, for this copolymer, the solubilization capacity does not change significantly. In contrast, the micelles of copolymer $E_{11}B_8$ are spherical at 25 °C but are cylindrical, even worm-like, at 40 °C [25] and the solubilization capacity is much increased. The determining factor is the value of the association number in relation to the average length of the hydrophobic block: a copolymer with a short hydrophobic block cannot form a large spherical micelle core.

Plots of s_h against v_c for solutions of E/S copolymers, both diblock and triblock, are shown in Fig. 2.3. Copolymer $E_{17}S_8$ has been shown to form cylindrical micelles at temperatures in the range 25–30 °C [26] and, no doubt, does so at 37 °C. In keeping with this, solubilization capacities measured for this copolymer are high compared with those found for the other copolymers, which form spherical (or near spherical) micelles. Even so, the results for the other diblock copolymers are of interest. The association numbers of their micelles are large and almost independent of temperature, characteristic of spherical micelles ap-

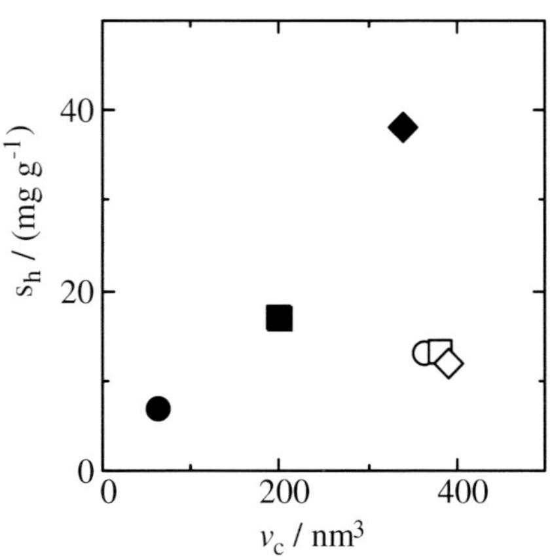

Fig. 2.2 Dependence on micelle core volume of solubilization capacity (s_h) for griseofulvin. Aqueous micellar solutions of copolymers $E_{11}B_8$ (filled symbols) and $E_{96}B_{18}$ (open symbols) at (●, ○) 25, (■, □) 37 and (◆, ◇) 40 °C.

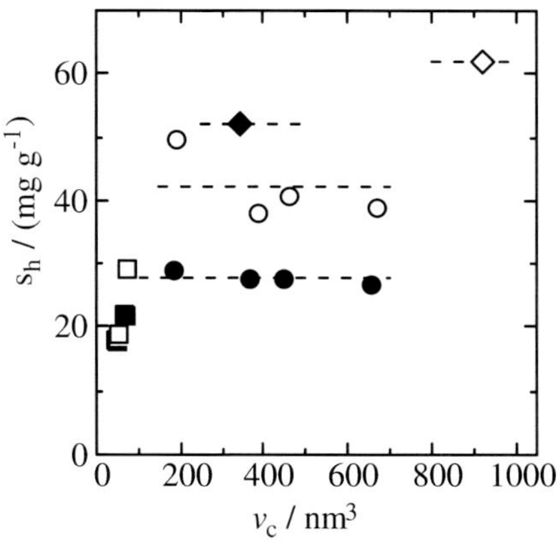

Fig. 2.3 Dependence on micelle core volume of solubilization capacity (s_h) for griseofulvin. Aqueous micellar solutions at (open symbols) 25 °C and (filled symbols) 37 °C of (●, ○) E_mS_n and S_nE_m and (■, □) $E_mS_nE_m$ copolymers forming spherical micelles and of (◆, ◇) $E_{17}S_8$.

proaching their upper size limit. However, unlike micellar solutions of copolymer $E_{96}B_{18}$, the solubilization capacity increases as temperature is increased: compare Figs. 2.2 and 2.3. The reason for this difference in behavior has not been established. As indicated by the difference in glass transition temperatures of lengthy S and B chains, $T_g \approx +40\,°C$ compared with –80 °C [29], S blocks in a micelle core at low temperatures will be less mobile than B blocks and it may be that a change in mobility between 25 and 37 °C affects the rate of uptake of the drug from solution.

2.5
Method 2

Values of the excess solubilization capacity of the copolymer solutions for griseofulvin obtained by method 2 are listed in Table 2.5. As for method 1, results for different copolymer concentrations (1 or 2 wt%) were averaged. For comparison, values of s_h obtained at 25 °C by method 1 are included. On average, the solubilization capacity is approximately doubled by incorporating griseofulvin into the melt before dissolving the copolymer in hot water. The effect is similar

Table 2.5 Solubilization of griseofulvin in block copolymer solutions (method 2).

Copolymer	s_{cp} (mg g^{-1}) (method 2)	s_h (mg g^{-1}) (method 2)	s_h (mg g^{-1}) (method 1)
$E_{96}B_{18}$	6.9	29	13
$E_{43}B_{14}E_{43}$	6.1	19	7
$E_{45}S_8$	10.7	32	22
$S_{10}E_{135}$	9.6	56	35
$E_{82}S_8E_{82}$	5.8	48	21
$E_{20}S_{10}E_{20}$	31.4	77	28

for both E/S and E/B copolymers, although at a lower level for the E/B copolymers.

The limiting conditions for microphase separation in E/B copolymer melts are well known and the melts of $E_{96}B_{18}$ and $E_{43}B_{14}E_{43}$ are certainly disordered [76]. Microphase separation has not been detected in the melts of E/S copolymers, no doubt because the charge-transfer interaction between ether oxygen and the phenyl ring greatly reduces the value of the Flory–Huggins χ parameter, much as observed for block copolymers of poly(ethylene oxide) and styrene [77]. We conclude that the marked effect seen in Table 2.5 is not a result of griseofulvin favoring less polar domains in an ordered melt, but rather a rapid and irreversible transfer to the micelle cores from the disordered melt at the point of micellization when the drug-loaded melt is transferred to the aqueous phase at 65 °C before equilibration at 25 °C and subsequent filtration. It is possible that method 2 is equivalent to a variant of method 1 carried out at 65 °C, albeit over a much shorter time period, 1 h rather than 3–5 days. We have not explored that possibility, being wary of the degradation of both copolymer and drug held in solution at high temperature.

A change in micelle shape from spherical to cylindrical caused by solubilization of a core-compatible substance has been predicted by Nagarajan [16]. A change in temperature from 25 to 65 °C will cause an increase in association number and favor a sphere-to-cylinder transition. Copolymer $E_{20}S_{10}E_{20}$ is an obvious case to consider. Looping of the S block in the core means that this copolymer is geometrically equivalent to an $E_{20}S_5$ diblock and might be compared with copolymer $E_{17}B_8$ (see Section 2.4.1.2) [26], in which case a substantial increase in N_w to a value approaching 100 will certainly result in formation of cylindrical micelles and so the very high value of s_h which is observed.

2.6 Gelation

A property of concentrated solutions of Pluronic F127 ($E_{106}P_{69}E_{106}$), which was recognized as early as 1972, is the formation of a hard gel on warming from ambient to body temperature, so-called 'cold gelation' [78]. This sol–gel transition is a consequence of a decrease in solubility of the copolymer in water on heating, which results in the formation of additional micelles and subsequent gel formation as the micelles become so tightly packed that their mobility is restricted. The potential application of this characteristic has been explored for a wide range of drug delivery systems, mainly using Pluronic F127 as the delivery vehicle, including for ophthalmic [79–82], rectal [83], nasal [84] and parenteral use [85–90]. Unfortunately, although it is possible to enhance significantly the solubilization capacity of block copolymer micelles beyond that of the $E_mP_nE_m$ copolymers by increasing the hydrophobicity of the cores as discussed in this chapter, none of these compounds in our experience exhibits cold gelation. Thus, although the $E_mS_nE_m$ and $E_mG_nE_m$ copolymers have greatly improved solubilization capability and readily form gels in concentrated solution, nevertheless they are unsuitable for use in delivery systems which depend on the cold gelation effect.

We have recently shown [91] that a solution with the desired gelation characteristics may be formed by mixing a triblock $E_mP_nE_m$ copolymer with a suitable $E_mS_nE_m$ copolymer. Our study has shown, for example, that 50:50 wt% mixtures of $E_{62}P_{39}E_{62}$ (Pluronic F87) with either $E_{137}S_{18}E_{137}$ or $E_{82}S_9E_{82}$ at a total copolymer concentration of approximately 30 wt% are fluids of low viscosity at temperatures below 22–25 °C and gels of high elastic modulus at body temperature. These mixed systems, which combine the desirable gelation characteristics of solutions of the $E_mP_nE_m$ copolymers with the greater solubilizing capacities of solutions of the $E_mS_nE_m$ copolymers, have potential as vehicles for the controlled delivery of solubilized drug from gels formed *in situ* following subcutaneous injection of a low-viscosity aqueous solution. Recently, it has been shown that micellar solutions of mixed copolymers $E_{137}S_{18}E_{137}$ (80 wt%) and $E_{62}P_{39}E_{62}$ (20 wt%) undergo cold gelation and have a high solubilization capacity for griseofulvin [74].

2.7 Conclusion

Our current interest is in the potential use of micellar solutions of block copolymers for the solubilization of aromatic drugs of low aqueous solubility. Comparison of the solubilization properties of a range of copoly(oxyalkylene)s shows that solubilization capacities in milligrams per gram of hydrophobe for griseofulvin rank in the order G > S > B >> P, where $G = OCH_2CH(CH_2OC_6H_5)$, $S = OCH_2CH(C_6H_5)$, $B = OCH_2CH(C_2H_5)$ and $P = OCH_2CH(CH_3)$. The results also show that solubilization of griseofulvin by diblock copolymers is enhanced compared with that of comparable triblock copolymers.

Acknowledgments

We thank Dr Nágila Ricardo and Ms Elenir Pinho (Federal University of Ceará, Brazil), Dr Chiraphon Chaibundit (Prince of Songkla University, Thailand) and Mr Zhengyuan Zhou (University of Manchester) for use of their results before publication.

Notation

The block copoly(oxyalkylene)s of interest combine hydrophilic poly(ethylene oxide) with hydrophobic poly(propylene oxide), poly(1,2-butylene oxide), poly(styrene oxide) or poly(phenyl glycidyl ether). To describe the repeat units we use the following notation: E = oxyethylene, OCH_2CH_2; P = oxypropylene, $OCH_2CH(CH_3)$; B = oxybutylene, $OCH_2CH(C_2H_5)$; S = oxyphenylethylene, $OCH_2CH(C_5H_6)$ from styrene oxide; G = oxy(phenyloxymethylene)ethylene, $OCH_2CH(CH_2OC_5H_6)$ from phenyl glycidyl ether. We use subscripts m and n to denote number-average lengths of the hydrophilic and hydrophobic blocks, respectively, in repeat units so that, for example, a triblock copolymer formed by sequential copolymerization of propylene oxide and ethylene oxide is denoted $E_mP_nE_m$.

References

1 M. Yokoyama, *Crit. Rev. Ther. Drug Carrier Syst.*, **1992**, *9*, 213–248.
2 M.L. Adams, A. Lavasanifar, G.S. Kwon, *J. Pharm. Sci.*, **2002**, *92*, 1343–1351.
3 C. Allen, D. Maysinger, A. Eisenberg, *Colloids Surf. B*, **1999**, *16*, 3–27.
4 A.V. Kabanov, E.V. Batrakova, V.Y. Alakhov, *J. Control. Release*, **2002**, *82*, 189–212.
5 G.S. Kwon, K. Kataoka, *Adv. Drug Deliv. Rev.*, **1995**, *16*, 295–309.
6 V.P. Torchilin, *Expert Opin. Ther. Pat.*, **2005**, *15*, 63–75.
7 J.H. Collett, E.A. Tobin, *J. Pharm. Pharmacol.*, **1979**, *31*, 174–177.
8 K.T. Oh, T.K. Bronich, A.V. Kabanov, *J. Control. Release*, **2004**, *94*, 411–422.
9 P.N. Hurter, T.A. Hatton, *Langmuir*, **1992**, *8*, 1291–1299.
10 P.N. Hurter, J.M.H.M. Scheutjens, T.A. Hatton, *Macromolecules*, **1993**, *8*, 5592–5601, 5030–5040.
11 X. Li, W.L. Mattice, *Macromolecules*, **1997**, *30*, 1711–1717.
12 R. Nagarajan, K. Ganesh, *Macromolecules*, **1989**, *22*, 4312–4325.
13 R. Nagarajan, M. Barry, E. Ruckenstein, *Langmuir*, **1986**, *1*, 210–215.
14 R. Nagarajan, K. Ganesh, *J. Chem. Phys.*, **1989**, *90*, 5843–5856.
15 R. Nagarajan, K. Ganesh, *J. Colloid Interface Sci.*, **1996**, *184*, 489–499.
16 R. Nagarajan, *Colloids Surf. B*, **1999**, *16*, 55–72.
17 C. Booth, D. Attwood, *Macromol. Rapid Commun.*, **2000**, *21*, 501–527.
18 I.W. Hamley, S.-M. Mai, A.J. Ryan, J.P.A. Fairclough, C. Booth, *Phys. Chem. Chem. Phys.*, **2001**, *3*, 2972–2980.
19 K. Mortensen, W. Brown, *Macromolecules*, **1993**, *26*, 4128–4135.
20 K. Mortensen, J.S. Pedersen, *Macromolecules*, **1993**, *26*, 805–812.
21 K. Schillen, W. Brown, R.M. Johnsen, *Macromolecules*, **1994**, *27*, 4825–4832.
22 O. Glatter, G. Scherf, K. Schillen, W. Brown, *Macromolecules*, **1994**, *27*, 6046–6054.
23 S. Hvidt, E.B. Jørgensen, W. Brown, K. Schillen, *J. Phys. Chem.*, **1994**, *98*, 12320–12328.
24 E.B. Jørgensen, S. Hvidt, W. Brown, K. Schillen, *Macromolecules*, **1997**, *30*, 2355–2364.
25 C. Chaibundit, N.M.P.S. Ricardo, C. Booth, M. Crothers, *Langmuir*, **2002**, *18*, 4277–4283.
26 Z. Yang, M. Crothers, D. Attwood, J.H. Collett, N.M.P.S. Ricardo, L.G.A. Martini, C. Booth, *J. Colloid Interface Sci.*, **2003**, *263*, 312–317.
27 C. Chaibundit, P. Sumanatrakool, S. Chinchew, P. Kanatharana, C.E. Tattershall, X.-F. Yuan, C. Booth, *J. Colloid Interface Sci.*, **2005**, *283*, 544–554.
28 G. Allen, C. Booth, S.J. Hurst, C. Price, F. Vernon, R.F. Warren, *Polymer*, **1967**, *8*, 406–413.
29 G. Allen, C. Booth, C. Price, *Polymer*, **1967**, *8*, 414–418.
30 A. Sunder, H. Türk, R. Haag, H. Frey, *Macromolecules*, **2000**, *33*, 7682–7692.
31 C.J. Rekatas, S.-M. Mai, M. Crothers, M. Quinn, J.H. Collett, D. Attwood, F. Heatley, L. Martini, C. Booth, *Phys. Chem. Chem. Phys.*, **2001**, *3*, 4769–4773.
32 T. Anarson, P.H. Elworthy, *J. Pharm. Pharmacol.*, **1981**, *33*, 141–144.

33 P.H. Elworthy, M.S. Patel, *J. Pharm. Pharmacol.*, **1982**, *34*, 543–546.

34 S.H. Yalkowsky, Y. He, *Handbook of Aqueous Solubilities*, CRC Press, Boca Raton, FL, **2003**.

35 R.H. Whitmarsh, in *Nonionic Surfactants, Poly(oxyalkylene) Block Copolymers*, Surfactant Science Series, Vol. 60, V.M. Nace (Ed.), Marcel Dekker, New York, **1996**, Ch. 1.

36 P.J. Flory, *Principles of Polymer Chemistry*, Cornell University Press, Ithaca, NY, **1953**, p. 336.

37 G.-E. Yu, A.J. Masters, F. Heatley, C. Booth, T.G. Blease, *Macromol. Chem. Phys.*, **1994**, *195*, 1517–1538.

38 S. Boileau, in *Comprehensive Polymer Science, Vol. 3, Chain Polymerization 1*, G.C. Eastmond, A. Ledwith, S. Russo, P. Sigwalt (Eds.), Pergamon Press, Oxford, **1989**, Chap. 32.

39 F. Heatley, G.-E. Yu, W.-B. Sun., E.J. Pywell, R.H. Mobbs, C. Booth, *Eur. Polym. J.*, **1990**, *26*, 583–592.

40 F. Heatley, G.-E. Yu, M.D. Draper, C. Booth, *Eur. Polym. J.*, **1991**, *27*, 471–478.

41 See, for example, N.K. Reddy, P.J. Fordham, D. Attwood, C. Booth, *J. Chem. Soc., Faraday Trans.*, **1990**, *86*, 1569–1572.

42 V.M. Nace, R.H. Whitmarsh, M.W. Edens, *J. Am. Oil Chem. Soc.*, **1994**, *71*, 777–781.

43 G.-E. Yu, M. Ameri, Z. Yang, D. Attwood, C. Price, C. Booth, *J. Phys. Chem. B*, **1997**, *101*, 4394–4401.

44 J.P.A. Fairclough, G.-E. Yu, S.-M. Mai, M. Crothers, A.J. Ryan, K. Mortensen, C. Booth, *Phys. Chem. Chem. Phys.*, **2000**, *2*, 1503–1507.

45 S.-M. Mai, J.P.A. Fairclough, K. Viras, P.A. Gorry, I.W. Hamley, A.J. Ryan, C. Booth, *Macromolecules*, **1997**, *30*, 8392–8400.

46 W. Mingvanish, S.-M. Mai, F. Heatley, C. Booth, D. Attwood, *J. Phys. Chem. B*, **1999**, *103*, 11269–11274.

47 S.-M. Mai, J.P.A. Fairclough, N.J. Terrill, S.C. Turner, I.W. Hamley, M.W. Matsen, A.J. Ryan, C. Booth, *Macromolecules*, **1998**, *31*, 8110–8116.

48 G.-E. Yu, Y.-W. Yang, Z. Yang, D. Attwood, C. Booth, V.M. Nace, *Langmuir*, **1996**, *12*, 3404–3412.

49 S.-M. Mai, C. Booth, A. Kelarakis, V. Havredaki, A.J. Ryan, *Langmuir*, **2000**, *16*, 1681–1688.

50 M. Crothers, Z.-Y. Zhou, N.M.P.S. Ricardo, Z. Yang, P. Taboada, C. Chaibundit, D. Attwood, C. Booth, *Int. J. Pharm.*, **2005**, *293*, 91–100.

51 M. Crothers, D. Attwood, J.H. Collett, Z. Yang, C. Booth, P. Taboada, V. Mosquera, N.M.P.S. Ricardo, L. Martini, *Langmuir*, **2002**, *18*, 8685–8691.

52 Z. Yang, M. Crothers, N.M.P.S. Ricardo, C. Chaibundit, P. Taboada, V. Mosquera, A. Kelarakis, V. Havredaki, L. Martini, C. Valder, J.H. Collett, D. Attwood, F. Heatley, C. Booth, *Langmuir*, **2003**, *19*, 943–950.

53 P. Taboada, G. Velasquez, S. Barbosa, Z. Yang, S.K. Nixon, Z.-Y. Zhou, F. Heatley, M. Ashford, V. Mosquera, D. Attwood, C. Booth, *Langmuir*, **2006**, *22*, 7465–7470.

54 P. Taboada, G. Velasquez, S. Barbosa, V. Castelletto, S.K. Nixon, Z. Yang, F. Heatley, I.W. Hamley, M. Ashford, V. Mosquera, D. Attwood, C. Booth, *Langmuir*, **2005**, *21*, 5263–5271.

55 S.K. Nixon, S. Hvidt, C. Booth, *J. Colloid Interface Sci.*, **2004**, *280*, 219–223.

56 S. Hvidt, C. Trandum, W. Batsberg, *J. Colloid Interface Sci.*, **2002**, *250*, 243–251.

57 P. Linse, M. Malmsten, *Macromolecules*, **1992**, *25*, 5434–5439.

58 W.F. Gadelle, W.J. Koros, R.S. Schechter, *Macromolecules*, **1995**, *28*, 4883–4892.

59 P. Alexandridis, J.F. Holzwarth, T.A. Hatton, *Macromolecules*, **1994**, *27*, 2414–2425.

60 E.V. Batrakova, S.-M. Li, S. Li, A. Venne, V. Alakhov, A.V. Kabanov, *Pharm. Res.*, **1999**, *16*, 1373–1379.

61 W.J. Harrison, G.J. Aboulgasem, F.A.I. Elathrem, S.K. Nixon, D. Attwood, C. Price, C. Booth, *Langmuir*, **2005**, *21*, 6170–6178.

62 G. Wanka, H. Hoffmann, W. Ulbricht, *Macromolecules*, **1994**, *27*, 4145–4159.

63 M. Bohorquez, C. Koch, T. Trygstad, N. Pandit, *J. Colloid Interface Sci.*, **1999**, *216*, 34–40.

64 A. Kelarakis, V. Havredaki, G.-E. Yu, L. Derici, C. Booth, *Macromolecules*, **1998**, *31*, 944–946.

65 A. Kelarakis, V. Havredaki, C.J. Rekatas, C. Booth, *Phys. Chem. Chem. Phys.*, **2001**, *3*, 5550–5552.

66 H. Altinok, S.K. Nixon, P.A. Gorry, D. Attwood, C. Booth, A. Kelarakis, V. Havredaki, *Colloids Surf. B*, **1999**, *16*, 73–88.

67 C. Chaibundit, unpublished work.

68 G.-E. Yu, Y.-L. Deng, S. Dalton, Q.-G. Wang, D. Attwood, C. Price, C. Booth, *J. Chem. Soc., Faraday Trans.*, **1992**, *88*, 2537–2544.

69 A.D. Bedells, R.M. Arafeh, Z. Yang, D. Attwood, F. Heatley, J.C. Padget, C. Price, C. Booth, *J. Chem. Soc., Faraday Trans.*, **1993**, *89*, 1235–1242.

70 A. Kelarakis, V. Havredaki, C.J. Rekatas, S.-M. Mai, D. Attwood, C. Booth, A.J. Ryan, I.W. Hamley, L.G.A. Martini, *Macromol. Chem. Phys.*, **2001**, *202*, 1345–1354.

71 Z.-Y. Zhou, unpublished work.

72 C. Booth, D. Attwood, C. Price, *Phys. Chem. Chem. Phys.*, **2006**, *8*, 3612–3622.

73 Q.-G. Wang, G.-E. Yu, Y.-L. Deng, C. Price, C. Booth, *Eur. Polym. J.*, **1993**, *29*, 665–669.

74 N.M.P.S. Ricardo, M.E.N. Pinho, unpublished results.

75 B. Chu, Z.-K. Zhou, in *Nonionic Surfactants, Poly(oxyalkylene) Block Copolymers*, Surfactant Science Series, Vol. 60, V.M. Nace (Ed.), Marcel Dekker, New York, **1996**, p. 67.

76 A.J. Ryan, S.-M. Mai, J.P.A. Fairclough, I.W. Hamley, C. Booth, *Phys. Chem. Chem. Phys.*, **2001**, *3*, 2961–2971.

77 H. Frielinghaus, N. Hermsdorf, R. Sigel, K. Almdal, K. Mortensen, I.W. Hamley, L. Messe, L. Corvazier, A.J. Ryan, D. van Dusschoten, M. Wilhelm, G. Floudas, G. Fytas, *Macromolecules*, **2001**, *34*, 4907–4916.

78 I.R. Schmolka, *J. Biomed. Mater. Res.*, **1972**, *6*, 571–582.

79 S.D. Desai, J. Blanchard, *J. Pharm. Sci.* **1998**, *87*, 226–230.

80 K. Edsman, J. Carlfors, R. Petersson, *Eur. J. Pharm. Sci.*, **1998**, *6*, 105–112.

81 A.H. El-Kamel, *Int.J. Pharm.*, **2002**, *241*, 47–55.

82 S. Miyazaki, C. Yokouchi, T. Nakamura, N. Hashiguchi, W.-M. Hou, M. Takada, *Chem. Pharm. Bull.* **1986**, *34*, 1801–1808.

83 S.S. Pisal, A.R. Paradkar, K.R. Mahadik, S.S. Kadam, *Int.J. Pharm.*, **2004**, *270*, 37–45.

84 M. Guzmán, F.F. García, J. Molpeceres, M.R. Aberturas, *Int.J. Pharm.*, **1992**, *80*, 119–127.

85 T.P. Johnston, M.A. Punjabi, C.F. Frölich, *Pharm. Res.*, **1992**, *9*, 425–434.

86 S. Miyazaki, Y. Ohkawa, M. Takada, D. Attwood, *Chem. Pharm. Bull.*, **1992**, *40*, 2224–2226.

87 E.A. Pec, Z.G. Wout, T.P. Johnston, *J. Pharm. Sci.*, **1992**, *81*, 626–630.

88 P.-L. Wang, T.P. Johnston, *Int.J. Pharm.*, **1995**, *113*, 73–81.

89 M. Katakam, W. R, Ravis, A.K. Banga, *J. Control. Release*, **1997**, *49*, 21–26.

90 A. Paavola, J. Yliruusi, P. Rosemberg, *J. Control. Release*, **1998**, *52*, 169–178.

91 N.M.P.S. Ricardo, M.E.N. Pinho, Z. Yang, D. Attwood, C. Booth, *Int. J. Pharm.*, **2005**, 300, *22*–31.

3
Controlling the Physical Stability of Liposomal Colloids

Ferenc Csempesz and István Puskás

Abstract

Uncharged homopolymers such as poly(vinyl alcohol), polyvinylpyrrolidone (PVP) and methylcellulose were used for the stabilization of small unilamellar vesicles in aqueous media of various composition. Polymer-free and polymer-bearing liposomes from dimyristoylphosphatidylcholine and dipalmitoylphosphatidylcholine were prepared under standardized conditions. Sterically stabilized liposomes were formed by incorporating dissolved macromolecules in the membrane bilayer of vesicles. The stability of liposomal dispersions, at physiological pH values and in the presence of natural cyclodextrins (CDs), were studied. The physical stability of vesicles was tested by measuring the size distribution of liposomes by means of a Malvern Zetasizer 4 apparatus. It was shown that most of the polymers are efficient steric stabilizers for the vesicles. PVP of high molecular mass exhibited the most efficient stabilizing effect at each pH studied, indicating that the formation of a relatively thick polymer layer around the lipid bilayers ensures enhanced and prolonged physical stability of liposomes. Also, the polymers proved to be suitable for preventing the destruction of phospholipid bilayers that dissolved natural CDs cause. Using these macromolecules, vesicle stability and interactions in dispersions of liposomal colloids can be well controlled.

3.1
Introduction

Colloidal delivery systems have many potential uses in science and technology since active materials can be bound to or incorporated into the carrier. In recent years, a key application of liposomes, microspheres, polymeric nanoparticles and nanocapsules has been in pharmaceuticals [1–6]. Most of the interest in using colloidal carriers stems from the belief that such systems with controlled stability and surface properties of dispersed particles may enhance the efficacy or safety of existing drugs. Also, the slow release properties of carriers may rep-

Colloids and Interface Science Series, Vol. 3
Colloid Stability and Application in Pharmacy. Edited by Tharwat F. Tadros

ISBN: 978-3-527-31463-8

resent a potential means to provide selective and prolonged therapeutic activity that may underlie optimized drug therapy.

Liposomes find perhaps the most extensive use as drug delivery vehicles of anticancer and antimicrobial therapeutics. Vesicles with alternating aqueous (hydrophilic) and lipid (hydrophobic) domains provide opportunities for the formulation of controlled release and site-specific drug-delivery systems [7–16]. Due to the biphasic nature of vesicles, both lipophilic and hydrophilic ingredients are accommodated depending on their solubility in the liposome components; consequently, almost any type of drug can be encapsulated. The medical utility of what are now called conventional liposomes (CLs) is limited by their rapid uptake by mononuclear phagocytic systems (MPS). Ensure prolonged therapeutic activity of active ingredients either bound or incorporated, while maintaining long-term vesicle stability, is also required. Interest in liposomes as drug carriers was rejuvenated by the introduction of sterically stabilized liposomes (SLs). Because of their reduced recognition and uptake by the immune system, these sophisticated liposomes have been referred to as "stealth" liposomes [12–14]. The enhanced biological stability of SLs is a result of the inhibition of interactions with plasma proteins such as opsonins and lipoproteins. Due to a suitable polymer coating, sterically stabilized liposomes also have a longer lifetime in the bloodstream. In addition, the polymer layer on the surface of liposomes reduces attractive forces and increases the hardcore repulsion between the possible reactants [9, 12].

A number of fundamental studies have demonstrated that liposomes containing phospholipid derivatized with hydrophilic polymer of relatively low molecular weight [e.g. poly(ethylene glycol) (PEG)] exhibit prolonged circulation. Much less attention has been paid to the use of high polymers in liposome stabilization. In the work described in this chapter, the effects which may lead to destruction of liposome membranes in aqueous dispersions on the one hand, and the influence of uncharged homopolymers on the physical stability of vesicles on the other, were studied. The aim of these investigations was to find suitable macromolecular stabilizers for colloidal vesicles of dimyristoylphosphatidylcholine (DMPC) and dipalmitoylphosphatidylcholine (DPPC). To this end, the mean size and size distribution of polymer-free and polymer-bearing vesicles and also the change in time of these characteristics were measured by photon correlation spectroscopy (PCS).

3.2 Experimental

3.2.1 Materials

L-α-DMPC and DPPC were used for the preparation of liposomes. These phospholipids, purchased from Sigma Chemical (St. Louis, MO, USA), were of analytical grade.

Water-soluble homopolymers such as methylcellulose (MC), hydrolyzed poly (vinyl alcohol) (PVA) and polyvinylpyrrolidone (PVP), which proved to be efficient steric stabilizers for several colloidal dispersions, were used for the preparation of polymer-bearing vesicles. The polymers were fractionated samples, prepared from the commercial products Tylose MH50 MC, Powal 420 PVA and GAF K-30 and K-90 PVP. The degree of polymerization of the MC and the PVA was 340 and 2450 and that of the K-30 and K-90 PVP was 280 and 8120, respectively.

Natural α-, β- and γ-cyclodextrins (CDs), well known to be capable of forming inclusion complexes with many amphiphilic and lipophilic compounds, were used. They were obtained from Cyclolab Cyclodextrin Research and Development Laboratories (Budapest, Hungary) and were of pharmaceutical grade.

The pH of the liposome dispersions was adjusted to a physiologically relevant value such as 1.2 and 8.2, using a mixture of hydrochloric acid and sodium chloride or mixtures with various ratios of 0.2 M sodium hydrogenphosphate and 0.1 M citric acid, as buffer solution.

All other reagents were of analytical grade and were purchased from Reanal (Budapest, Hungary). Aqueous solutions were prepared with doubly distilled water. For dynamic light scattering measurements, the distilled water was additionally purified by filtration (Minisart NML, pore size 0.2 µm; Sartorius, Göttingen, Germany).

3.2.2 Methods

3.2.2.1 Preparation of Liposomes

Liposomal dispersions were prepared from DMPC and DPPC in aqueous media in the presence and absence of a polymer by ultrasonication as described in detail elsewhere [17–20].

Dispersions of DMPC liposomes were formulated by twofold ultrasonication. First, a coarse lipid suspension was sonicated for 20 min at 20 kHz frequency with 9.0 ± 1.0 µm amplitude at 297 K, using an MSE 150 W Ultrasonic Disintegrator (MSE, Crawley, UK), then the suspension was diluted so that the final concentration of DMPC in the dispersion was 2.0 g dm^{-3} and the DMPC:polymer ratio, in each polymer-containing dispersion, was 10:1 (w/w). Following the second sonication, carried out in the same way, the dispersion was ultracentrifuged for 45 min at 19 000 rpm by a Beckman Model J2-21 ultracentrifuge (Beckman, Fullerton, CA, USA), in order to remove the possibly formed lipid aggregates or large vesicles.

Colloidal dispersions with DPPC vesicles were prepared by a modified sonication method [19]. DPPC (0.050 g) was dissolved in 2.50 mL of absolute ethanol and 240 µl of this solution were added to 4.50 mL aqueous solutions with a micropipette. The sample was then sonicated for 5 min using Realsonic 40SF ultrasound equipment (maximum electric power 300 W, frequency of ultrasound 37 kHz) Realtrade Ltd. (Budapest, Hungary). The bath of the sonicator was incubated at constant temperature in the range 25.0–45.0 °C. The homogeneous dispersion of the freshly prepared DPPC liposomes exhibited a slight opalescence.

3.2.2.2 Characterization of the Physical Stability of Liposomes

The change in time of the particle size, size distribution and polydispersity are sensitive indicators of the kinetic stability of colloidal dispersions. These parameters in the dispersions of DMPC and DPPC liposomes were determined by PCS at 25 °C, using a Zetasizer 4 instrument (Malvern Instruments, Malvern, UK). The intensity of the fluctuations of the laser light scattered from the vesicles allows the determination of an autocorrelation function that can be used to calculate the diffusion coefficient and size of liposomes undergoing Brownian motion. From the intensity distribution of light scattered from vesicles, the number and volume distributions can also be obtained and the corresponding mean sizes of vesicles and the polydispersity index (*PI*) can be calculated by PCS analysis.

3.3 Results

3.3.1 Particle Size and Size Distribution of Liposomes

Vesicle size is a relevant characteristics regarding both the organ distribution and encapsulation or adsorption efficiency of liposomal drug carriers. The dispersions used in these investigations were formulated at 25.0 °C. The size distribution functions, the number-, volume- or *Z*-average mean size of the vesicles and their polydispersity index were determined by PCS analysis. Since the loading capacity of vesicles closely correlates with the entrapped volume, in this study the volume-average diameter and/or the volume distribution function were usually given as characteristics for liposome dispersion.

Typical volume distribution functions for the DMPC and the DPPC liposomes are shown in Fig. 3.1. The number- and volume-average mean sizes of liposomes, calculated by multimodal analysis from the intensity distribution, are also given.

The results demonstrate that small unilamellar liposomes with a narrow size distribution were formed by the sonication methods used, from both DMPC and DPPC. Subsequent to the preparation, there were no meaningful differences in the mean sizes of the chemically different liposomes, but the DPPC vesicles were somewhat smaller and less polydisperse. The adequacy of the results of PCS analysis was confirmed by electron microscopy. In Fig. 3.2, a freeze-fracture (electron) micrograph of the DPPC liposomes is shown. The vesicles are spherical, with diameters in the range 40–50 nm.

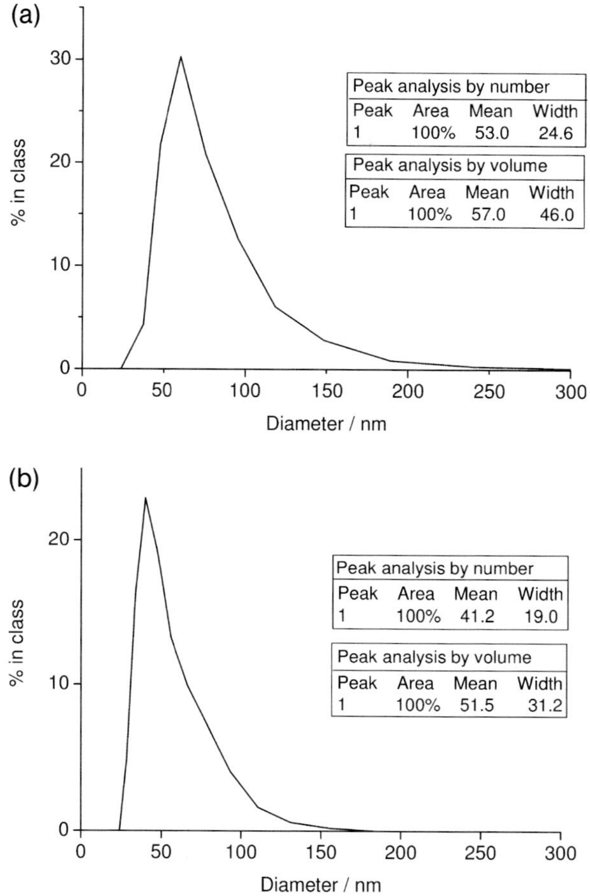

Fig. 3.1 Volume distribution of (a) DMPC and (b) DPPC liposomes in aqueous dispersion.

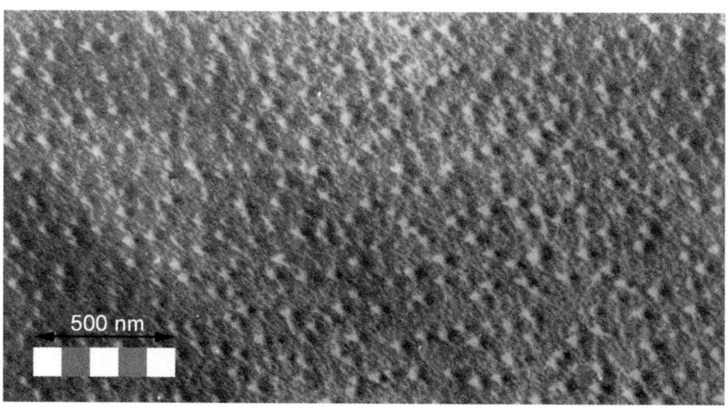

Fig. 3.2 Freeze-fracture (electron) micrograph of the DPPC liposomes.

3.3.2
Kinetic Stability of Liposome Dispersions

Vesicle stability, which can be divided into physical, chemical and biological stability of phospholipid bilayers, is one of the most important issues in applications of liposomal colloids. The kinetic stability of liposome dispersions is crucially influenced by the vesicle stability and also by the extent of electrical and/or steric repulsive forces operating between the vesicles. Breaking the cooperativity of phospholipid membranes, due to degradative effects or aggregation of liposomes, which may take place on storage of the dispersions, causes shifts in the mean size and size distribution towards higher values. Therefore, from the change in time of the size distribution, the aggregation state of vesicles and hence the kinetic stability of their dispersions can be estimated.

To illustrate the differences in the state of aggregation of polymer-free and polymer-containing vesicles several days after their preparation, the volume-average mean size (D) of DMPC liposomes and the standard deviation of D measured in different dispersion media are shown in Table 3.1.

There are no significant differences in the mean sizes between the DMPC and the DPPC liposomes after several days of storage. Nevertheless, some increase in the mean (hydrodynamic) diameter of the liposomes formulated in polymer solutions can be observed, indicating that the dissolved macromolecules are attached to the lipid membranes. These results also demonstrate that the dispersions of polymer-containing vesicles are kinetically more stable than those of the bare liposomes. During storage, aggregation and/or fusion of the polymer-free liposomes takes place, but the sorbed macromolecules decrease these processes. K90 PVP and MC proved to be effective stabilizers for the DMPC and the DPPC vesicles, respectively. It seems reasonable to assume that the enhanced stability of polymer-containing liposomes can be ascribed to the

Table 3.1 Mean size (*D*) and polydispersity (*PI*) of bare and polymer-bearing liposomes stored in aqueous dispersion.

Liposome	1 day		7 days	
	D (nm)	*PI*	*D* (nm)	*PI*
DMPC	57±23	0.41	137±41	0.68
DMPC + PVA	64±18	0.41	74±26	0.47
DMPC + K30 PVP	59±15	0.42	78±30	0.53
DMPC + K90 PVP	60±18	0.36	71±24	0.48
DPPC	52±16	0.34	101±33	0.70
DPPC + MC	58±20	0.32	68±15	0.47
DPPC + PVA	73±24	0.31	86±24	0.68
DPPC + K30 PVP	65±14	0.43	100±32	0.47
DPPC + K90 PVP	69±17	0.41	80±17	0.50

formation of a protective macromolecular sheath around the vesicles, which ensures steric stabilization against the aggregation of liposomes.

3.3.3
Effect of pH on the Stability of DMPC Liposomes

Most biomolecules, including lipids, may undergo different degradation reactions in both acidic and basic environments [3, 10, 14]. The degradative processes that may also take place on storage in liposomal phospholipids can affect the chemical stability of vesicles and hence the kinetic stability of liposome dispersions.

The effect of the pH of the dispersion medium on the long-term stability of liposomes is illustrated in Figs. 3.3 and 3.4, where the relative increase in the volume-average mean sizes (D/D_0) of DMPC liposomes stored for several weeks in buffered dispersion media at pH 1.2 and 8.2, respectively, is plotted as a function of the time of storage.

These plots illustrate a strong dependence of the kinetic stability of the liposome dispersions on the pH of the medium. At each pH studied, notable differences in the vesicle size between the bare and the polymer-containing DMPC liposomes can be observed. Also, the increase in the size of most polymer-bearing liposomes is significantly retarded in both weakly acidic and basic media. In strongly acidic medium (pH 1.2), the bare DMPC liposomes are practically unstable (Fig. 3.3). By a few days after preparation, a polydisperse dispersion with very large particles forms, indicating that extensive aggregation and/or fusion of the primary vesicles takes place. The polymers, however, exhibit a weak

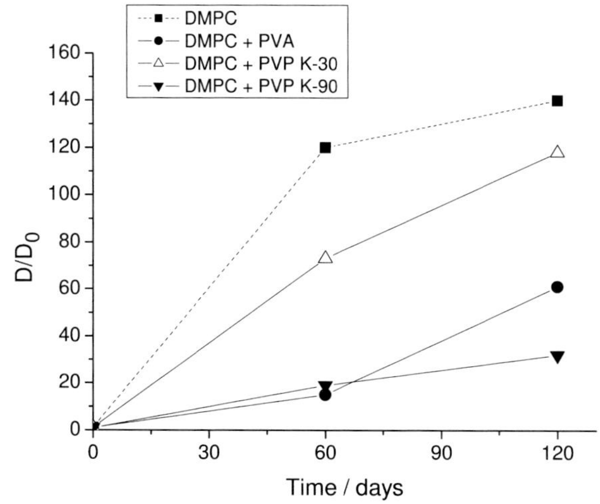

Fig. 3.3 Relative increase in the volume-average mean sizes of bare and polymer-bearing DMPC liposomes in buffered medium at pH 1.2 as a function of time.

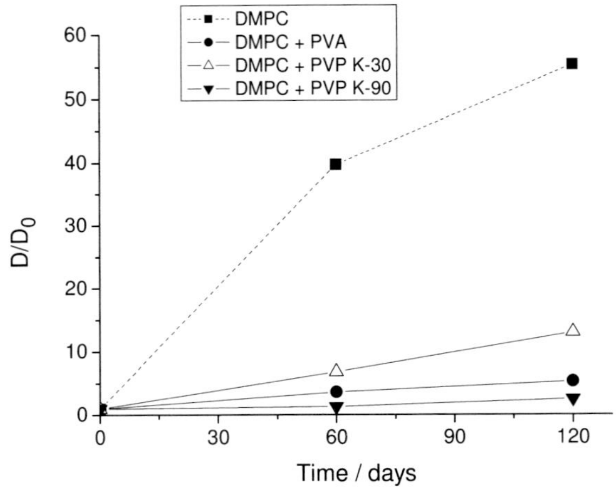

Fig. 3.4 Relative increase in the volume-average mean sizes of bare and polymer-bearing DMPC liposomes at pH 8.2 in buffered medium as a function of time.

stabilizing effect for the liposomes. Among the polymers, K90 PVP seems to be the best stabilizer for the DMPC vesicles at this pH.

After 1 week of storage in the strongly acidic medium, a many-fold increase in the size of the polymer-bearing liposomes can also be observed and in the meantime the dispersions have become completely polydisperse. Among the liposomes studied, the DMPC + K90 PVP vesicles are the only ones for which the polydispersity index remained below 1.0 after 120 days of storage.

In mild environments, all the liposomes are much more stable than when kept in the strongly acidic medium and the polymer-free vesicles exhibit the lowest stability. The stability of the bare DMPC liposomes kept either in the weakly acidic or in the weakly basic media was less than that when stored at neutral pH [18]. Almost the same trend holds for the polymer-bearing liposomes, but the vesicles of the same composition stored at pH 8.2 are much more stable (Fig. 3.4).

Nevertheless, it is worth pointing out that the DMPC + K-90 PVP liposomes also exhibit an enhanced stability in weakly acidic and basic dispersion media. These results suggest again that uncharged homopolymers may prove to be efficient steric stabilizers for the DMPC liposomes, at least in mild dispersion media.

To elucidate the observed alterations in the stability of liposomes in acidic and basic environments, the changes in the chemical stability of the vesicles at various pH values should certainly be taken into account. The most common pathways of chemical degradation that can lead to a change in the stability of vesicles are oxidation and, especially, hydrolysis. Oxidation is efficiently prevented by the use of saturated lipids. Hydrolysis normally leads to the cleavage of a C–O bond of the backbone and shows a strong pH dependence. The reaction rate of hydrolysis for more characteristic systems shows a minimum at about pH 6.5 and it steeply increases in both the acidic and basic regions

[3, 10]. Taking all these results into account, it is reasonable to assume that the significantly lowest stability of both the bare and polymer-bearing liposomes in strongly acidic media can be attributed to the more intense chemical degradation of the liposomal phospholipid at pH 1.2 than in milder environments.

3.3.4 Effect of Dissolved Cyclodextrins on the Stability of DPPC Liposomes

CDs are used in pharmaceuticals as specific "molecular or nanocapsules" [21]. The cyclic oligosaccharides are capable of forming inclusion complexes with drugs by taking up a molecule (or its hydrophobic part) into the cavity [22, 23]. The complexation will affect many of the physicochemical properties, e.g. the aqueous solubility, of sparingly soluble active ingredients. A major obstacle in the use of CDs, however, is that certain CD derivatives can interact with the cholesterol and the phospholipid components of the membranes of living cells, resulting in irreversible changes in the structure of the membrane. Such structural damage may lead to a considerable reduction in the physical stability and, for example, alterations in the permeability for a drug of the cell membrane [24–26].

The effect of dissolved CDs on the physical stability of DPPC liposomes was investigated also by measuring over time the average size of the vesicles formulated in CD solutions. As a typical example, Fig. 3.5 illustrates how α-CD affects the vesicle stability. The relative increase in the mean size of the liposomes (D/D_0) compared with the size of the vesicles prepared in the absence of the CD (D_0) is plotted as a function of time. The temperature of storage was 25 °C and the molar ratio of DPPC to CD ($n_{DPPC}:n_{CD}$) was 1:7.

For the pure DPPC liposomes in aqueous dispersion, a monomodal size distribution was detected after 7 days of storage. However, in presence of CD the

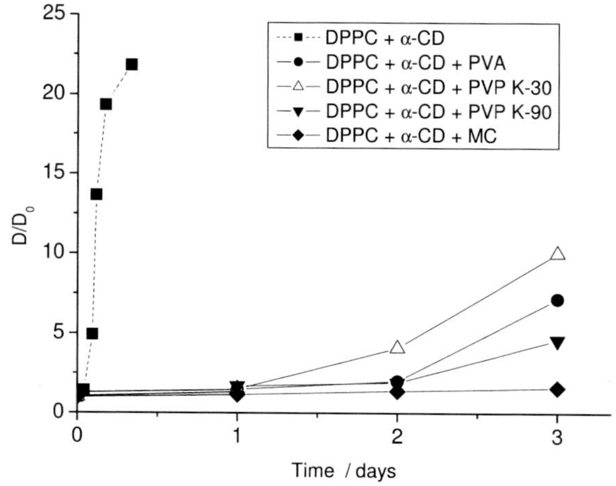

Fig. 3.5 Relative increase in the size of bare and polymer-bearing DPPC liposomes due to α-CD in aqueous dispersions.

average size of the vesicles increased significantly even after 8 h and the samples became highly polydisperse. These changes clearly show that dissolved CDs, above a critical CD:phospholipid ratio, may cause considerable damage to the structure of the phospholipid membranes. Due to this destructive effect of CD molecules, the physical stability of the liposomes is definitely lowered and aggregation and/or fusion of the vesicles induced.

The order of reducing the physical stability of the DPPC membranes (above a critical CD:DPPC ratio) is γ-CD < β-CD < α-CD. This order is the same as that of the affinity for the phospholipid membranes of the CDs, obtained from sorption measurements. In addition, it is demonstrated that the effect of the different CDs on the DPPC bilayers also depended on the phospholipid:CD ratio [19].

The lower stability of the polymer-free DPPC membranes in solutions of the CDs can very likely be ascribed to complexation between the phospholipid and CD molecules. Incorporation of the dissolved CD molecules into the phospholipid bilayers may result in a substantial decrease in the cooperativity of the DPPC membranes, which ultimately may lead to disintegration of the liposomes.

Nevertheless, it can also be seen that the polymer-bearing vesicles exhibit much higher physical stability than the bare vesicles. These differences demonstrate well that dissolved macromolecules can be effectively used for the prevention of the destruction of phospholipid bilayers for a longer time. In other words, vesicle stability and colloidal interactions between vesicles can be considerably modified and so the kinetic stability of liposome dispersions can also be well controlled by incorporation of uncharged polymers in the vesicle membranes.

3.4 Conclusions

Uncharged polymers attachable to phospholipid bilayers can be used as effective steric stabilizers for dispersions of liposomal colloids. Among the polymers used, PVP of high molecular mass exhibited the most efficient stabilizing effect, indicating that the formation of a relatively thick polymer layer around the lipid bilayers ensures enhanced and prolonged physical stability of the liposomes. These results provide experimental evidence that uncharged polymers can be efficient stabilizers of vesicles. Also, the colloidal interactions between the vesicles and hence the kinetic stability of liposome dispersions can be well controlled by using uncharged polymers.

References

1 A. D. Bangham, R.W. Horne, *J. Mol. Biol.*, *8*, 660–668 (**1964**).
2 R. H. Müller, *Colloidal Carriers for Controlled Drug Delivery and Targeting*, Wissenschlaftliche Verlagsgesellschaft, Stuttgart, **1991**.
3 J. Kreuter, *Colloidal Drug Delivery Systems*, Vol. 66, Marcel Dekker, New York, **1994**.
4 T. Landh, *FEBS Lett.*, *13*, 13–17 (**1995**).
5 M. Eddin, *Annu. Rev. Biomol. Struct.*, *32*, 257–283 (**2003**).
6 J. Barauskas, M. Johnson, F. Tiberg, *Nano Lett.*, *5*, 1615–1619 (**2005**).
7 G. Gregoriadis, A. C. Allison, *Liposomes in Biological Systems*, Wiley, New York, **1986**.
8 G. Gregoriardis, *Liposomes as Drug Carriers, Recent Trends and Progress*, Wiley, Chichester, **1988**.
9 M. Jamshaid, S. J. Farr, P. Kearney, I.W. Kellaway, *Int. J. Pharm.*, *48*, 125–131 (**1988**).
10 D. D. Lasic, *Liposomes – From Physics to Application*, Elsevier, Amsterdam, **1993**.
11 D. D. Lasic, F. J. Martin (Eds.), *Stealth Liposomes*, CRC Press, Boca Raton, FL, **1995**.
12 H. Ishiwata, A. Vertut-Doi, T. Hirose, K. Miyajima, *Chem. Pharm. Bull.*, *43*, 1005–1011 (**1995**).
13 D. D. Lasic, *Nature*, *380*, 561–562 (**1996**).
14 D. D. Lasic, D. Papahadjiopoulos, *Medical Applications of Liposomes*, Elsevier, Amsterdam, **1997**.
15 G.W. Boswell, D. Buell, I. Bekersky, *J. Clin. Pharmacol.*, *38*, 583–592 (**1998**).
16 J. Shah, Y. Sadhale, D. M. Chilikiru, *Adv. Drug Deliv. Rev.*, *47*, 229–250 (**2001**).
17 F. L. Grohmann, F. Csempesz, M. Szögyi, *Acta Pharm. Hung.*, *66*, 197–202 (**1996**).
18 F. L. Grohmann, F. Csempesz, *Magy. Kém. Foly.*, *104*, 53–59 (**1998**).
19 I. Puskás, L. Szente, F. Csempesz, *Magy. Kém. Foly. Közl.*, *3*, 33–37 (**2005**).
20 I. Puskás, L. Barcza, L. Szente, F. Csempesz, *J. Inclusion Phenom. Macrocycl. Chem.*, *54*, 89–93 (**2006**).
21 J. Szejtli, *Cyclodextrin Technology*, Kluwer, Dordrecht, **1988**.
22 Y. L. Loukas, P. Jayasekera, G. Gregoriadis, *Int. J. Pharm.*, *117*, 85–94 (**1995**).
23 K. H. Frömming, J. Szejtli, *Cyclodextrins in Pharmacy*, Kluwer, Dordrecht, **1993**.
24 L. Szente, *Cyclodextrin News*, *16* (4), 59–61 (**2002**).
25 J. Nishijo, S. Shiota, K. Mazima, Y. Inoue, H. Mizuno, J. Yoshida, *Chem. Pharm. Bull.*, *48*, 48–52 (**2000**).
26 T. Irie, K. Uekama, *J. Pharm. Sci.*, *86*, 147–162 (**1997**).

4
Mathematical Modeling of Coagulation and Flocculation of Colloidal Suspensions Incorporating the Influence of Surface Forces

Venkataramana Runkana and Ponisseril Somasundaran

Abstract

A general population balance model for the coagulation and flocculation of colloidal suspensions by inorganic electrolytes and polymers is presented. Since the rate of aggregation is determined primarily by the surface forces between colloidal particles, modern theories of surfaces are incorporated into the model. The efficiency of particle collisions is computed as a function of total interaction energy, which is assumed to be a sum of van der Waals attraction and electrical double layer repulsion when inorganic electrolytes are used. The model was extended for situations in which structural forces such as short-range hydration repulsion are active. An expression derived from the scaling theory of polymer adsorption was included to account for forces due to adsorbed polymer. The irregular and open structure of flocs is taken into account by computing the collision frequency factor as a function of floc mass fractal dimension. The model was tested with published experimental data for both electrolyte-induced coagulation and polymer-induced flocculation of metal oxide and polystyrene latex suspensions. The simulation results closely mimic experimental data for the evolution of mean aggregate size with time. The model can be utilized to study the influence of variables such as pH, electrolyte concentration, polymer concentration and temperature on flocculation kinetics.

4.1
Introduction

Colloidal suspensions are encountered in many industrial processes dealing with particulate solids. They can be classified as either dilute or concentrated, depending on the particle concentration. Several consumer products such as toothpastes, paints and food products are in the form of concentrated suspensions and many solid–liquid separation processes involve dilute suspensions, for example, water treatment and mineral beneficiation. Moreover, many inter-

Colloids and Interface Science Series, Vol. 3
Colloid Stability and Application in Pharmacy. Edited by Tharwat F. Tadros

ISBN: 978-3-527-31463-8

mediate process streams in the manufacture of pulp and paper, ceramic components, polymers and textiles are in the form of suspensions. It is desirable to disperse the particles uniformly within the solvent and to retain long-term stability or shelf-life in the case of products such as paints, inks and pharmaceutical suspensions. On the other hand, fast removal of suspended solids is a prerequisite for efficient solid–liquid separation. Inorganic electrolytes, polymers and surfactants are commonly employed as additives to induce coagulation or flocculation of suspended particles. The stability or rate of aggregation of a colloidal suspension depends strongly on the interaction or surface forces acting between particles. These forces include van der Waals attraction, electrical double layer repulsion/attraction, hydration repulsion, hydrophobic attraction and, in the presence of polymers, steric repulsion or bridging attraction (Israelachvili 1991). The nature and magnitude of these forces depend on variables such as particle size distribution, pH, temperature, type and concentration of added or dissolved electrolyte species and the concentration and properties of the polymer or surfactant used.

Coagulation and flocculation essentially involve aggregation of colloids and depend strongly on the interactions among colloidal particles, solvent or suspending medium, electrolyte species and polymers. These processes are complex in nature because several phenomena occur simultaneously or sequentially. These include the formation of clusters or aggregates or flocs, restructuring of flocs, adsorption of dissolved electrolyte species on particle surfaces and the subsidence, sedimentation or creaming of flocs. Additional phenomena such as adsorption of polymer chains on particle surfaces and stretching or reconformation of adsorbed chains occur in the presence of polymers. When shear is applied to improve the rate of flocculation, fragmentation of flocs, consolidation and restructuring of flocs, desorption of adsorbed chains and possibly scission of polymer chains in solution can also take place.

The effectiveness of coagulation or flocculation is usually represented in terms of readily measurable parameters such as settling rate of flocs, percentage of solids settled, supernatant turbidity, sediment volume or weight, moisture content and strength of flocs. These parameters are indirect measures of flocculation efficiency and the majority of them are actually functions of aggregate or floc size distribution (FSD), which is a direct and explicit indicator because flocculation is essentially an aggregation phenomenon. Population balances have been developed, starting from the classical work of Smoluchowski (1917), to predict the time evolution of aggregate size distribution during coagulation. Since the process is complex, several assumptions are usually made to simplify the mathematical treatment (Thomas et al. 1999). For example, it was assumed earlier that both primary particles and flocs are spherical in shape. However, with the advent of fractal geometry (Mandelbrot 1983) and the recognition that flocs are fractal in nature, the open and irregular structure of flocs is now routinely represented in terms of their mass fractal dimension in flocculation models. Similarly, initially the kinetics of floc fragmentation were not included in the population balance. However, later models incorporated simultaneous colloid aggregation and floc fragmentation

(Lu and Spielman 1985; Spicer and Pratsinis 1996; Somasundaran and Runkana 2003). However, the majority of flocculation models either assume that all particle collisions lead to aggregation or treat particle collision efficiency as an adjustable parameter. The probability or efficiency of colloid aggregation upon collision is actually a function of the surface forces between interacting particles. Recognizing this, we have incorporated fundamental theories of surface forces into the population balance framework to model flocculation (Runkana et al. 2004, 2005, 2006). This chapter summarizes our recent work on the modeling of coagulation and flocculation in the presence of inorganic electrolytes and polymers. Although coagulation and flocculation are used interchangeably nowadays, for the sake of convenience, aggregation of suspensions in the presence of inorganic electrolytes is termed coagulation whereas aggregation in the presence of polymers is referred to as flocculation here.

4.2 Mathematical Model

Flocculation is a complex process and involves several phenomena, as mentioned earlier. A comprehensive mathematical model should, in principle, incorporate the kinetics of particle and/or cluster aggregation, kinetics of floc fragmentation, heterogeneous fluid flow pattern in stirred tanks, interaction forces between particles and dynamics of polymer adsorption, if polymers are used. The size distribution and strength of flocs depend on numerous variables:

- *Properties of solids:* Primary particle size distribution, apparent density, surface potential or charge density of particles in the solvent and particle shape.
- *Properties of solvent/suspension:* Temperature, pH, solids concentration, electrolyte concentration and polymer concentration.
- *Properties of polymer/polyelectrolyte:* Average molecular weight or molecular weight distribution, charge density distribution and branching or nature of functional groups.
- *Flocculation tank geometry:* Tank height and diameter, number and dimensions of baffles, impeller type and diameter, number of blades/paddles and impeller speed.

Taking note of the various phenomena and the parameters that influence flocculation, it can be stated that it is necessary to combine the basic principles of chemical engineering with those of surface and colloid science to develop a realistic model for flocculation. We have developed a reasonably general population balance model (Fig. 4.1) for the flocculation of dilute colloidal suspensions in the presence of inorganic electrolytes and polymers. It will be noticed that several phenomena related to flocculation are incorporated except the heterogeneous structure of fluid flow in stirred suspensions and dynamics of polymer adsorption. The model can readily be extended by combining a suitable computational fluid dynamics model for fluid flow and a model for polymer adsorp-

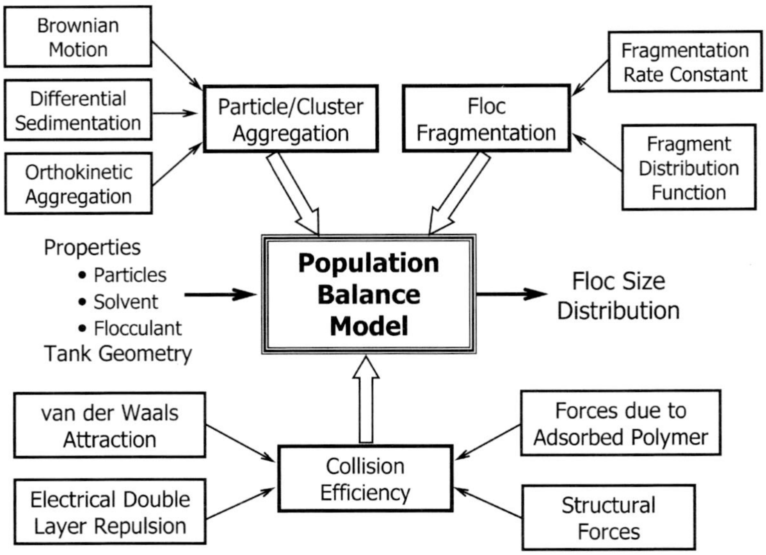

Fig. 4.1 General population balance framework for flocculation incorporating simultaneous aggregation–fragmentation kinetics and influence of surface forces.

tion dynamics. Although we have included kinetics of aggregation in shear flow and floc fragmentation kinetics in the population balance (Runkana 2003; Somasundaran and Runkana 2003), only results related to the aggregation of suspensions under quiescent flow conditions are discussed here.

4.2.1
Population Balance Equation for Aggregation

Population balances are commonly employed to model agglomeration processes such as coagulation and flocculation in order to simulate the evolution of aggregate size distribution with time. The size-continuous form of the population balance equation for aggregation of dilute colloidal suspensions in a batch process is given by (Smoluchowski 1917)

$$\frac{\partial n(v,t)}{\partial t} = -\int_0^\infty \alpha(v,u)\beta(v,u)n(v,t)n(u,t)\mathrm{d}u + \frac{1}{2}\int_0^v \alpha(v-u,u)\beta(v-u,u)n(v-u,t)n(u,t)\mathrm{d}u \tag{1}$$

where n is the number concentration of aggregates (or particles), v and u denote aggregate volume, t is aggregation time, β is the collision frequency factor and

α is the collision efficiency factor. The first term on the right-hand side accounts for the loss or disappearance of aggregates of size v due to their interaction with aggregates of all sizes. The second term represents the rate of formation of aggregates of size v due to interaction between aggregates of smaller sizes.

It is generally not possible to derive a closed-form analytical solution to the integro-partial differential equation of flocculation (Eq. 1). Analytical solutions can be derived only by making drastic simplifications in the kernels. For realistic collision kernels representative of aggregation of dilute suspensions, it becomes necessary to employ numerical solutions after discretizing Eq. (1) with respect to size into a set of non-linear ordinary differential equations (ODEs). The numerical solution of this set of ODEs is also not a trivial task, as the number of size classes required to represent the size spectrum adequately is very large and the set is fairly stiff. In order to overcome these problems, the size classes are usually lumped or grouped, either uniformly or geometrically, into a smaller number of sections. In the case of uniform grouping, the difference between the upper and lower bounds of a size interval is constant, whereas in geometric grouping, the ratio of the upper and lower bounds is fixed or allowed to change while solving the equations. Uniform grouping provides accurate solutions for the complete size distribution but at the expense of computational time, as a large number of sections are required to represent the full size distribution. Geometric grouping, on the other hand, results in somewhat less accurate solutions but it is computationally more efficient as only a small number of sections is required. The general framework of population balances and their solution techniques is described in an excellent manner elsewhere (Ramkrishna 2000).

The discretized population balance equation for aggregation, based on the method of geometric sections proposed by Hounslow et al. (1988), is adopted here to represent coagulation or flocculation processes that do not involve fragmentation of flocs. The rate of change of particle or floc number concentration due to aggregation alone is given by (Hounslow et al. 1988)

$$\begin{aligned}\frac{dN_i}{dt} &= N_{i-1}\sum_{j=1}^{i-2} 2^{j-i+1}\alpha_{i-1,j}\beta_{i-1,j}N_j + \frac{1}{2}\alpha_{i-1,i-1}\beta_{i-1,i-1}N_{i-1}^2 \\ &\quad - N_i\sum_{j=1}^{i-1} 2^{j-i}\alpha_{i,j}\beta_{i,j}N_j - N_i\sum_{j=i}^{\max}\alpha_{i,j}\beta_{i,j}N_j\end{aligned} \tag{2}$$

where N_i is number concentration of particles or aggregates in a section i. The frequency of collisions between aggregates is directly related to the collision volume of an aggregate. The characteristic volume V_i of an aggregate in section i is computed as the arithmetic average of its lower and upper bounds:

$$V_i = \frac{b_i + b_{i-1}}{2} \tag{3}$$

where b_{i-1} and b_i are lower and upper boundary volumes of section i, respectively. The upper bound is computed as a function of the lower one with a geometric sectional spacing factor of 2:

$$b_i = 2b_{i-1} \tag{4}$$

The primary particle volume is denoted V_1. An aggregate of volume V_i comprises 2^{i-1} primary particles.

4.2.2
Collision Frequency

Aggregation takes place as a result of particle–particle, particle–cluster and cluster–cluster collisions, which occur due to Brownian motion, applied shear and differential sedimentation. During the initial stages of flocculation, collisions involve mainly primary particles and lead to the formation of small clusters, which aggregate further to form larger aggregates or flocs. Colloid aggregation is usually characterized in terms of two regimes, diffusion-limited aggregation (DLA) and reaction-limited aggregation (RLA). DLA takes place when repulsive forces between particles are negligible. It is generally fast and follows power-law growth kinetics. On the other hand, RLA follows slow, exponential growth kinetics (Lin et al. 1989). The aggregate growth in these two regimes is shown schematically in Fig. 4.2. Although solid volume or mass is conserved, porosity and shape may alter when two aggregates combine. This implies that the collision radius of an aggregate depends on its structure. The aggregates are commonly referred to as mass fractals because aggregate mass M_i is related to its collision radius r_{C_i} through the mass fractal dimension d_F (Feder 1988):

$$M_i \propto r_{C_i}^{d_F} \tag{5}$$

The collision radius of an aggregate containing n_0 primary particles of radius r_0 is given by (Flesch et al., 1999)

$$r_{C_i} = r_0 \left(\frac{n_{0i}}{C_L} \right)^{1/d_F} \tag{6}$$

where C_L is the aggregate structure prefactor. The aggregate fractal dimension is related to particle size distribution, solids concentration, temperature, pH, ionic strength and polymer concentration (Klimpel and Hogg 1986; Amal et al. 1990b; Zhang and Buffle 1995).

Smoluchowski (1916) derived the collision kernel or frequency factor for the aggregation of spherical primary particles under the DLA regime. Since collisions mainly involve clusters or aggregates, except during the initial stages of flocculation, the equation for spherical particles derived by Smoluchowski needs to be modified to incorporate the collision radii of interacting aggregates (Eq. 6).

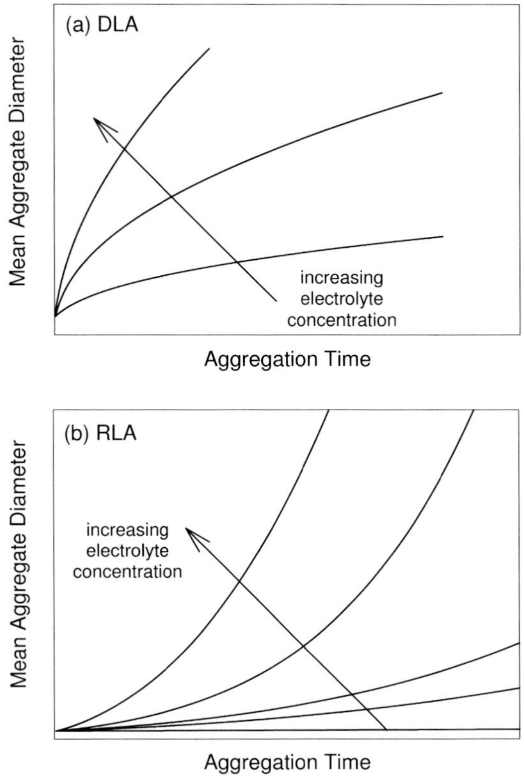

Fig. 4.2 Schematic representation of aggregation kinetics in (a) diffusion-limited aggregation–power law growth and (b) reaction-limited aggregation–exponential growth.

This can be derived easily and one obtains the following equation for the collision frequency factor for aggregation under the DLA regime, $\beta_{i,j}^{\mathrm{DLA}}$:

$$\beta_{i,j}^{\mathrm{DLA}} = \frac{2k_{\mathrm{B}}T}{3\mu}\left(\frac{1}{r_{\mathrm{c}_i}} + \frac{1}{r_{\mathrm{c}_j}}\right)(r_{\mathrm{c}_i} + r_{\mathrm{c}_j}) \tag{7}$$

where k_{B} is Boltzmann's constant, T is the suspension temperature, μ is the dynamic viscosity of the suspending fluid and r_{c_i} and r_{c_j} are the collision radii of particles or aggregates belonging to sections i and j, respectively. This equation, however, is not applicable for RLA because it occurs in the presence of repulsive forces and follows slow exponential growth kinetics. In the case of RLA, aggregation takes place only after several collisions have occurred between particles or aggregates due to the repulsive forces. Although several empirical relations have been proposed in the literature for the collision frequency factor for the RLA regime, none of them predicts the aggregation kinetics accurately. Recently, we derived the following kernel based on theoretical arguments proposed by Ball et al. (1987) for cluster–cluster collisions in RLA (Runkana et al. 2005):

$$\beta_{i,j}^{\mathrm{RLA}} = K_{\mathrm{RLA}} r_0{}^2 \left(\frac{1}{C_{\mathrm{L}}}\right)^{2/d_{\mathrm{F}}} (n_{0_i}{}^{1/d_{\mathrm{F}}} + n_{0_j}{}^{1/d_{\mathrm{F}}})^2 \tag{8}$$

where $\beta_{i,j}^{\mathrm{RLA}}$ is collision frequency factor for RLA kinetics, K_{RLA} is a proportionality constant or lumped parameter, which takes into account the effect of temperature and viscosity of the suspension, and n_{0_i} and n_{0_j} are the number of primary particles in aggregates belonging to sections i and j, respectively. This kernel has been found to be accurate for reaction-limited aggregation of γ-alumina suspensions in the presence of various electrolytes (Runkana et al. 2005).

As aggregation proceeds, flocs grow larger in size and settle at different velocities, depending on floc size, structure and density. The collision frequency factor for aggregation by differential sedimentation, $\beta_{i,j}^{\mathrm{DS}}$, is given by (Camp and Stein 1943)

$$\beta_{i,j}^{\mathrm{DS}} = \frac{2\pi g}{9\mu}(r_{\mathrm{c}_i} + r_{\mathrm{c}_j})^2 |r_{\mathrm{c}_i}{}^2(\rho_i - \rho_l) - r_{\mathrm{c}_j}{}^2(\rho_j - \rho_l)| \tag{9}$$

where ρ_i and ρ_l are the densities of aggregate and fluid, respectively, and g is the acceleration due to gravity. The aggregate density ρ_i, which decreases as size increases, can be estimated using the following equation (Jiang and Logan 1991)

$$\rho_i = \rho_0 \left(\frac{r_{\mathrm{c}_i}}{r_0}\right)^{d_{\mathrm{F}}-3} \tag{10}$$

This equation is applicable only when the aggregate density decreases linearly with size, when plotted on a logarithmic scale.

When polymers are used for flocculation, depending on polymer adsorption and its conformation at the solid–liquid interface, the collision radii of particles increase and the interaction forces between particle surfaces are altered, which, in turn, affect the rate of flocculation. Polymer adsorption is a complex phenomenon and depends on pH, temperature, nature of solvent, surface charge distribution on particles, charge density distribution of polymer chains, ionic strength, etc. Polymer chains may adsorb in the form of thin layers or trains, coils or loops and tails at the solid–liquid interface (Chandar et al. 1987; Fleer et al. 1993). Different chain conformations are shown schematically in Fig. 4.3. The composite polymer-coated particle radius is estimated by adding the adsorbed layer thickness to the solid particle radius. It is assumed here that polymer adsorption is very fast and polymer chains attain equilibrium conformation before a meaningful number of particle collisions have taken place. In addition, it is also assumed that polymer adsorbs uniformly and the adsorbed layer thickness is same for all the particles.

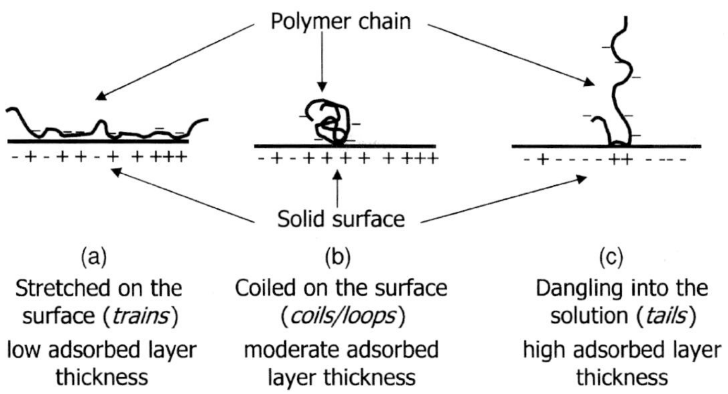

Fig. 4.3 Schematic representation of different polymer conformations at the solid–liquid interface.

Not all of the collisions between particles necessarily lead to aggregation if repulsive forces are stronger than attractive forces. The influence of these surface forces is taken into account by the collision efficiency factor in the population balance. The method of computing the collision efficiency factor and its relation to various surface forces is described in the next section.

4.2.3 Collision Efficiency

Depending on the surface and colloid chemistry of the system, the nature and magnitude of various interaction forces between particles will change and influence colloid stability and rate of aggregation. The collision efficiency factor for aggregates is computed as the reciprocal of the modified Fuchs' stability ratio W for two primary particles k and l (Fuchs 1934; Derjaguin and Muller 1967; McGown and Parfit 1967; Spielman 1970):

$$W_{k,l} = \frac{\int_{r_{0_k}+r_{0_l}}^{\infty} D_{k,l} \frac{\exp(V_T/k_B T)}{s^2} \mathrm{d}s}{\int_{r_{0_k}+r_{0_l}}^{\infty} D_{k,l} \frac{\exp(V_{vdW}/k_B T)}{s^2} \mathrm{d}s} \tag{11}$$

where $D_{k,l}$ is hydrodynamic correction factor, given by Honig et al. (1971):

$$D_{k,l} = \frac{6\bar{h}_0{}^2 + 13\bar{h}_0 + 2}{6\bar{h}_0{}^2 + 4\bar{h}_0} \tag{12}$$

V_T is the total energy of interaction and V_{vdW} is the van der Waals energy of attraction between two primary particles of radii r_{0_k} and r_{0_l} (assumed spherical), s is the distance between particle centers ($s = r_{0_k} + r_{0_l} + h_0$), h_0 is the distance of closest approach between particle surfaces and $\bar{h}_0 = 2h_0/(r_{0_k} + r_{0_l})$ is the normalized distance. It is assumed that the efficiency of aggregate collisions depends mainly on the interaction between particles lying on the surface of the aggregates. The forces between particles decrease rapidly with distance and, as such, the interaction between aggregates can be approximated by the interaction between the surface particles (Firth and Hunter 1976).

The total interaction energy is assumed to be sum of all the surface forces expected to be present between particles under the given conditions of pH, temperature, type and concentration of electrolyte species and type and concentration of coagulant or flocculant employed and its properties. These forces include van der Waals attraction, electrical double layer attraction/repulsion, hydration repulsion and hydrophobic attraction. When polymers are present, the electrochemical nature of the particle surfaces also becomes modified, which leads to a change in the nature and magnitude of van der Waals attraction and electrical double layer force. In addition, steric repulsion or bridging attraction has to be included when computing the total interaction energy. The equations for various interaction energies relevant for the results presented here are given in the following sections.

4.2.3.1 Van der Waals Attraction

The van der Waals energy of attraction between bare particles, $V_{\mathrm{vdW}}^{\mathrm{H}}$, is given by (Hamaker 1937)

$$V_{\mathrm{vdW}}^{\mathrm{H}} = -\frac{A}{6}\left\{\frac{2r_{0_k}r_{0_l}}{s^2-(r_{0_k}+r_{0_l})^2} + \frac{2r_{0_k}r_{0_l}}{s^2-(r_{0_k}-r_{0_l})^2} + \ln\left[\frac{s^2-(r_{0_k}+r_{0_l})^2}{s^2-(r_{0_k}-r_{0_l})^2}\right]\right\} \tag{13}$$

where A is the non-retarded Hamaker constant of solids in the solvent. The van der Waals attraction between particles covered with polymer layers differs from that between bare particles because the properties of the polymer solution such as dielectric constant and refractive index will be different from that of the pure solvent. Vold (1961) derived an equation for the van der Waals attraction between two identical particles having adsorbed layers of the same polymer. Subsequently, Vincent (1973) generalized Vold's expression for two dissimilar solids and polymers. The general expression of Vincent (1973), given below, is used in the present work.

$$\begin{aligned} -12V_{\mathrm{vdW}}^{\mathrm{V}} = {} & H_{s_k s_l}\left(A_{s_k}^{\frac{1}{2}} - A_m^{\frac{1}{2}}\right)\left(A_{s_l}^{\frac{1}{2}} - A_m^{\frac{1}{2}}\right) \\ & + H_{p_k p_l}\left(A_{p_k}^{\frac{1}{2}} - A_{s_k}^{\frac{1}{2}}\right)\left(A_{pl}^{\frac{1}{2}} - A_{s_l}^{\frac{1}{2}}\right) \\ & + H_{p_k s_l}\left(A_{p_k}^{\frac{1}{2}} - A_{s_k}^{\frac{1}{2}}\right)\left(A_{s_l}^{\frac{1}{2}} - A_m^{\frac{1}{2}}\right) \\ & + H_{p_l s_k}\left(A_{p_l}^{\frac{1}{2}} - A_{s_l}^{\frac{1}{2}}\right)\left(A_{s_k}^{\frac{1}{2}} - A_m^{\frac{1}{2}}\right) \end{aligned} \tag{14}$$

where $V_{\mathrm{vdW}}^{\mathrm{V}}$ is the van der Waals energy of attraction between particles with adsorbed polymer and A_p, A_m and A_s are non-retarded Hamaker constants of solids, solvent and polymer, respectively, across a vacuum. The non-retarded geometric function $H(x, y)$ is (Hamaker, 1937)

$$H(x,y) = \frac{y}{x^2+xy+x} + \frac{y}{x^2+xy+x+y} + 2\ln\left(\frac{x^2+xy+x}{x^2+xy+x+y}\right) \tag{15}$$

where $x = \Delta/2R_k$ and $y = R_l/R_k$. Expressions for Δ, R_k and R_l for different H functions in Eq. (15) are as follows (Vold 1961):

$$H_{s_k s_l}: \quad \Delta = h_0;\ R_k = r_{0_k} + \delta_k;\ R_l = r_{0_l} + \delta_l \tag{16 a}$$

$$H_{p_k p_l}: \quad \Delta = h_0 + \delta_k + \delta_l;\ R_k = r_{0_k};\ R_l = r_{0_l} \tag{16 b}$$

$$H_{p_k s_l}: \quad \Delta = h_0 + \delta_k;\ R_k = r_{0_k};\ R_l = r_{0_l} + \delta_l \tag{16 c}$$

$$H_{p_l s_k}: \quad \Delta = h_0 + \delta_l;\ R_k = r_{0_k} + \delta_k;\ R_l = r_{0_l} \tag{16 d}$$

where δ_k and δ_l are the thicknesses of the adsorbed polymer layers on primary particles k and l, respectively. It was assumed in deriving the above equations that the polymer concentration is uniform within the adsorbed layer.

The above expressions do not take into account the retardation phenomenon, which sets in at about 5 nm from the particle surface (Israelachvili 1991). The retardation effect is incorporated by multiplying the unretarded van der Waals attraction with a correction function proposed by Gregory (1981):

$$f_R(h_0) = 1 - \frac{b_R h_0}{\lambda_R} \ln\left(1 + \frac{\lambda_R}{b_R h_0}\right) \tag{17}$$

where f_R is the retardation correction function, λ_R is the characteristic wavelength of interaction and b_R is a fitting parameter. λ_R is typically 100 nm and the value of b_R is 5.32 (Gregory 1981).

4.2.3.2 **Electrical Double Layer Forces**

When two colloidal particles approach each other, they experience electrostatic attraction or repulsion depending on the nature of the electrical double layer. This interaction energy can be calculated accurately by rigorous solution of the Poisson–Boltzmann equation. Since this is a difficult task, approximate expressions have been proposed in the literature (Hogg et al. 1966; Bell et al. 1970; Sader et al. 1995). The analytical expression derived by Bell et al. (1970) based on the linear superposition approximation is used in the present work. The interaction energy due to the electrical double layers, V_{edl}, between two spherical particles of radii r_{0_k} and r_{0_l} and surface potentials ψ_{0_k} and ψ_{0_l} is given by

$$V_{\text{edl}} = 64\pi\varepsilon_{\text{r}}\varepsilon_0 \left(\frac{k_{\text{B}}T}{z_{\text{c}}e}\right)^2 \left(\frac{r_{0_k} r_{0_l}}{r_{0_k} + r_{0_l}}\right) \tanh\left(\frac{z_{\text{c}}e\psi_{0_k}}{4k_{\text{B}}T}\right) \tanh\left(\frac{z_{\text{c}}e\psi_{0_l}}{4k_{\text{B}}T}\right) \exp(-\kappa h_0) \tag{18}$$

where e is elementary charge, z_{c} is the valence of the counterion, ε_0 is the permittivity of free space and ε_{r} is the relative permittivity or dielectric constant of the solvent. The Debye–Hückel parameter κ is a function of electrolyte concentration, valence of electrolyte ions and temperature (Israelachvili 1991). The surface potential depends on pH, temperature, type and concentration of electrolyte species and concentration, molecular weight and charge density of the polymer added (Tjipangandjara et al. 1990; Zhang and Buffle 1995).

4.2.3.3 **Bridging Attraction/Steric Repulsion**

The scaling theory (de Gennes 1981, 1982) is used to compute forces due to adsorbed polymer layers. This theory is based on minimization of a surface free energy functional subject to the constraint that the total amount of polymer adsorbed is fixed in the region between two surfaces having adsorbed layers. The form of the surface energy functional depends on the nature of the solvent and polymer concentration. The scaling theory for good solvents is utilized in the present work. In the scaling theory, the surface free energy functional is assumed to be a sum of two contributions, a surface term to account for the interaction between polymer and surface and a bulk contribution due to variation of polymer concentration from the surface (de Gennes 1982):

$$\gamma^{\text{S}} - \gamma_0^{\text{S}} = -|\gamma_1^{\text{S}}|\Phi_{\text{s}} + a_{\text{Sc}}k_{\text{B}}T \int_0^{h_0} \frac{1}{\xi^3(\Phi)} \left\{1 + \left[m_0 \frac{\xi(\Phi)}{\Phi} \frac{\text{d}\Phi}{\text{d}z}\right]^2\right\} \text{d}z \tag{19}$$

where γ^{S} and γ_0^{S} are the surface free energy of the polymer solution and of the pure solvent, respectively, γ_1^{S} is the local solute–interface interaction energy per unit area, $\Phi(z)$ is the polymer volume fraction at a distance z from the surface, $\Phi_{\text{S}} = \Phi(z = 0)$, a_{Sc} and m_0 are numerical constants and ξ is the correlation length, the average distance between two successive contacts of a polymer chain with other chains.

The polymer concentration is maximum at the solid–liquid interface and decreases gradually to the bulk value. When two surfaces bearing adsorbed polymer are brought close to each other, it can be assumed that a symmetrical polymer concentration profile develops between the surfaces. If the intersurface gap becomes very narrow, it can be further assumed that the polymer concentration will be uniform throughout the gap. Making use of this simplification, Klein and Rossi (1998) derived an expression for the intersurface pressure between two flat surfaces. After making similar assumptions and applying the Derjaguin approximation (Derjaguin 1934), we derived the following equation for interaction energy between two unequal polymer-coated spherical particles, V_S (Runkana et al. 2006):

$$V_S = \left(\frac{2\pi r_{0_k} r_{0_l}}{r_{0_k} + r_{0_l}}\right)\left(\frac{a_{Sc} k_B T}{a_m{}^3}\right)\Phi_{S_0}{}^{\frac{9}{4}} D_{Sc}\left\{-\frac{16\Gamma D_{Sc}}{\Gamma_0}\ln\left(\frac{2\delta}{h_0}\right) + \frac{4 D_{Sc}{}^{\frac{5}{4}}}{2^{\frac{5}{4}}}\left(\frac{8\Gamma}{\Gamma_0}\right)^{\frac{9}{4}}\left[\frac{1}{h_0{}^{\frac{1}{4}}} - \frac{1}{(2\delta)^{\frac{1}{4}}}\right]\right\} \tag{20}$$

where Γ is the total amount of polymer adsorbed on a single surface, Γ_0 is the amount adsorbed at saturation, a_m is the effective monomer size, Φ_{S_0} is the polymer concentration at a single saturated surface and D_{Sc} is the scaling length, a measure of segment–surface attraction and related to γ_1^S (de Gennes, 1982). Both Φ_{S_0} and D_{Sc} can be obtained from polymer volume fraction profiles measured by neutron scattering and reflectometry. The first term in the braces in Eq. (20) represents short-range attraction while the second term accounts for excluded volume repulsion. The relative magnitudes of these two contributions determine whether polymer adsorption results in bridging attraction or steric repulsion.

4.2.3.4 **Hydration Repulsion**

Force measurements between surfaces of several materials such as mica, silica and alumina have shown the existence of a short-range repulsive force (Pashley 1981; Pashley and Israelachvili 1984; Ducker et al. 1994; Karaman et al 1997) under certain conditions of pH and electrolyte concentration. This force has been termed hydration or solvation repulsion because it supposedly arises due to the structure of water or solvent molecules at the solid–liquid interface. Although several theories have been proposed to describe the existence of this force, none of them has been able to explain experimental observations under different conditions (Johnson et al. 2000; Manciu and Ruckenstein 2001). The hydration force F_{hyd} decreases exponentially from the surface and usually a simple exponential decay function is fitted to experimental force–distance profiles (Israelachvili 1991; Ducker et al. 1994):

$$F_{hyd} = P_{hyd}\exp(-h_0/\lambda_{hyd}) \tag{21}$$

where P_{hyd} is a structural or hydration force constant and λ_{hyd} is the hydration decay length. Typical values of P_{hyd} and λ_{hyd} are in the ranges 1×10^6–5×10^8 N m^{-2} and 0.2–1.0 nm, respectively (Thompson and Collins 1994). By applying the Derjaguin approximation (Derjaguin 1934), the following equation was derived for the hydration interaction energy, V_{hyd}, between two unequal spherical particles (Runkana et al. 2005):

$$V_{hyd} = \left(\frac{2\pi r_{0_k} r_{0_l}}{r_{0_k} + r_{0_l}}\right) P_{hyd}\lambda_{hyd}{}^2\exp(-h_0/\lambda_{hyd}) \tag{22}$$

4.3
Results and Discussion

The mathematical model for coagulation and flocculation was implemented and validated against published data for different situations involving coagulation by inorganic electrolytes and flocculation by polymers. In the case of coagulation, the model was used for the simulation of both diffusion-limited and reaction-limited aggregation in the presence of different electrolytes. In the case of polymer-induced flocculation, results for two well-known mechanisms of flocculation, simple charge neutralization and bridging, are presented.

The collision frequency factor was calculated as a sum of contributions due to differential sedimentation (Eq. 9), and due to Brownian motion (Eq. (7) for DLA and Eq. (8) for RLA). The collision efficiency factor was computed as the reciprocal of the stability ratio (Eq. 11), which was evaluated by numerical integration using the Romberg technique (Press et al. 1992). Since it is assumed that the efficiency of aggregate collisions is a function of the interaction between primary particles, the properties of primary particles such as particle radius and surface potential were used for calculating the stability ratio. Moreover, if the suspension contains only one type of solid, the particle surface potential will be identical for all particles, that is $\psi_{0_k} = \psi_{0_l} = \psi_0$ in Eq. (18), which was used for calculating EDL repulsion in all the simulation results presented here. The effect of retardation was taken into account by multiplying the non-retarded van der Waals attraction by the correction function given by Eq. (17).

The population balance model involves a number of parameters, which can be measured experimentally or, in principle, calculated theoretically. These are mostly related to various particle interaction energies, except the mass fractal dimension of aggregates, which is required to evaluate the collision frequency factor. If model parameters were not available, representative values were taken from the literature for related systems or materials and adjusted by trial and error to match the simulation results with the experimental aggregate size distribution or time evolution of mean aggregate size data. Parameter estimation is commonly carried out by minimizing the error between simulation results and experimental data using a suitable optimization technique. However, this can result in a blind search without regard to practical values observed experimentally, and there is also a possibility of determining optimum but physically unrealistic parameter values due to mutual compensation in the highly non-linear environment of model equations. In order to avoid these, a systematic two-stage search strategy was followed. In the first stage, the interaction energy–surface separation diagram was used, where necessary, as a guide to arrive at physically realistic parameter values in various equations for surface forces, using reported values from experimental studies on similar systems as initial estimates. In the second stage, parameters such as aggregate fractal dimension were adjusted, if necessary, to achieve closer agreement between simulated aggregate growth curves or size distributions and experimental data.

We employed 30 geometric sections to represent the entire size spectrum, starting from the primary particles. That is, a set of 30 non-linear ordinary differential equations, each given by the discretized population balance equation (Eq. 2), were simultaneously solved numerically by Gear's predictor–corrector technique (Gear 1971) to predict the time evolution of the number concentration of aggregates in 30 size classes. During the integration of population balance equations, the computed aggregate size distribution was tested at each time step for conservation of solid volume. The loss of total volume was less than 1% for the results presented here.

The mass mean aggregate size is usually compared with the mean size measured experimentally by light scattering (Amal et al. 1990a; Flesch et al. 1999). In all the simulation results presented here, the mass mean aggregate radius was computed from the predicted aggregate size distribution using the following equation (Amal et al. 1990a):

$$\bar{r} = \frac{\sum_i y_{N_i} {r_{c_i}}^4}{\sum_i y_{N_i} {r_{c_i}}^3} \tag{23}$$

where y_{N_i} is number fraction of aggregates belonging to section i. The results for the coagulation of suspensions in the presence of inorganic electrolytes are presented first.

4.3.1 Electrolyte-induced Coagulation

4.3.1.1 Diffusion-limited Aggregation

The mathematical model was first tested with published data for the coagulation of hematite at different concentrations of KCl (Runkana et al. 2004). The data were taken from experimental studies by Amal et al. (1990a), which were conducted at 25 °C using a solids concentration of 2.25×10^{16} m^{-3} and reasonably monodisperse primary particles of mean diameter 100 nm.

The classical DLVO theory was used to compute the total interaction energy, with the van der Waals attraction given by the Hamaker expression for bare particles (Eq. 13) and EDL repulsion given by Eq. (18). The Hamaker constant of hematite in water was taken as 2.4×10^{-20} J (Amal et al. 1990a). The mass fractal dimension of aggregates was obtained by the principle of dynamic scaling (Runkana et al. 2004). The predicted evolution of mean aggregate diameter with time at different KCl concentrations is compared with experimental data in Fig. 4.4. The surface potential was used to match simulation results with experimental data and was found to be +29.5 mV, which is close to the hematite particle zeta potential of +31 mV, reported by Amal et al. (1990 a). The complete set of experimental data and model parameters is listed in Table 4.1. It can be seen from the table that the floc fractal dimension decreases as the electrolyte concentration increases. Higher electrolyte concentrations result in faster aggregation and lead to open floc structures.

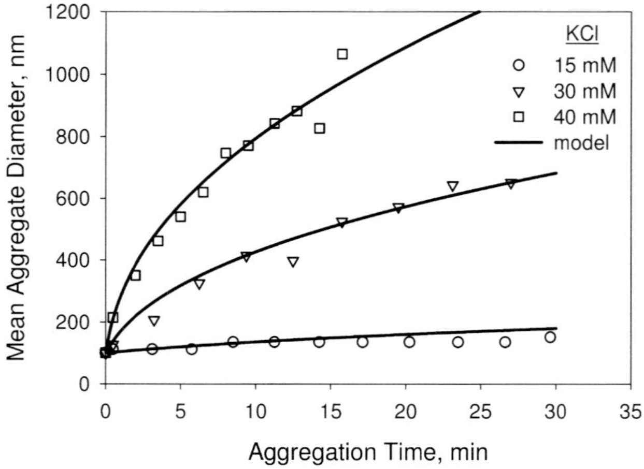

Fig. 4.4 Simulated and experimental time evolution of mean diameter of hematite aggregates in the DLA regime at different KCl concentrations. Experimental data and model parameters in Table 4.1. (Data from Amal et al. 1990a).

Table 4.1 Experimental data and model parameters for simulation results in Fig. 4.4.

Parameter	Value
Initial particle number concentration (m^{-3})	2.25×10^{16}
Mean primary particle diameter (nm)	100
Temperature (°C)	25
Hamaker constant of hematite in water (J)	2.4×10^{-20}
Particle surface potential (mV)	+29.5
KCl concentration (mM)	**Aggregate fractal dimension**
15	3.00
30	2.31
40	2.20

4.3.1.2 **Reaction-limited Aggregation**

RLA occurs when repulsive forces are stronger than attractive forces. In addition to the EDL force, under certain conditions of pH and electrolyte concentration, short-range repulsion could cause a reduction in collision efficiency between particles, which in turn leads to slow aggregation kinetics. The model was next implemented and validated with data from experimental studies of Beattie et al. (1996) and Waite et al. (2001) on the slow aggregation of γ-alumina suspensions in the presence of KCl, NaCl and KNO_3 electrolytes (Runkana et al. 2005). The mean radius of primary particles in both of these studies was 60 nm and the solids concentration in the suspension was 12.5 g L^{-1}, which corresponds to a primary particle number concentration of approximately 4.32×10^{18} m^{-3}. The experiments were conducted at 25 ± 1 °C and pH 4.5. The measured zeta potential of alumina particles was approximately 30 mV at pH 4.5 in KNO_3 solutions.

One of the main features of the experimental results was that aggregation did not occur at electrolyte concentrations below 0.5 M and above this the aggregation kinetics followed slow exponential growth instead of the usual power-law growth, which indicates that aggregation of γ-alumina in these experimental studies belongs to the RLA regime. Hence the collision frequency factor given by Eq. (8) was used, instead of the Smoluchowski kernel (Eq. 7) while solving the discretized PBE (Eq. 2). In addition, force measurements by atomic force microscopy (Karaman et al. 1997) between an aluminum-coated silica sphere and a flat aluminum substrate have shown the existence of short-range (< 3 nm) repulsive forces in 1 mM KCl at different pH levels. Similar short-range forces were also observed between sapphire surfaces in 0.1 M NaBr solution at pH 3 (Ducker et al. 1994). The DLVO theory could explain the interaction forces only up to a surface separation distance of about 3–5 nm. In order to describe the force–distance profiles at shorter separations, it was necessary to invoke an additional repulsive force, which was proposed to be hydration repulsion. Hence, in order to account for this, the DLVO theory was extended by including the hydration repulsion (Eq. 22) in the present work.

The RLA model contains three adjustable parameters, hydration force constant (P_{hyd}), hydration decay length (λ_{hyd}) and proportionality constant in the collision kernel (K_{RLA}). In order to obtain physically meaningful parameter values, a systematic two-stage search strategy was employed. In the first step, the interaction energy–surface separation diagram was utilized as a guide to determine hydration force parameters. In the second step, the proportionality constant K_{RLA} was adjusted to match as closely as possible the predicted time evolution of the mean aggregate diameter with experimental data.

The interaction energy–surface separation diagram for γ-alumina particles of radius 60 nm in water at pH 4.5 is shown in Fig. 4.5. The van der Waals attraction was calculated using the Hamaker theory (Eq. 13). A non-retarded Hamaker constant of 6.7×10^{-20} J, reported by Ducker et al. (1994) for sapphire surfaces in water, was assumed to be applicable for γ-alumina in water. The particle surface potential was assumed to be approximately equal to the measured zeta potential of 30 mV at pH 4.5, that is, $\psi_{0_k} = \psi_{0_l} = 30$ mV. It can be seen from Fig. 4.5 that the classical DLVO theory (curve d) predicts a strong attraction at 0.5 M NaCl. Hence it was necessary to include a short-range repulsion force (curve a) so as to preclude coagulation at this electrolyte concentration. The two hydration force parameters, P_{hyd} and λ_{hyd}, were adjusted by trial and error such that the suspension will be predicted to be stable. The values of P_{hyd} and λ_{hyd} were found to be approximately 1.9×10^7 N m^{-2} and 0.6 nm, respectively. The total interaction energy, incorporating hydration repulsion and shown as curve c in Fig. 4.5, is weakly repulsive, which indicates that the suspensions will remain stable at 0.5 M NaCl, as observed experimentally.

Finally, aggregation of γ-alumina suspension was simulated by solving the discretized PBE (Eq. 2), incorporating Eq. (8) for the collision frequency factor and

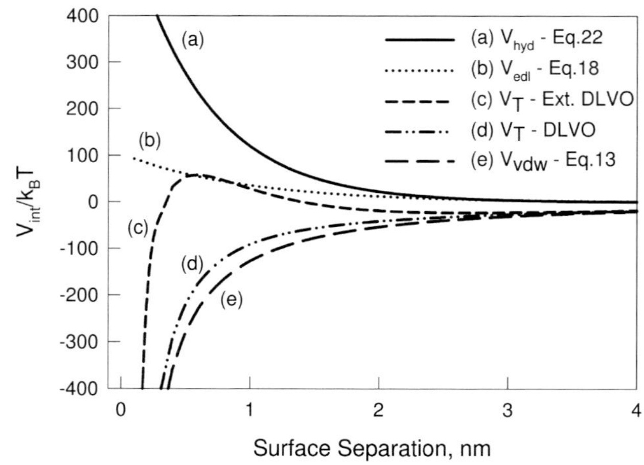

Fig. 4.5 Interaction energy–surface separation diagram for γ-alumina particles using DLVO and extended DLVO theories (particle diameter, 120 nm; pH, 4.5; ionic strength, 0.5 M).

calculating the total interaction energy as a sum of van der Waals attraction (Eq. 13), EDL repulsion (Eq. 18) and hydration repulsion (Eq. 22). The stability ratios, computed using Eq. (11) and K_{RLA}, obtained by fitting the population balance model to the experimental data, are given in Table 4.2, along with other input data and model parameters. Simulation results, obtained using these parameter values and an experimentally measured aggregate mass fractal dimen-

Table 4.2 Experimental data and model parameters for simulation results in Fig. 4.6.

Parameter		Value
Particle number concentration (m^{-3})		4.32×10^{18}
Mean primary particle diameter (nm)		120
pH		4.5
Temperature (°C)		25
Particle surface potential (mV)		+30
Mass fractal dimension of aggregate		2.1
Hamaker constant of γ-alumina in water (J)		6.7×10^{-20}
NaCl concentration (M)	**Stability ratio [a]**	**K_{RLA} ($m\ s^{-1}$)**
0.5	9.49×10^{6}	7.0×10^{-3}
0.6	1.78×10^{6}	5.0×10^{-3}
0.65	0.88×10^{6}	3.5×10^{-3}
0.7	0.46×10^{6}	3.4×10^{-3}
0.8	0.15×10^{6}	2.2×10^{-3}

a) The stability ratio was computed using Eq. (11).

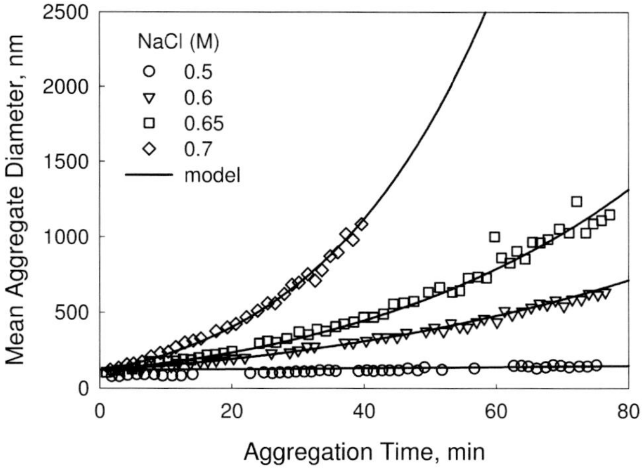

Fig. 4.6 Simulated and experimental time evolution of mean diameter of γ-alumina aggregates in the RLA regime at different NaCl concentrations. Experimental data and model parameters in Table 4.2. (Data from Waite et al., 2001).

sion of 2.1 (Waite et al. 2001), are compared with the measured evolution of mean aggregate diameter with time in Fig. 4.6. It can be seen that the computed growth curves are in reasonable agreement with the experimental data. It can be observed from Table 4.2 that the stability ratio decreases as the NaCl concentration increases, indicating destabilization of the suspension.

4.3.2 Polymer-induced Flocculation

Results obtained using the population balance model for polymer-induced flocculation by two well-known mechanisms, simple charge neutralization (Runkana et al., 2004) and bridging (Runkana et al. 2006), are presented in this section. As mentioned previously, it was assumed that polymer adsorption on oppositely charged particle surfaces is very fast and equilibrium conformation is achieved before collisions between particles take place. It was also assumed that the polymer adsorbs uniformly and the adsorbed layer thickness is same for all particles.

4.3.2.1 Simple Charge Neutralization

The experimental data of Zhang and Buffle (1995) for the flocculation of hematite in the presence of poly(acrylic acid) (PAA) by simple charge neutralization were used to test the model (Runkana et al. 2004). Fairly monodisperse primary particles of mean diameter 77 nm and a high molecular weight (1.36×10^6 g mol^{-1}) PAA were used in these experimental studies, which were conducted at pH 3 in HCl medium at 25±0.3 °C and using a solids concentration of 2.4×10^{16} m^{-3}. It was inferred from photon correlation spectroscopy (PCS) that the hydrodynamic layer thickness of the adsorbed layer was about 5±3 nm over a PAA concentration range of 5–71 mg L^{-1}.

The discretized PBE (Eq. 2) is now solved to simulate the aggregation of composite primary particles of radius 43.5 nm, consisting of solid particles of radius 38.5 nm covered with a 5-nm thick PAA layer. The collision frequency factors (Eqs. 7 and 9) and the stability ratio (Eq. 11) were calculated using the composite particle radius, that is $r_{0_k} = r_{0_l} = 43.5$ nm. The van der Waals attraction between particles with adsorbed layers was computed using Vincent's expression (Eq. 14). The Hamaker constants of hematite, water and PAA across a vacuum were taken as 1.52×10^{-19} J (Amal et al. 1990a), 3.7×10^{-20} J (Israelachvili 1991) and 8×10^{-20} J (Vincent 1973), respectively. The surface potentials were initially assigned measured zeta potential values, reported by Zhang and Buffle (1995). They were subsequently adjusted slightly to match the simulated time evolution of the mean floc diameter with experimental data. The measured zeta potentials and fitted surface potentials are given in Table 4.3, along with other input data and model parameters. The aggregate fractal dimensions, presented in Table 4.3, were obtained by dynamic scaling of mean floc size–time data (Runkana et al. 2004) and used for calculating collision frequency factors.

The predicted evolution of mean aggregate diameter with flocculation time at some typical PAA concentrations is compared with the experimental data of Zhang and Buffle (1995) in Fig. 4.7. It can be observed that the simulation results are reasonably close to the results of measurements. Moreover, it can be

Table 4.3 Experimental data and model parameters for simulation results in Fig. 4.7.

Parameter		Value
Particle number concentration (m^{-3})		2.4×10^{16}
Mean primary particle diameter (nm)		77
KCl concentration (mM)		1.0
pH		3.0
Temperature (°C)		25
Hamaker constant of hematite across a vacuum (J)		15.2×10^{-20}
Hamaker constant of water across a vacuum (J)		3.7×10^{-20}
Hamaker constant of PAA across a vacuum (J)		8.0×10^{-20}
Adsorbed layer thickness (nm)		5.0
PAA concentration (mg L^{-1})	**Floc fractal dimension**	**Surface potential (mV)**
0.129	2.41	+18.0
0.185	1.87	+17.0
0.200	1.89	+17.0
0.428	2.28	–18.0
0.443	2.74	–18.0

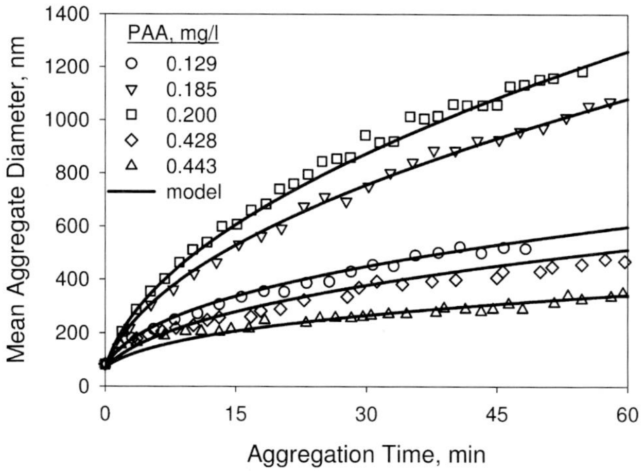

Fig. 4.7 Simulated and experimental time evolution of mean diameter of hematite aggregates at different PAA concentrations. Experimental data and model parameters in Table 4.3. (Data from Zhang and Buffle 1995).

seen that rate of aggregate growth initially increases with PAA concentration and then decreases at higher concentrations. It is well known that the rate of flocculation increases with polymer concentration, reaches a maximum and then decreases as the polymer concentration increases above the optimum value. It was possible to simulate this trend by the population balance model using the experimentally measured adsorbed layer thickness, zeta potentials and aggregate fractal dimensions derived from the experimental data, without adjusting any model parameters such as the Hamaker constant.

In the above simulations for flocculation by simple charge neutralization, it was not necessary to introduce any steric or bridging forces. The effect of polymer adsorption was taken into account by the simple modification of the van der Waals attraction for polymer-coated particles and by assigning surface potentials close to hematite particle zeta potentials in the presence of PAA. Moreover, since the adsorbed layer thickness is very small (3–5 nm), there will not be any polymer bridging. However, it would be necessary to include steric or bridging forces in the case of other polymer-induced flocculation mechanisms, namely charge patch neutralization and bridging. The latter case is discussed in the following section.

4.3.2.2 **Bridging Flocculation**

Bridging flocculation takes place when polymer chains adsorb on more than one particle and act as bridges. This generally happens when adsorbed polymer chains have long dangling tails protruding into the solution such that particle aggregation can occur even in the presence of repulsive forces. Moreover, the rate of flocculation is generally high because the collision radii of particles increase due to the large adsorbed layer thickness. Bridging flocculation was simulated by solving the discretized PBE (Eq. 2) and assuming the total interaction energy to be a sum of van der Waals attraction (Eq. 14), EDL repulsion (Eq. 18) and bridging attraction or steric repulsion due to adsorbed polymer (Eq. 20). The collision frequency factor was calculated as a sum of contributions due to Brownian motion (Eq. 7) and differential sedimentation (Eq. 9).

The model was tested (Runkana et al. 2006) with experimental data published by Biggs et al. (2000) for flocculation of anionic polystyrene latex particles by a high molecular weight (16×10^6 g mol^{-1}) cationic quaternary ammonium-based derivative of polyacrylamide. These experiments were carried out at pH 6 in 10^{-4} M KNO_3 solution using primary particles of mean radius 165 nm. The solids concentration was 0.05% w/w, which corresponds to a particle number concentration of approximately 2.53×10^{16} m^{-3}. The experiments were conducted following a two-stage procedure. In the first step, latex suspension was vigorously stirred with a magnetic stirrer for 60 s, after polymer addition. In the second step, the partially flocculated suspension was transferred to a light scattering cell and flocculation was allowed to proceed under perikinetic aggregation conditions. Since it is not possible to assign a suitable shear rate for magnetic stirring, only the second stage of the flocculation experiment is simulated here as perikinetic aggregation coupled with differential sedimentation. The primary aggregates, obtained after magnetic stirring, were assumed to be the starting material for simulating the second stage of flocculation. The mean size of these primary aggregates was computed using the published floc size distribution (FSD) data at 2 min and used as the initial condition to predict the evolution of FSD with time.

As discussed earlier, model parameters were obtained systematically in two steps. In the first step, parameters required for estimating various interaction energies were taken from relevant literature sources and the computed interaction energy–surface separation diagram was employed as a guide to obtain appropriate numerical values. In the second step, the floc fractal dimension alone was adjusted by trial and error to match computed and experimental floc size distributions. The Hamaker constants of polystyrene, water and polyacrylamide across a vacuum were taken as 6.6×10^{-20}, 3.7×10^{-20} and 8×10^{-20} J, respectively (Vincent 1973; Israelachvili 1991). The parameters in Eq. (20) for interaction due to adsorbed polymer can be obtained from forces between polymer-coated surfaces measured by the surface forces apparatus (SFA) or the atomic force microscope (AFM) and from polymer volume fraction profiles measured by neutron scattering. The term $a_{\mathrm{Sc}} k_{\mathrm{B}} T / a_{\mathrm{m}}{}^3$ in Eq. (20) was assigned 3×10^5 N m^{-2}, as reported by Klein and Rossi (1998) from forces measured between poly(ethylene oxide) (PEO)-coated mica surfaces. The scaling length D_{Sc} was assumed to be 4 nm, based on volume fraction profiles of PEO on polystyrene surfaces measured by neutron scattering (Cosgrove et al. 1990). The particle surface potential ψ_0, the polymer volume fraction at a single saturated surface ψ_{S_0}, the fractional polymer surface coverage Γ/Γ_0 and the adsorbed layer thickness δ were employed as adjustable parameters. Since we are interested in bridging flocculation, parameter values were searched by trial and error such that there is net attraction due to poly-

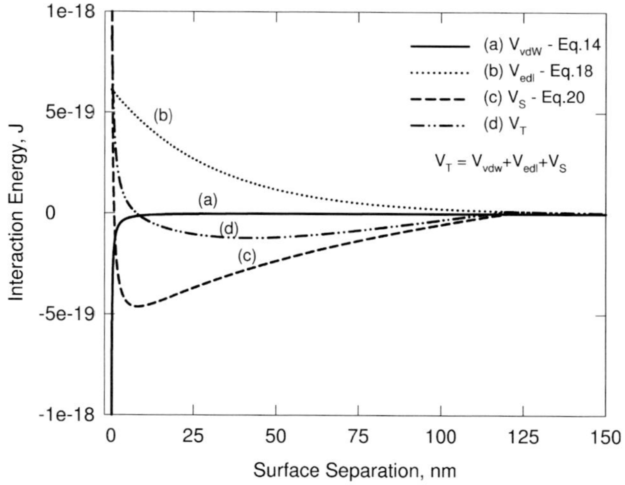

Fig. 4.8 Interaction energy–surface separation diagram for polystyrene latex particles in the presence of a polyacrylamide derivative (particle diameter, 330 nm; pH, 6.0; KNO_3 concentration, 0.0001 mM).

mer bridging. After some trials, the values of ψ_0, Φ_{S_0}, Γ/Γ_0 and δ were obtained as −31 mV, 0.21, 0.475 and 60 nm, respectively. The interaction energy curves for two polymer-coated spherical particles of radius 165 nm are shown in Fig. 4.8. It can be seen that there is a net attraction due to polymer bridging, under the experimental conditions and for the parameter values employed.

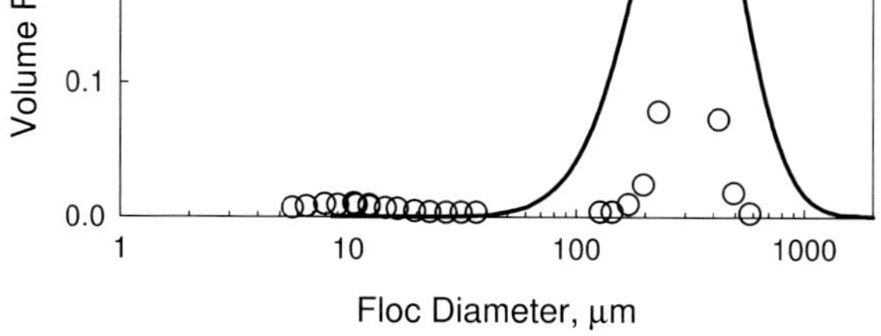

Fig. 4.9 Simulated and experimental size distribution of polystyrene latex flocs obtained after 30 min of flocculation by a polyacrylamide derivative. Experimental data and model parameters in Table 4.4. (Data from Biggs et al. 2000).

Table 4.4 Experimental data and model parameters for simulation results in Figs. 4.9 and 4.10.

Parameter	Value
Particle number concentration (m^{-3})	2.53×10^{16}
Mean primary particle diameter (nm)	330
KNO_3 concentration (mM)	10^{-4}
pH	6.0
Hamaker constant of water across a vacuum (J)	3.7×10^{-20}
Hamaker constant of polystyrene across a vacuum (J)	6.6×10^{-20}
Hamaker constant of polyacrylamide across a vacuum (J)	8.0×10^{-20}
Surface potential (mV)	–31.0
Mass fractal dimension of an aggregate	1.75
Polymer volume fraction at a single saturated surface	0.21
Scaling length (nm)	4.0
Lumped parameter in Eq. (20) ($a_{Sc}k_B T/a_m{}^3$) (N m^{-2})	3.0×10^5
Adsorbed layer thickness (nm)	60
Fractional polymer surface coverage (Γ/Γ_0)	0.475

Next, the discretized PBE (Eq. 2) was solved to simulate the evolution of the floc size distribution (FSD) using the parameters obtained above. The mass fractal dimension of flocs was used as the final adjustable parameter to obtain close agreement between the simulated and experimental FSD. A value of 1.85, reported by Glover et al. (2000) for the fractal dimension of alumina flocs produced by bridging flocculation with a mixture of polyacrylamide and PAA, was used as the initial guess and adjusted subsequently if required. The predicted FSD is compared

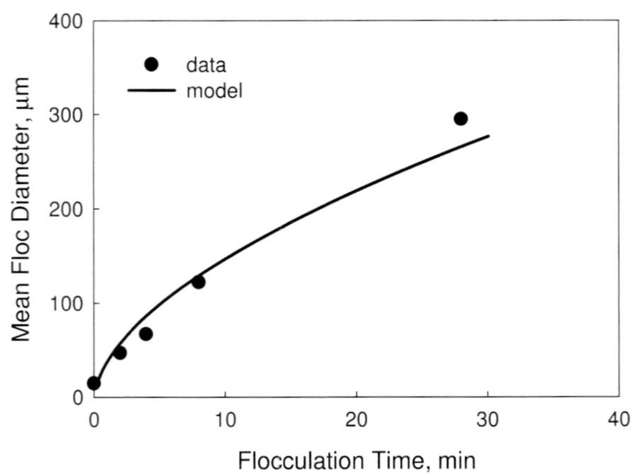

Fig. 4.10 Simulated and experimental time evolution of mean diameter of polystyrene latex flocs obtained by bridging flocculation with a polyacrylamide derivative. Experimental data and model parameters in Table 4.4. (Data from Biggs et al. 2000).

in Fig. 4.9 with experimental FSD obtained with a 20 ppm polymer concentration after 30 min of flocculation. The evolution of the mean floc diameter with time is shown in Fig. 4.10. A complete list of experimental data and model parameters is provided in Table 4.4. The predicted FSD is broader than the measured FSD, as can be seen in Fig. 4.9. However, the kinetics of bridging flocculation are captured reasonably well by the model, as is evident in Fig. 4.10.

4.4 Conclusions

The rate of coagulation or flocculation of colloidal suspensions is determined mainly by the surface forces between suspended particles, which depend on particle size distribution, pH, temperature, type and concentration of electrolyte species and concentration and molecular weight distribution of polymer added. A detailed and general population balance model, which incorporates modern theories of surface forces, for the coagulation and flocculation of colloidal suspensions by inorganic electrolytes and polymers is presented. The classical DLVO theory was used along with the Smoluchowski kernel for Brownian coagulation to simulate diffusion-limited aggregation in the presence of inorganic electrolytes. The model was adapted for reaction-limited aggregation kinetics by incorporating a novel collision kernel for slow, exponential growth and by extending the DLVO theory to include the effect of short-range hydration repulsion. For flocculation in the presence of polymers, a modified expression for the van der Waals attraction between particles with adsorbed polymer layers was used. An equation derived from the scaling theory of polymer adsorption was included in the model to account for the interaction energy due to adsorbed polymer.

The model was tested with experimental data published in the literature for diffusion-limited aggregation of hematite suspensions by KCl, reaction-limited aggregation of γ-alumina suspensions by different electrolytes, flocculation of hematite in the presence of PAA by simple charge neutralization and bridging flocculation of polystyrene latex by a polyacrylamide derivative. The simulation results are in reasonable agreement with experimental data for the evolution of the mean aggregate diameter or aggregate size distribution with time.

Acknowledgments

This work was supported by the National Science Foundation (NSF Grants INT-96-05197 and INT-01-17622) and the NSF Industry/University Cooperative Research Center (IUCRC) for Advanced Studies in Novel Surfactants at Columbia University (NSF Grant EEC-98-04618). The authors thank the management of Tata Research Development and Design Centre for permission to publish this chapter. V.R. thanks Professor E.C. Subbarao, Professor Mathai Joseph, Professor P.C. Kapur and Dr. Pradip for their advice and encouragement.

References

Amal, R., J. R. Coury, J. A. Raper, W. P. Walsh, T. D. Waite, Structure and kinetics of aggregating colloidal hematite. *Colloids Surf. A 46*, 1–19 (**1990a**).

Amal, R., J. A. Raper, T. D. Waite, Fractal structure of hematite aggregates *J. Colloid Interface Sci. 140*, 158–168 (**1990b**).

Ball, R. C., D. A. Weitz, T. A. Witten, F. Leyvraz, Universal kinetics in reaction-limited aggregation. *Phys. Rev. Lett. 58*, 274 (**1987**).

Beattie, J. K., J. K. Cleaver, T. D. Waite, Anomalous aggregation behavior of aluminium oxyhydroxides. *Colloids Surf. 111*, 131–138 (**1996**).

Bell, G. M., S. Levine, L. N. McCartney, Approximate methods of determining the double-layer free energy of interaction between two charged colloidal spheres. *J. Colloid Interface Sci. 33*, 335–359 (**1970**).

Biggs, S., M. Habgood, G. J. Jameson, Y.-d. Yan, Aggregate structures formed via a bridging flocculation mechanism. *Chem. Eng. J. 80*, 13–22 (**2000**).

Camp, T. R., P. C. Stein, Velocity gradients and internal work in fluid motion. *J. Boston Soc. Civ. Eng. 30*, 219–237 (**1943**).

Chander, P., P. Somasundaran, N. J. Turro, K. C. Waterman, Excimer fluorescence determination of solid–liquid interfacial pyrene-labeled polyacrylic acid conformations. *Langmuir 3*, 298–300 (**1987**).

Cosgrove, T., T. L. Crowley, K. Ryan, J. R. P. Webster, The effects of solvency on the structure of an absorbed polymer layer and dispersion stability. *Colloids Surfaces A 51*, 255–269 (**1990**).

de Gennes, P. G., Polymer solutions near an interface. 1. Adsorption and depletion layers. *Macromolecules 14*, 1637–1644 (**1981**).

de Gennes, P. G., Polymer solutions near an interface. 2. Interaction between two plates carrying adsorbed polymer layers. *Macromolecules 15*, 492–500 (**1982**).

Derjaguin, B. V., Friction and adhesion. IV. The theory of adhesion of small particles. *Kolloid-Z. 69*, 155–164 (**1934**).

Derjaguin, B. V., L. D. Landau, A theory of the stability of strongly charged lyophobic sols and of the adhesion of strongly charged particles in solutions of electrolytes. *Acta Physicochim. URSS 14*, 633 (**1941**).

Derjaguin, B. V., V. M. Muller, Slow coagulation of hydrophobic colloids. *Dokl. Akad. Nauk SSSR 176*, 738–741 (**1967**).

Ducker, W. A., Z. Xu, D. R. Clarke, J. N. Israelachvili, Forces between alumina surfaces in salt solutions: non-DLVO forces and the implications for colloidal processing. *J. Am. Ceram. Soc. 77*, 437–443 (**1994**).

Feder, J., *Fractals.* Plenum Press, New York (**1988**).

Firth, B. A., R. J. Hunter, Flow properties of coagulated colloidal suspensions. I. Energy dissipation in the flow units. *J. Colloid Interface Sci. 57*, 248–256 (**1976**).

Fleer, G. J., M. A. Cohen Stuart, J. M. H. M. Scheutjens, T. Cosgrove, B. Vincent, *Polymers at Interfaces.* Chapman and Hall, New York (**1993**).

Flesch, J. C., P. T. Spicer, S. E. Pratsinis, Laminar and turbulent shear-induced flocculation of fractal aggregates. *AIChE J. 45*, 1114–1124 (**1999**).

Fuchs, N., Über die Stabilität und Aufladung der Aerosole. *Z. Phys. 89*, 736–743 (**1934**).

Gear, C. W., *Numerical Initial Value Problems in Ordinary Differential Equations.* Prentice-Hall, Englewood Cliffs, NJ (**1971**).

Glover, S. M., Y.-d. Yan, G. J. Jameson, S. Biggs, Bridging flocculation studied by light scattering and settling. *Chem. Eng. J. 80*, 3–12 (**2000**).

Gregory, J., Approximate expressions for retarded van der Waals interaction. *J. Colloid Interface Sci. 83*, 138–145 (**1981**).

Hamaker, H. C., The London–van der Waals attraction between spherical particles. *Physica IV*, 1058–1072 (**1937**).

Hogg, R., T. W. Healy, D. W. Fuerstenau, Mutual coagulation of colloidal dispersions. *Trans. Faraday. Soc. 62*, 1638–1651 (**1966**).

Honig, E. P., G. J. Roebersen, P. H. Wiersema, Effect of hydrodynamic interaction on the coagulation rate of hydrophobic colloids. *J. Colloid Interface Sci. 36*, 97–109 (**1971**).

Hounslow, M. J., R. L. Ryall, V. R. Marshall, A discretized population balance for nucleation, growth and aggregation. *AIChE J. 34*, 1821–1832 (**1988**).

Israelachvili, J. N., *Intermolecular and Surface Forces*, 2nd edn. Academic Press, New York (**1991**).

Jiang, Q., B. E. Logan, Fractal dimensions of aggregates determined from steady-state size distributions. *Environ. Sci. Technol. 25*, 2031–2038 (**1991**).

Johnson, S. B., G. V. Franks, P. J. Scales, D. V. Boger, T. W. Healy, Surface chemistry–rheology relationships in concentrated mineral suspensions. *Int. J. Miner. Process. 58*, 267 (**2000**).

Karaman, M. E., R. M. Pashley, T. D. Waite, S. J. Hatch, H. Bustamante, A comparison of the interaction forces between model alumina surfaces and their colloidal properties. *Colloids Surf. A 129/130*, 239–255 (**1997**).

Klein, J., G. Rossi, Analysis of the experimental implications of the scaling theory of polymer adsorption. *Macromolecules 31*, 1979–1988 (**1998**).

Klimpel, R. C., R. Hogg, Effects of flocculation conditions on agglomerate structure. *J. Colloid Interface Sci. 113*, 121–131 (**1986**).

Lin, M. Y., H. M. Lindsay, D. A. Weitz, R. C. Ball, R. Klein, P. Meakin, Universality in colloid aggregation. *Nature 339*, 360–362 (**1989**).

Lu, C. F., L. A. Spielman, Kinetics of floc breakage and aggregation in agitated liquid suspensions. *J. Colloid Interface Sci. 103*, 95–105 (**1985**).

Manciu, M., E. Ruckenstein, Role of the hydration force in the stability of colloids at high ionic strengths. *Langmuir 17*, 7061–7070 (**2001**).

Mandelbrot, B. B., *The Fractal Geometry of Nature*. Freeman, New York (**1983**).

McGown, D. N., G. D. Parfitt, Improved theoretical calculation of the stability ratio for colloidal systems. *J. Phys. Chem. 71*, 449–450 (**1967**).

Pashley, R. M., DLVO and hydration forces between mica surfaces in Li^{+1}, Na^{+1}, K^{+1} and Cs^{+1} electrolyte solutions in the range 1–100 nm. *J. Colloid Interface Sci. 83*, 531–546 (**1981**).

Pashley, R. M., J. N. Israelachvili, DLVO and hydration forces between mica surfaces in Mg^{+2}, Ca^{+2}, Sr^{+2} and Ba^{+2} chloride solutions. *J. Colloid Interface Sci. 97*, 446–455 (**1984**).

Press, W. H., S. A. Teukolsky, W. T. Vetterling, B. P. Flannery, *Numerical Recipes in FORTRAN: the Art of Scientific Computing*, 2nd edn. Cambridge University Press, Cambridge (**1992**).

Ramkrishna, D., *Population Balances: Theory and Applications to Particulate Systems in Engineering*. Academic Press, New York (**2000**).

Runkana, V., Mathematical modeling of flocculation and dispersion of colloidal suspensions. *DESc Thesis*, Columbia University, New York (**2003**).

Runkana, V., P. Somasundaran, P. C. Kapur, Mathematical modeling of polymer-induced flocculation by charge neutralization. *J Colloid Interface Sci. 270*, 347–358 (**2004**).

Runkana, V., P. Somasundaran, P. C. Kapur, Reaction-limited aggregation in presence of short-range structural forces. *AIChE J. 51*, 1233–1245 (**2005**).

Runkana, V., P. Somasundaran, P. C. Kapur, A population balance model for flocculation of colloidal suspensions by polymer bridging. *Chem. Eng. Sci. 61*, 182–191 (**2006**).

Sader, J. E., S. L. Carnie, D. Y. C. Chan, Accurate analytical formulas for the double-layer interaction between spheres. *J. Colloid Interface Sci. 171*, 46–54 (**1995**).

Smoluchowski, M. v., Drei Vorträge über Diffusion, Brownsche Molekularbewegung und Koagulation vor Kolloidteilchen. *Phys. Z. 17*, 557–571 (**1916**).

Smoluchowski, M. v., Versuch einer mathematischen Theorie der Koagulationskinetik kolloider Lösungen. *Z. Phys. Chem. 92*, 129–168 (**1917**).

Somasundaran, P., V. Runkana, Modeling flocculation of colloidal mineral suspensions using population balances. *Int. J. Miner. Process. 72*, 33–55 (**2003**).

Spicer, P. T., S. E. Pratsinis, Coagulation and fragmentation: universal steady-state particle-size distribution. *AIChE J. 42*, 1612–1620 (**1996**).

Spielman, L. A., Viscous interactions in Brownian coagulation. *J. Colloid Interface Sci. 33,* 562–571 (**1970**).

Thomas, D. N., S. J. Judd, N. Fawcett, Flocculation modeling: a review. *Water Res. 33,* 1579–1592 (**1999**).

Thompson, D. W., I. R. Collins, Electrolyte-induced aggregation of gold particles on solid surfaces. *J. Colloid Interface Sci. 163,* 347–354 (**1994**).

Tjipangandjara, K., Y.-B. Huang, P. Somasundaran, N. J. Turro, Correlation of alumina flocculation with adsorbed polyacrylic acid conformation. *Colloids Surf. A 44,* 229–236 (**1990**).

Verwey, E. J. W., J. Th. G. Overbeek, *Theory of the Stability of Lyophilic Colloids.* Elsevier, Amsterdam (**1948**).

Vincent, B., The van der Waals attraction between colloid particles having adsorbed layers. II. Calculation of interaction curves. *J. Colloid Interface Sci. 42,* 270–285 (**1973**).

Vold, M. J., The effect of adsorption on the van der Waals interaction of spherical colloidal particles. *J. Colloid Sci. 16,* 1–12 (**1961**).

Waite, T. D., J. K. Cleaver, J. K. Beattie, Aggregation kinetics and fractal structure of γ-alumina assemblages. *J. Colloid Interface Sci. 241,* 333–339 (**2001**).

Zhang, J., J. Buffle, Kinetics of hematite aggregation by polyacrylic acid: importance of charge neutralization. *J. Colloid Interface Sci. 174,* 500–509 (**1995**).

5
Emulsion Stability and Interfacial Properties – Application to Complex Emulsions of Industrial Interest

Danièle Clausse, Delphine Daniel-David, François Gomez, Ljepsa Komunjer, Isabelle Pezron, Christine Dalmazzone, and Christine Noïk

Abstract

A lot of interest has been devoted to colloid stability principles and their application to emulsion stability. However, when dealing with opaque and concentrated multi-component industrial emulsions, theoretical concepts are much more difficult to apply. This chapter presents experimental correlations between emulsion stability and interfacial properties in the case of complex emulsions of industrial interest. Methods of characterization specific to these systems are described. Special focus is placed on emulsions in the food and petroleum industries. Although these fields of investigation can be perceived as being far removed from each other, common features can be identified. Food emulsions generally contain mixtures of surface-active molecules and proteins which compete for adsorption at the interface. In many cases, emulsion stability depends on protein–surfactant interactions at the interface. The formation of a protein–surfactant interfacial complex generally has a positive influence on emulsion stability whereas non-interactive surfactants will partially displace the protein from the interface, which usually has a destabilizing effect. Similarities with the action of chemical emulsion breakers used in the petroleum industry are pointed out. Water-in-crude oil emulsions are also multi-component systems stabilized by natural emulsifiers such as asphaltenes, resins and small solid particles such as clays. Similarly to proteins, asphaltene molecules are strongly adsorbed at the interface, where they aggregate as a network. Efficient demulsifiers rapidly adsorb at the oil/water interface and modify the viscoelastic properties of the film. It appears that the evolution of emulsions with time is generally accompanied by observed changes in interfacial properties. However, theoretical models to establish these correlations are still lacking.

Colloids and Interface Science Series, Vol. 3
Colloid Stability and Application in Pharmacy. Edited by Tharwat F. Tadros

ISBN: 978-3-527-31463-8

5.1 Introduction

Emulsions are dispersed systems consisting of at least two immiscible liquids, commonly water and a non-polar oil. Emulsions are found in a large range of manufactured products, including food, agrochemicals, inks and paints, pharmaceuticals and cosmetic products. These emulsions are formulated and produced using various techniques, but they can also occur naturally as the result of particular conditions favorable to their formation. This is the case in the petroleum industry, as will be seen in the section devoted to these products. The control of emulsion stability is of great importance regardless of the mode of preparation. Stability is meant here as kinetic stability since these systems are thermodynamically unstable. For manufactured emulsions, long-term stability is generally required. Conversely, for natural emulsions, destabilization of the system and separation of the two phases are usually needed. In the last stage of emulsion evolution, the rupture of the interfacial film induces droplet coalescence and breaking of the emulsion. Therefore, it clearly appears that the properties of the interface between the two immiscible liquids and the role of adsorbed surface-active molecules are of utmost importance. Numerous independent studies have been conducted in order to obtain information on the one hand about interfacial properties of the water/air, oil/air and water/oil interfaces and on the other hand on the emulsion stability itself. The corresponding references will be given in the following sections. A much smaller number of studies has been devoted to the relationship between interfacial properties and emulsion stability. The problems encountered result from the rather complex nature of emulsion formulations. Many components are present, such as emulsifiers, resins, natural polymers, proteins or solid particles, themselves characterized by a complex structure. Any compound, present even in a minute amount, may play a determinant role in the emulsion stability.

This chapter is an attempt to present the experimental work that has been undertaken to correlate interfacial properties with emulsion stability. More theoretical developments and experiments to be performed in ground and microgravity conditions are being carried out in the frame of a more general project supported by ESA and referred to as FASES (Fundamental and Applied Studies of Emulsion Stability), in which the authors of this chapter are involved [1, 2].

In this chapter, the methods used for characterizing complex emulsions and their stability will first be presented, followed by those used to characterize interfacial properties. Two particular examples where complex emulsions are present, in the food and petroleum industries, are subsequently described in detail.

5.2 Macroscopic Characterization of Complex Emulsions: Properties and Stability

Various types of emulsion can be obtained depending on their formulation and mode of preparation. The next section presents a synthetic description of these emulsions. Then, the relevant parameters characterizing the emulsions are given and finally the various known mechanisms leading to the separation of the emulsion into at least two phases are discussed.

5.2.1 Emulsion Definitions

Simple emulsions are referred as water-in-oil (W/O) or oil-in-water (O/W) emulsions. W/O emulsions consist of water droplets dispersed in a continuous oil phase and O/W emulsions are made of oil droplets dispersed in a continuous aqueous medium. To be comprehensive, the formulation of such a medium should also include all additives present. They play a determinant role in the formation and aging of emulsions, as will be described later. Due to their double affinity for water and for oil, they are preferentially found at the interfaces between the droplets and the continuous medium. Their choice depends on the kind of emulsion wanted. These molecules are commonly designed as surface-active molecules, surfactants or more specifically here as emulsifiers. More details about emulsions can be found in general books dealing with these systems [3–5].

Multiple emulsions have been formulated and studied for their capability to encapsulate active products [6]. They have also been considered as media for separating compounds [7]. W/O/W emulsions can be described as being formed of oil globules containing tiny water droplets, globules that are themselves dispersed in water. The distribution of the oil and water phases can be different: water globules containing tiny oil droplets are dispersed within oil and the corresponding emulsion is referred as an O/W/O emulsion. The material composing the globules can be assimilated to a liquid membrane and mass transfers have been observed in these media [7].

Mixed emulsions may also be classified as W/O or O/W emulsions. They have been specially formulated to study mass transfers between droplets [8–10]. For that purpose, two primary emulsions containing droplets of different composition were mixed gently so that the resulting emulsion contained two types of droplet population. Should a mass transfer occur between the droplets, the resulting emulsion will change towards an emulsion containing only one type of droplets. The final droplet composition can be determined from the formulation of the original emulsions, assuming that no other process is involved. As mass transfer between droplets can be a cause of instability, examples will be given in the section devoted to emulsion stability. Recent studies have been undertaken to transfer laboratory knowledge of emulsion formulation to large-scale plant production [11, 12] and they appeared to be very helpful both for emulsion engineering and for setting up the appropriate processes.

5.2.2
Relevant Parameters

Due to their complex formulation and structure, it is crucial to identify the parameters that are the most appropriate to characterize an emulsion. Water and oil phases may contain surfactants, salts and solids that could be very different in nature, as will be seen in the sections devoted to emulsions in the food and petroleum industries. Therefore, all the parameters relevant to emulsion formulation have to be precisely identified. The mode of preparation is also of primary importance, especially the sequence of introduction of the various compounds, the conditions of mixing, mixing devices and temperature, time of mixing, etc. These factors are generally not detailed enough in the literature and a real effort has to be made in order to improve the transfer of information between research groups working in this field. The physicochemical parameters allowing the characterization of the evolution of the emulsion versus time are other important factors. These parameters will be defined in the next section. Nevertheless, from now on, we will consider the droplet size as the main parameter while keeping in mind that composition, viscosity and interfacial tension also have to be taken into account. Other parameters specific to the techniques used for following the stability could also be relevant. In this case, dielectric constant, freezing and melting temperatures and enthalpies could also be parameters that need to be known.

5.2.3
Emulsion Stability

As early as 1957 [13], it was stated that "it is an unfortunate fact that there exists no single coherent theory of emulsion stability" and it can be said that the situation is still nearly the same. This is not surprising considering that various mechanisms can create an evolution of the emulsion and many studies have been undertaken to give a coherent theory for each mechanism. Therefore, as emulsions are thermodynamically unstable, the term stability should first be understood as a kinetic stability and furthermore linked to the mechanism considered. The main mechanisms of destabilization are represented schematically in Fig. 5.1. It is not within the scope of this chapter to describe these mechanisms thoroughly; readers can find detailed developments in basic articles devoted to emulsions [13–17].

These mechanisms can be briefly described as follows.

Creaming or sedimentation is not a breaking of the emulsion but can be considered as a separation into two emulsions, one of which is richer in dispersed phase than the original emulsion, and the other one poorer. The emulsion which is more concentrated is called the "cream". Creaming is due to the difference between the densities of the droplets and the continuous medium in which they are dispersed. A way to avoid this mechanism would be to use a gel as suspending medium or two immiscible liquids of identical density. This

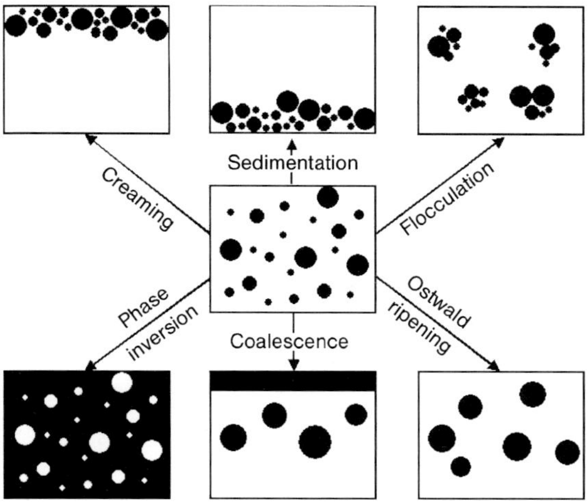

Fig. 5.1 Schematic representation of the breakdown processes in emulsions. (Redrawn from Tadros [18]).

approach is suitable when information on model emulsions is required. It does not work when emulsion stability studies are dealing with real systems for which the formulation cannot be changed at will. In that case, the gravitation effect has to be cancelled by another means and is the reason why studies of emulsion stability need to be performed in microgravity conditions [1].

Ostwald ripening leads to mass transfer from the smallest to the largest droplets without any contact between the droplets. It results from the fact that the solubility of the droplet components depends on their size due to the excess pressure inside the droplets (Laplace equation). The size concerned is commonly less than 1 μm and therefore it is generally not the main mechanism leading to the breaking of the emulsion, but it could be the initiator.

Flocculation is defined as the aggregation of droplets to give three-dimensional clusters without coalescence occurring. Therefore, the droplets retain their individual identities in clusters named flocs. The emulsion is not broken but its state and therefore its properties could be very different from the original ones.

Coalescence results from the rupture of the thin liquid film formed between two droplets in contact and ultimately leads to the complete separation of phases.

The mechanisms mentioned above are the most obvious and the best known. Less well known are those dealing with mass transfer due to composition gradients within the emulsion and partial or full solidification of the dispersed phase. These mechanisms and their consequences for emulsion stability are thoroughly described in papers dealing with multiple emulsions and emulsions for which the water droplets contain a solute [7, 19, 20].

5.2.4 Methods Used to Characterize Emulsion Stability

An emulsion stability index may be defined. Some emulsions may be stable (without significant separation of phases) over a long period. It may then be difficult to characterize the stability of the emulsion by a direct analysis of the evolution of its properties (local concentrations, creaming or sedimentation, change in droplet size distribution) which will give rise to phase separation. Hence it is of interest to test the emulsion at the beginning of its lifetime. One widely used analytical test consists in mechanically perturbing the sample (using centrifugation [21] or classical agitation [22]) and then characterizing the modified sample (observation of the level of creaming, sedimentation or phase separation). This characterization may be achieved by visual observation or by one of the techniques described below. The emulsion stability index obtained may be related to the long-term stability of the initial sample [23].

It is also possible to characterize local composition. If one considers an alteration of the emulsion due to a modification of local concentrations only, the size of the droplets may remain unchanged. Even if the global concentration of species present in the system is not modified, the local amount of a substance may vary with time. This is the case when the system is not at thermodynamic equilibrium. Of course, most emulsions are not in equilibrium by definition. Diffusion of chemical species is possible inside the divided medium such as surfactant diffusion [24] (cyclic dimpling or osmotic swelling), ion diffusion or molecular diffusion (this process is used to perform mass transfers inside emulsions, for example). The local modifications of concentration induce a change in interparticular forces. This may favor sedimentation, creaming or coalescence and is therefore related to emulsion stability. Kinetic factors such as surface and bulk diffusion fluxes must be determined for each molecule in order to estimate the local changes of concentration. Segregation phenomena (sedimentation and creaming) may be characterized by most of the techniques listed below [25].

The bottle test is a standard method used in petroleum science to characterize water-in-crude oil emulsion stability [26]. It is simply defined by the volume ratio of the separated water phase, measured in a glass vessel, against time (from hours to months depending on the system). This test has the advantage of being very simple and it allows an easy comparison of the stability of several samples [27]. A very similar test exists (ASTM method D1479 and DIN method 51367) for determining the stability of metalworking fluids (MWF) [28].

The consistency of an emulsion is a function of several parameters of the system, for example the phase ratio of dispersed droplets, size distribution of these droplets and the presence of additives (polymers, solids, etc.). Rheological measurements (steady-state, controlled stress and oscillatory measurements) provide information which may be related to the parameters cited above and to emulsion stability. Tadros [18] illustrated the different possibilities of rheology in the field of emulsions and his paper should be consulted for more details.

Optical microscopy is the most direct technique for observing emulsions [29]. In practice, reliable measurements are possible above 1 μm. For smaller droplets, Brownian motion causes images to appear blurred. Using this technique, it is possible to determine local heterogeneity of a sample (creaming or sedimentation) and phase separation. It is also possible to access the droplet size distribution of the emulsion. Nevertheless, optical microscopy is time consuming and is limited to dilute samples. To provide correct results, several precautions must be taken: the sample to be analyzed must be fully representative of the emulsion, a large number of particles must be measured to ensure statistical accuracy and the effects of sectioning must be taken into account. On the other hand, it is a very simple way to characterize an emulsion and it is often used as a standard method.

Traditional microscopy may be advantageously substituted by confocal laser scanning microscopy (CLSM) to follow microstructure changes inside emulsions, such as aggregation or gelation [30]. This technique permits an enhancement of image contrast [31] by reducing out-of-focus noise and the generation of three-dimensional images of structures. In this context, CLSM is used in food science to characterize emulsions [32, 33]. Limitations are the same as those of classical microscopy: samples have to be diluted and image analysis must be performed to obtain quantitative data about the size distribution of the sample.

Electron microscopy [scanning electron microscopy (SEM) and transmission electron microscopy (TEM)] is another microscopic technique which permits the direct observation of emulsions, particularly those for which droplets are too small for an optical observation (below 1 μm). It is widely used, notably in food science [23]. A particularity of this technique is the necessity to maintain high vacuum inside the instrument. Consequently, samples must be free of volatile components. Using specific apparatus, the temperature may be lowered to allow the analysis of classical emulsions (cryogenic electron microscopy). The resolving power of electron microscopy is on the nanometer scale, depending on the technique. Limitations and precautions are the same as those for classical microscopy.

Atomic force microscopy (AFM) is able to give information about structure at the molecular level. Hence it permits the characterization of interface changes in emulsions. It is complementary to the other microscopic techniques. It may be used, for example, to determine wax morphology at the oil/water interface and the role of the emulsifier, which may influence emulsion stability [34], or to observe the structure of aggregates [35].

To obtain information about the droplet size distribution of an emulsion and its evolution over time, techniques involving light scattering are probably the most widely used [36] from food science [37] to mineral processes [38]. Of course, various commercial systems based on light scattering exist and they are simple to utilize. A variety of methods are also available, varying from simple turbidity measurement to treatment of diffraction data using complex algorithms. Among these techniques, we can cite the measurements of turbidity [39, 40], reflectance [41], color [42] and light scattering [23, 43] such as photon correlation spectroscopy and Doppler shift spectroscopy. Even though these methods seem to be very rapid and simple, commercial instruments have to make some assumptions about the shape of the droplet size distribution and the optical properties of emulsion droplets (their real and imaginary refractive indices) must be known in order to calculate a most probable distribution. Therefore, the data obtained must be treated with caution. Nevertheless, comparisons between several emulsions and, of course, the study of emulsion stability are possible. It should be noted that dilution of a sample is obligatory and it is therefore necessary to verify that the studied sample (and its stability) is not perturbed by dilution.

It is possible to access the droplet size distribution of an emulsion by counting the drops while they pass through a calibrated hole. As the sample is subjected to a voltage gradient, the measurement of current between the electrodes permits the detection of droplets passing through the hole and the calculation of their size. This technique is capable of determining droplet size distributions between about 0.6 and 400 μm but presents some limitations [18, 44]. The continuous phase of the emulsion has to be electrically conductive, which precludes the study of inverse emulsions (i.e. W/O). The sample may also be diluted in such a way that only one droplet at a time passes through the hole.

Both particle size and concentration may be determined in emulsions by accurate acoustic or electroacoustic measurements [45–47]. The theory is valid for a wide range of droplet sizes, varying from nanometers to 1 mm. Taking into account drop–drop interactions for the treatment of analytical data, it is possible to characterize concentrated and opaque systems, but the presence of gas bubbles could cause problems. Commercial instruments based on ultrasonic spectrometry are now available and may be used for real systems, without any dilution, in several fields such as the food industry and cosmetology.

The dielectric permittivity of an emulsion may be measured over a range of electromagnetic frequencies. For emulsions containing charged particles, it is then possible to determine a droplet size distribution, even in concentrated media, but this technique is not widely utilized yet [48, 49].

Nuclear magnetic resonance (NMR) and in particular the methods which determine molecular self-diffusion are now used to characterize emulsion properties and lately their stability [50]. This technique may provide useful information about emulsions and may be used for concentrated samples. The main restriction is the analytical instrumentation and so only research studies are conducted using NMR. Nevertheless, the recent literature reports emulsion stability studies by NMR [25, 34, 51].

Recently, Clausse et al. [52] developed a method based on the undercooling phenomenon, allowing the determination of droplet size distribution in concentrated and/or opaque emulsions. When a sample is cooled at a constant rate, the solidification temperature of the droplets is a function of their size. Using a simple scanning calorimeter to detect the crystallization, it is possible to correlate a thermogram with a size distribution histogram without complex algorithms or restrictive theory. This technique does not need any dilution and a statistical response dealing with a large number of droplets, around 10^6, is obtained. This method has been proposed for studying emulsion stability in microgravity conditions [1].

Various other techniques may be used in particular systems in order to characterize emulsion properties and stability. The list is extensive and here we cite just measurements using conductimetry [53], microwave attenuation [54] and voltammetry [55].

5.3 Methods Used to Characterize Interfacial Properties

5.3.1 Experimental Methods

As has been said previously, in order to stabilize the oil/water interface when an emulsion is formed, introduction of surfactant molecules, polymers or fine particles is necessary. Otherwise, rapid coalescence of drops leading to the energetically favorable diminution of the total interfacial area takes place. It is therefore clear that the destabilization of such systems is connected with the properties of adsorbed layers at the interface.

In the last few decades, a number of reviews on the experimental methods concerning interfacial properties related to the stability of dispersed systems have been published [56–60].

5.3.2 Adsorption

Adsorption of surface-active molecules at the interface gives rise to the formation of more or less ordered two-dimensional (2D) structures and results in diminished surface tension, γ. Gibbs was the first to relate the lowering of surface tension $d\gamma$ to the variation of chemical potential of the bulk solution of surfactant, $d\mu$, and the surface excess concentration, Γ. In this way, from the γ versus $\ln c$ curve (where c is the bulk concentration), one can determine the average area occupied per molecule at the interface, $a = 1/\Gamma$, but the information about the 2D structure of the film has to be obtained by other methods.

5.3.2.1 Structure of Adsorbed Layers

In the case of adsorbed layers subjected to deformation, the monolayer response to it can be characterized by 2D analogues of elasticity (or compressibility) and viscosity. It was again Gibbs who first introduced the notion of elasticity of a monolayer of small surface-active molecules, but it was only in 1960s and 1970s that the subject of interfacial rheology really started to develop from both experimental and theoretical points of view [61–63]. Recent renewal of interest in better comprehension of adsorption phenomena has given significant impetus to the development of experimental techniques allowing precise measurements of interfacial properties. Furthermore, new interpretations of classical adsorption models and propositions for new ones are also often found in the literature ([64] and references cited therein).

By analogy with the deformation of a three-dimensional solid, Gibbs defined surface dilatational elasticity as the variation of the surface tension of a monolayer when its area is changed (dilated or compressed). In order to meet quasi-equilibrium conditions, the monolayer should be insoluble in the liquid subphase and the imposed film deformation should be quasi-reversible.

According to Exerowa and Kruglyakov [65], there should be no direct relationship between the stability of a monolayer and its elasticity. Fruhner et al. [66] examined the rheological properties of films of several well-known surface-active molecules and their experimental results confirmed this theoretical prediction. Furthermore, Exerowa and Kruglyakov suggested that the film stability should be related to the change in the elasticity of the film with the change in its thickness. Lucassen-Reynders et al. [67], who studied the dependence of elasticity of protein films on the concentration of the protein at the interface, showed that the main factor affecting this dependence is the variation of the mean molar area of protein with surface excess.

5.3.2.2 Principal Surface/Interface Parameters

It is generally accepted that low interfacial tension is a necessary condition for the formation of the oil/water interface. One can therefore deduce that the destabilization of such an interface will be accompanied by increased interfacial tension. Although many different experimental techniques allow the measurement of the liquid/air surface tension, not all of them are adapted for oil/water interfacial tension determination. Amongst those allowing the measurement of even very low values of γ, the classical spinning drop technique is probably the best known. It is based on the measurement of the elongation of an oil droplet subjected to rotation in a continuous aqueous phase. The main drawback to using this technique routinely is the relatively important quantity of continuous phase required. The same is true for the Wilhelmy plate and Du Noüy ring techniques where, by means of a sensitive electrobalance, the force is continuously measured while a platinum ring or plate is driven through the oil/water interface. Both methods give access to the equilibrium value of interfacial tension. However, it is commonly accepted nowadays that dynamic surface/interfacial tension (variation of γ with time) is an important parameter for the understanding of the stability of interfaces.

The simplest method for the determination of dynamic interfacial tension, γ_{dyn}, is the classical drop volume technique. In this method, drops of the oil phase are formed at the tip of a capillary of known radius immersed in the aqueous phase. The rate of creation of the oil/water interface can be varied while the corresponding values of interfacial tension are deduced from the quantity of oil and number of drops created.

Shape analysis of objects with axial symmetry became easily accessible mainly due to the low cost of powerful calculators. Recent techniques based on the analysis of a contour of a droplet of one liquid immersed in another liquid are promising tools for studying interfacial properties. The method for the determination of γ is based on the equilibration of gravitational force and surface tension force which governs the shape of drops and bubbles. At the origin of this method is the remarkable pioneering work of Bashforth and Adams published in 1883 [68]. By comparing the photographed profiles of several sessile drops of different sizes with the calculated contour obtained by solving (by hand!) a differential equation based on the Laplace law and gravitational law, they successfully determined the surface tension of liquid mercury.

Recent experimental devices use various configurations, such as a liquid drop in air, an air bubble in liquid or a drop of one liquid in another liquid. Two main difficulties with the technique are the precise determination of the contour of the interface, a crucial parameter for the calculation of the correct value of γ, and the impossibility of maintaining the droplet in position for low γ values. The latter drawback can prohibit the use of the method in the case of an oil/water interface, especially in the presence of surface-active substances. In spite of these limitations, the technique is rapidly gaining in importance. In addition to the small quantity of sample needed, there are other considerable advantages in comparison with classical methods, e.g. by using the bubble configuration the problem of evaporation of small drops and the danger of contamination of the drop surface are eliminated or significantly reduced. The other noteworthy advantage is the possibility of imposing variations of droplet/bubble volume, inducing the variation of its area and simultaneously measuring the resulting variation of γ. In this way, a whole new domain of interfacial rheology becomes experimentally accessible [69–71]. The same parameters are also accessible by another important interfacial technique, the Langmuir balance. This involves the measurement of surface pressure, π (the difference in the interfacial tensions before and after addition of native surfactants), in a monolayer while its surface area is varied, i.e. contracted or expanded (the apparatus can be seen as a 2D analogue of the classical cylinder and piston set-up in the case of a 3D gas). The π versus area A isotherm obtained gives access to the structural changes in a monolayer and it can be very rich in the sense that many different 2D phases can be observed. In addition, the oscillatory movement of barriers and the simultaneous measurement of π (via measurement of instantaneous values of γ) give access to the rheological properties of the monolayers [62, 70]. Deposited insoluble monolayers (or Langmuir monolayers) are much more often studied, but lately also Gibbs monolayers adsorbed from the subphase have become of interest, although a consensus concerning the relevance of the data obtained has not yet been arrived at in the literature.

In addition to the isotherms, a number of research group uses fluorescence and/or Brewster angle microscopy to visualize the two-dimensional structures obtained while a monolayer is compressed. A large number of beautiful and sometimes intriguing photographs have been published in the last 20 years [72, 73].

In order to measure quantitatively the height of a surface layer, ellipsometry has been introduced. By analyzing the changes in reflected light of an elliptically polarized light beam striking the surface, one can obtain information about the thickness and the refractive index of the monolayer. In this way it is possible to quantify the amount of adsorbed species, for example, to deduce the surface excess Γ at different states of a 2D structure [74].

5.4 Impact of Interfacial Phenomena on the Stability of Complex Emulsions

5.4.1 Oil-in-Water Food Emulsions

Most manufactured foodstuffs are food colloids composed of multi-phase dispersions including an oil phase, an aqueous phase, often solid particles as fat crystals or starch granules and sometimes gas bubbles. The objective of the food manufacturer is to obtain a homogeneous mixture which is not only stable but also has the required specific properties (such as nutritional, organoleptic and rheological properties). O/W emulsions (mayonnaise, homogenized milk, salad dressing, ice-cream, cream liqueurs) and less common W/O emulsions (margarine, spreads) are important classes of food colloids. Considerable work has been devoted to food emulsions during the last 50 years and many books [23, 59, 75–77] and reviews [78–81] have been written on this subject, and it is not possible to quote all of them. This section will show how the system complexity and multi-component interactions between food emulsion ingredients make the development of theoretical models very challenging.

5.4.1.1 Surface-active Food Emulsion Components

Surface-active components of food emulsions include mainly low molecular weight surfactants and proteins. Small surfactants are characterized by distinct polar and non-polar parts which efficiently reduce the oil/water interfacial tension. They may be water soluble [ethoxylated sorbitan esters (Tween®), ethoxylated alcohols (Brij®), sucrose esters, alkylpolyglucosides, etc.) or oil soluble (mono- and diglycerides, phospholipids, sorbitan esters (Span®), etc.) [57]. Most food emulsifiers are non-ionic, a few are anionic or amphoteric, but not cationic. Such surfactants are particularly efficient during the emulsification process as they diffuse rapidly to the interface and promote short-term stability against droplet coalescence. They form a condensed layer stabilized by electrostatic repulsions or through the Gibbs–Marangoni mechanism [82]. In this case, rapid lateral diffusion of emulsifiers in response to surface concentration gradients will drag the adjacent continuous phase and restore the liquid film between the droplets, thus preventing the rupture of the interfacial film [83].

Proteins are the most common class of food emulsifiers and stabilizers as they are natural, non-toxic and easily available ingredients [82]. They are also surface-active molecules although they do not have a classical amphiphilic structure. The spatial arrangement of a protein influences its interfacial properties. Examples of proteins used in food emulsions are lysozyme (globular and rigid), bovine serum albumin (globular and flexible) and β-casein (random coil) [84]. Proteins diffuse to the interface more slowly than small surfactants but their affinity for the interface is higher. Adsorption is considered to be almost irreversible, although proteins can be displaced from the interface by changing the environmental conditions (pH, temperature, ionic strength) or by addition of surfactants. This irreversible character has been attributed to a gradual increase in physical and chemical bonds between the adsorbed proteins [81]. The effect is more pronounced with globular proteins than with random coil proteins. Proteins form a rigid interfacial layer characterized by viscoelastic properties which usually provide a strong mechanical barrier to coalescence. The protein layer ages and its mechanical properties slowly change with time.

Table 5.1 illustrates the differences between interfacial properties of proteins and small surfactants [57]. Because of these differences, proteins and surfactant are often referred to as incompatible ingredients [82, 85, 86], although they are often simultaneously present in real food systems, as in ice-cream or whipped toppings.

In food emulsions, stability is not the only required property. Specific properties of the products, in particular control of texture and of rheology where other components of food emulsion also play role, are often sought after [75]. Hydrocolloids, such as polysaccharides, are water-soluble biopolymers added to control the rheological properties of the continuous phase, for instance to increase its viscosity or to gel it [79]. In such a way the emulsion stability is favored. Modified polysaccharide, such as hydrophobically modified cellulose [85, 87, 88] or dextran [89], are surface-active molecules and may even be used alone as emulsifiers [90]. Emulsion stability in this case has been related to the conformation of the polymer at the interface, more precisely to the ratio of loop–train and train–tail segments [87]. The apparent surface activity of unmodified polysaccharides has not been clearly explained and is still under debate [79, 91]. In some instances, it has been attributed to the presence of a small amount of proteins, either as contaminants or as a part of the molecular structure [79].

The next section will illustrate the complex behavior observed when these components are mixed, particularly when proteins are present. Important factors related to proteins are conformation changes at the interface due to adsorption or induced by some change in environmental conditions. In addition, surfactant–protein and polysaccharide–protein interactions may become important at the interface and/or in the bulk.

Table 5.1 Comparison of some typical interfacial properties of proteins and surfactants. (Adapted from M.A Bos and T. van Vliet [57]).

Properties	Proteins	Surfactants
Molecular weight (g mol^{-1})	~20 000	~200
Surface excess (mol m^{-2})	~10^{-7}	~10^{-5}
Adsorbed amount (mg m^{-2})	~2	~2
Equilibrium surface tension upon adsorption (A/W) (mN m^{-1})	47–57	22–42
Equilibrium interfacial tension upon adsorption (O/W) (mN m^{-1})	8–28	10^{-3}–25
Reversibility of adsorption	Quasi-irreversible	Reversible
Rate of surface tension lowering	Slow	Fast
Conformation change upon adsorption	Yes	No
Ability to build up surface tension gradients	Low	High
Binding to polysaccharides	Yes	No

5.4.1.2 **Relationship Between the Properties of the Protein Interfacial Film and the Stability of Food Emulsions**

Food emulsions are stabilized by complex interfacial structures ranging from protein membranes to gel networks, liquid crystals, multilayers or solid particles [75]. Different methods, described in Section 5.2.4, have been implemented in order to characterize the emulsion stability. They involve mainly droplet size analysis, by optical microscopy, light scattering techniques (dynamic light scattering, laser diffraction) or Coulter Counter multisizer, creaming observations either visually or by scattering techniques and bulk viscosimetry. The rate of coalescence is also directly studied by measuring the contact time between an oil droplet and the oil/water interface before coalescence occurs [82]. Table 5.2 gives some examples of the different techniques used to correlate emulsion stability with interfacial properties. It can be seen that there is no unique way to define emulsion stability, as was been stressed in the Introduction. As a consequence, the results obtained in different studies may sometimes be difficult to compare and controversial conclusions may arise.

Most systems involved in food emulsions include protein as a main component. The rheological and mechanical properties of protein films have attracted a lot of attention in an attempt to explain emulsion destabilization mechanisms [57]. However, a complication encountered in this kind of study is that changing the composition of the system will modify the droplet size distribution and that alone has a pronounced effect on the emulsion stability.

Upon adsorption, proteins partially unfold to expose their hydrophobic groups to the non-polar oil surface. The extent of unfolding, which can be considered as

Table 5.2 Some examples of the various techniques used to correlate emulsion stability to interfacial properties (emulsions properties were generally measured regularly over a period of time indicated in parentheses).

System content	Methods used to measure the interfacial properties	Methods used to characterize the emulsion stability	Ref.
Oil phase: *n*-hexadecane *Aqueous phase*: mixture of proteins and polysaccharides in buffer (pH 7)	Surface tension (equilibration time 30 min) Interfacial shear viscosity (equilibration time 1 h)	Conductivity measurements (0–280 min) Droplet size (laser diffraction/0–7 days) Bulk viscosimetry	85
Oil phase: soybean oil *Aqueous phase*: solution of proteins in buffer (pH 4–7)	Interfacial shear viscosity Surface pressure (Langmuir trough)	Droplet size (Coulter Counter/0–8 days) Turbidity (0–100 h) Time of coalescence of single droplets (0–600 min)	92
Oil phase: *n*-tetradecane *Aqueous phase*: solution of proteins in buffer (pH 7) – proteins cross-linked by enzymes	Surface dilational viscosity	Droplet size (laser diffraction/0–7 days) Back scattering (Turbiscan/0–60 min) Bulk viscosimetry (2–4 h)	102
Oil phase: triglycerides *Aqueous phase*: solution of polysaccharides in water	Dynamic interfacial tension (pendant drop)	Droplet size (laser diffraction/0–6 months) Bulk viscosity (0–6 months) Temperature stability testing (0–6 months)	87
Oil phase: *n*-dodecane *Aqueous phase*: solution of polysaccharides in water	Surface and interfacial tension (Wilhelmy plate and ring) (equilibration time 15–60 min)	Creaming (visual observation/0–60 days) Droplet size (photon correlation spectroscopy/0–60 days) Zeta potential	89
Oil phase: *n*-tetradecane-containing emulsifiers *Aqueous phase*: solution of proteins in buffer (pH 7)	Interfacial tension (pendant drop)	Droplet size (laser diffraction/0–500 min) Orthokinetic stability (droplet size distribution of the emulsion kept under stirring/0–500 min)	86
Oil phase: sunflower and soybean oil containing emulsifiers *Aqueous phase*: solution of proteins in water	Interfacial tension (Wilhelmy plate)	Droplet size (laser diffraction/1–7 days) Bulk viscosimetry (1–7 days)	103

partial denaturation, depends on the native structure of the protein and influences the interfacial properties. It was observed that the shear viscosity of the protein interfacial film is much higher for globular proteins, such as lysozyme, than for random coil proteins, such as β-casein, because of stronger molecular interactions in the film [92]. Enhanced emulsion stability can be correlated in most cases with higher interfacial shear viscosity [93]. However, it has been shown recently that globular protein films may become brittle and fracture, leaving holes in the layers, and this may explain why in some cases a rigid film may lead to less stable emulsions [57]. Optimum behavior of the protein film would be characterized by high interfacial modulus but also the ability to respond rapidly to external deformations. However, no theory so far can reliably relate the interfacial properties to the stability of emulsions stabilized by a protein film [57].

Modification of secondary and tertiary conformations of globular proteins by thermal treatment prior to emulsification can induce enhanced emulsion stability [92]. In the case of β-lactoglobulin, thermal treatment resulted in protein partial denaturation, as illustrated by a decrease in α-helix content and an increase in random coil conformation [92]. The improvement in rheological properties was attributed to increased flexibility of the treated chain and a tendency for more loop formation.

In some cases, the effect of protein structural modification on the emulsifying ability of proteins also gave rise to controversial results. Studies on high-pressure induced aggregation of soy proteins demonstrated either positive [94] or negative [95] effects on their emulsifying properties. Recent report on high-pressure treatment of lupin proteins [96] revealed improved rheological properties and emulsion stability, in contrast to milk proteins [97] subjected to the same treatment.

Understanding the relationship between interfacial properties and emulsion stability is even more difficult in real food emulsions for which different components compete for the interfacial area. Most real systems include mixtures of emulsifiers (low molecular weight surfactants) and proteins. It is often stated that, in real systems, mixtures of surfactants and proteins result in maximum emulsion stability against coalescence [57]. The effect can actually be positive or negative depending on the surfactant concentration and the nature of the protein–surfactant interaction [75]. The formation of protein–surfactant interfacial complex generally has a positive influence on emulsion stability whereas non-interactive surfactants will partially displace the protein from the interface, which usually has a destabilizing effect [75]. Combination of computer simulation and AFM images evidenced that the displacement of protein films by surfactants occurs in a non-homogeneous way [98]. Surfactant molecules adsorb in defects of the protein layer. Surfactant-rich domains grow progressively, compressing the protein-rich regions until their collapse and the disruption of the protein interfacial network [82, 98]. This result, illustrated in Figure 5.2 for the Tween 20-β-lactoglobulin system [82], evidences phase separation between protein and surfactant at the interface and gives a new insight into the displacement of protein films by surfactants.

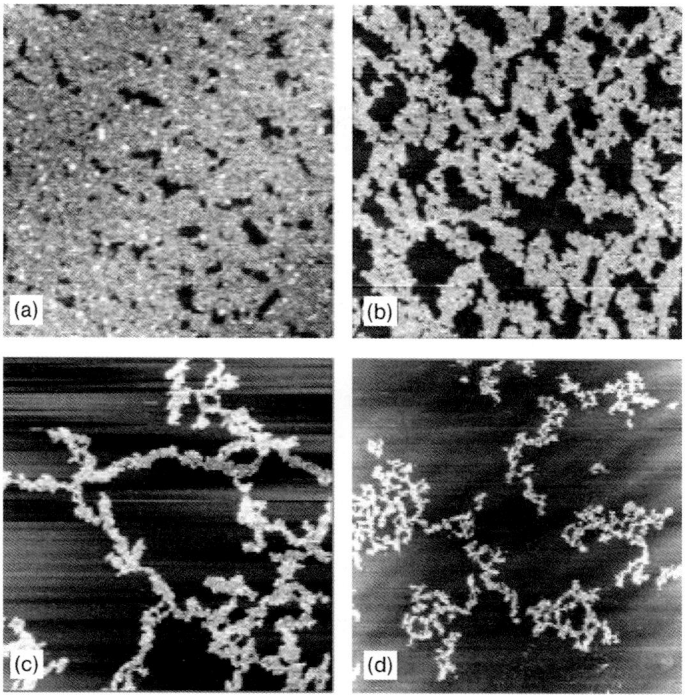

Fig. 5.2 AFM images of a β-lactoglobulin film progressively displaced by Tween 20. Surfactant molecules adsorb in holes of the protein film. As more surfactant is added, the protein film is compressed until the network fails. The light regions represent the protein network and the dark regions represent surfactant domains. (Reprinted with permission from Elsevier [82]).

Proteins also interact with other components of food emulsions, in particular with hydrocolloids such as polysaccharides [79]. Interactions between proteins and polysaccharides can be attractive or repulsive depending on the molecular structure and environmental conditions. Repulsive interactions due to thermodynamic incompatibility will be a destabilizing factor resulting in phase separation and depletion flocculation. Conversely, favorable interactions will lead to the formation of an interfacial complex resulting in increased emulsion stability due to the combination of steric and electrostatic repulsive forces [79]. Most studies carried out on protein–polysaccharide complexes concern electrostatic complexes, mainly between proteins (at pH below the isoelectric point) and anionic polysaccharides [99–101]. Electrostatic interactions favoring interfacial coadsorption of both components has a positive effect on emulsion stabilization [99–101]. In the presence of surface-active polysaccharides, competitive adsorption can take place, as was observed with low molecular weight surfactants. Synergistic interactions were observed between β-lactoglobulin and hydroxypropylmethylcellulose, resulting in even greater values of surface elasticity than the protein film alone [88]. However, the origin of this effect is not fully explained.

Understanding and controlling multi-component interactions involving proteins is very important because even a small change in weak attractive or repulsive interactions can have a drastic effect on the emulsion stability and specific properties [79].

Microencapsulation processes have been designed to reinforce the interface, for instance by cross-linking the protein film [102], by forming interfacial liquid crystalline layers [103] or by creating multiple-layer membranes using an electrostatic deposition method [104]. Adsorption of cationic chitosan on a negative protein membrane by interfacial engineering technology had a positive effect on emulsion stability and could lead to a new type of emulsion with improved properties.

Despite the inherent difficulty in developing theoretical models to correlate interfacial properties with emulsion stability, this research field is very active and innovative formulations designed to improve the product properties are developed on a daily basis.

5.4.2 Water-in-Crude Oil Emulsion: Formation and Destabilization

Emulsions are encountered throughout the petroleum industry, from drilling, production and transportation to refining. Among all the types of emulsions, water-in-crude oil emulsions that form during oil production are the subject of many studies because of their extreme stability, which complicates the operations of separation between oil and water.

Crude oil exploitation is always accompanied by water production. Water, which is naturally present in the oil reservoir, mixes with crude oil to form very stable emulsions. High shear rates and zones of turbulence encountered at the wellhead in the choke-valve cause their formation. W/O emulsions are preferentially created because of the oil-soluble nature of the stabilizing compounds. Nevertheless, one can find O/W emulsions when the water content in the production effluent becomes very high or even multiple emulsions such as W/O/W or O/W/O, when the water content is close to the inversion point of the emulsion. W/O emulsions can be very stable due to the presence of asphaltenes and resins, which play the role of natural emulsifiers, and also because of small solids (crystallized waxes and clays, for example) [105–108]. This kind of emulsion is undesirable because its volume and viscosity, higher than those of crude oil, raise operational costs. Therefore, before refining, the water content has to be reduced in order to produce high-quality oil and also to eliminate chloride salts which corrode equipment. Dehydration of crude oil is usually achieved by gravitational separation or electrocoalescence. The times required for demulsification are often very long and chemical agents are needed to hasten it. Blends of surfactants are commonly used, such as amines, copolymers of ethylene oxide and propylene oxide, ethoxylated phenols, etc.; they are formulated in solvents such as short-chain alcohols or aromatic or heavy aromatic naphtha [109–111]. The most useful method to quantify emulsion stability is the bottle test, which allows the determination of the water volume separation with time. Differential scanning calorimetry has also been used to study the evolution of the water droplet size distribution and to measure accurately the water content [52, 112].

First, we will examine how emulsion stability can be affected by physicochemical parameters, then the mechanisms of emulsion destabilization by chemical demulsifiers proposed in the literature will be presented.

5.4.2.1 **Correlation Between Water-in-Crude Oil Emulsion Stability and Physicochemical Parameters**

Water-in-crude oil emulsions are strongly stabilized by native crude oil surfactants, which prevent water droplet coalescence by forming a rigid film or skin around the droplets, but the mechanism is not well understood. Among the native surfactants, asphaltenes and resins are often considered to be responsible for emulsion stability. Asphaltenes and resins have the same structure but their molecular weight is different (Fig. 5.3).

Asphaltenes are defined as the crude oil fraction insoluble in *n*-heptane or *n*-pentane but soluble in toluene, whereas resins are soluble in aliphatic and aromatic solvents (except in methanol). They both contain hydrophobic and hydrophilic moieties, which confer them their surface-active properties. Asphaltenes tend to self-associate in the form of aggregates whose size depends on solvent conditions, temperature and pressure. Yen described asphaltenes as stacks of flat sheets of condensed aromatic sheets, which are linked to each others by

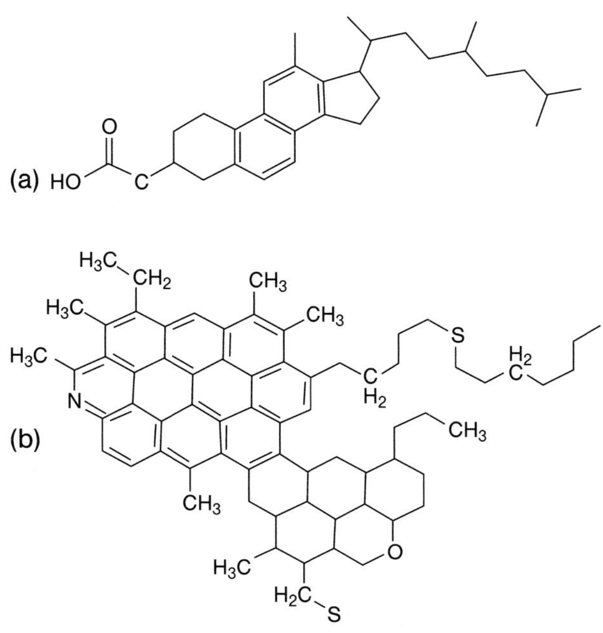

Fig. 5.3 Example of (a) resin and (b) asphaltene molecules [113].

ether, sulfide and aliphatic chains [114]. Therefore, asphaltene aggregation is the association of aromatic compounds caused by π–π interactions and hydrogen bonding; in this case, asphaltene molecules are said to form colloids. However, if the association of asphaltene molecules is driven by hydrophobic–hydrophilic interactions, asphaltene aggregates can be called micelles. In crude oil, asphaltene aggregates are stabilized by resins and lighter polar components by adsorption mechanisms. In fact, resins act like surfactants in colloidal suspension [115]. When crude oil is brought in contact with water, asphaltenes and resins compete for the water/oil interface [110]. Resins, which are lighter than asphaltenes, are the first entities to reach the water/oil interface. As a consequence, the solubility conditions for the asphaltenes change, inducing the precipitation and accumulation of the asphaltene particles around the water droplets. It is suggested that in this case the emulsion stabilization is assured by steric repulsion between asphaltene particles at the water/oil interface or by presence of other solid particles [110]. According to McLean and Kilpatrick, emulsion stability is also related to the solubility state of the asphaltene aggregates in the bulk [116]. The resin-to-asphaltene ratio, the aromaticity ratios of the crude oil and resins with respect to asphaltenes and the concentration of polar functional groups in resins and asphaltenes influence emulsion stability. High values of these parameters mean unstable emulsions. The interactions between asphaltenes and resins are a preponderant factor in emulsion destabilization. Indeed, resins are able to solvate asphaltene aggregates because of their polar and dispersive nature [117]. They disrupt π–π and polar interactions existing within asphaltene aggregates and thus inhibit the surface-active nature of asphaltenes. Resins can replace the asphaltenes at the oil/water interface and induce emulsion destabilization if their concentration is high enough in the bulk, but they cannot alone stabilize W/O emulsions [118]. Furthermore, according to the same reference, beyond the asphaltene precipitation point, the emulsions become unstable because precipitated asphaltenes would not participate in the stabilization mechanism. Eley et al. proposed that the most stable emulsions exist when the asphaltenes are near the precipitation point [119]. Bouriat et al. [120] used static and dynamic tensiometry to determine the mean area occupied by asphaltene molecules at a newly prepared water/asphaltened cyclohexane interface, showing that asphaltene molecules and not large aggregates adsorb at the interface. From rheological studies, they stated that 2-day-old interfaces exhibit a glass transition zone, behave like a gel near their gelation point and are built by a universal process of aggregation. Obviously, the form in which asphaltenes stabilize W/O emulsions still remains debatable.

Although asphaltenes and resins are the main surface-active molecules that compose the rigid skin preventing water droplet coalescence, other surface-active molecules such as naphthenic acids and porphyrins or small particles such as wax crystals and clays can enhance emulsion stability. It has also been proved that bulk viscosity can play a preponderant role. Johansen et al. studied the ability of crude oils from 32 different wells to form stable emulsions [121]. The crude oils contained a large amount of saturated hydrocarbons, a varying content of waxes (2–15 wt%) and a low content of asphaltenes (0–1.5 wt%). They were characterized by density, viscosity, surface and interfacial tension measurements. It was found that the emulsion viscosity, the crude viscosity and the content of waxes were the most significant parameters for predicting the stability of these emulsions.

It is worth noting that the lowering of interfacial tension is not sufficient to predict W/O emulsion stability [122, 123]. Nevertheless, a decrease in interfacial tension with time was observed [124], which corresponds to the mass transport of the crude oil native surfactants to the oil/water interface followed by a steric arrangement. These results are in agreement with current observations made on emulsion stabilization with time in oil fields. Moreover, the effect of aging on crude oils that have been exposed to air and light and thus to oxidation and photo-oxidation has also been studied [125]: the most stable emulsions show very low interfacial tension values due to the presence of oxidized aromatic compounds.

The stability of water-in-decane emulsions stabilized by surface-active fractions from North Sea crude oils has been investigated by Sjöblom et al. [126]. They found a good correlation between interfacial pressure and the stability of the emulsions. Indeed, stable emulsions are characterized by high interfacial pressure values. The effect of the oil aromaticity has also been studied. The addition of benzene in decane leads to unstable emulsions because of the solubility of the native surfactants in aromatic solvents. They become less surface active and leave the oil/water interface. These observations strengthen the hypothesis of steric stabilization of water-in-crude oil emulsions.

We saw that the oil/water interfacial film plays an important role in the stabilization mechanism. When only native surfactants are present in crude oil, an interfacial tension gradient sets up, with higher interfacial tension within the film and lower interfacial tension outside the film. The resulting mass transport inhibits the phenomenon of film drainage and therefore prevents the water droplets from coalescing. Interfacial tension gradients induce film viscosity and film elasticity. Hence many interfacial rheological studies have been performed in order to correlate the behavior of the interfacial film with the emulsion stability.

High interfacial viscosities often characterize stable emulsions and, surprisingly, low interfacial viscosities do not mean unstable emulsions: it is rather the compressibility of the interfacial film and hence its elasticity, which are involved in the mechanism of resistance to droplet coalescence. Acevedo et al. studied the interfacial behavior of asphaltene and crude oil systems by using a planar rheological method [127]. Viscoelastic and elastic behavior have been observed. They concluded that interfacial viscosities are not relevant for predicting emulsion stability and that only interfacial elasticity should be considered. They showed that the high values of elastic modulus leading to high stability of W/O emulsions were attributable to flocculation of asphaltene–resin aggregates at the interface.

By using a Langmuir-type oil/water interfacial film balance, Jones et al. showed that stable, incompressible, non-relaxing films act like a mechanical barrier to coalescence [122]. Bhardwaj and Hartland measured the rheological properties of oil/water interfaces using a drop-volume apparatus. They measured the compressibility of the oil/water interface by decreasing the drop size after allowing the interface to age. They concluded that crude oil with a high asphaltene and resin content presents highly incompressible interfacial films [124]. Bauget et al. performed drop oscillations using the pendant drop method and showed that resins diminish the resistance of the interfacial film [128]. They also found that oil/water interfaces in the presence of asphaltenes are characterized by long adsorption times and the formation of a rigid film. pH has a strong influence on the interfacial properties of asphaltenes at the oil/water interface. Recent studies, using the pendant drop method, have shown that the surface activity and the hydrophilic behavior of the asphaltenes are enhanced at low or high pH values because their functional groups become charged [129]. Indeed, asphaltenes have more affinity with the interface when they are charged. They tend to accumulate more easily at the interface and the resulting emulsions are more stable. The consequences are a decrease in the interfacial tension and an increase in the dilatational elastic modulus. The observed effect is stronger at high pH because asphaltenes have more acidic than basic groups.

The relevance of correlations between the viscoelastic properties of the water/oil interface and emulsion stability are also confirmed by many studies performed on the action of emulsion breakers (demulsifiers), as will be presented in detail in the next section. For instance, Kim et al. measured the rheological properties of an oil film separating two aqueous brine phases [111]. They added poly(propylene oxide)-type demulsifiers and found that unstable emulsions are characterized by low film dilatational moduli.

Some typical examples of the techniques that are used to correlate emulsion stability to interfacial properties can be found in Table 5.3. One can see that the diversity of the chosen systems and the methods employed makes comparisons between the different studies difficult.

5.4.2.2 Destabilization of Water-in-Oil Emulsions by Chemical Demulsifiers

Cocktails of demulsifiers are added to break water-in-crude oil emulsions in oil production. Ese et al. [130] showed that the addition of demulsifiers to asphaltene monolayers caused the opening of the rigid asphaltene structure.

Chemical demulsifiers interact with the oil/water film so that they modify the interfacial properties. Demulsifiers invert the interfacial tension gradient due to their superior surface activity, inducing film drainage, which leads to emulsion destabilization. Bhardwaj and Hartland [131] have shown that efficient demulsifiers reduce the interfacial tension between oil and water but also displace the native surfactants in order to adsorb at the interface. Djuve et al. asserted that high surface activity is also very important [132]; nevertheless, decreasing interfacial tension and adsorption at the interface are essential properties of demulsifiers, but they are not sufficient. Thus, demulsifiers have to show a high diffusion/adsorption rate and be able to act on viscoelastic properties of the oil/water film. They can decrease the surface dilatational modulus of the interfacial film

Table 5.3 Some examples of the techniques used to correlate emulsion stability with interfacial properties.

System under study	Methods used to measure the interfacial properties	Methods used to characterize the emulsion stability	Ref.
Crude oil [a]/brine (3.5% NaCl) Model oil [b]/brine (3.5% NaCl) Aqueous phase = 5–60% High and low molecular weight chemical additives	Liquid–liquid Langmuir trough: Surface pressure–area isotherms Langmuir–Blodgett films studied by means of AFM	Measurement of the water volume resolved after 30 min at 50 °C	110
Crude oil [c]/brine (33 800 ppm of NaCl) Aqueous phase = 30% (v/v) Poly(propylene oxide)-type chemical additives (100 ppm)	Film tensiometer: Dynamic film tension (dynamic film elasticity) and film stress relaxation measurements Drop volume method: Interfacial activity and adsorption kinetic measurements	Measurement of the water volume resolved as a function of time at 70 °C	123
Model oil [d]/brine [e]	Planar rheological method: Elasticity and viscosity surface moduli	Measurement of the water volume resolved after 24 h at 25 °C (bottle tests)	127
Model oil [f]/brine (5 g L^{-1} NaCl)	Dynamic surface tension instrument: Dynamic interfacial tension and dilational viscoelastic properties	Measurement of the water volume resolved after 30 min of centrifugation at 4000 rpm by means of optical microscopy (25 °C) (bottle tests)	129
Crude oil/water Model oil [g]/water Aqueous phase = 30% Chemical additives [h]: 100–1000 ppm in crude oil 0.1–50 ppm in model oil	Tensiometer (Wilhelmy plate method): Interfacial tension Viscous traction deep channel viscometer: Interfacial shear viscosity Drop volume method: Dynamic interfacial tension	Measurement of the water volume resolved as a function of time	133

Table 5.3 (continued)

System under study	Methods used to measure the interfacial properties	Methods used to characterize the emulsion stability	Ref.
Model oil[i)]/brine (0.01 M NaCl) Aqueous phase = 50% Chemical additives: ethylene oxide and propylene oxide copolymers (200 ppm)	Tensiometer (Du Nouy ring): Interfacial tension	Measurement of the water volume resolved after 1 h at 60 °C using gravitational settling	139
Crude oil[j)]/ultra pure water Aqueous phase = 30% (v/v) Chemical additives: polysiloxane copolymers (100 and 200 ppm)	Liquid–liquid Langmuir trough: Surface elasticity	Measurement of the water volume resolved after 1 h at 23 °C by means of optical microscopy	142

a) From Venezuela, Nigeria, North Sea and European continent.
b) Asphaltenes dissolved in toluene–decane (30:70).
c) From Mississippi.
d) Crude oil extra heavy Cerro Negro diluted in xylene [30% (v/v)] or asphaltenes dissolved in xylene [3.21% (w/v)].
e) Triply distilled water or brine [2% (w/v) NaCl].
f) Asphaltenes dissolved in toluene (0.01–1%).
g) Asphaltenes dissolved in heptane–toluene (70:30).
h) Alkoxylated resins, ethoxylated phenol, blend of phenol resins.
i) 5% bitumen diluted in toluene.
j) From a French field.

and must have a high diffusivity [111, 133]. Zhang et al. [134] studied mixed monolayers of asphaltenes and a demulsifier (a blend of oxyalkylated alkylphenol–formaldehyde resins) using a Langmuir interfacial trough at the air/water and heptol/water interface (heptol is a mixture of heptane and toluene). They showed that monolayers are more compressible when the demulsifier is added at both interfaces; the demulsifier reduces the strength of the asphaltene films. Jones et al. [122] have shown, using a Langmuir-type liquid–liquid film balance, that a mixture of an alkoxylated phenol with an alkoxylated ester can displace incompressible crude oil/water films and also inhibit film build-up. A soap of carboxylic acid with an alkoxylated ester can only slow the rate of film build-up. The Langmuir technique has also been used by Ese et al. [135] to study the influence of chemical additives of high molecular weight at both the air/water and oil/water interfaces. They concluded that some of the demulsifiers increase the compressibility of the interfacial film by altering the asphaltene structure, as resins can do.

The ability of demulsifiers to interact with the interface depends on their structure and their composition. Singh [136] correlated the efficiency of commercial demulsifiers with their influence on oil/water film surface pressure–area isotherms, by using a liquid–liquid Langmuir trough. He found that efficient demulsifiers reduce interfacial tension, indicating their fast migration at the oil/water interface, and that their action is determined by their structure and their orientation at the interface. Djuve et al. compared the efficiency of low and high molecular weight demulsifiers and found that demulsifier action is improved due to high surface activity when the molecular weight is high [132]. Sun et al. [137] studied the action of straight- and branched-chain demulsifiers containing ethylene oxide on the viscoelastic properties of a model oil/water film (the model oil contains the surface-active fractions of the crude oil, which are dissolved in *n*-decane). They demonstrated that the branch-chain demulsifier is the most able to replace the native surfactants at the interface because the space between the adsorbed molecules at the interface is smaller than that for the straight-chain demulsifier: native surfactants cannot reach the interface any longer.

Some authors stated that the hydrophilic lipophilic balance (HLB) values of demulsifiers could determine their ability to break W/O emulsions [138, 139], that is, that high HLB values would facilitate emulsion destabilization. Nevertheless, Al-Sabagh et al. investigated the action of 10 demulsifiers and found no relation between the HLB values and the demulsification efficiency. They explained the high efficiency of demulsifiers containing poly(ethylene oxide) by their partition coefficients close to unity rather than by their HLB values [140]. Demulsifiers that are only soluble in water cannot be good W/O emulsion breakers. Indeed, studies have shown that molecules with equal partitioning between the aqueous and the oil phase are the most efficient demulsifiers [131, 133]. Therefore, amphiphilic demulsifiers should be preferred. Recently, the action of three linear silicone copolymers on the crude oil/water interface has been studied using a liquid–liquid Langmuir trough, after being first characterized at the air/water interface [141, 142]. The copolymers were based on poly (alkylene oxide)-modified polydimethylsiloxane chains. The results confirmed that amphiphilic character is a major factor in destabilization efficiency. Indeed, the two amphiphiles were very efficient demulsifiers whereas the third, which was completely hydrophobic, was a poor emulsion breaker. The isotherms emphasized the interactions of the two amphiphilic demulsifiers with the oil/water interface during the compression at very low surface pressures whereas the hydrophilic demulsifier had almost no influence.

Referring to W/O emulsion stabilization, a strong correlation exists between the emulsion stability and the viscoelastic properties of the interfacial film. Asphaltene aggregates and resins are the main components of a viscoelastic film, which acts like a mechanical barrier to water droplet coalescence. Resins can dissolve the asphaltene aggregates and diminish the film resistance when their concentration in the bulk is high enough.

Concerning emulsion destabilization by chemicals, it is worth noting that efficient demulsifiers can rapidly adsorb at the oil/water interface in order to invert the interfacial tension gradient. Their action is to increase the film compressibility and to reduce its elasticity. They should have equal partitioning between the aqueous and the oil phases.

5.5 Conclusions

The aim of this chapter was to review current experimental data dealing with interfacial properties and the stability of complex emulsions. Particular attention was devoted to emulsions encountered in food science and in the petroleum industry. These systems can be highly concentrated and consequently display strong opacity and high viscosity. Furthermore, as they are manufactured products such as food materials or natural materials such as petroleum fluids, they contain a great number of components. Surface-active components are often characterized by a complex structure, sometimes with ramifications, and can be irreversibly adsorbed at the oil/water interface. Therefore, interfacial properties are not easy to determine as equilibrium is difficult to reach and even to define. Another factor is the type of emulsion under study. Referring to food science, emulsions are essentially O/W whereas in the petroleum industry emulsions are preferentially W/O. Nevertheless, provided that the droplets are large enough for the droplet curvature not to be involved and even if the mechanisms responsible for the destabilization can be different, the ultimate destabilization process, namely coalescence, will occur in both cases. Therefore, it appears that the interfacial properties of the film obtained when droplets are close to each other are very important. It was not within the scope of this chapter to study the mechanism of film breakdown, involving nucleation of holes as in the case of foams, but it can be stated that interfacial properties must play an important role in this mechanism. Special attention was given to the various experimental techniques used to characterize the interfacial properties of liquid/liquid interfaces. Referring to emulsion stability, the problem is to define accurately what indeed is "stability". "Stability" does not represent a specific property of the emulsion and very often is meant as an "evolution" of the emulsion versus time. Even so, various relevant parameters can be defined. Rapid assessment of emulsion stability can be obtained by the "bottle test", which consists in monitoring the extent of phase separation following the partial breakdown of the emulsion. For more precise characterization, the evolution of droplet size distribution with time is considered. However, measurement of the droplet size very often requires dilution of the emulsion and the interpretation of the results must take this factor into account. More recently, ripening composition due to composition gradients within the emulsions has also been considered. In this context, it is very difficult to establish a correlation between interfacial properties and emulsion stability and, from the data found in the literature, no general rule can be applied. There is a deficiency in this field, even at the experimental level, and an effort must be made in order to quantify emulsion stability properly and to make correlations with well-defined interfacial properties of the liquid/liquid interface. Also, a better link must be made between single-component model systems and multi-component real systems. Micro- or nanoparticles are often present in real systems and their role in emulsion stability has attracted a lot of attention in the past few years [143]. Future studies on complex emulsions will have to include systematically particles, combined or not with surfactants, to understand better the specific role of each component. To progress in this area, closer contact between the scientific communities working on emulsions and foams should be very fruitful. Numerous studies have been performed to characterize the interfacial properties of foam films and there are some similarities that should be considered when dealing with emulsion films made of droplets close to coalescence.

References

1 A. Passerone, L. Liggieri, R. Miller, D. Clausse, A. Steinchen, G. Loglio, A. Di Lullo, The FASES project for the investigation of emulsion stability in microgravity, in *Third World Congress on Emulsions, Lyon, September 24–27, 2002*, abstracts 1-B/233, **2002**, 115.

2 L. Liggeri, A. Passerone, F. Ravera, M. Ferrari, R. Miller, G. Loglio, D. Clausse, A. Steinchen, J.D. Sylvain, A. Di Lullo, Adsorption dynamics, interfacial elasticity and emulsion stability: the scientific basis of the FASES MAP project, in *1st International Symposium on Microgravity Research and Application in Physical Science and Biotechnology, Sorrento, September 10–15, 2000*, ESA SP-454, **2001**, 133.

3 P. Sherman (Ed.), *Emulsion Science*, Academic Press, New York, **1968**.

4 P. Becher (Ed.), *Encyclopedia of Emulsion Technology*, Marcel Dekker, New York, **1983**.

5 J. Sjöblom (Ed.), *Emulsion – A Fundamental and Practical Approach*, NATO ASI Series, Kluwer, Dordrecht, **1992**.

6 J.L. Grossiord, M. Seiller, *Multiple Emulsions: Structure, Properties and Applications*, Editions de Santé, Paris, **1998**.

7 J.R. Avendano-Gomez, J.L. Grossiord, D. Clausse, *J. Colloid Interface Sci.* **2005**, *290*, 533.
8 J. Avendano-Gomez, J.L. Grossiord, D. Clausse, *Entropie* **2000**, *224/225*, 110.
9 L. Taisne, *PhD Thesis*, Université Paris 6, **1997**.
10 D.I. Fletcher, B.P. Binks, J.H. Clint, S. Rippon, *Langmuir* **1999**, *15*, 4495.
11 J.L. Salager, R.E. Anton, M.A. Briceno, L. Choplin, L. Marquez, A. Pizzino, M.P. Rodriguez, *Polym. Int.* **2003**, *52*, 471.
12 J.L. Salager, in F. Nielloud, G. Marti-Mestres (Eds.), *Pharmaceutical Emulsions and Suspensions*, Drugs and Pharmaceutical Science Series, Vol. 105, Marcel Dekker, New York, **2000**, 73–126.
13 P. Becher, *Emulsions: Theory and Practise*, American Chemical Society monograph series No. 162, Am. Chem. Soc., Washington, DC, **1957**.
14 F.T. Tadros, B. Vincent, in P. Blecher (Ed.), *Encyclopedia of Emulsion Technology*, Marcel Dekker, New York, **1983**.
15 P.I. Fletcher, in D. Möbius, R. Miller (Eds.), *Drops and Bubbles in Interfacial Research*, Elsevier, Amsterdam, **1998**, *8*, 129–286.
16 J. Bibette, F. Leal Calderon, P. Poulin, *Prog. Phys.* **1999**, *62*, 563–592.
17 J. Sjöblom (Ed.), *Emulsions and Emulsion Stability*, Surfactant Science Series, New York, **1996**.
18 T. Tadros, *Adv. Colloid Interface Sci.* **2004**, *108/109*, 227.
19 D. Clausse, B. Fouconnier, J. Avendano-Gomez, *J. Dispers. Sci. Technol.* **2002**, *23*, 379.
20 D. Clausse, I. Pezron, L. Komunjer, *Colloids Surf. A* **1999**, *152*, 23.
21 A.L. Smith, D.P. Mitchell, in A.L. Smith (Ed.), *Theory and Practice of Emulsion Technology*, Academic Press, London, **1976**, 61–74.
22 M. Britten, H.J. Giroux, *J. Food Sci.* **1991**, *56*, 792
23 J. McClements, *Food Emulsions: Principles, Practices and Techniques*, 2nd ed, CRC Press, Boca Raton, FL, **2005**.
24 I.B. Ivanov, P.A. Kralchevsky, *Colloids Surf. A* **1997**, *128*, 155.
25 M.M. Robins, *Curr. Opin. Colloid Interface Sci.* **2000**, *5*, 265.
26 K.J. Lissant, *Demulsification – Industrial Applications*, Marcel Dekker, New York, **1983**.
27 P.C. Schorling, D.G. Kessel, I. Rahimian, *Colloids Surf. A* **1999**, *152*, 95.
28 J.P. Byers, in J.P. Byers (Ed.), *Metalworking Fluids*, Marcel Dekker, New York, **1994**, 147–174.
29 R.J. Mikula, in L.L. Schramm (Ed.), *Emulsions – Fundamental and Applications in the Petroleum Industry*, Vol. 231, American Chemical Society, Washington, DC, **1992**, 79–129.
30 S. Kerstens, B.S. Murray, E. Dickinson, *Food Hydrocolloids* **2005**, *19*, 625.
31 T. Wilson, *Confocal Microscopy*, Academic Press, London, **1990**.
32 P.W. de Bont, G.M.P. van Kempen, R. Vreeker, *Food Hydrocolloids* **2002**, *16*, 127.
33 M. Langton, E. Jordansson, A. Altskar, C. Sorensen, A. Hermansson, *Food Hydrocolloids* **1999**, *13*, 113.
34 D. Rousseau, S.M. Hodge, *Colloids Surf. A* **2005**, *260*, 229.
35 A.R. Kirby, A.P. Gunning, V.J. Morris, *Trends Food Sci. Technol.* **1995**, *6*, 359.
36 D.G. Dalgleish, in J. Sjöblom (Ed.), *Emulsions and Emulsion Stability*, Vol. 61, Surfactant Science Series, New York, **1996**, 287–326.
37 V. Kontogiorgos, C.G. Biliaderis, V. Kiosseoglou, G. Doxastakis, *Food Hydrocolloids* **2004**, *18*, 987.
38 D. Beydoun, G. Guang, R.P. Chhabra, J.A. Raper, *Powder Technol.* **1998**, *97*, 72.
39 M.G. Song, S.H. Jho, J.Y. Kim, J.D. Kim, *J. Colloid Interface Sci.* **2000**, *230*, 213.
40 J. Deluhery, N. Rajagopalan, *Colloids Surf. A* **2005**, *256*, 145.
41 A.M. El-Hamouz, A.C. Steward, in *Proceedings of SPE Annual Technical Conference and Exhibition, Denver, CO*, **1996**, 785–796.
42 D.J. McClements, *Curr. Opin. Colloid Interface Sci.* **2002**, *7*, 451.
43 R.J. Hunter, *Intoduction to Modern Colloid Science*, Oxford University Press, Oxford, **1993**.
44 R.J. Hunter, *Foundations of Colloid Science*, Vol. 1, Oxford University Press, Oxford, **1986**.

45 A. S. Dukhin, P. J. Goetz, *Ultrasound for Characterizing Colloids, Particle Sizing, Zeta Potential, Rheology*, Elsevier, New York, **2002**.
46 D. J. McClements, *Adv. Colloid Interface Sci.* **1991**, *37*, 33.
47 R. J. Hunter, *Colloids Surf.* **1998**, *141*, 37.
48 M. Clausse, in P. Blecher (Ed.), *Encyclopedia of Emulsion Technology*, Marcel Dekker, New York, **1983**, 77–157.
49 H. Fordedal, Y. Schildberg, J. Sjöblom, J. L. Volle, *Colloids Surf.* **1996**, *106*, 33.
50 O. Söderman, B. Balinov, in J. Sjöblom (Ed.), *Emulsions and Emulsion Stability*, Vol. 61, Surfactant Science Series, New York, **1996**.
51 J. P. Hindmarsh, K. G. Hollingsworth, D. I. Wilson, M. L. Johns, *J. Colloid Interface Sci.* **2004**, *275*, 165.
52 D. Clausse, F. Gomez, C. Dalmazzone, C. Noik, *J. Colloid Interface Sci.* **2005**, *287*, 694.
53 S. Raynal, I. Pezron, L. Potier, D. Clausse, J. L. Grossiord, M. Seiller, *Colloids Surf. A* **1994**, *91*, 191.
54 C. Thomas, J. P. Perl, D. T. Wasan, *J. Colloid Interface Sci.* **1990**, *139*, 1.
55 F. Marken, R. D. Webster, S. D. Bull, S. G. Davies, *J. Electroanal. Chem.* **1997**, *437*, 209.
56 D. Langevin, *Adv. Colloid Interface Sci.* **2000**, *88*, 209.
57 M. A. Bos, T. van Vliet, *Adv. Colloid Interface Sci.* **2001**, *91*, 437.
58 D. Langevin, S. Poteau, I. Hénaut, J. F. Argillier, *Oil Gas Sci. Technol. – Rev. IFP* **2004**, *59*, 511.
59 P. Walstra, *Physical Chemistry of Food*, Marcel Dekker, New York, **2003**.
60 D. Möbius, R. Miller (Eds.), *Novel Methods to Study Interfacial Layers*, Elsevier Science, Amsterdam, **2001**.
61 J. Lucassen, M. van der Tempel, *Chem. Eng. Sci.* **1972**, *27*, 1283.
62 J. Lucassen, D. Giles, *J. Chem. Soc., Faraday Trans. 1* **1975**, *71*, 217.
63 P. R. Garrett, P. Joos, *J. Chem. Soc., Faraday Trans 1* **1976**, *72*, 2161.
64 V.B. Fainerman, E. H. Lucassen-Reynders, R. Miller *Adv. Colloid Interface Sci.* **2003**, *106*, 237.
65 D. Exerowa, P. Kruglyakov, *Foam and Foam Films, Theory, Experiment, Application*, Studies in Interface Science, Vol. 5, Elsevier, Amsterdam, **1998**.
66 H. Fruhner, K.-D. Wantke, K. Lunkenheimer, *Colloids Surf. A* **1999**, *162*, 193.
67 E. H. Lucassen-Reynders, V. B. Fainerman, R. Miller, *J. Phys. Chem. B* **2004**, *108*, 9173.
68 F. Bashforth, J. C. Adams, *An Attempt to Test the Theories of Capillary Action by Comparing the Theoretical and Measured Forms of Drops of Fluid*, Cambridge University Press, Cambridge, **1883**.
69 J. Benjamins, E. H. Lucassen-Reynders, in D. Möbius, R. Miller (Eds.), *Proteins at Liquid Interfaces*, Elsevier Science, Amsterdam, **1998**, 341–384.
70 E. H. Lucassen-Reynders, A. Cagna, J. Lucassen, *Colloids Surf. A* **2001**, *186*, 63.
71 P. Saulnier, F. Boury, A. Malzert, B. Heurtault, Tz. Ivanova, A. Cagna, I. Panaïotov, J. E. Proust, *Langmuir* **2001**, *17*, 8104.
72 X. Zhai, G. Brezesinski, H. Mohwald, J. Li, *J. Phys. Chem. B* **2004**, *108*, 13475.
73 C. Gourier, M. Alba, A. Braslau, J. Daillant, M. Goldmann, C.M. Knobler, F. Rieutord, G. Zalczer, *Langmuir* **2001**, *17*, 6496.
74 T. Nylander, A. Hamraoui, M. Paulsson, *Int. J. Food Sci. Technol.* **1999**, *34*, 573.
75 E. Dickinson, *An Introduction to Food Colloids*, Oxford University Press, Oxford, **1992**.
76 S. E. Friberg, K. Larsson, J. Sjoblöm (Eds.), *Food Emulsions*, 4th edn., Marcel Dekker, New York, **2004**.
77 E. Dickinson, J. M. Rodríguez Patino (Eds.), *Food Emulsions and Foams: Interfaces, Interactions and Stability*, Royal Society of Chemistry, Cambridge, **1999**.
78 M. M. Robins, A. D. Watson, P. J. Wilde, *Curr. Opin. Colloid Interface Sci.* **2002**, *7*, 419.
79 E. Dickinson, *Food Hydrocolloids* **2003**, *17*, 25.
80 D. G. Dalgleish, *Trends Food Sci. Technol.* **1997**, *8*, 1.
81 G. A. van Aken, T. B. J. Blijdenstein, N. E. Hotrum, *Curr. Opin. Colloid Interface Sci.* **2003**, *8*, 371.
82 P. Wilde, A. Mackie, F. Husband, P. Gunning, V. Morris, *Adv. Colloid Interface Sci.* **2004**, *108/109*, 63.

83 W. E. Ewers, K. L. Sutherland, *Aust. J. Sci. Res. Ser. A5* **1952**, 697.

84 D. E. Graham, M. C. Phillips, *J. Colloid Interface Sci.* **1979**, *79*, 403.

85 D. K. Sarker, M. Axelos, Y. Popineau, *Colloids Surf. B* **1999**, *12*, 147.

86 M. Cornec, A. R. Mackie, P. J. Wilde, D. C. Clark, *Colloids Surf. A* **1996**, *114*, 237.

87 C. Wollenweber, A. V. Makievski, R. Miller, R. Daniels, *Colloids Surf. A* **2000**, *172*, 91.

88 J. C. Arboleya, P. J. Wilde, *Food Hydrocolloids* **2005**, *19*, 485.

89 C. Rouzes, A. Durand, M. Leonard, E. Dellacherie, *J. Colloid Interf. Sci.* **2002**, *253*, 217.

90 A. G. Gaonkar, *Food Hydrocolloids* **1991**, *5*, 329.

91 J. Leroux, V. Langendorff, G. Schick, V. Vaishnav, J. Mazoyer, *Food Hydrocolloids* **2003**, *17*, 455.

92 D. A. Kim, M. Cornec, G. Narsimhan, *J. Colloid Interf. Sci.* **2005**, *285*, 100.

93 E. Dickinson, B. S. Murray, G. Stainsby, *J. Chem. Soc., Faraday Trans.* **1988**, *84*, 871.

94 A. Denda, R. Hayashi, in C. Balney, R. Hayashi, K. Heremans, P. Masson (Eds.), *High Pressure and Biotechnology*, John Libbey Eurotext, Montrouge, France, **1992**, 333–335.

95 E. Molina, A. Papadopoulo, D. A. Ledward, *Food Hydrocolloids* **2001**, *15*, 263.

96 N. Chapleau, M. de Lamballerie-Anton, *Food Hydrocolloids* **2003**, *17*, 273.

97 E. Dumay, M. Kalichevsky, J. C. Chetfel, *J. Agric. Food Chem.* **1997**, *45*, 713.

98 P. A. Pugnaloni, E. Dickinson, R. Ettelaie, A. R. Mackie, P. J. Wilde, *Adv. Colloid Interface Sci.* **2004**, *107*, 27.

99 E. Dickinson, in A. M. Stephen (Ed.), *Food Polysaccharides and Their Applications*, Marcel Dekker, New York, **1995**, p. 501.

100 S. Laplante, S. L. Turgeon, P. Paquin, *Carbohydr. Polym.* **2005**, *59*, 425.

101 Y. S. Gu, E. A. Decker, D. J. McClements, *Food Hydrocolloids* **2005**, *19*, 83.

102 R. Sharma, M. Zakora, K. B. Qvist, *Food Chem.* **2002**, *79*, 493.

103 J. Holstborg, B. V. Pedersen, N. Krog, S. K. Olesen, *Colloids Surf. B* **1999**, *12*, 383.

104 U. Klinkesorn, P. Sophanodora, P. Chnachhoti, E.A. Deker, D. J. McClements, *Food Hydrocolloids* **2005**, *19*, 1044.

105 E. Papirer, C. Bourgeois, B. Siffert, H. Balard, *Fuel* **1982**, *61*, 732.

106 G. P. Canevari, *Mar. Pollut. Bull.* **1982**, *13*, 49.

107 A. L. Bridié, T. H. Wanders, W. Zegveld, H. B. van der Heide, *Mar. Pollut. Bull.* **1980**, *11*, 343.

108 A. J. McMahon, in J. Sjöblom (Ed.), *Emulsions – A Fundamental and Practical Approach*, Kluwer, Dordrecht, **1992**, p. 135.

109 C. W. Angle, in J. Sjöblom (Ed.), *Encyclopaedic Handbook of Emulsion Technology*, Marcel Dekker, New York, **2001**, p. 541.

110 J. Sjöblom, E. E. Johnsen, A. Westvik, M. H. Ese, J. Djuve, I. H. Auflem, H. Kallevik, in J. Sjöblom (Ed.), *Encyclopaedic Handbook of Emulsion Technology*, Marcel Dekker, New York, **2001**, 595–619.

111 Y. H. Kim, A. D. Nikolov, D. T. Wasan, D. T. Diaz-Arauzo, C. S. Shetty, *J. Colloid Interface Sci.* **1996**, *17* (1), 33.

112 C. Dalmazzone, H. Seris, *Rev. Inst. Fr. Pét.* **1998**, *53*, 463.

113 J. Murgich, J. Rodríguez M., Y. Aray, *Energy Fuels* **1996**, *10*, 68.

114 T. F. Yen, *Energy Sources* **1974**, *1*, 447.

115 J. P. H. Pfeiffer, R. N. Saal, *J. Phys. Chem.* **1940**, *44*, 139.

116 J. D. McLean, P. K. Kilpatrick, *J. Colloid Interface Sci.* **1997**, *189*, 242.

117 P. M. Spiecker, K. L. Gawrys, C. B. Trail, P. K. Kilpatrick, *Colloids Surf. A* **2003**, *220*, 9.

118 O. V. Gafanova, W. Yarranton, *J. Colloid Interface Sci.* **2001**, *241*, 469.

119 D. D. Eley, M. J. Hey, J. D. Symonds, *Colloids Surf.* **1988**, *32*, 87.

120 P. Bouriat, N. E. Kerri, A. Graciaa, J. Lachaise, *Langmuir* **2004**, *20*, 7459.

121 E. J. Johansen, I. M. Skjärvö, T. Lund, J. Sjöblom, H. Söderlund, G. Boström, *Colloids Surf.* **1988–89**, *34*, 353.

122 T.J. Jones, E.L. Neustadter, K.P. Wittingham, *J. Can. Pet. Technol.* **1978**, 100.

123 Y.H. Kim, D.T. Wasan, P.J. Breen, *Colloids Surf. A* **1995**, 235.

124 A. Bhardwaj, S. Hartland, *J. Dispers. Sci. Technol.* **1998**, *19*, 465.

125 M. Desmaison, C. Piekarski, S. Piekarski, J.P. Desmarquet, *Rev. Inst. Fr. Pét.* **1984**, *5*, 603.

126 J. Sjöblom, L. Mingyuan, A.A. Christy, T. Gu, *Colloids Surf.* **1992**, *66*, 55.

127 S. Acevedo, G. Escobar, L.B. Gutiérrez, H. Rivas, X. Gutiérrez, *Colloids Surf. A* **1993**, *71*, 65.

128 F. Bauget, D. Langevin, R. Lenormand, *J. Colloid Interface Sci.* **2001**, *239*, 501.

129 S. Poteau, J.F. Argillier, D. Langevin, F. Pincet, E. Perez, *Energy Fuels* **2005**, *19*, 1337.

130 M.H. Ese, J. Sjöblom, J. Djuve, R. Pugh, *Colloid Polym. Sci.* **2000**, *278*, 532.

131 A. Bhardwaj, S. Hartland, *J. Dispers. Sci. Technol.* **1993**, *14*, 541.

132 J. Djuve, X. Yang, I.J. Fjellanger, J. Sjöblom, E. Pellizetti, *Colloid Polym. Sci.* **2001**, *279*, 232.

133 M.A. Krawczyk, D.T. Wasan, C.S. Shetty, *Ind. Eng. Chem. Res.* **1991**, *30*, 367.

134 L.Y. Zhang, Z. Xu, J.H. Masliyah, *Langmuir* **2003**, *19*, 9730.

135 H.H. Ese, L. Galet, D. Clausse, J. Sjöblom, *J. Colloid Interface Sci.* **1999**, *220*, 293.

136 B.P. Singh, *Energy Sources* **1994**, *16*, 377.

137 T. Sun, L. Zhang, Y. Wang, S. Zhao, B. Peng, M. Li, J. Yu, *J. Colloid Interface Sci.* **2002**, *255*, 241.

138 N.S. Ahmed, A.M. Nassar, *J. Polym. Res.* **2001**, *8*, 191.

139 J. Wu, Y. Xu, T. Dabros, H. Hamza, *Colloids Surf. A* **2004**, *252*, 79.

140 A.M. Al-Sabagh, A.M. Badawi, M.R. Noor El-Den, *Pet. Sci. Technol.* **2002**, *9/10*, 887.

141 D. Daniel-David, I. Pezron, D. Clausse, C. Dalmazzone, C. Noïk, L. Komunjer, *Phys. Chem. Chem. Phys.* **2004**, *6*, 1570.

142 D. Daniel-David, I. Pezron, C. Dalmazzone, C. Noïk, D. Clausse, L. Komunjer, *Colloids Surf. A* **2005**, *270/271*, 262.

143 B. Binks, *Curr. Opin. Colloid Interface Sci.* **2002**, *7*, 21–41.

6
Wetting Film Dynamics and Stability

Boryan Radoev, Klaus Werner Stöckelhuber, Roumen Tsekov, and Philippe Letocart

Abstract

Although wetting films in many aspects are similar to other (e.g. soap and emulsion) thin liquid films, there are also some differences in their behavior. First is the stability of wetting films and the factors on which they depend. In contrast to soap and emulsion films, the surfaces of which are homogeneous, solid substrates of wetting films are heterogeneous as a rule, unless special measures for their homogenization are taken. Here we mean primarily a heterogeneous distribution of the surface energy, i.e. the existence of hydrophobic domains in the hydrophilic surface phase or vice versa. As is known, such hydrophobic domains could play the role of gas-phase nucleation centers and nowadays it is widely accepted that nano-bubbles can be formed therein. This chapter considers the effect of nano-bubbles adhering to a solid surface on the stability of wetting films. It is shown that the existence of nano-bubbles is crucial for the lifetime of wetting films. Another peculiarity typical of hydrophobic solid surface, the so-called slippage effect, is also investigated and its contribution to the dispersion equation of capillary waves in wetting films is taken into account.

6.1
Introduction

Thin films, building blocks of dispersed systems, manifest themselves mainly in three variants: soap, emulsion and wetting films. There are many equilibrium and dynamic characteristics typical of all three kinds of films, e.g. the surface (DLVO) forces acting therein, the type of hydrodynamics in the films (the so-called lubrication flow), the equilibrium and stability conditions, film rupture, etc. However, in addition to their common nature, each kind of film possesses specific features, which should be taken into account when interpreting the corresponding experimental data and their influence on the behavior of dis-

Colloids and Interface Science Series, Vol. 3
Colloid Stability and Application in Pharmacy. Edited by Tharwat F. Tadros

ISBN: 978-3-527-31463-8

persed systems in general. The aim of this chapter is to review some of the peculiarities related to the rheology, stability and rupture of wetting films. As is shown, all observed phenomena are due to effects characteristic of solid–liquid surfaces. Section 6.2 is devoted to the hydrodynamics of wetting films with special attention paid to a new experimental method (see Section 6.2.2) for exciting and measuring surface waves on wetting films. Among the most interesting results obtained by this method is the anomalous high slippage observed at solid surfaces. The presence of nano-bubbles adhering to a solid–liquid surface is suggested as a possible origin of this effect. The idea of the existence of nano-bubbles in wetting films appears again in Section 6.3.3, this time as a cause of film rupture. The detailed analysis shows that the region above a nano-bubble resembles more a foam film, which is unstable as a rule, thus explaining why wetting films with repulsive interactions could also rupture.

6.2 Dynamics of Wetting Films

6.2.1 Drainage of Liquid Films and Dimple Relaxation

During the past century, the hydrodynamic behavior of thin liquid films has been extensively studied in relation to the stability of disperse systems. The film drainage is usually described by the Navier–Stokes equations, applied to a liquid film of thickness much smaller than its radius. An interesting effect for thinning films is the coupling of the film drainage and the unstable hydrodynamic modes [1–3], the theoretical description of which requires correct calculation of the film drainage rate. The latter depends substantially on the shape of the film interfaces; usually a dimple [4–6] or wimple [7, 8] form with maximum or minimum of the film thickness in the film center, respectively. Classically, the rate of thinning is modeled by the classical Reynolds expression:

$$V_{\text{Re}} = \frac{2h^3 \Delta p}{3\eta R^2} \tag{1}$$

first applied to thin liquid films by Scheludko [9]. This equation is valid for films with tangentially immobile and plane-parallel surfaces, where h is the average film thickness, R is the film radius, η is the liquid viscosity and $\Delta p = p_\sigma - \Pi$ is the driving pressure (p_σ is the capillary pressure in the meniscus and Π is the disjoining pressure).

Deviations from Eq. (1) due to tangential surface mobility have been reported [10–14] and the theory [15–18] predicts a thinning rate according to the Reynolds equation with a reduced effective viscosity μ, $V_{\text{Re}} = 2h^3\Delta p/3\mu R^2$ [compare with Eq. (1)]. In the case of a wetting film μ can be expressed as [10]

$$\mu = \eta \frac{Ma + Na + Ap + MaNa + ApNa/2}{12 + 4Ma + 4Na + MaNa} \tag{2}$$

where $Ma = haE_G/\eta(D_s a + Dh)$, $Na = h\beta/\eta$ and $Ap = -ah^2\Gamma(\partial_\Gamma \Pi)/\eta(aD_s + Dh)$ are the so-called *Marangoni*, *Navier* and *adsorption-pressure* numbers.

In *Ma* (see also *Ap*) $a = \partial_c \Gamma$ is the adsorption length (Γ is the adsorption and c is the surfactant concentration), $E_G = -\Gamma\partial_\Gamma\sigma$ is the surface Gibbs elasticity (σ is the surface tension on the film/air interface) and D_s and D are the surface and bulk diffusion coefficients of the surfactant. The Marangoni number accounts for the effect of surfactants on the surface mobility, i.e. on the tangential velocity of fluid film/gas interface [15]. As can be seen, *Ma* depends on two factors, potential interactions (Gibbs elasticity) and dissipative characteristics (viscosity, diffusion). At a pure (single-component) liquid/gas interface there is no adsorption, $Ma = 0$, the tangential surface velocity is maximum and we have the so-called "free" surface. In this case, Eq. (2) takes the form ($Ap = 0$ for pure liquid, see below):

$$\mu = \eta \frac{Na}{12 + 4Na} \tag{2 a}$$

The other limit, $Ma >> 1$, corresponds to a practically immobilized surface (zero tangential surface velocity). It is realized at a sufficient amount of surfactant (high enough Gibbs elasticity) and Eq. (2) takes the form

$$\mu = \eta \frac{1 + Na}{4 + Na} \tag{2 b}$$

The Navier number, *Na*, accounts for the slippage on the film/solid interface. Discussions about the slippage phenomenon began [20] at the same time as the formulation of the Navier–Stokes equations. They are strongly related to the rheological models of ideal (not viscous) and real (viscous) fluids. According to Stokes, the adherence of fluids at interfaces (non-slip condition) is a straight consequence of their viscous nature. Ideal fluids slip along the interface because no adhesion arises there. Feynman, in his famous lecture in physics, called ideal fluids "dry", in order to emphasize their non-adhering nature. The current revival of the slip problem is related to advanced methods of experimental studies [21, 22] and to computer modeling, which make it possible to describe the motion of a liquid along a solid from first principles. In *Na*, β is the slip coefficient on the film/solid interface, but some authors use slip-length $b \equiv \beta/\eta$ ($Na = h/b$) to characterize the adhesion [23].

Usually for hydrophilic surfaces, the non-slip boundary condition holds, which corresponds to infinite slip coefficient (zero slip-length), i.e. infinite Navier number ($Na \to \infty$). Finite slippage (i.e. $Na < \infty$) is detected experimentally for hydrophobic surfaces [23]. The limit $Na = 0$ describes the boundary condition of an ideal fluid at a solid surface (zero friction, unlimited slip-length,

$b \to \infty$ and from a rheological point of view is equivalent to $Ma = 0$ at a liquid/gas surface [compare Eq. (2a) at $Na = \to \infty$ and Eq. (2b) at $Na = 0$]. These two cases ($Ma = 0$, $Na >> 1$ and $Ma >> 1$, $Na = 0$) correspond to a film with an immobilized and a "free" interface, with $\mu = \eta/4$. The trivial case of two immobilized surfaces is equivalent to $Na >> 1$, $Ma >> 1$, with $\mu = \eta$.

The slippage effect is especially important for the hydrodynamics of extremely narrow regions, thin films, three-phase contact zones [23], etc., since $Na \sim h$, i.e. the thinner the film, the greater the slippage.

Finally, the adsorption–pressure number *Ap* takes into account the dependence of the disjoining pressure on the adsorption on the film surface. From this point of view, it is not directly correlated with the dissipative nature of the film surfaces, but rather with the driving forces in thin films.

Due to the surface convection (non-zero surface velocity, $u \neq 0$, Fig. 6.1), a gradient of the adsorption ($\nabla \Gamma$) at the liquid–gas surface appears. If the disjoining pressure is a function of Γ, $\Pi(\Gamma)$, then it leads to a corresponding pressure gradient, $\nabla \Pi = (\partial_\Gamma \Pi) \nabla \Gamma$.

This effect is pronounced in the case of ionic surfactants since their adsorption gradient is equivalent to the surface charge density gradient, $\nabla q_s = F \sum_k z_k \nabla \Gamma_k$ (F is the Faraday constant, z_k is the valency of the adsorbed ion), and to electrostatic disjoining pressure gradient, $\nabla \Pi_{el} = (\partial_{q_s} \Pi) \nabla q_s$. In contrast to *Ma* and *Na*, *Ap* could be either positive or negative.

Some experiments [24] have detected rates of thinning depending much more weakly on the film radius as compared with the Reynolds equation. Such a behavior cannot be explained by the surface mobility, the effect of which is not expected to change the functional dependence of the thinning rate on the film radius. Equation (1) requires strictly symmetrical drainage between two parallel flat surfaces, which is supplied in the Reynolds case by the rigidity of the solid interfaces. In contrast, the shape of wetting films is determined via the balance of the viscous, capillary and surface forces. Experimental investigations above have shown that large films are non-homogeneous in thickness and the thinning rate of dimpled films is always larger than the prediction of the Reynolds law.

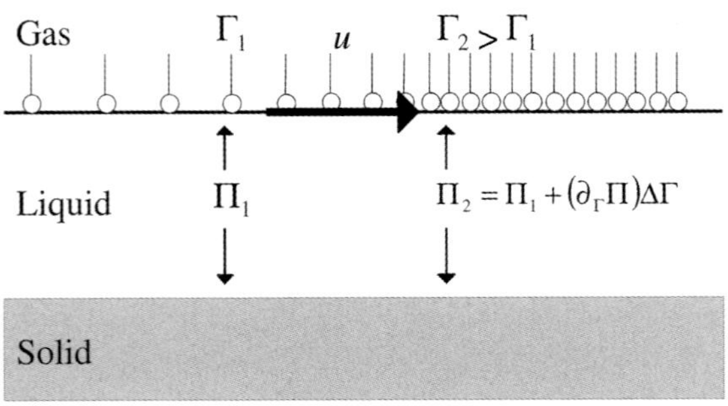

Fig. 6.1 Distribution of disjoining pressure Π due to adsorption distribution Γ; u, surface velocity; $\Delta\Gamma = \Gamma_2 - \Gamma_1$.

Recently, a classification of the film drainage has been proposed [25] by introducing a dynamic fractal dimension, α, which takes into account the effective degree of freedom for relaxation of the film interfaces. In this way, a generalized drainage law is derived:

$$V = \frac{2h^3 \Delta p}{3\mu R^2} \left(\frac{R^2 \Delta p}{16\sigma h} \right)^{\frac{2-\alpha}{2+\alpha}} \tag{3}$$

Equation (3) discriminates between different types of drainage by the corresponding value of α. For instance, if the film possesses two solid interfaces the latter induce strong correlation in both the two directions and $\alpha = 2$. Hence, according to Eq. (3), the film drains with the Reynolds velocity. If the drainage is strictly axisymmetric but not confined between solid interfaces, the radial direction is free for relaxation and a dimple occurs. In this case $\alpha = 1$ and Eq. (3) reduces to

$$V = \frac{1}{3\mu} \sqrt[3]{\frac{h^8 \Delta p^4}{2\sigma R^4}} \tag{4}$$

One can imagine a completely random film broken down to many uncorrelated sub-domains. In this case $\alpha = 0$ and Eq. (3) acquires the form

$$V = \frac{h^2 \Delta p^2}{24\mu\sigma} \tag{5}$$

Note that the rate of drainage of such stochastically corrugated films does not depend on the film radius since any spatial correlation of the dynamics is missing. Finally, one can imagine that a film is broken down to sub-domains, which are not completely uncorrelated. In this case $\alpha = 1/2$, if the pressure fluctuations in the film obey the thermodynamic law [26], i.e. their amplitude is inversely proportional to the square root of the number of sub-domains in the film. Hence Eq. (3) changes to the following rate of drainage:

$$V = \frac{1}{6\mu} \sqrt[5]{\frac{h^{12} \Delta p^8}{4\sigma^3 R^4}} \tag{6}$$

This equation predicts that the rate of drainage is inversely proportional to the film radius to the power 4/5. Such a behavior was observed experimentally [24, 26]. The non-linear dependence of the thinning rate V on the driving pressure Δp from Eq. (6) has also been experimentally proven [25].

In fact, the thinning of wetting films is a more complicated process passing through several stages. Since initially the drainage at the film border is much faster than that in the film center [4–6, 27, 28], the thickness at the barrier rim quickly reaches a value close to the equilibrium thickness h_e. This dimpled structure is a non-equilibrium one and the next stage in the film thinning is the dimple relaxation. Under the action of capillary and disjoining forces, the dimple shrinks until the film reaches a flat shape at equilibrium. Sometimes, however, under non-equilibrium conditions the dimples can grow driven by external flows [29].

In the frame of the lubrication approximation, the evolution of the local thickness H of an axisymmetric wetting film is governed by the following equation [19, 30]:

$$\partial_t H = -\frac{1}{r}\partial_r\left\{\frac{H^3}{12\mu(H)} r\partial_r\left[\frac{\sigma(H)}{r}\partial_r\left(r\frac{\partial_r H}{\sqrt{1+(\partial_r H)^2}}\right)+\Pi(H)\right]\right\} \equiv \hat{L}\{H\} \quad (7)$$

where t is time, r is radial coordinate and $\hat{L}$ is the evolution operator. Due to the complexity of this equation, it could be solved only numerically [29]. Hereafter we present an approximate procedure for the description of the dimple relaxation which elucidates the physical aspects and is more transparent. A rigorous treatment of the dimple relaxation via Eq. (7) requires relevant boundary and initial conditions. The exact boundary conditions can be written only far away in the meniscus, where the lubrication theory, and Eq. (7), are no longer applicable. Moreover, Eq. (7) is valid for the latest stage of the dimple relaxation and the relevant initial profile can only be experimentally specified. Hence it is impossible to close the mathematical problem and further treatment requires empirical modeling. For this reason, an approximate method was developed [30, 31] for calculating the evolution under the assumption that the film shape is known.

The film profile can be generally represented as an r^2-power expansion. Since $r \leq R$, one is able to approximate the dimple in the film region by a finite series, e.g. a biquadratic form. It can be additionally specified by application of the two known conditions at the barrier rim: equilibrium thickness $H(R) = h_e$ and minimum of the film profile $(\partial_r H)_R = 0$. Thus, the biquadratic polynomial acquires the form

$$H = h_e + [h_0(t) - h_e][1 - (r/R)^2]^2 \quad (8)$$

The thickness h_0 in the film center is the only unknown function of time in Eq. (8). Suppose that at time t the film profile is given by Eq. (8). The profile at time $t + \tau$ can be calculated from Eq. (7), which in the limit of small τ can be rewritten as

$$\tilde{H}(r, t+\tau) = H(r, t) + \tau\hat{L}H(r, t) \quad (9)$$

Hence the consequent profile $\tilde{H}$ can be generated, which satisfies Eq. (7) but does not obey the necessary boundary conditions. To apply the boundary conditions, we postulate that the real profile $H(r, t+\tau)$ is the best fit of the profile $\tilde{H}(r, t+\tau)$. Hence, to calculate the film profile, a minimization of the square of the deviation of the two functions all over the film is required. This criterion combined with Eqs. (8) and (9) leads to the following recurrent relation:

$$h_0(t+\tau) = h_0(t) + 10\tau \int_0^1 (1-x^2)^2 \hat{L}[h_0(t)(1-x^2)^2 + h_e x^2(2-x^2)]x\mathrm{d}x \qquad (10)$$

where $x = r/R$. It is only a matter of integration to calculate the evolution of the thickness in the film center, which introduced in Eq. (8) will provide the whole film profile evolution.

We applied the method described above to wetting films from a 1 mM aqueous solution of KCl on a glass surface [32]. In order to check the theory, we compared it with experimental results of dimple relaxation and very good agreement was observed for the thinning in the film center, which was monitored by the classical light interference method. Assuming that the OH^- ions adsorb of the water/air interface [33], we estimated their adsorption length as $a = 0.14$ mm. This is a reasonable value since the concentration of OH^- ions is very low and it is well known that the slope of the adsorption isotherm is large when the concentration approaches zero. The value of a corresponds to adsorption $\Gamma = 14$ nmol m^{-2} and surface charge density $q_s = -1.35$ mC m^{-2} on the water/air surface. The distribution of the OH^- ions on the water/air surface was also simulated. Initially, the adsorption in the film center is about 30% less than the equilibrium value due to film drainage. With advancing time the thinning rate of the film decreases and the adsorption increases to reach its equilibrium value at the end of the process. Thus for film thicknesses larger than 80 nm, *Ap* is small and the reduced viscosity μ practically equals η. If the film thickness decreases below 80 nm, the adsorption–pressure number becomes negative, which decreases μ and close to the equilibrium thickness it is twice as small as η. It is important to note that the Marangoni number *Ma* of the OH^- surfactant in water is about 1000. Hence the water/air surface is tangentially immobile even in the case of pure water. This is an important novelty since in hydrodynamics the water/air surface is usually considered as free with zero stress.

Our next aim was to investigate further the effect of ionic surfactants on the film dynamics [34]. For this reason, wetting films were formed on a glass surface from a 1 mM aqueous solution of sodium dodecyl sulfate (SDS) [30], which is an anionic surfactant. SDS contributes to the dimple evolution either through the Marangoni effect or through the disjoining pressure; the adsorption of DS^- ions generates an electrostatic disjoining pressure that is much larger than the van der Waals component at the concentration considered [35]. The evolution of the thickness at the film center and the thickness at $r = 0.625R$ was monitored by the classical light interference method. By comparing the theory with experimental data, the surface diffusion coefficient of DS^- ions is determined as a single fitted parameter. Hence important information about the interfacial properties of the DS^- adsorption layer is obtained. It is shown that the interfacial electrostatics decrease dramatically the Marangoni number by suppressing the surface elasticity. This effect, enhanced by the effect of the adsorption dependence of the disjoining pressure, leads to an important increase in the thinning rate at small film thickness. For film thickness larger than 100 nm, *Ap* is small and negative and *Ma* is large and positive. However, if the film thickness drops below 100 nm, these numbers decrease strongly. Therefore, the coupling between the disjoining pressure and adsorption is important and leads to a substantial increase in the thinning rate at the end stage of the film drainage. This could be a plausible explanation of the discrepancy in the mobility determined by drainage experiments and by the dynamics of artificial waves [35] on an equilibrium film from an ionic surfactant solution.

6.2.2
Acoustically Excited Surface Waves

Capillary surface waves on liquids have been investigated for a long time, beginning with the first prediction by Smoluchowski in 1908 [36] that a liquid surface must scatter light, mainly near its critical temperature. The first quantitative theory was developed in 1913 by Mandelstam, who on the basis of the equipartition theorem obtained for the root mean square amplitude A_ζ of a surface wave on deep water the well-known relation [37]

$$A_\zeta \sim \sqrt{k_B T/\sigma} \tag{11}$$

where k_B is the Boltzmann constant, T the absolute temperature and σ the surface tension.

Systematic studies of capillary waves on thin-film surfaces began in the late 1960 and since then a large number of relevant publications have appeared (e.g. [38–43]). Artificial wave propagation in foams was studied specifically by Sun et al. [44], the bending mode in free films by Bergmann [45] and wetting films by Schulze and coworkers [19, 35].

The core method presented here, the so-called DSFB (Derjaguin–Scheludko force balance), for exciting and measuring surface waves on wetting films, is shown in Figs. 6.2 and 6.3 [35].

The equilibrium thickness of the film is adjusted by a piston and monitored by a baratron gauge. The dynamic pressure variations are generated by a loudspeaker, as seen in Fig. 6.2. In this way, a wave with a circular wave front is propagating from the film edge towards the film center (Fig. 6.3).

As is known, a dynamic fingerprint of a wave process $\zeta \sim \exp(i\omega t)$ is its dispersion relation $\omega(k)$, with $\omega = 2\pi\nu$ (ν=wave frequency) and $k = 2\pi/\lambda$ (λ=wavelength). For wetting films, in the case of long waves ($\lambda >> h$), the following dispersion relation is derived in [19]:

$$\omega = ih^3k^2(\sigma k^2 - \partial_h \Pi)/3\mu \tag{12}$$

Actually, Eq. (12) is an analogue of Eq. (1), and noting that ω is proportional to the velocity $\partial_t\zeta = i\omega\zeta$, it is clear that $(\sigma k^2 - \partial_h \Pi)$ plays the role of driving pressure [compare with Δp, Eq. (1)]. Essential in Eq. (12) is the fact that μ is the effective viscosity defined in Eq. (2), which provides the principle opportunity to

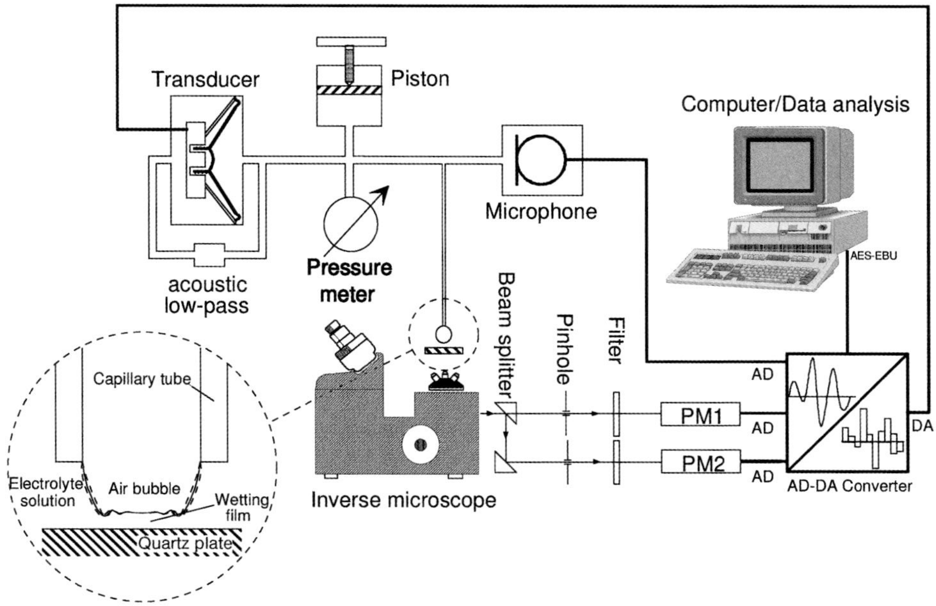

Fig. 6.2 Overview of the Derjaguin–Scheludko force balance (DSFB) apparatus for studying acoustically excited waves on wetting films. Frequency range of oscillation, 10–200 Hz. Inset: oscillating film with the adjacent meniscus (bubble surface).

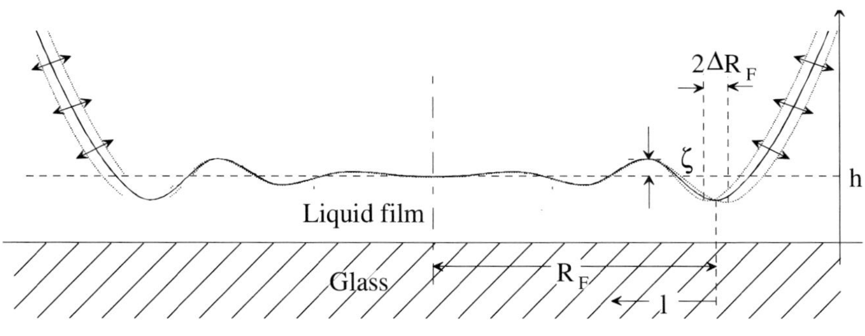

Fig. 6.3 Cross-section of oscillating bubble (meniscus) and wetting film. Note that the meniscus and the film are mechanically coupled.

obtain information about the thermodynamics and rheology of wetting films. As seen below, this possibility can be realized by measuring λ vs. ν via the DSFB method (Fig. 6.4).

The results for this model system are summarized in Fig. 6.4:

- Due to the low concentration of surfactant, both Marangoni and adsorption–disjoining pressure effects are negligible.

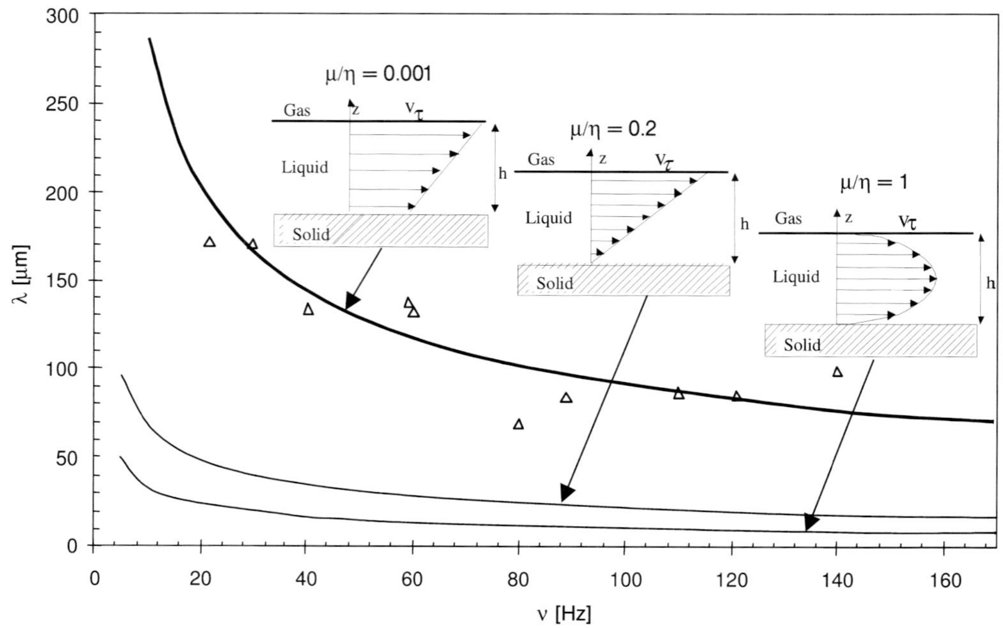

Fig. 6.4 Dispersion equation $\lambda(\nu)$, where λ = wavelength and ν = frequency. Δ = Experimental data for a glass–liquid–gas model system with $h_{eq} = 80$ nm, $\partial_h \Pi = -1.9 \times 10^{10}$ N m^{-3}, $\psi_{bubble} = -80$ mV and $\psi_{SiO_2} = -100$ mV. Solid lines: theoretical dependence ($\lambda \sim 1/\sqrt{\nu}$ at $\sigma k^2 \ll |\partial_h \Pi|$) for three different μ values (see Eq. 12). The insets illustrate the slip conditions for the corresponding μ values.

- The relation between the frequency and the wavelength of this model system corresponds qualitatively with the theory (Eq. 12) at $\sigma k^2 << |\partial_h \Pi|$.
- As seen, the traditional assumption of a solid–liquid surface as a fully blocked surface (i.e. $Na \rightarrow \infty$) cannot be maintained. On the contrary, a value of $Na = 0.2$ is required in order to fit the experimental results.

Experiments were carried out for wetting films of different aqueous solutions of potassium chloride (0.1–1 mM) on quartz, with a small amount (0.1 mM) of SDS as an air–electrolyte surface potential stabilizer. The results (Fig. 6.4) show that this method reveals interesting surface rheology effects [19, 35]. The experiment shows that there is a quantitative discrepancy with the traditional non-slip condition at solid surfaces, which can only be explained by a more generalized rheological model.

A possible origin of the established slippage could be the nano-bubbles adhering at the solid interface (Fig. 6.5). As discussed in Section 6.3.3, there is much evidence for the existence of nano-bubbles at the not perfectly wettable water/quartz interface. Figure 6.5 shows schematically how the role of the finite bubble liquid–gas surface mobility leads to a pseudo-slip effect. The consequence is the anomalously low Na value obtained.

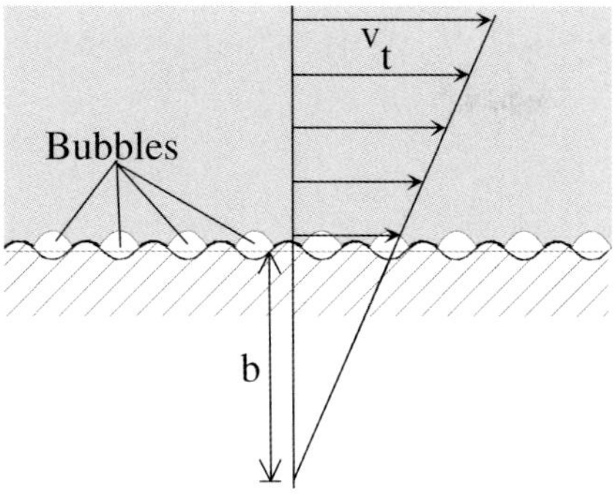

Fig. 6.5 Model of enhanced solid–liquid mobility due to adhering nano-bubbles; *b* is the slip-length (see Section 6.2.1); v_t = tangential velocity.

6.3 Stability of Wetting Films

6.3.1 Stable Films

The stability and rupture behavior of aqueous wetting films on solid surfaces is an important step in many coagulation processes of colloidal systems. In particular, the interaction of an air bubble with a solid particle in the industrial flotation process [46], which is widely used in mineral processing, in paper recycling and waste water treatment, the forming, thinning and rupture of thin intervening water films play a crucial role. Whether the film rupture takes place in the short contact time of the particle and the bubble (millisecond scale) or not is of decisive consequence for flotation success and yield.

In general, a positive sign of the disjoining pressure ($\Pi > 0$) is the condition for equilibrium (non-thinning) liquid films (see Figs. 6.6 and 6.7), whereas their stability depends on the sign of the disjoining pressure derivative ($\partial_h \Pi < 0 \leftrightarrow$ stable films) [9]. Usually disjoining pressures increase (in their absolute values)

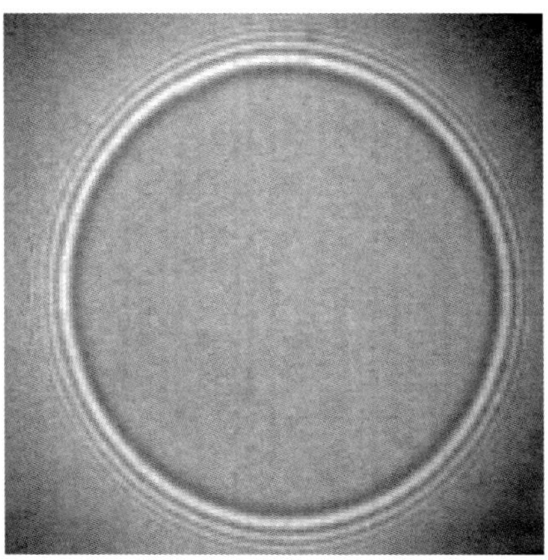

Fig. 6.6 Flat, stable equilibrium film of 10^{-3} M KCl solution on hydrophilic silica.

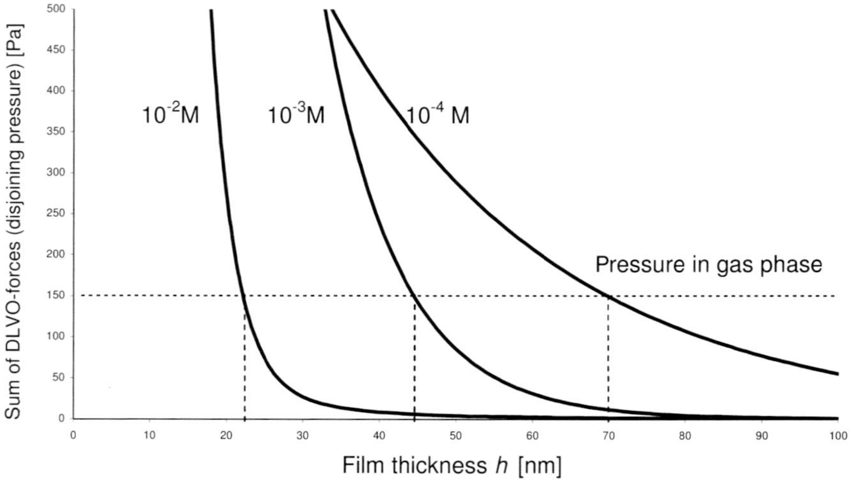

Fig. 6.7 Force–distance curves for wetting films on silica with different ionic strength (10^{-4}–10^{-2} M KCl). ($A = 1 \times 10^{-20}$ J, $\psi_1 = -35$ mV, $\psi_2 = -30$ mV). The line at 150 Pa gives a typical experimental bubble pressure of the DSFB. The abscise coordinates of the intersection points between the disjoining pressure Π and the capillary pressure p_σ in the meniscus are the corresponding equilibrium thicknesses h_{eq}.

with decreasing thickness ($\partial_h|\Pi| < 0$) and then both the equilibrium and the stability conditions can be determined only by the sign of Π: $\Pi > 0$ (equivalent to $\partial_h \Pi < 0$) guarantees equilibrium and stable films and, vice versa, $\Pi < 0$ (equivalent to $\partial_h \Pi > 0$) corresponds to unstable films (see Figs. 6.11–6.13). Experiments with foam and emulsion films mainly [16] have shown excellent agreement with this theory, whereas wetting films have revealed some peculiarities. In Section 6.3.3 it will be shown clearly that part of these peculiarities is due to substantial heterogeneity typical of solid surfaces.

The most investigated systems in this field are aqueous wetting films on silica surfaces; i.e. quartz, glass or the oxide layer on silicon wafers. In this asymmetric system, all DLVO forces are repulsive, $\Pi_{vdW} > 0$; $\Pi_{el} > 0(\Pi_{vdW} = -A/6\pi h^3$, $A_{silica/water/air} = -1 \times 10^{-20}$ J [47]; the potential of the clean water/air interface is negative, $\psi_1 = -35$ mV [48], and the silica/water interface at neutral pH is also negatively charged, $\psi_2 = -30$ mV). Hence stable wetting films can be formed on meticulously cleaned silica surfaces.

Figure 6.6 presents a typical picture of a stable equilibrium film formed in DSFB of 10^{-3} M KCl solution on hydrophilic silica. The uniform gray inside the wetting film shows the homogeneous thickness of the film. The Newton fringes at the film edge are due to the meniscus profile at the bubble film contact.

In Fig. 6.7 are represented DLVO forces versus distance curves for three different electrolyte solutions. As can be seen, the corresponding equilibrium thicknesses h_{eq} are simply the abscissa coordinates of the intersection point between the total disjoining pressure Π and the capillary pressure p_σ in the meniscus, i.e. equivalent to the pressure balance $\Pi(h_{eq} = p_\sigma$ i.e. $\Delta p = 0$ (see Eq. 1).

It should be pointed out again that the case of both repulsive van der Waals forces and repulsive electrostatic double layer forces is limited to some asymmetric films and can therefore only be found in ternary systems such as gas–water–solid or oil–water–solid [47]. In colloidal systems where two identical media interact through a second phase, the van der Waals forces are always attractive, so only metastable films can be found.

The above-mentioned technological flotation process does not succeed if stable water films are formed between solid particles and gas bubbles.

6.3.2 Rupture of Wetting Films (Unstable Wave Mechanism)

If the resultant DLVO force in the film is attractive ($\Pi<0$), the most important consequence is that the film becomes unstable, i.e. its drainage ends inevitably with a rupture. We have studied this rupture process in systems of repulsive van der Waals force ($\Pi_{vdW}>0$) and attractive electrostatic interactions ($\Pi_{el}<0$) so that the total interaction forces remains attractive ($\Pi=\Pi_{vdW}+\Pi_{el}<0$). The occurrence of attractive electrostatic forces can be realized experimentally by overcharging the silica surface by Al^{3+} ions [49, 50] or by applying TiO_2 on glass-surfaces in a pH range more acidic than the isoelectric point of this system (our unpublished measurements). Figure 6.8 shows $\Pi(h)$ isotherms of a silica sur-

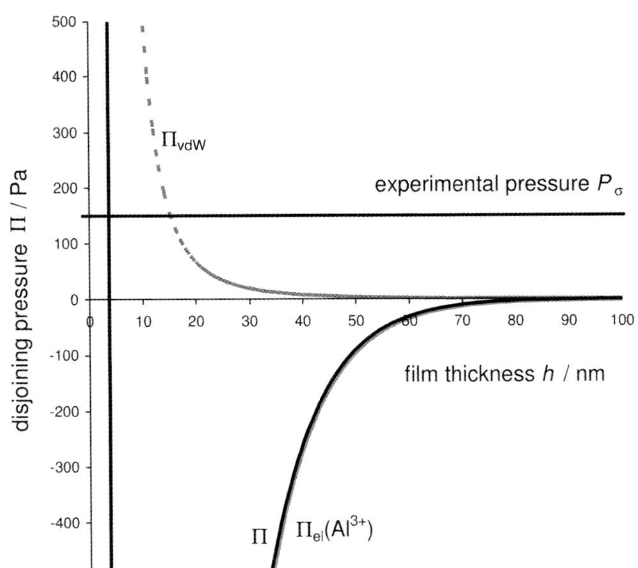

Fig. 6.8 Force–distance curves for films containing 10^{-4} M $AlCl_3$ and 10^{-3} M KCl. The solid surface is positively charged due to the Al^{3+} ions ($\psi_2=+35$ mV), the air/liquid interface remains negatively charged ($\psi_1=-35$ mV). $\Pi=\Pi_{vdW}+\Pi_{el}$, total disjoining pressure.

face overcharged by Al^{3+} ions. As can be seen, the electrostatic interaction for this particular system predominates in the entire interval above rupture thicknesses ($h_{\rm rupt} \approx 35$ nm, see Fig. 6.9) and the resultant disjoining pressure Π is practically equal to $\Pi_{\rm el}$ [51]:

$$\Pi_{\rm el} = 2\varepsilon\varepsilon_0\kappa^2\left(4\frac{k_{\rm B}T}{ze}\right)\tanh\left(\frac{ze\psi_1}{4k_{\rm B}T}\right)\tanh\left(\frac{ze\psi_2}{4k_{\rm B}T}\right)\exp(-\kappa h) \qquad (13)$$

As already mentioned (see Section 6.2.2), fluid surfaces are corrugated (due to thermal fluctuations) but, in contrast to "deep" water, where the waves are stable [with finite amplitudes, Eq. (11)], surface waves in thin films could be unstable (with increasing amplitude). In the case of attractive DLVO forces, a principle possibility arises that part of the wave spectrum may become unstable. According to the theory [9, 40], a wave (named critical wave) with a length $\lambda_{\rm cr}$ separates the Fourier spectrum into stable and unstable parts (Fig. 6.9). The critical wavelength is proportional to the square root of the film surface tension σ divided by the first derivative of the disjoining pressure $\partial_h\Pi$ with respect to the film thickness h [9]:

$$\lambda_{\rm cr} \sim \sqrt{\sigma/\partial_h\Pi} \qquad (14)$$

Relation (14) follows directly from Eq. (12) by substituting $(\sigma k^2 - \partial_h\Pi) = 0$ therein.

Figure 6.9 represents a typical trend of the function $\lambda_{\rm cr}(h)$ calculated from Eq. (14) with Π from Eq. (13). As can be seen, the "short waves" spectrum is stable ($\lambda < \lambda_{\rm cr}$), whereas the "long waves" spectrum is unstable ($\lambda > \lambda_{\rm cr}$).

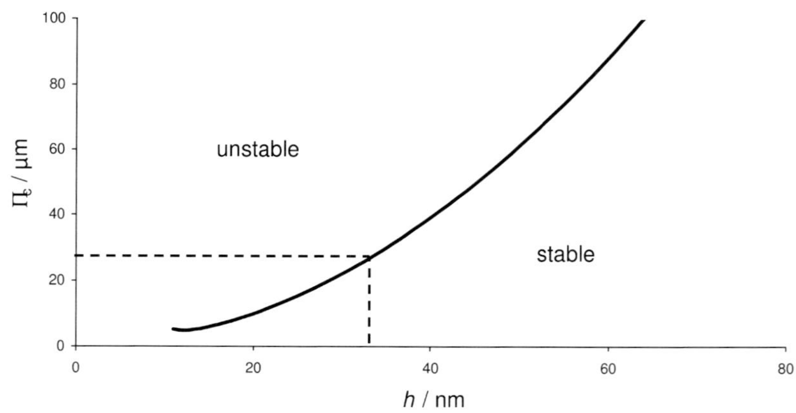

Fig. 6.9 Stable and unstable capillary wave spectrum regions; $\lambda_{\rm cr}(h)$ = curve calculated by Eqs. (13) and (14) for the experimental conditions of Fig. 6.8. The particular coordinates ($h \approx 35$ nm; $\lambda_{\rm cr} \approx 25$ µm) are in good agreement with some experimental observations (see Fig. 6.10).

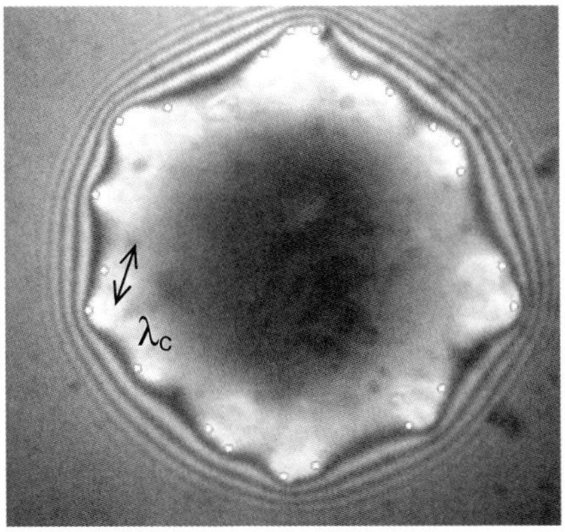

Fig. 6.10 Ruptured film on silica, positively charged by 10^{-4} M Al^{3+} ions; 10^{-2} M KCl electrolyte. The holes in the film marked as white dots for better visibility are nearly equidistant.

It is worth noting that the unstable modes rupture mechanism considered here is equivalent to a spontaneous (non-equilibrium) process, i.e. unstable films will rupture at any stationary thickness (h_f=constant) for which a critical mode $\lambda_{cr}(h)$ exists. In the next section we will see that for heterogeneous wetting films there is also the possibility of another mechanism, where rupture occurs only at a definite thickness [non-spontaneous mechanism, Eq. (18)].

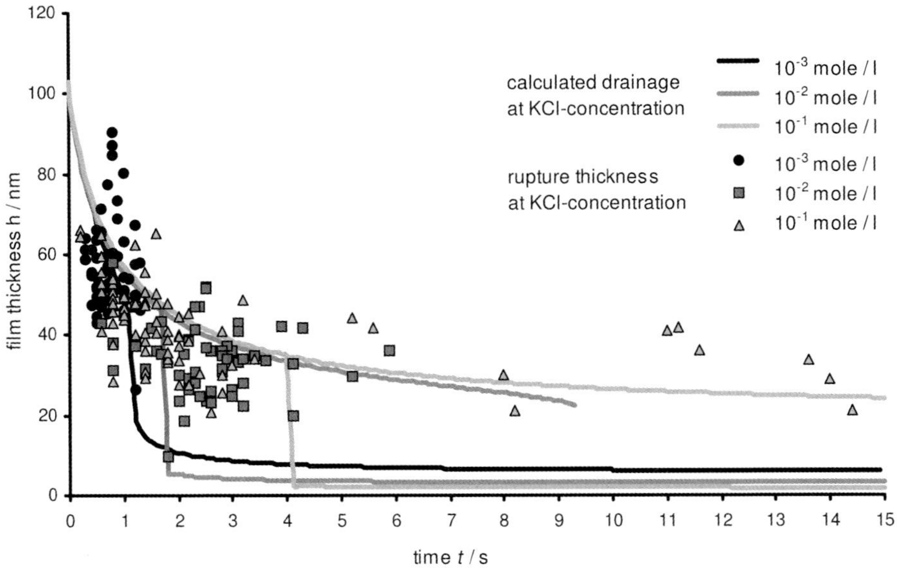

Fig. 6.11 Drainage and rupture of an aqueous wetting film of $AlCl_3$ (10^{-4} M) solution on glass at different KCl concentrations.
Lines: calculated curves according to the Reynolds equation (Eq. 1).
Symbols: rupture thickness/lifetime.

When the amplitudes of the unstable waves are increasing, they will reach the opposite film surface at a given moment. For soap films, this contact point usually leads to rupture [40], whereas for wetting films this is followed by hole formation, known in the literature also as the dewetting process [49, 50, 52].

If one assumes that every touch leads with large probability to the formation of a hole, the distance between them should be scaled by the wavelength. The experimental data from Fig. 6.10 are in a good agreement with this prediction. The evaluated λ_c value from Fig. 6.9 for the experimentally measured rupture thickness h_{rupt} of 35 nm leads to a critical wavelength λ_c of approximately 25 μm, which is of the same order as the measured distance of the holes in Fig. 6.10.

Further evidence for the existence of the capillary wave rupture mechanism can be seen in the results in Fig. 6.11. The curves $h_{film}(t)$ represent film drainage calculated by Eq. (1) at three KCl concentrations, i.e. at three different electrostatic disjoining pressures (see, e.g., Fig. 6.7). As can be seen, the experimentally measured rupture thicknesses (the different types of dots) lie in the range of the corresponding Reynolds law curves [51, 53]. The high scattering of the rupture data is typical for metastable systems.

6.3.3
Film Rupture Caused by Nano-bubbles (Nucleation Mechanism)

Let us consider an aqueous wetting film on a pure silica surface, i.e. not recharged with Al^{3+} ions. As already discussed, such films are stable with equilibrium thicknesses depending on the KCl concentration and the experimental pressure p_σ (see Fig. 6.7). However, if the silica surface is made hydrophobic (e.g. via methylation), wetting films become unstable and rupture, although all DLVO forces remain repulsive. It should be recalled that the hydrophobization does not change the repulsive nature of DLVO forces, provided that neither the Hamaker constant ($A = -1 \times 10^{-20}$ J) nor the surface potential ($\psi_2 \approx -30$ mV) at neutral pH values of the silica surface are changed significantly [54, 55]. The rupture thickness in this system can reach very high values, up to hundreds of nanometers, whereas the lifetimes are very short, of the order of parts of a second. To explain this problem, some groups introduced a "long-range hydrophobic force" of unknown physical origin [56, 57]. Nowadays it is widely accepted that the rupture of these films is caused by small gas bubbles (in the sub-micrometer range) adhering to the hydrophobic solid surface [58]. A number of papers have been published that give clear evidence for the existence of such nano-bubbles, by means of IR spectroscopy [59], force measurements [60–62] and also by image-producing methods such as tapping-mode atomic force microscopy [63–65]. As already mentioned, the rupture mechanism in this case is non-spontaneous, in contrast to the spontaneous unstable wave mechanism (see Section 6.3.2). However, due to the mode of bubble formation, it is known in the literature as the nucleation rupture mechanism [50–53].

Figure 6.12 shows rupture thicknesses h_{rupt} (and lifetimes) of wetting films at different hydrophobized silica surfaces. The degree of hydrophobization is char-

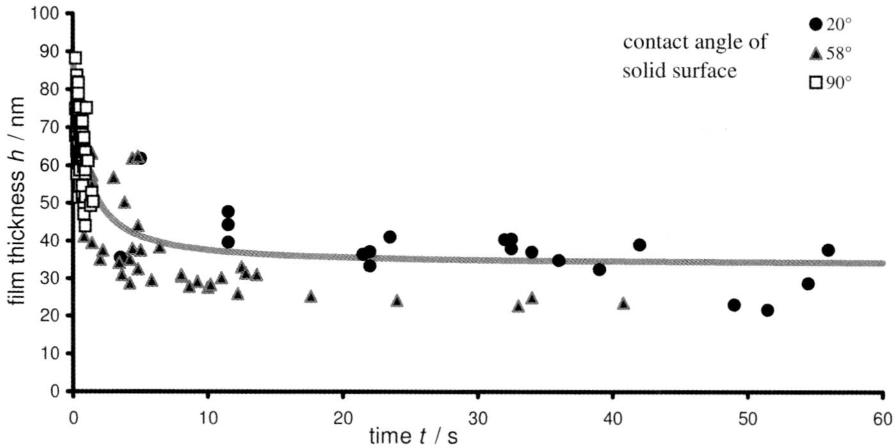

Fig. 6.12 Drainage and rupture of wetting films on methylated silica, with different contact angles. Line, calculated curves according to the Reynolds equation (Eq. 1) (KCl concentration 10^{-3} M); symbols, rupture thickness.

acterized by the corresponding advancing angle, 20, 58, 90° in this particular case. The following conclusions can be drawn from the experimental data presented in Fig. 6.12:

- The greater the degree of hydrophobization (the larger the advancing contact angle), the larger is the rupture thickness (the shorter the lifetime).
- The ruptures take place along the theoretical drainage curve (Eq. 1) where only repulsive DLVO forces are taken into account.

The last observation strongly supports the assumption mentioned above that the hydrophobization does not change the repulsive character of the surface forces in the wetting film on a silica surface and that no additional attractive forces (such as "long-range hydrophobic forces") are acting in this case.

Also, the microscopically observed rupture (Fig. 6.13) gives an indication of a nucleation process: the rupture starts from a single hole somewhere in the film – in all probability at the place where the largest nano-bubble is located – and leads to complete dewetting of the whole film area (radius ca. 100 μm) in a few milliseconds.

Note the drastic difference from the film evolution in the case of hydrophilic silica surfaces where ruptures occur at several points but the contact spots do not expand (Fig. 6.10).

The detailed analysis [51, 53, 66, 67] shows that gas bubble formation actually creates non-homogeneities in the wetting film. In the region above the bubble, the surface forces are much closer to surface forces in foam (air–water–air) films and differ from those in wetting films (Fig. 6.14). Wetting films (silica–water–air) have both van der Waals and electrostatic forces repulsive (Fig. 6.12), whereas in foam films the van der Waals interactions are attractive. Figure 6.14

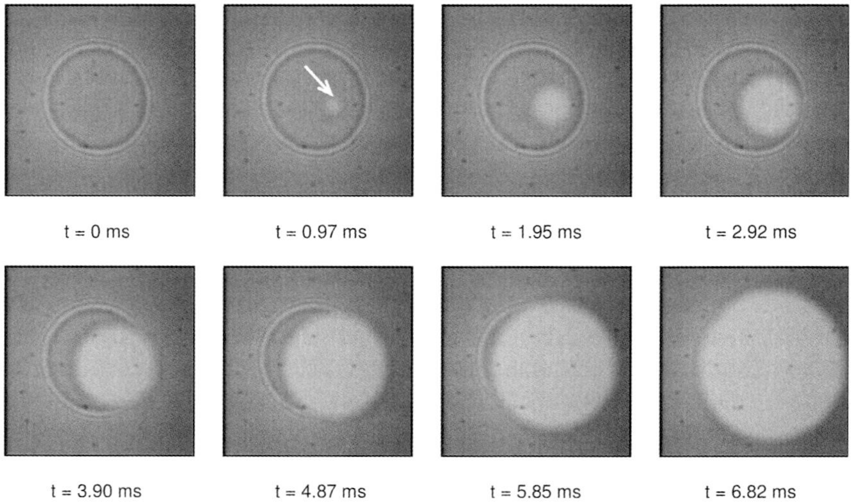

Fig. 6.13 Microinterferometric high-speed video frames of the rupture of an aqueous wetting film on methylated silica. The arrow on the second frame shows the point where nucleation starts.

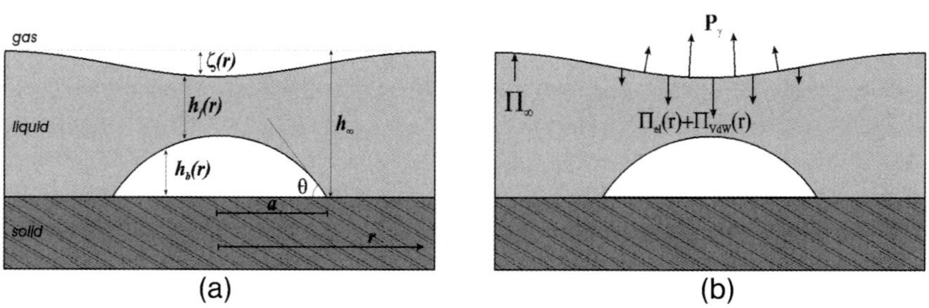

Fig. 6.14 Equilibrium profile and pressure acting on a film surface in the presence of a bubble. (a) Surface deformation due to local attractive forces ($\Pi<0$); h_∞, wetting film thickness far from the bubble. (b) Pressures acting at the film surface. Π_∞, disjoining pressure far from the bubble (repulsive, $\Pi_\infty>0$).

shows schematically the equilibrium profile and the pressure balance of a wetting film caused by a bubble adhering to the solid surface.

Quantitative results for a film surface deformation above the bubble apex $\zeta(r=0)$ are demonstrated in Fig. 6.15. The equilibrium form $\zeta(r)$ is the solution of the pressure balance equation [67]:

$$\Pi(r,\zeta) + p_\sigma(\zeta) = \Pi_\infty(h_\infty) \tag{15}$$

where $\Pi(r,\zeta)$ and $\Pi_\infty(h_\infty)$ are the resultant disjoining pressures inside and outside the bubble region, respectively, and $p_\sigma(\zeta)$ is the local capillary pressure. In

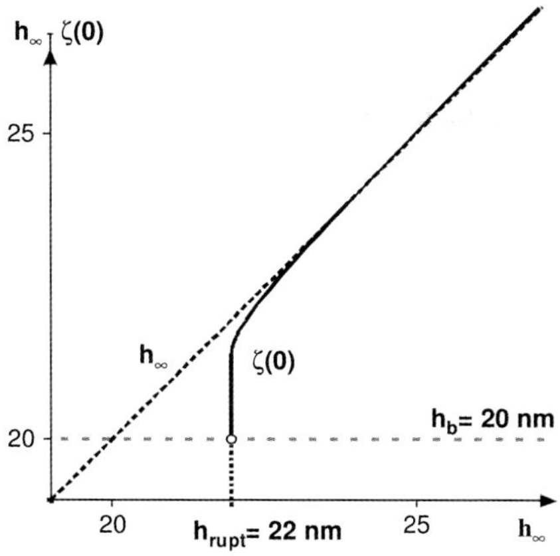

Fig. 6.15 Equilibrium deformation $\zeta(r=0)$ of a film surface above a bubble apex (see Fig. 6.14). The rupture occurs when $\zeta(0)$ reaches the bubble apex. Rupture thickness value $h_{\text{rupt}}=22$ nm calculated with Eq. (15) is close to the bubble height $h_b=20$ nm.

contrast to the disjoining pressure $\Pi_\infty(h_\infty)$, which depends solely on the wetting film thickness h_∞, $\Pi(r,\zeta)$ is a function of the radial coordinate r. Actually, in this case we have a multilayer (gas–liquid–gas–solid) with variable thicknesses of the liquid phase $h_f(r)$ and of the gas (bubble) phase $h_b(r)$ (Fig. 6.14), which reflects on the radial dependence of $\Pi(r,\zeta)$. For example, the corresponding van der Waals disjoining pressure, $\Pi_{\text{vdW}}(r,\zeta)$, becomes

$$\Pi_{\text{vdW}}(r,\zeta) = -\frac{1}{6\pi}\left\{\frac{A_{\text{LL}}}{h_f{}^3(r)} - \frac{A_{\text{SL}}}{[h_\infty - \zeta(r)]^3}\right\} \tag{16}$$

with $A_{\text{SL}}=4.7\times10^{-20}$ J $>$ $A_{\text{LL}}=3.7\times10^{-20}$ J as Hamaker constants for solid (glass)–liquid (water) media [68]. Obviously, with a sufficiently thin liquid layer $[h_f < (h_\infty - \zeta)]$ above the bubble, the second term in Eq. (16) prevails and $\Pi_{\text{vdW}}(r,\zeta)<0$, i.e. we can expect local attractive van der Waals interactions. In contrast, far away from the bubble in the wetting film, the van der Waals interactions are repulsive:

$$\Pi_{\text{vdW}}(h_\infty) = -\frac{1}{6\pi}\frac{A_{\text{LL}} - A_{\text{SL}}}{h_\infty{}^3} > 0 \tag{17}$$

Figure 6.15 represents an equilibrium deformation of a film above a bubble apex. As can be seen, the surface deformations are negligible ($\zeta<1$–3 nm) except for heights shortly before the moment of contact with the bubble, i.e. just before the film rupture. According to the mechanism discussed in Section 6.3.2, films rupture when their two surfaces touch each other at least at a single point. In the system considered here, this condition reads

$$\zeta(r=0, h_{\text{rupt}}) = h_{\text{rupt}} - h_b(0) \tag{18}$$

Note that, according the above definition, h_{rupt} is the wetting film thickness (far away from the bubble), i.e. $h_{\text{rupt}}=h_\infty$. For the particular case in Fig. 6.15, the theoretically estimated rupture thickness h_{rupt} is close to the bubble height, i.e. $h_{\text{rupt}} \approx h_{\text{b}}(0)$ [67].

Condition (18) defines an equilibrium (non-spontaneous) rupture mechanism, but as already mentioned (see Section 6.3.2), the complete solution of a rupture problem needs an additional stability analysis. Generally, stability depends on the evolution of perturbations of the equilibrium characteristics, i.e. whether the perturbation amplitudes grow or diminish with time [69]. For thin liquid films the thickness perturbation evolution $\delta h(r,t)$ is significant, i.e. at $\partial_t|\delta h| < 0$ the film is stable, whereas at $\partial_t|\delta h| > 0$ it is unstable. Obviously, the behavior of $\delta h(r,t)$ is determined by the surface forces, i.e. by the pressure balance perturbation $\delta[\Pi(r) + p_\sigma]$ (see Eq. 15). The relation between $\partial_t \delta h$ and $\delta(\Pi + p_\sigma)$ obeys Eq. (7), discussed in Section 6.2.1. The evolution equation defining $\delta h(r,t)$ follows directly in explicit form by substituting $H \equiv h_{\text{f}}(eq) + \delta h(r,t)$ in Equation (7). After expanding the terms of the sum in a series $(p_\sigma + \Pi) \approx (p_\sigma + \Pi)_{\text{eq}} + p_\sigma(\delta h) + (\partial_h \Pi)\delta h$ by taking into account the equilibrium condition (15), one obtains the following expression for the evolution equation of $\delta h(r,t)$ [16]:

$$\partial_t \delta h = -\nabla\{h_f{}^3(r)\nabla[\sigma\nabla^2\delta h + (\partial_h \Pi)_{\text{eq}}\delta h]/12\mu\} \tag{19}$$

For the sake of brevity, the standard nabla (∇) operator notation is used in Eq. (19). Note that in the frame of the linear perturbation approach the capillary pressure operator is also linearized $[p_\sigma = \sigma\nabla^2\delta h(r,t)]$. For homogeneous films, i.e. at h_{f}=constant, Π=constant, Eq. (19) (after applying the standard Fourier transformation technique) leads directly to the dispersion condition (12). The main problem with heterogeneous films [either because of uneven thicknesses, h_{f} (r), or because of variable disjoining pressure, $\Pi(r)$], is that the Fourier image of Eq. (19) is no longer an algebraic but rather an integral equation. The solution of this more complicated (from the computational point of view) problem will be the goal of a future study. We will conclude this chapter with a comment that may be instructive for the qualitative analysis of Eq. (19). The main point here is that if the zone where attractive forces act in a heterogeneous wetting film has dimensions smaller than the critical wavelength (Eq. 14), this film will remain stable with respect to linear perturbations. In the system considered here, the attractive zone dimensions are of the order of the bubble contact radius a (Fig. 6.14), which is equivalent to the fact that linear instability (spontaneous rupture) could be expected if $a \geq \lambda_{\text{cr}}$. Noting that for glass–water–air films λ_{cr} values are of the order of tens of micrometers (see Fig. 6.9), i.e. 10^3 times larger than that of the nano-bubbles, one could hardly expect a substantial impact (shortening) on the rupture thicknesses due to instability effects.

Acknowledgment

B.R. is grateful for the financial support of the Bulgarian Ministry of Education and Science (Project Y-X-2/2003).

References

1 A. Sharma, E. Ruckenstein, *J. Colloid Interface Sci.*, *119*, 14 (**1987**).
2 A. Sharma, E. Ruckenstein, *Langmuir*, *3*, 760 (**1987**).
3 R. Tsekov, E. Ruckenstein, *Langmuir*, *9*, 3264 (**1993**).
4 D. Platikanov, *J. Phys. Chem.* *68*, 3619 (**1964**).
5 J. Joye, C.A. Miller, G.J. Hirasaki, *Langmuir*, *8*, 3083 (**1992**).
6 J.N. Connor, R.G. Horn, *Faraday Discuss.*, *123*, 193 (**2003**).
7 L.Y. Clasohm, J.N. Connor, O.I. Vinogradova, R.G. Horn, *Langmuir*, *21*, 8243 (**2005**).
8 R. Tsekov, O.I. Vinogradova, *Langmuir*, *21*, 12090 (**2005**).
9 A. Scheludko, *Adv. Colloid Interface Sci.*, *1*, 391 (**1967**).
10 B. Radoev, E. Manev, I.B. Ivanov, *Kolloid-Z.*, *234*, 1037 (**1969**).
11 D.F. Cooper, B.H. Davey, J.W. Smith, *Can. J. Chem. Eng.*, *54*, 631 (**1976**).
12 A.K. Malhotra, D.T. Wasan, *Chem. Eng. Commun.*, *55*, 95 (**1987**).
13 S.A.K. Jeelani, S. Hartland, *J. Colloid Interface Sci.*, *164*, 296 (**1994**).
14 A. Sonin, A. Bonfillon, D. Langevin, *J. Colloid Interface Sci.*, *162*, 323 (**1994**).
15 B. Radoev, D.S. Dimitrov, I.B. Ivanov, *Colloid Polym. Sci.*, *252*, 50 (**1974**).
16 I.B. Ivanov (Ed.), *Thin Liquid Films*, Marcel Dekker, New York, **1988**.
17 R.H. Davis, J.A. Schonberg, J. M. Rallison, *Phys. Fluids A*, *1*, 77 (**1989**).
18 A.K. Chesters, *Chem. Eng. Res. Des.*, *69*, 259 (**1991**).
19 R. Tsekov, H.J. Schulze, B. Radoev, P. Letocart, *Colloids Surf. A*, *142*, 287 (**1998**).
20 H. von Helmholtz, G. Pictowski, *Sitzber. Kaiserl. Akad. Wissensch., Math. Nat. Klasse, Abt. 2 (Wien)* 40 (**1868**) [cited in E. Ruckenstein, P. Rajora, *J. Colloid Interface Sci.*, *96*, 488 (**1983**)].
21 K.B. Migler, H. Hervet, L. Leger, *Phys. Rev. Lett.*, *70*, 287 (**1993**).
22 B.Z. Newby, M.K. Chaudhury, *Langmuir*, *13*, 1805 (**1997**).
23 O. Vinogradova, *Langmuir* *11*, 2213 (**1995**).
24 E. Manev, R. Tsekov, B. Radoev, *J. Dispers. Sci. Technol.*, *18*, 769 (**1997**).
25 R. Tsekov, E. Evstatieva, *Prog. Colloid Polym. Sci.*, *126*, 93 (**2004**).
26 R. Tsekov, *Colloids Surf. A*, *141*, 161 (**1998**).
27 J.K. Angarska, E.D. Manev, *Colloids Surf. A*, *223*, 27 (**2003**).
28 R. Tsekov, E. Ruckenstein, *Colloids Surf. A*, *82*, 255 (**1994**).
29 K.D. Danov, T.D. Gurkov, T. Dimitrova, I.B. Ivanov, D. Smith, *J. Colloid Interface Sci.*, *188*, 313 (**1997**).
30 R. Tsekov, P. Letocart, E. Evstatieva, H.J. Schulze, *Langmuir*, *18*, 5799 (**2002**).
31 D.J. Mitev, R. Tsekov, C.S. Vassilieff, *Colloids Surf. A*, *10*, 67 (**1997**).
32 R. Tsekov, P. Letocart, H.J. Schulze, *Langmuir*, *16*, 8206 (**2000**).
33 K.G. Marinova, R.G. Alargova, N.D. Denkov, O.D. Velev, D.N. Petsev, I.B. Ivanov, R.P. Borwankar, *Langmuir*, *12*, 2045 (**1996**).
34 D. Valkovska, K. Danov, *J. Colloid Interface Sci.*, *241*, 400 (**2001**).
35 P. Letocart, B. Radoev, H.J. Schulze, R. Tsekov, *Colloids Surf. A*, *149*, 151 (**1999**).
36 M. Smoluchowski, *Ann. Phys.*, *25*, 205 (**1908**).
37 L. Mandelstam, *Ann. Phys.*, *41*, 609 (**1913**).
38 W. Eisenmenger, *Acustica*, *9*, 327 (**1959**).
39 A. Scheludko, *Proc. K. Akad. Wettensch.*, *B65*, 87 (**1962**).
40 A. Vrij, *Colloid Sci.*, *19*, 1 (**1964**); *Discuss. Faraday Soc.*, *42*, 44 (**1966**).

41 J. Lucassen, M. van den Tempel, *J. Colloid Interface Sci. 41*, 491 (**1972**).
42 C. Stenvot, D. Langevin, *Langmuir, 4*, 1179 (**1988**).
43 Q. Wang, A. Feder, E. Mazur, *J. Phys. Chem., 98*, 12720 (**1994**).
44 Q.A. Sun, J.P. Buttler, B. Suki, D. Stamenovic, *J. Colloid Interface Sci., 163*, 269 (**1994**).
45 L. Bergmann, *Acustica, 9*, 185 (**1959**).
46 H.J. Schulze, *Physico-Chemical Elementary Processes in Flotation*, Elsevier, Amsterdam, **1984**; A.V. Nguyen, H.J. Schulze, *Colloidal Science of Flotation*, Marcel Dekker, New York, **2003**.
47 H. Lyklema, *Fundamentals of Interface and Colloid Science*, Vol. 1, Academic Press, London, **1991**.
48 Chr. Cichos, *Freiberger Forschungshefte, 7A*, 513 (**1973**).
49 H.J. Schulze, *Colloid Polym. Sci., 253*, 730 (**1975**).
50 H.J. Schulze, K.W. Stöckelhuber, A. Wenger, *Colloids Surf. A, 192*, 61 (**2001**).
51 K.W. Stöckelhuber, B. Radoev, A. Wenger, H.J. Schulze, *Langmuir, 20*, 164 (**2004**).
52 K.W. Stöckelhuber, *Eur. Phys. J. E, 12*, 431 (**2003**).
53 K.W. Stöckelhuber, H.J. Schulze, A. Wenger, *Chem. Eng. Tech. 24*, 619 (**2001**).
54 J. Laskowski, J.A. Kitchener, *J. Colloid Interface Sci., 29*, 670 (**1969**).
55 J. Sonnefeld, A. Göbel, W. Vogelsberger, *Colloid Polym. Sci., 273*, 926 (**1995**).
56 J.N. Israelachvili, R.M. Pashley, *Nature, 300*, 341 (**1982**).
57 J.N. Israelachvili, R.M. Pashley, *J. Colloid Interface Sci., 98*, 500 (**1984**).
58 J. Ralston, D. Fornasiero, N. Mishchuk, *Colloids Surf. A, 192*, 39 (**2001**).
59 W. Gong, J. Stearnes, D. Fornasiero, R. Hayes, J. Ralston, *Phys. Chem. Chem. Phys., 1*, 2799 (**1999**).
60 J. Mahnke, J. Stearns, J. Ralston, *Phys. Chem. Chem. Phys., 1*, 2793 (**1999**).
61 N. Ishida, M. Sakamoto, M. Miyahara, K. Higashitani, *Langmuir, 16*, 5681 (**2000**).
62 W.G. Tyrell, P. Attard, *Langmuir, 18*, 160 (**2002**).
63 N. Ishida, M. Sakamoto, M. Miyahara, K. Higashitani, *J. Colloid Interface Sci., 253*, 112 (**2002**).
64 J.W.G. Tyrell, P. Attard, *Phys. Rev. Lett., 87*, 176104 (**2001**).
65 S. Lou, J. Gao, X. Xiao, X. Li, G. Li, Y. Zhang, M. Li, J. Sun, J. Hu, *Chin. Phys., 10*, 108 (**2001**).
66 K.W. Stöckelhuber, H.J. Schulze, A. Wenger, *Prog. Colloid Polym. Sci., 118*, 11 (**2001**).
67 R. Slavchov, B. Radoev, K.W. Stöckelhuber, *Colloids Surf. A, 261*, 135 (**2005**).
68 J.N. Israelachvili, *Intermolecular and Surface Forces*, Academic Press, London, **1991**.
69 P. Glansdorff, I. Prigogine, *Thermodynamic Theory of Structure, Stability and Fluctuations*, Wiley-Interscience, London, **1972**.

7
Measuring Forces in the Colloidal Regime with the Atomic Force Microscope*

J. Mieke Kleijn, Marcel Giesbers, and Martien A. Cohen Stuart

Abstract

Since the early 1990s, atomic force microscopy (AFM) has become a key technique for measuring interaction forces in the colloidal regime (nanometer to micrometer scale), as testified by the continuous growth in the number of papers published. In this chapter, we focus on this technique and illustrate its versatility by describing a number of applications. First, we provide a general introduction and a short historical overview on surface force measurements. The principles of force measurements using AFM and the interpretation of such measurements and pitfalls are explained. Finally, we consider various fields of application (without striving for completeness), in which there is some emphasis on examples of work from our own laboratory, but ample attention to the work of others.

7.1
Introduction

7.1.1
General Introduction

Forces in the colloidal regime, often referred to as "surface forces", determine the behavior of a great variety of materials. They dictate the interactions between colloids and between colloids and (macroscopic) surfaces. Ensembles of molecules and ions affect these interactions, which is reflected in osmotic stress, thermal fluctuations, hydrophobic attraction and so on. The forces in this regime are of great importance in both natural and synthetic systems. For instance, the structure of soils and possibilities for their cultivation are to a large degree determined by the interactions between sand, clay and other particles. Physiological processes are dominated by interactions between protein mole-

* A list of symbols and abbreviations can be found at the end of this chapter.

Colloids and Interface Science Series, Vol. 3
Colloid Stability and Application in Pharmacy. Edited by Tharwat F. Tadros

ISBN: 978-3-527-31463-8

cules and membrane systems. In industry, knowledge and control over colloidal forces is essential, for example, for manufacturing paper, pharmaceuticals and paints, and for the production of particle layers on macroscopic surfaces. Such layers are used in many applications, such as protective coatings, television and computer screens and fluorescent lamps. On the one hand the colloidal dispersions used in the production processes should be stable for easy handling and obtaining homogeneous layers, whereas on the other hand attraction between surface and particles is necessary for good adhesion. For an extensive overview on surface forces, we refer in particular to the book by Israelachvili [1].

With the introduction of the atomic force microscope (AFM; also referred to as the scanning force microscope, SFM), a new tool became available to measure directly the interaction forces that are sensed by colloids. In comparison with the few techniques that were already available to measure surface forces, the AFM soon proved to be more versatile and relatively easy to use. Nowadays, by far the majority of data on interactions between colloids and surfaces that can be found in the literature has been obtained with the use of the AFM. In this chapter, we will review the technique of AFM force measurements, its possibilities and limitations and applications to elucidate interactions that occur in the colloidal regime. With respect to these applications, we do not strive for completeness and there is some emphasis on examples of work from our own laboratory, but ample attention to the work of others. When appropriate, we will refer to more focused or extensive reviews than given here. We start with a concise historical overview on surface force measurements.

7.1.2
Measuring Surface Forces – a Historical Overview

Measurement of interactions in the colloidal regime is an experimental challenge. In the past, information was for the most part gathered indirectly, for example by measurements on the stability of dispersions and soap films, aggregation kinetics and diffraction studies of the ordering in liquid suspensions. Over the years, several techniques have been developed to measure directly interaction forces between surfaces as a function of their separation distance. In 1954, Derjaguin et al. [2] measured the van der Waals attraction between a glass sphere and a flat glass plate in vacuum. For this they used a force-feedback technique in which the separation between the glass surfaces was determined optically and the interaction force was measured using an electrobalance. Satisfactory results, reasonably in line with predictions of the Lifschitz–van der Waals theory [3], were obtained in the separation range 100–700 nm.

Around the same time, Overbeek and Sparnaay [4, 5] measured van der Waals interactions between two parallel quartz plates in air using a spring balance. The surfaces were positioned and displaced using an air flow. The distance between the surfaces was determined optically, for separations over 100 nm by using interference fringes and for shorter distances via the intensity of the reflected light.

Other early techniques for measuring interactions in colloidal systems include the thin-film balance [6–8] to determine the interaction between liquid interfaces and the osmotic stress method [9] to measure repulsive forces between dispersed colloids. From both methods the disjoining pressure, i.e. the force per unit area needed to balance the interaction force between the surfaces, at a particular (equilibrium) distance can be obtained [10]. Whereas in other surface force methods the balancing force is often exerted by a spring, in the film balance it is the capillary pressure in a free film of (aqueous surfactant) solution in air that equals the interaction between its two interfaces. The capillary pressure is determined by measuring the thickness of the film using a microinterferometric method. Stable films can only be obtained if the disjoining pressure is repulsive. In the osmotic stress method an (osmotic) pressure is exerted on an aqueous dispersion, resulting in the removal of solvent until the volume of the sample is such that the repulsive interactions between the colloids just counterbalance the external pressure. There are various methods to generate pressure, with different pressure ranges [9], such as placing the sample in contact with a polymer solution of known osmotic pressure using a semipermeable membrane (1 kPa–10 MPa), exerting a physical pressure via a piston acting on the sample from which the water is allowed to escape through a semipermeable membrane (pressures up to 50 MPa) or exposing the sample to an atmosphere of known water vapor pressure, attained in the presence of saturated salt solutions ($\sim$300 MPa or higher). Alternatively, one can use hydrostatic pressure in a so-called osmomanometer (1–30 kPa) [11].

In 1968, Tabor and Winterton [12, 13] introduced a force balance in which two thin sheets of mica were placed in a crossed-cylinder geometry. One of the sheets was mounted on a piezoelectric device and the other on a flexible lever spring with known spring constant. The separation between the mica surfaces was determined using an interferometer. The deflection of the lever spring – and thus the interaction force – in a particular separation was derived from the difference between the distance traveled by the piezo element and the actual displacement of the surfaces relative to each other. The apparatus of Tabor and Winterton was further developed by Israelachvili and coworkers [14, 15] into the well-known *surface force apparatus* (SFA, see Fig. 7.1). In the SFA, the determination of both force and distance between the surfaces depends on the interferometric technique. The SFA has provided a wealth of excellent information (see, e.g. [10]). However, the technique is somewhat cumbersome in operation and limited to smooth, transparent and/or reflective-surface materials.

From around 1990, a number of surface force techniques were developed more or less in parallel. First, the development of bimorph force sensors [16, 17] offered new possibilities. A bimorph consists essentially of two layers of piezoelectric ceramic mounted on top of each other and electrically polarized in opposite directions. When the bimorph is bent, one side expands and the other contracts and the resulting piezoelectric voltages add together to give an output voltage that is linear with the deflection. Bimorphs can be used as the lever spring in the SFA, as an addition to the interferometric measurement of the in-

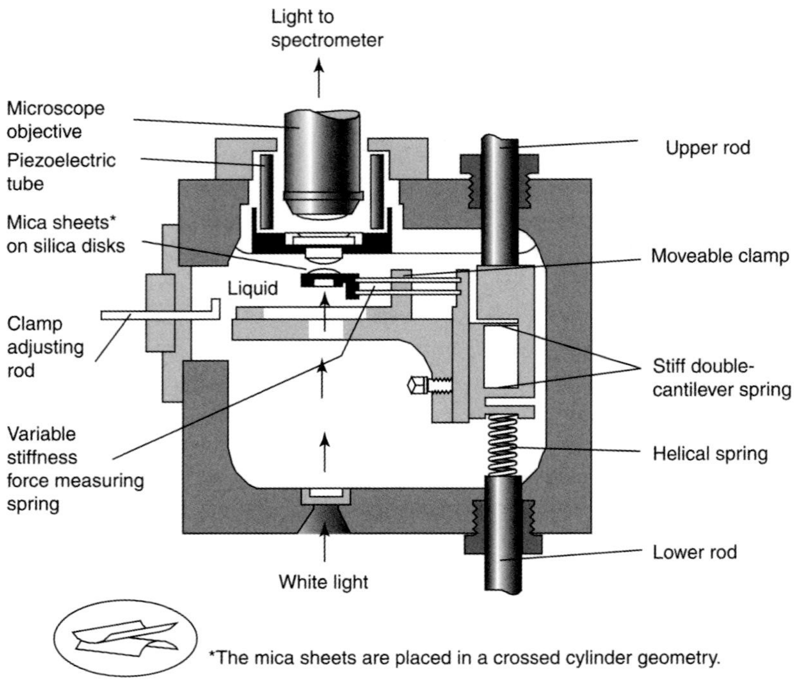

Fig. 7.1 Diagram of the surface force apparatus of Israelachvili and coworkers [14, 15].

teraction force [16]. The bimorph force sensor is also used in another force-measuring device, the so-called MASIF, which stands for *measurements and analysis of surface interaction and forces* [10]. In this apparatus, presented schematically in Fig. 7.2, one of the substrates is mounted on a piezoelectric tube and the other on the bimorph sensor and the tube moves the substrates towards and away from each other. The interacting surfaces have a spherical shape. The separation of the surfaces is not directly measured, but deduced from the piezo travel and the bimorph deflection. (In fact, the principles are very much the same as for AFM force measurements, which will be treated in detail later.) The MASIF can be used to measure forces between non-transparent substrates and it allows for rapid force measurements permitting non-equilibrium and time-dependent measurements. However, the interaction areas are relatively large and therefore the surfaces have to be smooth over areas of about 100 µm [2]. Mostly glass substrates have been used, made by heating the end of a glass rod until it has a spherical shape.

With total internal reflection microscopy (TIRM) [18, 19], one measures directly the potential energy of interaction between a particle immersed in a liquid and a transparent substrate beneath it (i.e. it does not concern a force balance). The principle is as follows: a laser beam is used to generate an evanescent wave at the substrate/liquid interface by total internal reflection. Any particle located close enough to the substrate will scatter the evanescent wave. Since

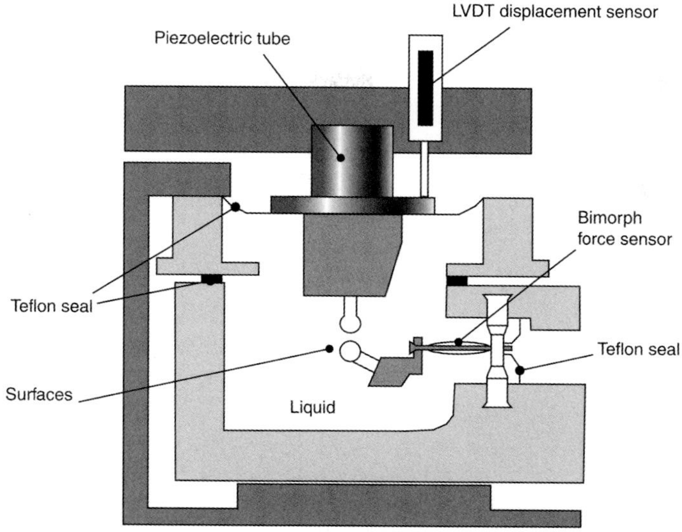

Fig. 7.2 Scheme of the MASIF force apparatus.

the evanescent field falls off exponentially with distance to the surface, the scattered intensity is related to the separation between the particle and substrate. It is possible to select and hold one particle by using optical tweezers and flow other particles out of the cell. Measuring the scattered intensity as a function of time provides a distribution of the separation distances. Since the probability of finding the particle at a particular position is related to the local potential energy through the Boltzmann equation, this distribution can be converted into a potential energy profile. In Fig. 7.3, an example of typical results is given.

The sensitivity of TIRM makes the technique very suitable for measuring very weak interactions. Standard TIRM probes the forces that keep a heavy Brownian particle elevated above a glass substrate. Drawbacks are that energetically unfavorable particle positions are poorly probed and that strongly attractive interactions are problematic. However, several strategies have been developed to overcome these problems, for example by varying the particle position using optical trapping [20]. Behrens [21] turned the TIRM set-up upside down, creating the evanescent field at the top of the sample cell, and studied the repeated sedimentation of a heavy colloidal sphere released from an optical trap near the surface. Every time the particle falls down from the surface, all particle–surface separations in the evanescent field are probed.

Optical tweezers have not only become popular for manipulation of small particles in other techniques (e.g. as mentioned above), but can also be used for force measurements [22, 23]. The principle involves the trapping and tracking of particles ranging in size from tens of nanometers to tens of micrometers. The force exerted by the optical trap on a probe particle is to a good approximation harmonic [22] and the position of the trapped particle fluctuates around the

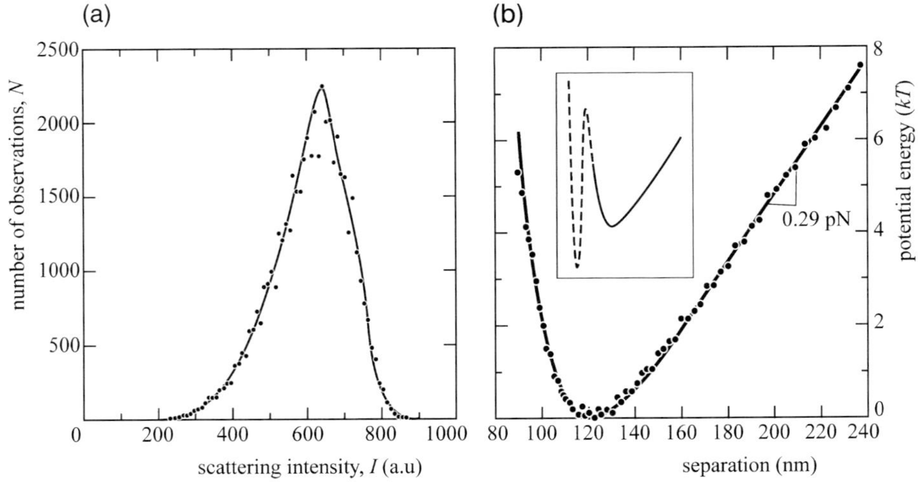

Fig. 7.3 Typical TIRM results. (a) Histogram of scattered intensities from a polystyrene latex sphere (ca. 10 µm in diameter) in 0.5 mM NaCl solution above a glass substrate. (b) The potential energy profile deduced from the histogram. It shows the secondary minimum due to gravity and electric double-layer repulsion. The inset shows the expected potential energy profile for smaller separations (deep primary minimum resulting from van der Waals attraction). (Redrawn from Prieve [19], with permission from Elsevier).

point where the force is zero. The shape of the optical trap (steepness of the interaction potential or "stiffness") is tuned by the optical system, i.e. optical lens and spatial filters. The probe movements can be tracked in two or three dimensions using, for example, video microscopy [24] or interferometrically in a so-called photonic force microscope [25]. The thermal fluctuations in the trap change when an external force acts on the probe and in the new equilibrium position the force exerted by the trap balances this external force. A histogram of particle positions gives the interaction potential and corresponding forces that act on the particle. It is clear that application of this method is not restricted to measurements near a surface. When the interaction with a wall is studied, one can make use of a second particle tightly anchored to the wall as a reference for zero separation or deduce the separation from the diffusion rates that decrease sharply near the wall. The maximum force in the trap is generally below 1 pN (depending on the laser power, size of the probe and optical contrast between probe and medium), so the method is especially useful for measuring weak interactions. An obvious advantage is that the technique does not involve a mechanical force transducer.

The AFM was initially developed to provide images and surface analysis with high spatial resolution, in the ideal case even atomic resolution [26]. In contrast to the scanning tunneling microscope (STM), which was invented a few years earlier [27], it is capable of imaging both conducting and non-conducting samples. The first force measurements with the AFM were performed primarily to improve the quality of AFM images: lowering the force gives better image resolution and reduces the risk of deforming or damaging the sample. In particular, Weisenhorn et al. [28] put considerable effort into studying the effect of the forces during image scanning.

In the early 1990s, several studies were published [29–31] in which the force between the tip and a flat substrate was measured as a function of their separation. However, since the geometry and other properties of AFM tips are not exactly known (and may vary with the tip), quantitative comparison with existing theory was not possible. This problem can be solved by attachment of a micrometer-sized particle of known geometry and well-defined surface properties, a so-called 'colloidal probe', at the end of an AFM cantilever spring [32, 33]. An additional advantage is that the forces between the probe and a flat substrate are much larger than those between the AFM tip and such a substrate and therefore can be measured more accurately (higher force resolution). Due to this development, nowadays AFM is not only a versatile imaging technique, but also has grown into an important surface force technique. With AFM it is possible to measure dynamic interactions, i.e. the role of approach rate can be examined. However, AFM does not have good sensitivity for extremely weak interactions. In the following sections we will go into the details of AFM and AFM force measurements.

7.1.3 The Atomic Force Microscope

Since its introduction in 1986 [26], the development and use of the AFM have boomed, not in the least because it can be applied to all kinds of sample surfaces in a wide range of media (air, gases, liquids). Therefore, surface processes such as adsorption and electrochemical metal deposition can be followed *in situ*. Moreover, as opposed to optical microscopes and scanning electron microscopes, the AFM (and other SPM techniques) measures surface structures in three dimensions (x, y and z) with sub-nanometer resolution. Considerable progress has been made with respect to instrumentation and probes, imaging techniques and sample preparation.

The AFM is surprisingly simple in its concept. The heart of the apparatus is a piezoelectric scanner, a common feature of each SPM, which allows a sharp probe (commonly referred to as the 'tip') to be moved with extreme precision over a sample surface. The scanner can be controlled to provide motion increments of less than 0.1 nm in three dimensions. Depending on the design of the instrument, the scanner either moves the sample or scans the tip over the (static) sample surface. In the first case, depicted in Fig. 7.4, measurements are restricted to samples of area less than ca. 1 cm^2. However, this design is more suitable to obtain images with high (molecular) resolution. The force between the sample surface and the tip is used as a feedback signal for the movements of the tip relative to the sample. Generally, the tip is mounted on a lever spring and the deflection of this spring represents the interaction between tip and sample. In the most common design,

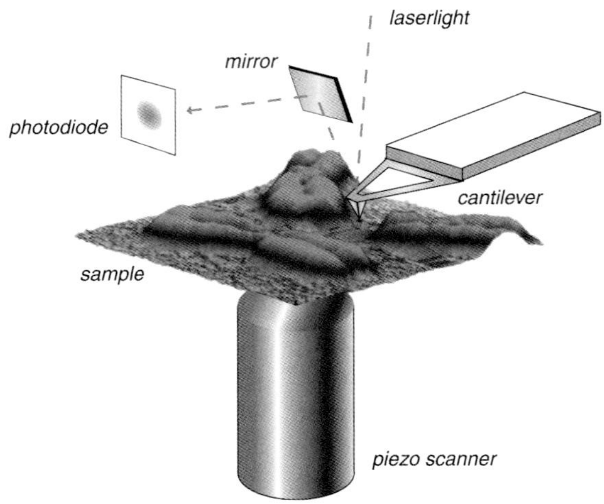

Fig. 7.4 Diagram of the most common AFM set-up.

a laser beam is reflected off the back of the spring, which is coated with a reflective (e.g. gold) layer, on to a segmented photodiode. The signal from this photodiode is simply the difference between the voltage output of its top and bottom segments. To convert a change in voltage into a change of deflection in nanometers, the response of the photodiode is recorded while moving the tip towards and away from a solid (non-deformable) surface. In the resulting plot of photodiode output versus piezo displacement, there is a linear region where the tip is in contact with the sample surface. In this so-called constant-compliance region, the piezo movement and the movement of the tip are the same. Therefore, the slope ($V\ nm^{-1}$) in this region provides the relation between the photodiode output and the lever spring deflection.

A large choice of AFM probes of variable spring constants, tip apex, shapes and coatings, etc., is commercially available. The most commonly used springs have a V-shape (cantilevers), making them less prone to lateral twisting and rolling as the tip scans the sample surface.

The geometry and size of the tip determine the resolution of the images. A standard AFM tip is made of silicon nitride and has a pyramidal shape, about 5 μm at the base and with a tip apex radius of curvature of the order of 10–30 nm. Any AFM image is a convolution of the surface morphology and the size and shape of the tip. This leads to a "blurring" effect which is, of course, particularly important when imaging very small features or relatively large height differences. To improve the resolution, sharper tips (with a high aspect ratio) may be used. These are made, for example, by sharpening standard tips by oxidation or by growing thin filaments on standard tips by means of focused ion beam deposition. Carbon nanotubes, consisting of fullerenes, seem to have ideal properties for AFM tips [34, 35]: they are only a few nanometers in diameter, yet they can be several micrometers long. Therefore, they can probe deep crevices and give images with high lateral resolution. Carbon nanotubes are relatively strong and are able to buckle elastically, which limits the force that is applied to the sample and ensures that they can survive a crash into the sample surface.

AFM lever springs with or without tips are produced by micro-fabrication techniques and are generally made of silicon nitride or single-crystal silicon with lengths from less than 1 μm to a few hundred micrometers. For special applications such as electric force microscopy (EFM), conductive AFM and magnetic force microscopy (MFM), tips made of other materials (e.g. tungsten, antimony-doped silicon) are employed.

In the basic mode of operation of the AFM, the so-called contact mode, the spring is deflected off the sample, thus exerting a force to the sample surface in the range 1–10 nN, depending on the conditions. During imaging, the deflection and thus the force exerted by the spring can be kept constant by adjusting the z-position of the piezo element. This implies that when the tip encounters higher features on the surface, the piezo moves the sample away from the tip and when scanning lower features the piezo moves the sample towards the tip. When the feedback is relatively strong, the sample is scanned under constant force and the z-movement of the piezo reflects the topography of the sample. This is more or less the standard way of imaging in the AFM and is commonly known as the "constant force" or "height" mode. When the feedback is relatively weak, the z-position of the piezo hardly changes during scanning. Now, the deflection of the spring depends on the local height of the sample. In this "deflection mode", the deflection of the cantilever is plotted as a function of the x,y position on the surface rather than the z-movement of the piezo.

For most applications, the force on the sample can be kept low enough to avoid deformation or damage of the surface. For imaging of soft (organic) surfaces or molecular layers that are not firmly attached to the surface, special modes of operation have been developed, such as the non-contact mode (hardly used any more because of the relatively low resolution) or tapping mode, also referred to as the intermittent contact mode or resonant mode. In the tapping mode, the tip – mounted on a lever spring that is generally stiffer than used for contact mode imaging – is forced to oscillate vertically with a frequency near the resonance frequency of the spring, while scanning laterally. Near the sample surface the amplitude varies: when the tip–sample distance decreases, the amplitude of the oscillation decreases and vice versa. The feedback system then adjusts the piezo so that the tip–sample distance is restored and the spring recovers its original amplitude of oscillation (the amplitude set point). In the tapping mode, forces acting laterally to the sample are virtually eliminated and the frequency of oscillation is so high that the collisions between tip and sample are completely elastic, provided that the amplitude set point is not chosen too small. Very soft and delicate objects, such as protein networks, surface micelles and lipid vesicles, have been studied in this way (see, for example, [36–38]).

In frictional force microscopy (FFM), also called lateral force microscopy (LFM), the lateral deflection or twist of the cantilever is monitored by the horizontal set of segments of the photodiode and the fast scan direction is normal to the long axis of the cantilever. In this mode of measuring, regions of different material can be distinguished even if there is no height difference [39]. LFM with a functionally modified tip is a way of chemical imaging, i.e. regions of different chemical nature can be identified. Quantification of the friction forces involved is complicated. In this chapter, we do not elaborate on this subject, but refer to the paper by Ralston et al. [40] and references therein.

New applications and variations on the AFM continue to be developed. Nice examples are near-field scanning optical microscopy (NSOM or SNOM), a combination of AFM and optical techniques, and force modulation microscopy (FMM or force modulation imaging, FMI). In NSOM the probe consists of a very thin (sub-wavelength size) glass fiber. The other end of this fiber is connected to a light source. Using force feedback, the probe maintains constant separation from the sample (a few nanometers). As light emanates from the tip, it only illuminates an area of the sample approximately equal to the aperture size, which may be less than 25 nm in diameter. Any collected optical contrast, either reflected from the sample or emitted by excited fluorophores, originates from this small area. Thus, NSOM overcomes the optical diffraction limit that restricts the resolution of even confocal microscopy. The FMM technique identifies differences in surface stiffness or elasticity. The tip is oscillated with small vertical amplitude and a frequency that is significantly faster than the scan rate. The average force is kept constant, equivalent to that in simple contact mode. The variation in cantilever deflection amplitude, reflecting local elasticity differences, is collected simultaneously with topographical information (non-oscillatory deflection). The same kind of information can be obtained by phase imaging in tapping mode, i.e. by recording the phase shift in the oscillation as a function of location on the sample. The phase shift is only small on a very stiff surface and increases at places where the surface is softer. Tapping mode and NSOM have been combined (TM–NSOM) to probe the compliance properties of phase-separated lipid monolayers doped with a fluorescent dye [41].

In colloid and interface science, AFM has become a standard technique for surface analysis and for the study of colloidal structures at interfaces. In the biosciences, cells, cellular structures and biomolecules can be imaged, characterized and manipulated *in situ*. For detailed information on methodologies and applications we refer to [41–43].

7.2
Force Measurements Using the AFM

7.2.1
Principles

To measure the interaction between the tip – or a colloidal probe particle glued to the tip – and a sample surface as a function of their separation, they are moved towards and away from each other. Usually, this is done in a continuous motion, but some home-built and commercial AFMs allow for a programmed mode of approach and retraction. (For example, pause for a particular time when in contact; different velocities on approach and retract; perform only one cycle of approach and retraction.)

The raw data obtained by the AFM consist of the measured deflection as a function of the z-displacement of the piezoelectric tube, as depicted in Fig. 7.5. The deflection versus z-displacement curve has to be converted into a curve in which the interaction force is plotted against the actual distance between the surfaces of probe and substrate. In order to achieve this, first the point of zero separation needs to be defined. This is derived from the constant-compliance region in which the surfaces are in contact (region III in Fig. 7.5). The onset of the constant-compliance region is taken as the origin of the separation axis. It should be noted that if there is an adsorption layer present at one or both of the surfaces, there is still a finite distance between the surfaces in the constant-compliance zone. In that case it is not possible to determine the absolute zero point of separation. Furthermore, if one or both of the surfaces is/are compressible, the first point of contact is not represented by the onset of the constant-compliance zone (in fact, there may not even be such a region). We will return later to the problems that may arise when trying to establish of the point of zero separation.

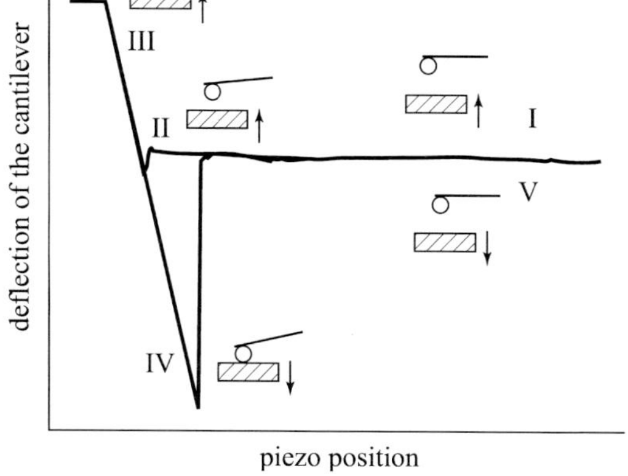

Fig. 7.5 A typical force graph in which the deflection of the cantilever is plotted against the piezo position. The position of the colloidal probe relative to the flat surface on the piezo is shown for several points of the curve, indicated by Roman numerals.

Subsequently, for each point of the force curve, the separation distance is calculated from the displacement of the piezo and the change in cantilever deflection relative to the onset of the constant-compliance region. This procedure is illustrated in Fig. 7.6.

In the range of deflection that is measured in the AFM, the relation between deflection and force F follows from Hooke's law:

$$F = -k\Delta z \tag{1}$$

where k is the spring constant and Δz the change in deflection with respect to its value at sufficiently large separations (where the interaction force is zero).

Usually, the force is displayed as a force divided by an effective radius R, in units of newtons per meter. This quantity is related to the interaction energy per unit area $U(D)$, between two flat surfaces at separation D, by

$$F(D)/R = 2\pi U(D) \tag{2}$$

This relation is known as the *Derjaguin approximation* [44] and is applicable when the range of interaction and the distance (D) between the surfaces is much less than the effective radius (R). For the interaction between two spheres, an effective radius is defined as

$$R = \frac{R_1 R_2}{R_1 + R_2} \tag{3}$$

From this equation, it follows directly that when the radius of one sphere is much larger than the other ($R_2 >> R_1$), the effective radius equals R_1. Obviously, this also holds for a sphere and a flat interface ($R_2 = \infty$). Normalization of the interaction force in this way, i.e. presenting it as F/R, facilitates comparison of force measurements performed with (colloidal) probes of different sizes or in systems with different geometries.

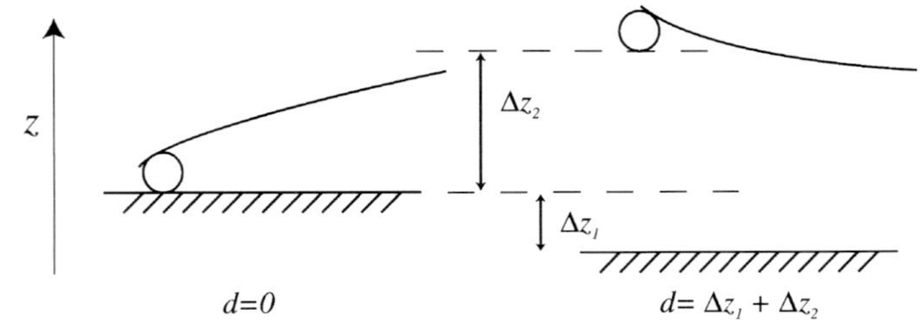

Fig. 7.6 Calculation of the distance between the surfaces. In the diagram on the left, the surfaces are in contact, i.e. their separation $d=0$ (start of the constant-compliance region during approach). In the diagram on the right, the surfaces are no longer (or not yet) in contact. Relative to the left diagram the piezo has traveled over a distance Δz_1 and the vertical deflection of the cantilever has moved over a distance Δz_2. The distance between the surfaces is now $d = \Delta z_1 + \Delta z_2$.

7.2.1.1 **Dynamic Measuring Methods**

Dynamic force measurements essentially involve the application of a low-amplitude oscillation in the vertical direction on top of the normal ramp movement, with a frequency that is much higher than that of the approach–retract cycle. The phase change and amplitude attenuation of the oscillation are monitored as a function of surface separation. These quantities together reflect the viscoelastic behavior of the medium between the surfaces. Notley et al. [45] have shown that this method can be used to measure the restricted drainage of solvent between overlapping polyelectrolyte layers and to determine the hydrodynamic thickness of adsorbed polymer layers.

7.2.2
Colloidal Probe AFM

In colloidal probe AFM, a probe particle is glued to an AFM cantilever spring. The cantilever is placed in the AFM and the interaction between the probe and a substrate is determined as a function of distance as described earlier. Usually a spherical probe particle is used, but sometimes it has another shape, for example a rod or fiber [46]. Particles of less well-defined geometry may be used, but this makes quantitative comparison with theory, and with measurements that have not been performed with the same particle, impossible. However, provided that one can measure with the same probe particle, trends can be determined, for example, differences in interaction with substrates of different nature or the effect of additives to the medium. In this way, in our laboratory the adhesion between biosulfur particles and various surfaces in air and in aqueous solution has been studied [47 a]. Instead of synthetic particles one can use biological probes like bacterial cells [47 b].

Attaching a particle to a cantilever is in principle simple (but may take some training): a sharp wire is used to put glue on the cantilever and subsequently a single particle is picked up with another sharp wire (to which it generally readily adheres due to capillary adhesion between the thin films of water present on wire and particle in ambient conditions) and placed on the glue. These actions are performed under an optical microscope and using a micromanipulator. Usually the glue is an epoxy resin that melts at elevated temperature (80–100 °C) and gluing is done on a heating stage. Of course, the glue should be inert and insoluble in the medium in which the measurements will be performed. More details can be found elsewhere [40, 48]. Most often particles several micrometers in diameter are used (see, for example, Fig. 7.7), although in some studies particles less than 1 μm in diameter have been attached to a cantilever [49]. For such small particles, it is necessary to use tipless lever springs.

The majority of colloidal probe studies have addressed the interaction between the probe and a flat substrate. However, it is also possible to measure interactions between, for example, two particles [50–52], two fibers [46] or a particle and a hollow polymeric capsule [53]. For this it is necessary to immobilize the second object on a substrate or to use closely packed particle layers. Further-

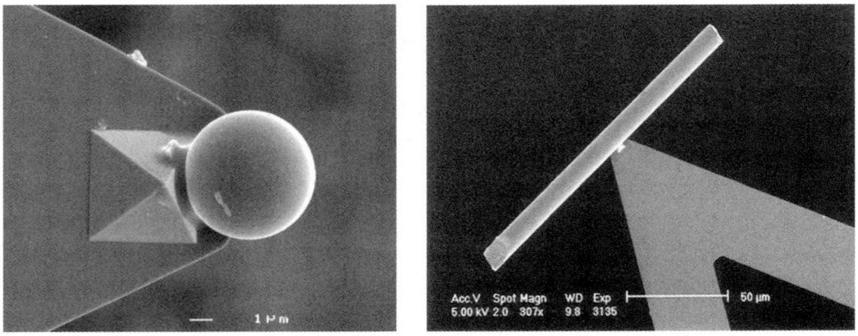

(a) (b)

Fig. 7.7 Scanning electron micrographs of colloidal probes glued to AFM cantilevers. (a) A 6-μm silica particle (from Giesbers [48]). (b) An α-alumina fiber attached to an AFM cantilever (from Meagher et al. [46], reproduced by permission of Elsevier).

more, a concentric location of the probe with respect to the second object is crucial and can be achieved by using the AFM imaging facility prior to the force measurements. Drift of the scanner can make this an awkward job and it is much more convenient to use an x,y closed-loop scanner, i.e. a scanner with a position sensor, the signal of which is used in a closed-loop algorithm to determine the voltage needed for correct movement of the scanner.

7.2.3 Calibration of the Piezoelectric Scanner

For force measurements, calibration of the displacement of the z-scanner is particularly important, for calculation of both the force and the separation between the surfaces. Calibration of x, y and z movements of the scanner is usually done by procedures using calibration grids with well-defined features, described and provided by the manufacturers of the AFM equipment. Following such procedures, the degree of expansion of the piezo for a given applied voltage is determined for movements in each direction, while the non-linearity of the movement is accounted for.

If a monochromatic laser is used to detect the deflection of the cantilever, an interferometric method of calibration for the z-movements can also be used [54]. In this method, the laser beam is directed in such a way that it is split between the end of the cantilever and a reflective substrate. By moving the cantilever and substrate relative to each other, an interference pattern is obtained. Since the laser wavelength and geometry of the set-up are known, the exact z-displacements of the piezo can be calculated. However, since such an interference pattern may also be a nuisance in force measurements (it may conceal the force curve), a number of commercial AFMs are provided with an incoherent laser, which makes this method of calibration difficult.

7.2.4
Determination of the Cantilever Spring Constant

To quantify the force between the tip and the sample, it is necessary to know accurately the spring constant k of the cantilever spring [see Eq. (1)]. Manufacturers of AFM probes provide rough nominal values (typically ranging from 0.01 to 100 N m^{-1}), but these are only a guide for choosing a particular type of lever spring and far from being accurate enough to calculate forces. Typically, even cantilevers that are made in the same batch (on the same wafer) vary significantly (by a factor of 1.5–2) in their spring constants [40, 48]. Therefore, it is necessary to determine the spring constant of each individual cantilever used in force measurements.

Various methods to obtain cantilever spring constants have been proposed and applied. In principle, the spring constant can be calculated from the shape, dimensions and material properties of the cantilever. For a single-beam lever spring [55]:

$$k = \left(\frac{t}{2l}\right)^3 Ew \tag{4}$$

where E is the elastic (Young's) modulus of the bulk material, t is the thickness of the cantilever and w and l are the width and length, respectively, of the cantilever beams. V-shaped (cantilever) springs are much more difficult to model mathematically and the simplest way to treat this problem is to describe the cantilever by two rectangular parallel beams. To avoid the use of a too simple model, Sader and White [56] performed a finite-element analysis of the static deflection of V-shaped springs and presented exact numerical results for the spring constant for a variety of cantilever dimensions.

The width and length of the cantilever beams can be measured fairly accurately, but it is difficult to obtain an exact value for the thickness and this may vary significantly even between cantilevers from one wafer. Since the spring constant depends on the thickness cubed [see Eq. (4)], this may lead to large deviations from the actual spring constant. Furthermore, the elastic modulus of the cantilever also varies due to (local) variations in the stoichiometry of the silicon nitride and that of the reflective coating, which are unavoidable in micro-fabrication techniques. These problems restrict the applicability of any model calculation of the spring constant. In practice, measurement of the spring constant of the individual spring constant is the best option to obtain quantitatively reliable force data.

A number of investigators have devised experimental methods to determine the spring constant. For example, Senden and Ducker [57] described a procedure in which a tungsten sphere (10–50 μm in diameter) of known mass is attached to the end of a cantilever. The resulting deflection of the cantilever is measured for the standard orientation and after placing the cantilever (i.e. the AFM head piece, including the cantilever holder and the optical detection system) upside down. The difference between the two measurements is twice the deflection due to gravity, from which the spring constant can be calculated. In many modern commercial AFM heads the laser is shut off when the head is turned upside down (for safety reasons), so this method cannot be used.

In several methods, the spring constant is derived from the resonance frequency of the unloaded cantilever [58] or the shift of the resonance frequency of the cantilever when loaded with particles of known mass. The latter is referred to as the Cleveland method [59] and is commonly recognized as a reliable method to obtain cantilever spring constants for force measurements for practically all AFMs without further equipment [40, 48]. Therefore, we describe it here in some detail.

The resonance frequency ν of a cantilever loaded with an added mass M is

$$\nu = \frac{1}{2\pi}\sqrt{\frac{k}{M+m}} \tag{5}$$

with m the effective mass of the cantilever. Rearrangement of this equation shows that

$$M = \frac{k}{(2\pi\nu)^2} - m \tag{6}$$

Thus, measuring the resonance frequencies for several attached masses and plotting M as a function of $(2\pi\nu)^2$ gives a line with slope equal to the spring constant of the cantilever and the intercept gives its effective mass. The added masses are spherical particles of known density (usually tungsten spheres) and various sizes, which are determined by optical microscopy. These are attached to the cantilever one by one in much the same way as a colloidal probe, except that no glue is used as in air the particles stick due to capillary adhesion. The accuracy of this method depends on the error in M (depending on the accuracy with which the diameter of the particles has been determined and whether they are indeed spherical), but also on the exact and same positioning of the particles on the cantilever. If the procedure is conducted carefully, the error in the spring constant is around 10%. A disadvantage is that it requires a time-consuming micromanipulation of particles for each cantilever that has been used for quantitative force measurements.

A very good compromise between accuracy and convenience is the "thermal noise" method, in which the spring constant is obtained from a power spectral density analysis of the thermal vibrations of the free cantilever, i.e. not interacting with a surface [60]. By modeling the cantilever as a harmonic oscillator with one degree of freedom and neglecting higher modes of oscillation, it can be simply derived that

$$k = \frac{k_B T}{\langle z^2 \rangle} \tag{7}$$

where $\langle z^2 \rangle$ is the mean square deflection of the cantilever in vertical direction due to thermal motion, k_B the Boltzmann constant and T the temperature.

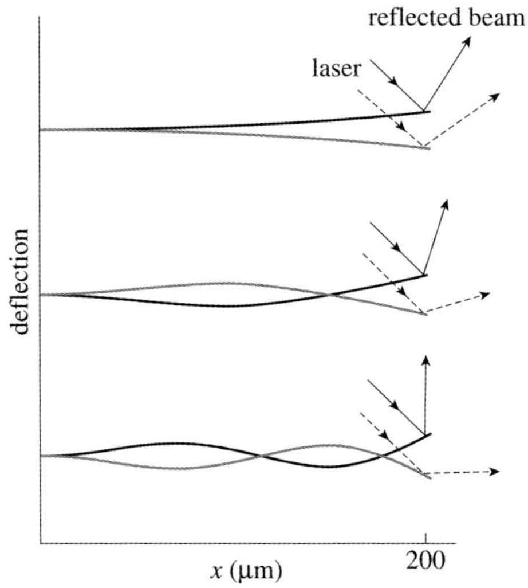

Fig. 7.8 The first three eigenmodes of a freely vibrating rectangular cantilever. This diagram illustrates that the direction of the reflected beam depends on the inclination of the cantilever (dz/dx) at the point of reflection. It should be noted that the effect is highly exaggerated, since the dimensions in the z-direction are not in scale with the x-axis (the deflection at the end of the cantilever resulting from thermal noise is in the order of only a few tens of nanometers).

Equation (7) follows from the equipartition theorem, stating that the thermal energy stored in the vibration equals $1/2k_BT$ and the fact that a bending of the cantilever by a small amount z corresponds to a potential energy of $1/2kz^2$. By calculation of the thermal noise of a rectangular cantilever with one free end, considering all possible vibration modes, Butt and Jaschke [61] showed that the relationship given in Eq. (7) is also true for the total thermal noise. The value of $\langle z^2 \rangle$ is equal to the area under the power spectrum of all thermal vibrations in the z-direction after subtraction of the white noise. Furthermore, Butt and Jaschke pointed out that if the fluctuations are measured optically using the reflection of a laser beam on the back of the cantilever, it is not the vertical displacement z, but rather the inclination (angle) (dz/dx), that is obtained (x being the coordinate in the longitudinal direction of the cantilever). The relation between inclination dz/dx and displacement z varies for the different modes of vibration since it depends on the shape of the deformations. Figure 7.8 shows schematically that oscillating shapes of a lever spring differ from its static shape. According to the analysis given by Butt and Jaschke, a correction factor of 4/3 has to be included:

$$k = \frac{4k_B T}{3\left\langle z^{*2} \right\rangle} \qquad (8)$$

where z^* is the (virtual) displacement as measured using the optical lever technique. A further complication is that because of bandwidth limitations only the first modes are experimentally accessible. For a rectangular (one-beam) lever spring, the contribution of the first peak in the noise power spectrum (first harmonic oscillation) to the virtual deflection is about 70% [61]. The findings of Butt and Jaschke have been verified experimentally by Lévy and Maaloum [62].

Stark et al. [63] calculated the thermal noise of a V-shaped cantilever by means of a finite element analysis and showed that the thermal noise of V-shaped cantilevers differs systematically from that of rectangular cantilevers. For the cantilever for which the analysis was performed, the contribution of the first harmonic oscillation to the noise power spectrum is approximately 78%. Therefore, using the simplifying assumption of a rectangular beam model in the determination of the spring constant for such a V-shaped cantilever from the first peak in the noise power spectrum would give rise to an error of about +8%.

The thermal vibration method gives fairly accurate results provided that the contributions of the various vibration modes for the particular type of cantilever to the noise power spectrum are taken into account. The software of modern commercial AFMs often allows for a sufficiently detailed spectral analysis of the thermal noise, which makes the method experimentally simple to perform. The viscosity of the medium or the effective mass of the cantilever does not affect determination of the spring constant using this method [64], so that calibration can be performed directly in the system in which force measurements are conducted.

7.2.5
Points of Attention

Nowadays, commercial AFMs are provided with user-friendly software to obtain force curves, to calibrate the spring constant of the lever spring, to convert the raw data into force-separation curves and so on. Hence force measurements are relatively simple to perform and no longer the domain of AFM experts and pioneers. However, although in some systems force measurements may be almost routine, in many cases it is not straightforward to obtain really reliable and accurate results. Here, we briefly discuss some points of attention and limitations of the technique. Details on good practice can be found in various papers, e.g. in the IUPAC technical report by Ralston et al. [40] and in a recent, extensive review by Butt et al. [65].

7.2.5.1 Scanner Peculiarities

As already pointed out, the correct calibration of the piezoelectric scanner, in particular with respect to the z-movement, is crucial for quantitative force measurements. The accuracy of not only the separation distance but also of the interaction force depends on this. Since the response of the piezo scanner to voltage is dependent on the age of the piezo and ambient humidity, regular calibration is necessary. Most of the change in the sensitivity occurs at the beginning of a scanner's lifetime.

In addition, one should be aware of several issues concerning the movements of the piezo-scanner, such as hysteresis, drift and creep. Hysteresis between the two scan directions is caused by non-linearity of the scanner: it is more sensitive (i.e. more movement per applied volt) at the end of a scan line than at the beginning. This causes differences in the forward and reverse scan directions. The same is true in force measurements: the piezo moves less per applied volt at the beginning of the approach or retraction than near the end. The non-linearity should be corrected in the calibration procedure, resulting in a non-linear voltage in real-time measurements to produce a linear scan in all directions.

During measurements, the average piezo position in the x, y and/or z direction may gradually change due to drift. For force measurements this may imply that the area of interaction gradually shifts over the substrate. Drift in the z-direction may be observed as a change in the length of the constant-compliance region. Creep is the drift of the piezo displacement after a relatively large offset voltage has been applied to it. In an image it appears as an elongation and stretching of features in the offset direction for some time after the offset. Systems with so-called closed-loop scanners automatically correct for drift in the x,y plane and/or in the z-direction. With closed-loop scanners the maximum scan rate is somewhat lower.

7.2.5.2 **Absolute Distance**

Lack of a direct method to determine the separation between the probe and the substrate surface is often mentioned as an inherent shortcoming of the AFM in force measurements. Indeed, in many cases this limits interpretation of force data, in particular in the case of soft, deformable surfaces. For adsorption layers of, for example, polymers of biomolecules on a solid substrate, the constant-compliance region may represent either the contact position of the solid surfaces (the true zero separation) if the layer is pushed away from the contact region or the position at which the elastic modulus of the adsorbed layer exceeds the spring constant of the AFM cantilever.

However, even for bare, non-deformable surfaces with strong repulsive forces, there may be a distance error, in particular when a cantilever with low spring constant is used. The determination of the zero of separation is based on the assumption that the constant-compliance zone in the raw data is caused by the direct coupling of the probe and the surface in hard contact. As demonstrated by Biggs et al. [66], this is not always true. From a simple semi-quantitative analysis of electric double-layer forces they showed that it may happen that an apparent constant-compliance regime is reached while the surfaces are still interacting through their double layers, i.e. before they really come into contact. Lever springs with a low spring constant are more sensitive to weak forces, but this also implies that – in the case of repulsion – the piezo has to travel a relatively long distance before constant compliance is reached. Before that is the case, the deflection can easily pass the limit of the photodiode detector. Together this may result in observing an apparent constant-compliance region. Evidence for the apparent constant-compliance behavior arising from electric double-layer repulsion was obtained by Piech and Walz [67], who examined depletion forces and, since such forces are weak, used a cantilever with low spring constant. Biggs et al. [66] investigated structural and depletion forces between silica surfaces in aqueous solutions of poly(styrene sulfonate) (PSS) by colloidal probe AFM and also by TIRM. Comparison between the data sets was used to calibrate the AFM data with respect to absolute separation distances. It was found that except for the highest PSS concentration used (1000 ppm), the AFM data have a significant error in the zero separation (in the order of 10–30 nm) and this was attributed to false constant-compliance regions prior to real contact between the surfaces.

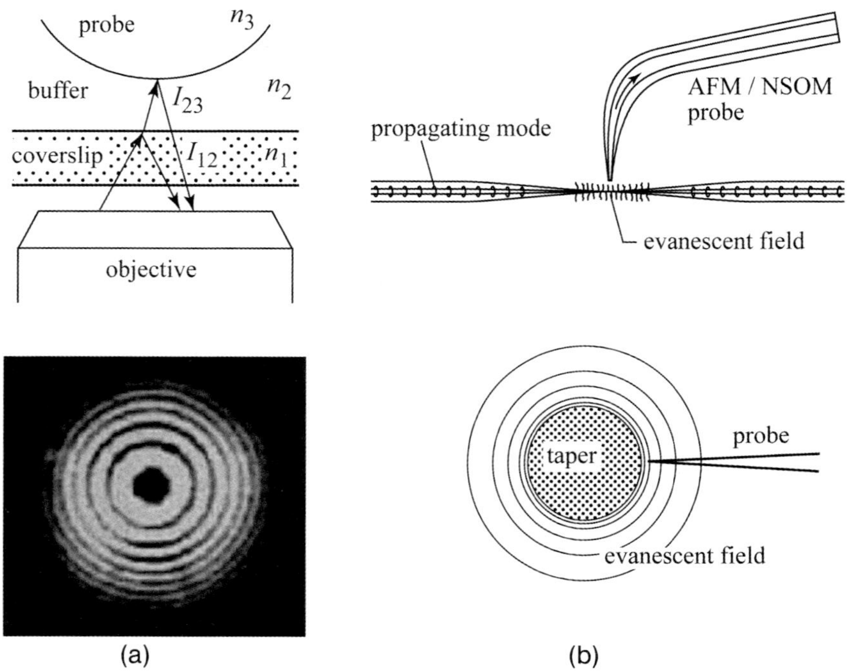

Fig. 7.9 (a) Application of reflection interference microscopy to measure the separation between a spherical AFM probe and sample. The interference pattern formed by the object beam I_{23} and the reference beam I_{12} is observed as circular interference fringes, which depend on the distance between probe and surface. (From Stuart and Hlady [68], redrawn with permission from the Biophysical Society). (b) Combination of NSOM and AFM force measurements. The near-field optical probe detects the intensity of the evanescent field generated by an optical fiber taper sensor. The intensity falls off exponentially with the distance between probe and surface. (From Huntington et al. [69], redrawn with permission from Elsevier).

The problems encountered in determining absolute distances in AFM force data may be overcome by the invention of new techniques and combinations of techniques. For example, integration of a reflection interference contrast microscope and an AFM makes it possible to measure independently the separation distance between the AFM probe and the sample surface [68] (see Fig. 7.9). However, to apply this method the substrate has to be optically transparent. Using a combined AFM–NSOM (near-field optical microscope) system, Huntington et al. [69] measured the separation between a probe and a compliant sample. This sample was a tapered optical fiber from which an evanescent field was generated (Fig. 7.9). The decay of the evanescent field was monitored using a near-field optical probe on a lever spring, while simultaneously measuring the deflection of the spring and thus the interaction force. Drawbacks of this approach are that NSOM is an expensive technique and the limited choice of both probes and substrates.

When one or both of the surfaces deform during force measurements (for example, between a probe particle and an air bubble or oil droplet), it becomes very difficult to interpret the interaction curves. The zero of separation is hard to determine, since surface contact does not give a linear region of constant compliance. Furthermore, in case of long-range attraction or repulsion, the shape of the surface(s) and therefore the relative separation may already change on approach. As an additional complication, deformations of viscoelastic bodies depend on the rate of which the force curves are recorded.

7.2.5.3 Surface Roughness

An advantage of colloidal probe AFM force measurements over the SFA is that the interactions between all kind of materials can be measured, since the surfaces do not have to be smooth over a large area (or transparent or reflective); the interaction area between the colloidal probe and the sample surface is typically in the range 100 nm^2–0.1 μm^2. However, surface roughness complicates the interpretation of force curves, in particular with respect to short-range forces. The surface roughness of both the substrate and the probe can be evaluated by AFM imaging [52, 70] and reverse imaging, i.e. scanning with the probe particles over a surface with sharp features [71].

For incompressible asperities, the result of surface roughness is that at minimum separation only a fraction of the opposing surfaces are in real contact. Forces of which the range is comparable to or shorter than the thickness of the roughness layer do not or only partly manifest themselves. As a consequence, the relative contribution of forces between colloids, for example van der Waals forces and electrostatic interactions, can be different than expected, leading to anomalous colloidal stability behavior [69]. It is obvious that also in colloidal probe AFM measurements roughness of the interacting surfaces attenuates short-range forces, such as van der Waals forces and solvent structural forces; contact forces that lead to adhesion develop for only a fraction of the interaction areas. Furthermore, surface roughness affects electrostatic double-layer interactions, since there is no well-defined plane of charge and the potential already falls off to a lower value over the roughness layer. We will return to these issues in Section 7.3.

7.2.5.4 **Hydrodynamics/Hydrodynamic Drag**

In AFM force measurements with a colloidal probe, hydrodynamic effects should be taken into account. When the probe particle approaches the other surface, the liquid between the surfaces has to be squeezed out of the gap. The repulsive hydrodynamic force involved increases linearly with the rate of approach and increases as the separation between the surfaces decreases. On retraction, liquid has to fill the gap again and then the hydrodynamic force is attractive. The forces can be described using fluid mechanics [65]. One should be aware that the frequently used assumption that adjacent to a solid wall liquid molecules have no velocity relative to the solid (no-slip condition) may generally be applicable for macroscopic systems, but on the nanometer scale it is questionable, in particular in the case of very smooth and lyophobic surfaces.

A really disturbing factor can be the hydrodynamic drag by the cantilever, which has much larger dimensions that the tip or colloidal probe. This can be a problem, especially when studying structural forces due to some kind of delicate molecular organization in the liquid. We will return to this in later sections. A way to assess the degree of hydrodynamic perturbation by the cantilever, and to minimize it, is to do force measurements at different rates and extrapolate to zero velocity.

7.3 Forces in the Colloidal Regime

In this section, we discuss typical examples of forces in the colloidal regime as measured with the AFM. Before doing so, we first consider some general properties of relevant forces. We then take a look at experimental data, arranged according to the composition of the systems in which they were measured.

Traditionally, "colloidal" forces between particles dispersed in a liquid medium are divided into many different kinds: double-layer forces, van der Waals forces, hydrophobic forces, polymer bridging and depletion forces, etc. In some cases, forces are detected which cannot be put under any of these headings, but are somehow caused by the structure of the medium between the surfaces; these forces have been called "structural forces". In this scheme, forces are often denoted by terms suggesting some sort of molecular mechanism: undulation, protrusion, steric, bridging, etc.; the list grows as more experimental data are reported. What seems to be lacking here is a clear conceptual framework. For example, "depletion forces" have much in common with "structural forces" and can indeed be understood in very similar terms. We therefore primarily distinguish between just two general classes of colloidal forces: electromagnetic forces and structural forces. In addition, one often detects and measures forces in the

AFM that arise as the result of the emergence of a new phase between the interacting surfaces. Essentially, such forces derive from the interfacial tension between the fluid phases and their selective wetting of the solid; we will therefore refer to these as "wetting forces".

Three electromagnetic forces are relevant in the colloidal regime: electrostatic forces due to the electric field present in the neighborhood of charged particles (the so-called Maxwell stress), magnetic forces as occur between magnetic particles and forces due to coupling between rapidly fluctuating (10^{-16} s) electronic dipoles (van der Waals dispersion forces). Typically, electromagnetic forces are fast; they propagate at the speed of light and can therefore be considered as instantaneous on the time-scale of colloidal events. For the van der Waals forces, however, the time it takes for signals to travel between particles at distances of the order of 20 nm becomes comparable to the fluctuation time, leading to "retardation", i.e. a faster decay of the force with distance.

Structural forces generally originate from modifications in the structure of the medium between two approaching surfaces. "Structure" here refers to variations in the density, orientation or shape of components in the liquid phase (solvent, solutes). Such variations occur in bulk liquids also, as fluctuations in relevant properties (density, orientation); each of these has its characteristic correlation length scale and amplitude. For example, pure, simple liquids far from the critical point have density fluctuations on the scale of a typical molecular diameter. Dipolar molecules such as water also have fluctuations in molecular orientation. In addition, solutions may have fluctuations in the concentration of the solute; for example, very pronounced concentration fluctuations are encountered in aqueous polyelectrolyte solutions of moderate concentration, due to the strong electrostatic repulsion between these highly charged objects. In Fig. 7.10, this is illustrated for partially sulfonated polystyrene (PSS) of various charge densities.

Any surface that interacts with a component in the liquid will cause changes in the structure of the adjoining fluid. The density oscillations of a pure liquid near a hard wall are one example; the "depletion zone" of negatively adsorbing polymers is another. It is the surface free energy stored in these modified density profiles which gives rise to a force; we propose to call all such forces "structural forces". Hence the depletion force is one example of a structural force; the steric force coming from an adsorbed block copolymer is another. The distance over which the structural perturbations propagate from the surface into the liquid gives the range of structural forces; in most cases this is a few correlation lengths. Hence depletion effects typically extend over distances of the order of the size of the polymer coils (or the mesh size of the transient polymer network), whereas low molecular weight liquids give rise to oscillatory force curves with a period that is much smaller. Examples of forces due to structure in simple liquids are known for water, liquid alkanes and spherical silane molecules [1].

Structural forces have time-scales associated with molecular rearrangement (diffusion, reorientation, reconformation) times in the system. Particularly for macromolecules at solid/liquid interfaces, such times may be long to very long. Therefore, structures may sometimes rearrange completely on experimental

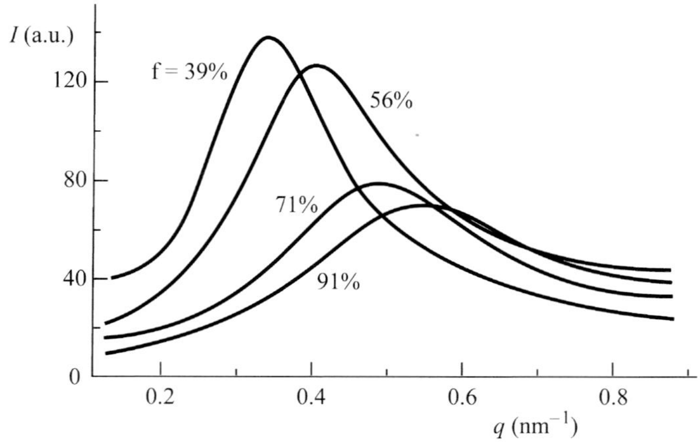

Fig. 7.10 Small angle X-ray scattering (SAXS) intensity (arbitrary units) as a function of wavevector q for aqueous solutions of differently charged PSS (charge fraction given by f). Concentration of polymer: 0.1 M. The broad peak is characteristic of salt-free polyelectrolyte solutions and its position q^* gives the correlation length of the bulk density fluctuations (i.e. $2\pi/q^*$). (Results of Qu et al. [72], redrawn with permission from the American Chemical Society).

time-scales, leading to forces that correspond to full thermodynamic equilibrium, whereas in other cases rearrangement may be virtually suppressed, so that the forces do not correspond at all to complete thermodynamic equilibrium. In general, incomplete relaxation always leads to additional repulsion upon approach, because there is additional work of a dissipative nature that needs to be done to bring surfaces together. For example, the repulsive 'lubrication force' which arises as a consequence of viscous flow of liquid from the gap between two approaching surfaces (see Section 7.2.5) is a force with such a dissipative origin, caused by the fact that a tangential pressure gradient cannot relax sufficiently rapidly. When layers of adsorbed polymers in a solvent are pressed together, the solvent can leave far more rapidly than the polymer, so that the polymer is concentrated. Upon approach, this leads to an unrelaxed osmotic pressure gradient and, hence, repulsion. Partial relaxation during contact may result in a new structure, which produces an attractive force upon retraction. It is important to be aware of such possibilities when dealing with experimental data. In specific cases, it may be difficult to ascertain to what extent (partial) relaxation has occurred in a given experiment.

7.3.1 Interactions Between Solid Surfaces in Aqueous Electrolyte

In aqueous solution, practically all solid surfaces bear an electric charge due to the dissociation of surface groups or specific adsorption of ions from solution. In addition, some surfaces bear a more or less constant charge either as a result of isomorphic substitution of ions in the material itself (e.g. clay minerals) or due to covalently bound functional groups (e.g. sulfonic acids). The surface charge is compensated by a countercharge in solution, together forming an electric double layer. When electric double layers overlap, surfaces attract or repel each other. These are relatively large forces, well within the range of AFM sensitivity. Colloidal probe AFM is a particularly suitable method for measuring such electric double-layer interactions. An example of force curves dominated by electric double-layer interactions is given in Fig. 7.11. Many other consistent force data exist for the interaction between a silica plate and a spherical silica particle [49, 50, 73–77]. For a large part due to the vast amount of data on electrostatic interactions in solution that has become available from AFM force measurements, electric double-layer modeling again receives a lot of attention [78–80].

The interaction between the diffuse parts of two electric double layers can be adequately described using the non-linear Poisson–Boltzmann (PB) equation. For planar surfaces in a symmetrical (z:z) electrolyte solution:

$$\frac{d^2\psi(z)}{dx^2} = \frac{2zen_0}{\varepsilon}\sinh\left[\frac{ze\psi(x)}{k_B T}\right] \tag{9}$$

where ψ is the electric potential at a distance x from the surface, e the unit charge, n_0 the bulk concentration of the electrolyte ions, ε the dielectric constant of the solution, k_B the Boltzmann constant and T the absolute temperature. The

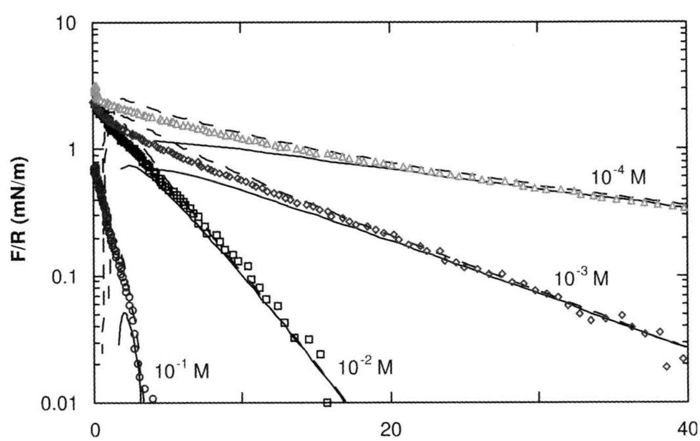

Fig. 7.11 Interaction between a spherical silica particle and a silica wafer in aqueous solutions of NaCl (pH ≈ 6). The data points are represented by the symbols. The curves were calculated using the DLVO theory and solving the Poisson–Boltzmann equation using boundary conditions of constant charge (dashed lines) and constant potential (solid lines). The force normalized to the particle radius (F/R) is plotted on a logarithmic scale, so that the slope of the curves at larger separations corresponds to the Debye length. (From Giesbers et al. [81], reproduced by permission of Elsevier).

interaction free energy can be calculated from the expression for the disjoining pressure Π:

$$\Pi = 2k_B T n_0 \left\{ \cosh\left[\frac{ze\psi(x)}{k_B T}\right] - 1 \right\} - \frac{\varepsilon}{2}\left[\frac{d\psi(x)}{dx}\right]^2 \quad (10)$$

The first term on the right-hand side represents the ion osmotic pressure and the second the Maxwell stress. Hence the double-layer force is partly electromagnetic and partly structural; the structural part tends to relax rapidly, but changes in surface charge may occur on experimental time-scales [82]. The expression holds for any position x in the overlapping diffuse double layers between the surfaces. The free energy per unit area of the diffuse double-layer interaction is obtained by integration over the distance (boundaries being the onsets of the diffuse parts of the double layers). Combination of the PB equation, Eq. (10) and the Derjaguin approximation (Eq. 7) and choosing suitable boundary conditions for integration allows calculation of the force profile $F(D)$ in the region of the diffuse parts of the overlapping double layers. For details see, for example, the book by Lyklema [83].

The most extreme boundary conditions that may be chosen to solve the PB equation are that of "constant charge" or "constant potential". The former implies that the charge in the diffuse part of the double layers, σ^d, is taken to be constant and the latter that the potentials at the onset of the diffuse double layers, ψ^d, are assumed to be constant. In the original papers by Derjaguin and Landau [84] and Verwey and Overbeek [85] (the basis of the DLVO theory), only diffuse double layers are treated. In such a Gouy–Chapman approach to the double layer, the above boundary conditions imply constant charge densities (σ^0) or constant potentials (ψ^0) at the actual surfaces of the interacting bodies. In some special cases one can assume validity of either of the two boundary conditions, e.g. σ^0 constant for clay plates when the surface charge is practically completely determined by isomorphic ion substitution or for very fast approach, so that the charges in the double layer cannot adjust. However, intermediate situations, in which both σ^0 and ψ^0 are adjusted to some extent, are more abundant. The process by which equilibrium is re-established upon a change in the overlap of double layers is called "regulation".

The simple model of a surface charge and a diffuse countercharge, as considered in DLVO theory, holds only for low surface potentials and low electrolyte concentration. There are a number of arguments, also based on experimental data, to refine the double-layer model with one (Stern layer) or two extra layers (inner and outer Helmholtz layers), adjacent to the surface [83]. In this way, specific adsorption of ions can be accounted for, and also the finite size of (hydrated) ions. Introduction of such layer(s) means that the onset of the diffuse double layer is no longer at the surface. One of the consequences is that there are no longer plausible physical arguments for choosing between one of the two sets of boundary conditions, constant charge or constant potential, to solve the PB equation. For a more sophisticated, i.e. more realistic, approach, a model for regulation is needed. In such a model σ^d and ψ^d are related as a function of separation or overlap. Since the formation of the double-layer charge is determined by the balance of specific (chemical) binding of ions to the surface and the counteracting effect of electrostatic repulsion, regulation involves both chemical and electric work (for a detailed discussion see Lyklema and Duval [79]). Furthermore, when the characteristic time-scales of adsorption and desorption of ions are larger than that of the change in separation of the surfaces, interaction dynamics start to play a role. This issue has been reviewed by Lyklema et al. [86].

A problem is that charge regulation characteristics only show up in the force curves at relatively small separations, say below 10–15 nm (for interactions between dissimilar surfaces this may be more), while in this region other interactions, such as van der Waals and hydration forces and surface roughness effects, can play a significant role. Therefore, it is rather pointless to fit the data using sophisticated double-layer models, except for illustrating that double-layer theory itself may already account for unexpected features in force curves.

In practice, the tail end of double-layer interaction curves is generally easily fitted to the PB equation (or to DLVO theory, i.e. also including van der Waals interactions) for constant charge (CC) as well as constant potential conditions (CP). To do this numerically, algorithms are available for symmetric systems, i.e. two identical surfaces interacting ("homo-interaction") [87, 88] and for asymmetric systems ("hetero-interaction") [89, 90]. As can be expected, it is often found that the measured curves are between these two extremes, as is also the case in Figs. 7.11 and 7.12.

For homo-interaction, theoretical force–distance curves can be fitted to the experimental curves with the double-layer potential ψ^d at infinite separation as the only fitting parameter. As a rule, the values obtained for ψ^d from CC or CP are not very different [91]. Furthermore, they are comparable to zeta potentials obtained from electrokinetic techniques [76, 77, 81, 91]. As an example, some data for silica from different authors are shown in Table 7.1.

Insight into hetero-interaction comes almost exclusively from AFM work. It is not possible to obtain unique values for both the double-layer potentials of the two different surfaces from a fit of the interaction curves with theory. To obtain meaningful information it is required that for at least one of the surfaces the potential at infinite separation is known. For this, ψ^d values can be used that are obtained from force measurements on symmetric systems or values obtained using other techniques.

The interaction between double layers of two identical surfaces is always repulsive. In the case of hetero-interaction, however, double-layer theory predicts either attraction or repulsion, which may change on approach due to charge reversal of one of the surfaces, depending on the charging mechanisms of both surfaces [79, 83, 93–95]. The experimental and calculated curves in Fig. 7.12 illustrate the complexity of electrostatic interactions between dissimilar surfaces. For surfaces of the same charge sign, both kept at a constant charge, the interaction is always repulsive (dashed curves in Fig. 7.12 a and b); if constant poten-

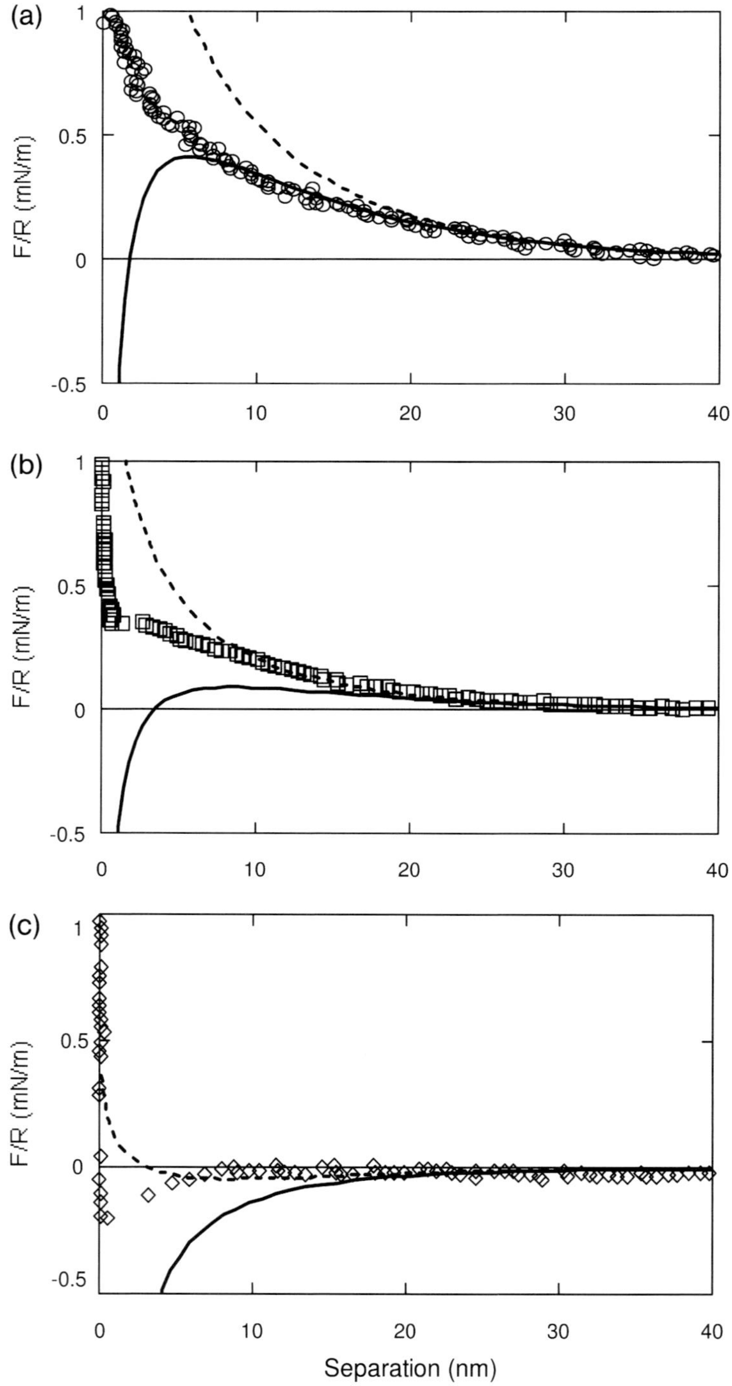

Fig. 7.12 (legend see p. 201)

Table 7.1 Double-layer potentials of silica (in mV) at neutral pH (~6) from DLVO fits of silica–silica interaction curves and zeta potentials (mV) of silica obtained by electrokinetic measurements.

Ionic strength (M)	ψ^d			ζ		
	AFM [a)]	AFM [b)]	SFA [c)]	Streaming potential [d)]	Streaming potential [e)]	Electrophoresis [f)]
10^{-1}	–21	–18	–23	–22		
10^{-2}	–34	–33	–28	–40		
10^{-3}	–53	–46	–32	–67	–47	–64
10^{-4}	–61	–88	–40	–83	–85	–74

a) Colloidal probe AFM; interaction between oxidized silicon wafer and silica particle; NaCl solutions [75].
b) Colloidal probe AFM; oxidized silicon wafer and silica particle; NaCl solutions [81].
c) SFA measurements; thin, smooth silica sheets (Suprasil); NaCl solutions [74].
d) Streaming potential measurements; vitreous silica; KNO_3 solutions [92].
e) Streaming potential measurements; vitreous silica (Suprasil); $NaNO_3$ solutions [76].
f) Electrophoresis; silica particles prepared using a modified Stöber process; $NaNO_3$ solutions [76].

tial is maintained, the repulsion turns into an attraction on approach (solid curves in Fig. 7.12 a and b) as a result of charge reversal on the surface with the lower potential. For surfaces of opposite charge signs maintaining constant potential, the interaction is always attractive (solid curve in Fig. 7.12 c); for constant charge surfaces of opposite sign, the attraction turns into a repulsion on approach (dashed curve in Fig. 7.12 c). The more extreme the ratio between the surface charge densities at infinite separation, the larger is the separation distance at which sign reversal of the interaction occurs. For the experimental curve measured at pH 4, where the ratio of the charge densities of the gold and silica surfaces is small, the bump at 3–4 nm probably reflects the onset of a change of repulsion into attraction; before the sign of the electrostatic interaction on approach really changes from positive to negative, other effects (such as

Fig. 7.12 The interaction on approach between a silica sphere and a gold-coated surface in 10^{-3} M solutions at three pH values, as indicated. The data are fitted to the PB equation for the boundary conditions of constant charge (dashed curves) and constant potential (solid curves). For each data set the potential of the silica surface (obtained from silica–silica interaction curves) was entered as a parameter and the potential of the gold surface was obtained from the best fit to the data. These input and fitted values are (a) ψ^d (silica) = –58 mV, ψ^d (gold) = –26 mV, (b) ψ^d (silica) = –35 mV, ψ^d (gold) = –13 mV; (c) ψ^d (silica) = –22 mV, ψ^d (gold) = +10 mV. (From Giesbers [81], reproduced by permission of Elsevier).

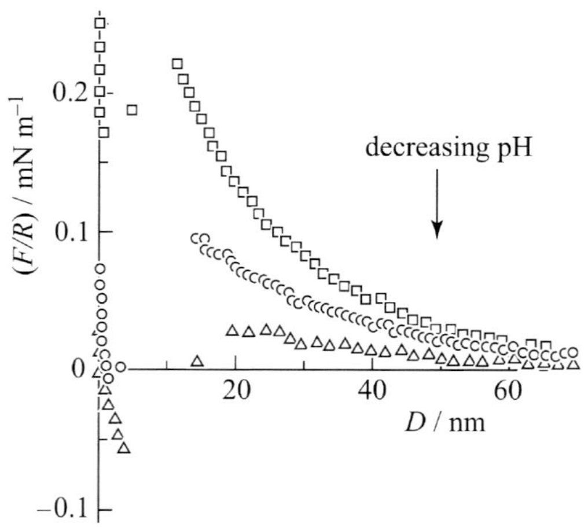

Fig. 7.13 Interaction between ZnS spherical particles and PbS cubes at pH 4, 5.8 and 10. Electrolyte: 2×10^{-4} M NaCl. (Results of Toikka et al. [96], redrawn with permission from the Royal Society of Chemistry).

hydration) start to dominate the force curve. The small maximum in the curve is not likely to be due to van der Waals forces, since the authors did not observe these in any of the other curves for silica–silica, gold–gold or silica–gold interactions [81]. Complete charge reversal has been observed experimentally by Toikka et al. [96], as shown in Fig. 7.13.

According to DLVO theory, van der Waals forces always exceed the double-layer forces at small distances, leading to an attractive minimum as the separation approaches zero. However, for a range of systems (silica, mica, metal and metal sulfide surfaces, homo- and hetero-interaction), the experimental force curves frequently either lack any indications of a van der Waals interaction [49, 73, 74, 81, 91, 96–99] or an attraction is found that is much smaller than expected [75] (see for example Figs. 7.11 to 7.13). In Fig. 7.14, another example is given for the interaction between two gold-coated surfaces. Fitting the data with the DLVO theory, using the calculated value for the non-retarded Hamaker constant (from Lifschitz theory) for the gold–water–gold system [100], 2.5×10^{-19} J, gives satisfactory results only for separations larger than about 20 nm. The fit is much better if only double-layer interactions are taken into account, the experimental points lying between the solutions for CC and CP.

There are several factors that can account for the deviation of measured interaction curves from the DLVO theory at small separations. The most important ones are surface roughness and solvent structural forces, i.e. repulsion between hydration layers. Sometimes the effect of an apparent constant-compliance region plays a role, as discussed before (see Biggs et al. [66] and references therein). Since van der Waals forces are relatively short-range and arise from differences in material properties between interacting bodies and medium [3] (i.e. they are not really surface forces), these are much smaller for two objects with rough surfaces in contact

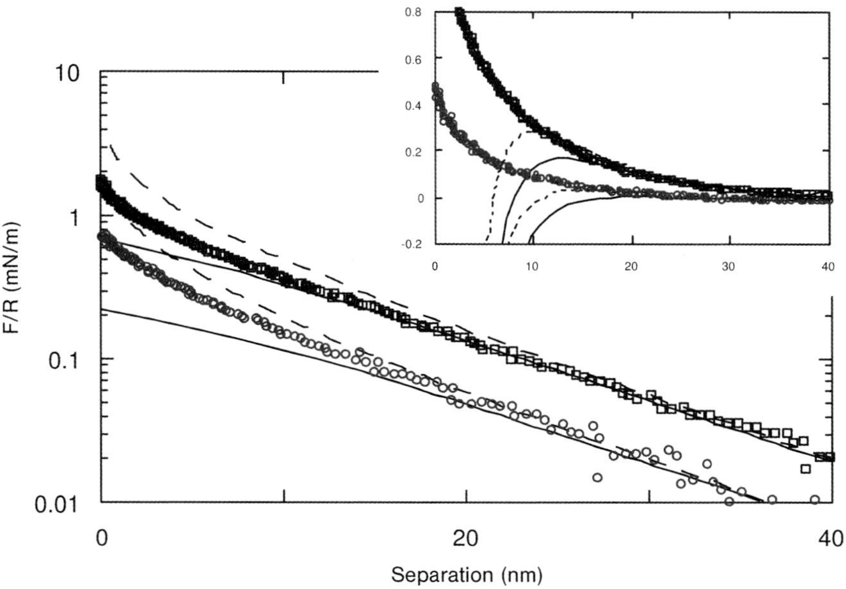

Fig. 7.14 Interaction between a gold-coated silica sphere and a gold-coated silicon wafer at pH 7 (○) and 11 (□). Total ionic strength 10^{-3} M. The curves were calculated using the non-linear PB equation and boundary conditions of constant charge (dashed lines) and constant potential (solid lines). Inset: the same interaction curves fitted using DLVO theory, i.e. including van der Waals interaction. (From Giesbers et al. [81], reproduced by permission of Elsevier).

than in the case of ideally smooth surfaces [97, 98]. It has been shown [99] that even when electrostatic repulsion between silica surfaces is practically zero, i.e. close to the isoelectric point of silica and at high ionic strength, there is no sign of attraction, which is explained by supposing that van der Waals attraction is completely obscured by a strong short-range repulsion due to the solvent structuring effect of the surface silanol groups (–OH). This hydration repulsion is greatly reduced when the silica surfaces are treated with (γ-aminopropyl)dimethylethoxysilane [101] or when the protons of the silanol groups are replaced by sodium ions [102], making it possible to observe van der Waals attraction, albeit significantly weaker than predicted by theory.

The Hamaker constant for interaction between metal objects is relatively high; compare, e.g., the calculated Hamaker constant for gold–water–gold [95], 2.5×10^{-19} J, with the calculated value for silica–water–silica [102], 8.5×10^{-21} J. The high calculated value for the gold–water–gold system was confirmed by AFM force measurements between a small silica sphere (ca. 2 μm in diameter) covered with a relatively thick layer of gold (0.6 μm) and a massive gold plate, which were in line with the high Hamaker constant [96]. However, if this value were applicable for the system described in Fig. 7.14, van der Waals forces would dominate the last 10–15 nm of the approach, as is shown in the inset, and should therefore be clearly visible in the force curves – despite any surface roughness and hydration. The fact that this is not the case is in line with results of coagulation experiments on gold sols [103, 104]. The phenomenon that metal sols are often much more stable than expected is caused by the fact that the particle sizes in these sols are generally of the order of only 10 nm. For such small metal particles, the Hamaker constant is much lower than for the bulk metals. This has been shown experimentally by Frens [105] and can be explained theoretically [106–108]. The strong dispersion forces between pieces of metal originate from the coupling of surface modes of free electrons and this mechanism becomes less effective when the dimensions of the interacting objects shrink down to the penetration depth of these modes or to the mean free path of the electrons. For the interaction curves displayed in Fig. 7.14, the gold coatings of colloidal probe and wafer were only 15 nm thick and therefore the van der Waals forces are much lower than expected on the basis the calculated (bulk) gold–water–gold Hamaker constant.

To return to the issue of surface roughness, it should be noted that it also affects the determination of the double-layer potential by a PB fit. First, there is no well-defined plane in which the surface charge is located; this charge is divided over the surface roughness layer. Second, in contact only the larger surface asperities are touching the opposite surface. Therefore, the double-layer overlap is smaller than for smooth surfaces in contact. Provided that the Debye length is considerably greater than the thickness of the roughness layer, the effect of roughness on the fit of the double-layer interaction is small. For short Debye lengths (i.e. high ionic strengths), however, the fitted ψ^d will be significantly lower than the actual value [70, 91]. The effects of surface roughness and other surface inhomogeneities on the interaction between electric double layers have been modeled by numerous groups (see the paper by Duval et al. [80] and references therein).

AFM force measurements with a silica probe particle have been used to study the double layer on a gold electrode in aqueous electrolyte solutions [90, 91, 103]. The interaction between probe and electrode is a strong function of externally applied potential, although the variation in double-layer potential at the gold/electrolyte interface is much smaller (only ca. 10%) [91, 103] than the variation in applied potential (Fig. 7.15). Apart from the effect of the externally applied potential, Barten et al. determined the double-layer potential as a function the pH of the solution. The effect of the pH on the double layer of gold was found to be less pronounced than that of the external potential and is ascribed to the presence of a limited number of proton binding sites at the gold surface, resulting from adsorption and partial discharge of hydroxyl ions. A comparable study has been performed on the double layer of a semiconducting TiO_2 electrode [104]. As might be expected, for this system the effect of pH is much stronger. The behavior of both the gold/electrolyte and the TiO_2/electrolyte interfaces is well described by an amphifunctional double-layer model [78, 91], which combines the features of reversible and polarizable interfaces.

The results of Barten et al. have been used for the interpretation of the adsorption of various charged macromolecules to the gold electrode [109]. As an example, Fig. 7.16 gives the adsorbed amounts of the protein lysozyme as a

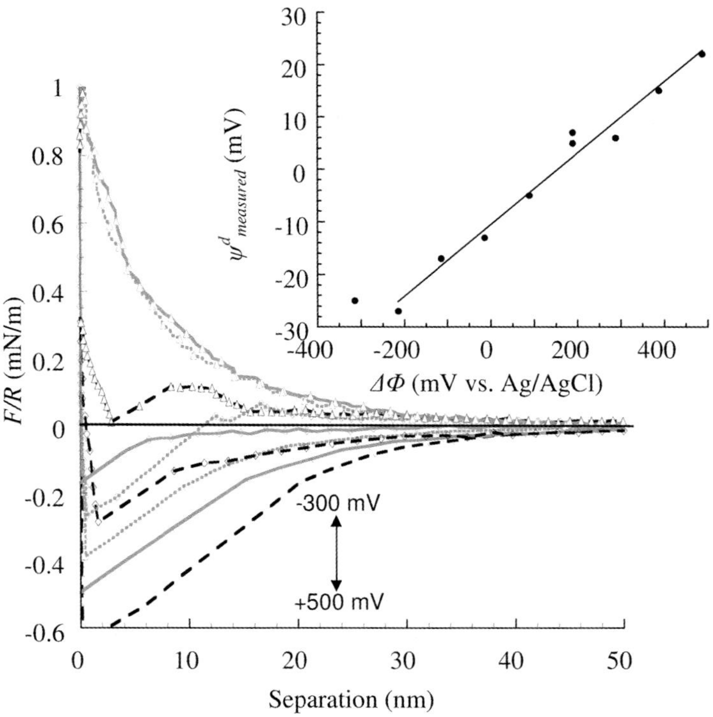

Fig. 7.15 Interaction between a gold substrate and a silica probe in aqueous electrolyte solution, at different externally applied potentials to the gold; 1 mM KNO_3, pH 4.7. The inset shows the corresponding double-layer potentials of the gold substrate obtained from fits to the PB equation and using the double-layer potential of the silica probe as an input parameter. (Results of Barten et al. [91], reproduced by permission of the American Chemical Society).

function of double-layer potential. The data were obtained as a function of pH (at open circuit potential) and as a function of applied potential (at two pH values). Under all experimental conditions, lysozyme is positively charged. The graph shows that the polarity – and not the electrostatic attraction or repulsion – is determining for the adsorption behavior. The extent of "spreading" of the protein molecules decreases with increasing polarity, resulting in a minimum in adsorption around the point of zero potential of the gold electrode. The deviations of the points measured at pH 3 and 11 from the curve reflect the effect of pH on the structural stability of the protein: at pH 3 this stability is low and at pH 11, near the isoelectric point of the protein, it is maximal. These findings illustrate that knowledge of the double-layer potential – as can be obtained from AFM colloidal probe measurements – is very useful for interpreting the effect of the electric potential of the solid/electrolyte solution interface and contributions of intrinsic changes in the adsorbing molecules.

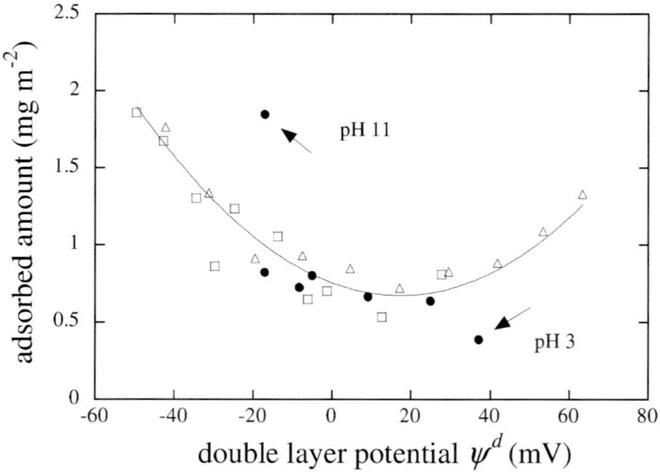

Fig. 7.16 Adsorption of lysozyme on a gold electrode as a function of the double-layer potential. Protein concentration in solution, 3 mg L^{-1}; background electrolyte, 1 mM KNO_3. ● Data from measurements obtained at different values of the solution pH, under open circuit conditions. The other data points result from measurements in which the double-layer potential was varied using an external source. △ pH 5; □ pH 6.4. (Data from Kleijn et al. [109], redrawn with permission from the American Chemical Society).

In SFA and AFM force measurements between hydrophobic surfaces in water, very long-range attractive forces have been observed (of rather variable range, from tens to several hundreds of nanometers; see, for example, [106, 107]). The origin of these forces, initially assumed by various authors to result from hydrophobic interactions between the surfaces, has been a topic of much debate [108], not least because of the variability in reported data. It has become clear that these forces were in fact wetting forces, caused by the emergence of very small gas bubbles (cavitation) and their subsequent coalescence between the interacting surfaces [110]. The effect can be suppressed by using extremely carefully degassed water. "Real" hydrophobic interactions are the result of the decrease in entropy (orientational freedom) of water molecules near a hydrophobic surface. Hence they can be categorized under structural forces and only extend over distances of a few (say five) layers of water molecules, i.e. in the order of 10 nm at the most.

7.3.2
Adhesion

In AFM force measurements, it is often observed that during withdrawal the probe adheres to the substrate up to some distance beyond the initial contact point in the approach curve, as depicted schematically in Fig. 7.5. As the piezo continues to retract, the lever spring is bent further until the spring force overcomes the adhesive force, and at that moment contact is lost and the lever spring jumps to its non-contact position. Inherent to the technique is that in case of attraction the pull-off deflection of the lever spring and the jump-off distance are always larger than the deflection and distance at which the probe and substrate jump into contact during approach, even if there are no specific contact forces (forces that develop only when the surfaces are in contact). This is due to the finite stiffness of the lever spring and can be made intelligible by constructing a force displacement curve in which at each distance the deflection counterbalances the probe–sample interaction [111]. Notwithstanding this, adhesion forces are frequently significantly stronger than forces encountered on approach. As a result, lever spring deflections may be so large that they result into a cut-off signal because of the limited detection range of the segmented photodiode detector (laser spot completely on the lower photodiode segment). Linear position-sensitive detectors have been employed to optimize measurements of strong adhesion forces [112].

Adhesion may arise from of all sorts of interatomic or intermolecular forces. For example, in vacuum or a dry atmosphere (nitrogen gas), adhesion is mainly determined by dispersion (van der Waals) forces. Under ambient conditions, capillary forces between the microscopically thin water films generally present on objects in air may result in large adhesive wetting forces (capillary bridges). Adhesion can also have a chemical origin, i.e. during contact chemical bonds are formed, or it may result from polymer bridging between the surfaces.

As an example, we present some data of Giesbers et al. [113], who measured the adhesion force between acid- and base-functionalized silica and gold surfaces immersed in water (Fig. 7.17). Many other examples can be found in the review of Kappl and Butt [114]. The pronounced effect of pH in Fig. 7.17 is explained by hydrogen bonding between protonated and deprotonated surface groups, resulting for the cases of amine–amine and carboxyl–carboxyl interaction in a maximum in the adhesion around the corresponding pK value. For the amine–carboxyl case the adhesion is stronger due to the formation of acid–base bonds, with a maximum in the intermediate pH range. Not only does the pH have a large effect on the adhesion, but so also has the background electrolyte concentration. From this, it was concluded that the number of bonds decreases with electrolyte concentration (screening of sites or counterion condensation) and/or that the adhesion is partly due to electrostatic attraction between parts of the surfaces directly adjacent to the areas of real physical contact.

The theories that are available to analyze adhesion are based on the principle that two bodies pushed together into contact will deform. For purely elastic deformations either the theory of Johnson, Kendall and Roberts (JKR) [115] or that of Derjaguin, Muller and Toporov (DMT) [116] are applicable. Both theories give the interaction area between a sphere (radius R) and a plate as a function of loading force and predict the force required to pull the sphere from the plate:

$$F_{\mathrm{adh}} = \frac{3}{2}\pi R W (\mathrm{JKR}) \tag{11}$$

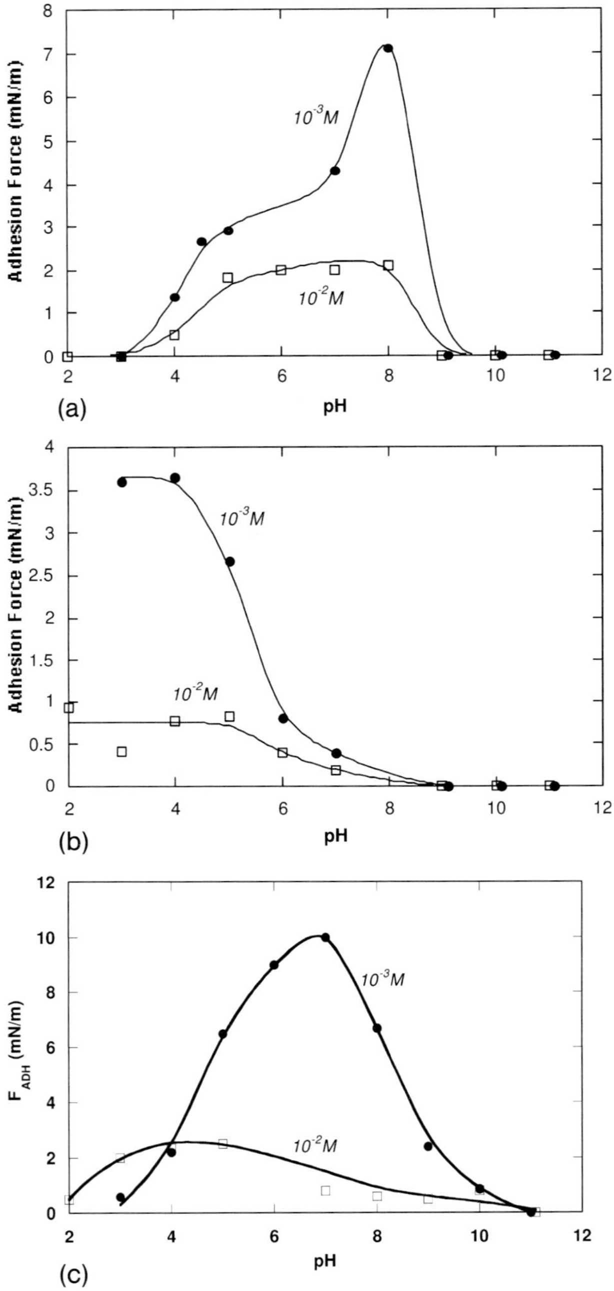

Fig. 7.17 Adhesion forces measured upon separating functionalized surfaces at two concentrations of NaCl as indicated. (a) Two amine-functionalized silica surfaces; (b) two carboxyl-functionalized gold surfaces; (c) an amine-functionalized silica colloidal probe and a carboxyl-functionalized gold-coated wafer. (From Giesbers et al. [113], reproduced by permission of Elsevier).

$$F_{\text{adh}} = 2\pi RW(\text{DMT}) \tag{12}$$

where W is the work of adhesion. Equation (12) is essentially the Derjaguin approximation (Eq. 2) applied to the pull-off point. The two equations differ only by a numerical factor of 4/3. Whereas the JKR approach is suitable for relatively high adhesive forces and soft materials (i.e. high energy surfaces and low elastic or Young's moduli), the DMT theory is a better approximation for low adhesion and hard surfaces. Furthermore, the JKR theory assumes that forces act only in the particle–substrate contact area, whereas the DMT theory includes long-range interaction forces outside the contact area. Various authors have described the transition between the two models, which are in fact limiting cases in more recently developed continuum contact mechanics theories (for example, by Maugis [117]). For details we refer to more extensive reviews [65, 111] and the review of Leite and Herrmann [118], which focuses on adhesion studies using the AFM.

Values reported in the literature for the work of adhesion between (functionalized) surfaces may be compared with theoretical values based on the Dupré equation [119]. According to this equation, the reversible work of adhesion, i.e. the free energy change per unit area in creating an interface between two bodies, is given by

$$W = \gamma_1 + \gamma_2 - \gamma_{12} \tag{13}$$

where γ_1 and γ_2 are the surface free energy of each of the surfaces in contact with the medium and γ_{12} is the free energy of the surfaces in contact. Giesbers et al. [113] have made such a comparison for their own results (depicted in Fig. 7.17) and for other data reported in the literature (see also the review by Drelich et al. [120]). Generally, for hard (rather non-deformable) surfaces the values obtained for W are lower than expected. The most obvious explanation is that the Dupré equation holds for physically ideal (i.e. smooth) surfaces. Of course, real surfaces have a certain degree of roughness and this decreases adhesion, since it reduces the physical contact area. It is also clear that adhesion is very sensitive to contamination of the surfaces. It should be noted, however, that measuring pull-off forces is not a very accurate method to determine surface energies because of the finite stiffness of the measuring system (in particular the lever spring) [111], apart from the fact that many adhesion processes are not reversible. Nevertheless, the pull-off process reflects a wide range of adhesive material properties.

Since the JKR and DMT theories are for purely elastic deformations, they do not comprise the dependence of adhesion force on contact time, rate of separation or applied force. The applied force or loading force can be readily calculated by multiplying the piezo travel after the point of first contact with the spring constant of the lever spring. Applied force, contact time and rate of separation will play a role in the case of plastic or viscoelastic deformations, for which an increasing contact time will result into a larger contact area while energy is dissipated along the contact line. The contact area may also increase as a result of particular probe–substrate interactions, such as hydrophobic attraction, which may lead to progressive coverage of the probe by soft material containing hydrophobic groups. For instance, strong adhesion between a probe and biological material in water may occur, which increases with contact time. Applying a higher loading force increases the deformation or indentation of the sample and therefore the contact area, which may also give rise to stronger adhesion in the case of non-conservative forces (dissipative interactions).

Considerable work on adhesion and particle deformation has been done by Vakarelski and coworkers [121–124] and much of it has been included in another review [114]. It suffices here to mention a study on the adhesion between a silica particle and mica surface in water and electrolyte solutions [121]. It was observed that when the particle is pushed towards the mica, the adhesive force increases with contact time, but this force does not depend on the load force. The explanation given is that on first contact there will still be water between the two surfaces and it takes some time to break the hydration layers. With increasing electrolyte concentration, the adhesion force was found to decrease. At a fairly high salt concentration (1 M), the remaining weak adhesion is independent of contact time for cations of low hydration enthalpy, but for cations of high hydration enthalpy it increases with contact time. Apparently, this difference is due to highly hydrated ions forming a thick but weakly adsorbed layer that can be broken relatively easily, whereas poorly hydrated ions form a thin but strongly adsorbed layer that cannot be broken by the forces applied in this study.

Dependence of adhesion on contact time and loading force can reveal much about the contact mechanics and mechanical properties (such as the stiffness, elastic modulus or viscoelastic parameters) of the materials. For the analysis of results in the case of purely plastic deformations, the theory developed by Maugis and Pollock [125] may be used. Mechanical properties can also be deduced from the slope of the force curve after contact (by comparison with a hard reference surface). This has been done, for example, for polyelectrolyte capsules by compressing them using a colloidal probe [126].

By measuring the pull-off force at many points on a sample, one can produce an adhesion map. Using AFM tips modified with, for example, thiols of various functionalities (such as NH_2, COOH or CH_3) or specific biomolecules, the local chemistry of the surface can be probed in terms of adhesion (see, e.g., Berger et al. [127]). This way of imaging is referred to as chemical force microscopy (CFM). Alternatively, as already mentioned, one can use lateral force microscopy and image the frictional forces between a modified tip and sample [39]. In so-called layered imaging or force–volume imaging, complete force curves are collected together with surface topography data [128]. A 3D data set is obtained giving the deflection of the cantilever at position (x, y, z), hence the names for this technique. In this way, local adhesion, hardness or deformability of samples can be imaged.

When the adhesion arises from chemical bonds between probe and tip, the single-bond force can be determined from a statistical analysis of adhesion val-

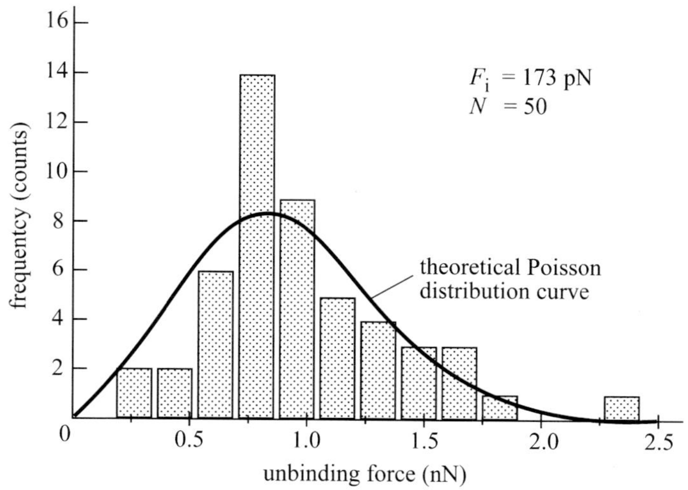

Fig. 7.18 Histogram of the forces needed to pull off a biotinylated AFM tip from an avidin-modified surface (50 measurements) in phosphate buffer (pH 7). F_i is the average individual bond rupture force as determined from a Poisson statistical analysis of the data. (Results of Lo et al. [130], redrawn with permission from the American Chemical Society).

ues [129]. Since both the tip and the surface can be modified, studies of a variety of interactions have been possible. For example, the binding strengths between biotin and avidin and biotin and streptavidin, typical examples of specific biological ligand–receptor systems, were found to be 173 ± 19 and 326 ± 33 pN, respectively [130] (Fig. 7.18).

7.3.3
Forces Between a Probe Particle and a Fluid Interface

Measurements of the interactions between solid particles and bubbles or droplets are challenging and of fundamental interest, since a number of aspects are different from measurements between two solid materials. Furthermore, these interactions are technologically of special interest, e.g. for the separation of mineral particles by flotation and for water purification. As shown by several groups [131–137], it is possible to measure the interaction between an air bubble or an oil droplet and a solid probe particle in aqueous solution using an AFM, although the study of such systems has many practical difficulties. Of course, the deformation of the liquid interface, with the surface tension as restoring force, is an important feature of these systems. Deformations occur throughout the interaction, even before contact is made. Because of this deformability, it is difficult to determine the separation distance. Furthermore, the local radius of curvature (which affects the calculation of the interaction energy using the Derjaguin approximation) changes in a way that is not always readily derivable. In principle, the shape of the interface can be calculated by use of the Laplace equation [138], but in many cases this is not straightforward and often practically impossible. Dagastine and coworkers [139–141] have put a lot of effort into describing the interaction force as a function of the distance of the probe from the rigid substrate (AFM stage) on which the bubble or droplet resides, taking into account various attractive and repulsive interactions.

As there is only one solid phase in the system, the liquid film adjacent to it can be discussed in terms of wetting. When the probe approaches the bubble, the aqueous film becomes thinner and less stable and will finally collapse (transition from complete wetting to partial wetting) except for very hydrophilic surfaces. Partial wetting occurs when the so-called spreading parameter $S=-(\gamma_{SW}+\gamma_{WF}-\gamma_{SF})$ is negative [142], where γ_{SW}, γ_{WF} and γ_{SF} are the interfacial tensions of the solid/aqueous solution, the aqueous solution/fluid (air, gas or oil) and the solid/fluid interfaces, respectively. Once the film is broken, a three-phase line is formed and a capillary force F_{cap} completely dominates the interaction. This force is given by

$$F_{cap} = 2\pi R\gamma_{WF} \sin a \sin(\theta_r - a) \tag{14}$$

where R is the particle radius, a the immersion angle as shown in Fig. 7.19 and θ_r the receding contact angle of the aqueous solution on the particle. The capillary force decreases on moving further into the bubble. Note that the contact angle is related to the interfacial tensions or to the spreading parameter by Young's law: $S=\gamma_{sw}(\cos\theta-1)$. In the absence of hysteresis, the value of S is an equilibrium quantity related to the average properties of the solid surface; the retraction and approach forces should then be equal. For simple fluids and solid surfaces, this is rarely the case, however.

From the immersion depth at which the force is zero and the radius of the (spherical) probe particle, the receding contact angle can be calculated ("microsphere tensiometry") [65, 143]. However, if the particle jumps into the bubble or droplet before the particle reaches the original position of the liquid interface, it is a problem to determine the immersion depth accurately from the force curve. Measurements between differently modified silica particles and air bubbles

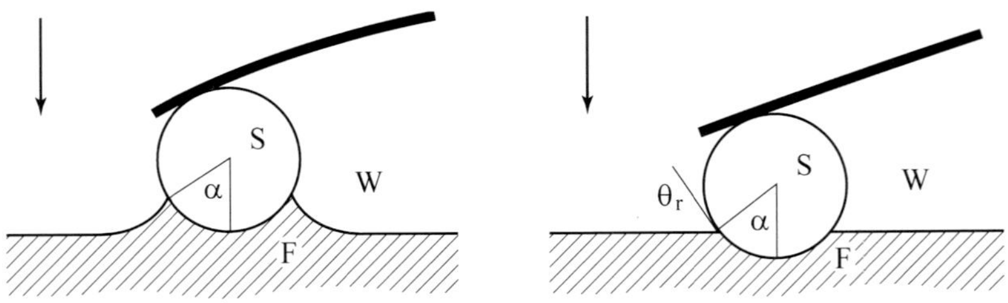

Fig. 7.19 Colloidal probe moving downwards at the interface between an air bubble and an aqueous solution. When the immersion angle a becomes equal to the receding contact angle θ_r, the capillary force is zero.

showed, as expected, that a hydrophobic particle jumps into the bubble and a hydrophilic particle is repelled [131]. Upon retraction, the capillary force is dependent on the advancing contact angle of the aqueous solution on the particle. For hydrophobic particles this leads to a large adhesion.

Electrostatic double-layer interactions between a silica probe and an air bubble were measured already by Ducker et al. [132] and Mulvaney et al. [134] and Hartley et al. [135] measured these interactions between silica and a decane droplet. Both the air/water and decane/water interfaces are negatively charged at neutral pH and, in line with this, repulsive double-layer interactions were found. The major difference between oil–particle and air–particle interactions in water is that the Hamaker constant for the silica–water–air bubble is strongly negative, leading to a van der Waals repulsion between the surfaces, whereas it is positive for the silica–water–decane system.

7.3.4 Forces Between Two Fluid Interfaces

The interactions between two oil (decane) drops in aqueous surfactant [sodium dodecyl sulfate (SDS)] solutions have to date – as far as we are aware – only been measured using an AFM by Dagastine et al. [144]. Such studies can provide insight into emulsion stability and drop coalescence. The authors attached one of the drops (diameter ca. 35 μm) to a hydrophobized gold-coated cantilever; the second drop (ca. 50 mm) was immobilized on a polymer film. The authors studied pseudo-static and dynamic interactions and the effect of changing the electric double-layer interaction and interfacial tension by adding an anionic surfactant. A complete and quantitative interpretation of the results has not yet been possible.

7.3.5 Surfaces Carrying End-grafted or Adsorbed Polymers

The presence of adsorbing or end-grafted polymers on a surface leads to a variety of structural forces in colloidal dispersions, ranging from steric stabilization to bridging flocculation, depending on the polymer conformation at the surface [145]. Polymers at interfaces play an important role in many industrial products and processes. For example, they may act as stabilizers in paints and in pharmaceutical and cosmetic creams and play an important role in the production of magnetic tapes, fluorescent lamps and television screens. Flocculation by polymers is exploited in, for example, water purification, mineral processing and paper manufacturing. Polymers at interfaces are also applied in adhesives, coatings and (polymer) composites. In these applications, the strength of the polymer surface contact is a decisive factor.

Several groups have investigated surfaces carrying layers of densely (and irreversibly) end-grafted polymer [47, 146–151]; these systems are often denoted "polymer brushes". The case of two identical surfaces (both carrying the same brush) has been amply studied by means of SFA methods [152]; the studies mentioned here deal with the interaction between a bare probe surface and a brush, a case for which AFM is a very suitable tool. As an example, in Fig. 7.20 the interaction between a standard silicon nitride AFM tip and a polystyrene (PS) surface with two densities of grafted poly(ethylene oxide) (PEO) chains is presented. Between the bare PS surface and the AFM tip, large adhesion is observed. At a low grafting density of PEO, where there is no excluded volume interaction between the polymer chains (mushroom regime), this adhesion is still present, but significantly lower. In the brush regime, there is a repulsive interaction as a result of compression and extension of the brush layer.

Yamamoto et al. [148, 149] studied the effects of chain length and grafting density on the resistance to compression of (polydisperse) poly(methyl methacrylate) (PMMA) brushes ($M_w/M_n \approx 1.5$) in toluene. In this solvent and at the grafting densities employed, no attractive force could be detected. The repulsive force curves measured for long chains were higher than those for short chains (at the same distances) and denser brushes extended further in solution, as ex-

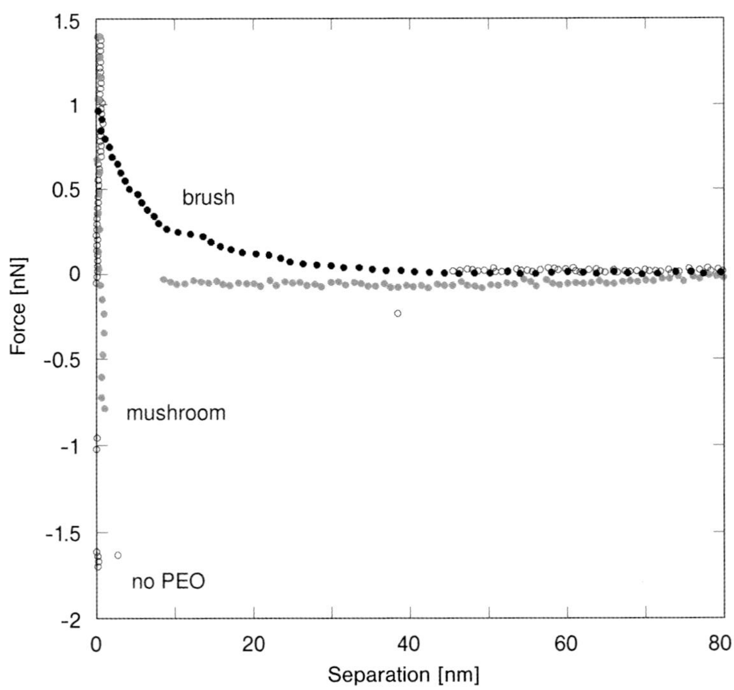

Fig. 7.20 Force between a silicon nitride AFM tip and a PS surface on which PS–PEO block copolymers have been deposited using the Langmuir–Blodgett technique [152], in aqueous 50 mM $NaNO_3$ solution. The length of the PEO chains is 770 monomers ($M_w = 33\,880$ g mol^{-1}). Open circles: bare PS surface; gray circles: grafting density PEO $\sigma = 0.013$ nm^{-1} (mushroom regime); black circles: $\sigma = 0.075$ nm^{-1} (brush regime). Retraction curves. (Unpublished results of W. de Vos, Laboratory of Physical Chemistry and Colloid Science, Wageningen University).

pected from current brush models. In order to establish an absolute value for the distance between the colloidal probe and the substrate on which the PMMA was grafted, the authors carried out a "scratch test" in which polymer was locally removed to expose bare substrate. Similar experiments were carried out by Butt et al. [147] with PS in cyclohexane; here the repulsion could be increased by raising the temperature, which improves the solvency of the PS–cyclohexane system and, hence, its osmotic pressure.

Of course, grafted chains can also adsorb on a bare (probe) surface. This may lead to both attractive "bridging" forces and repulsive "steric" forces; the balance between these for a given grafting density may depend on the strength of monomer–probe interaction. Such forces were studied as a function of grafting density and for two different substituted polyacrylamides [polymethoxyethylacrylamide (PMEA) and poly-*N*-isopropylacrylamide (PNIPAM)] by Goodman et al. [150]. At very low grafting density, PMEA gave a long-range attraction, which disappeared as soon as the grafting density reached values of the order of 0.02 nm^{-2}. With PNIPAM, the attractive forces were higher and more persistent. In another study [151], Goodman et al. further showed that a force–distance retraction curve displays "snap-off" peaks (Fig. 7.21), which can be used to characterize polydisperse brushes in terms of average contour length and chain length distribution.

Razatos et al. [47] managed to fix bacteria to an AFM tip by means of charge-driven deposition and with these as a "colloidal probe" investigated the forces between a PEO brush and bacteria and reported a monotonically repulsive force

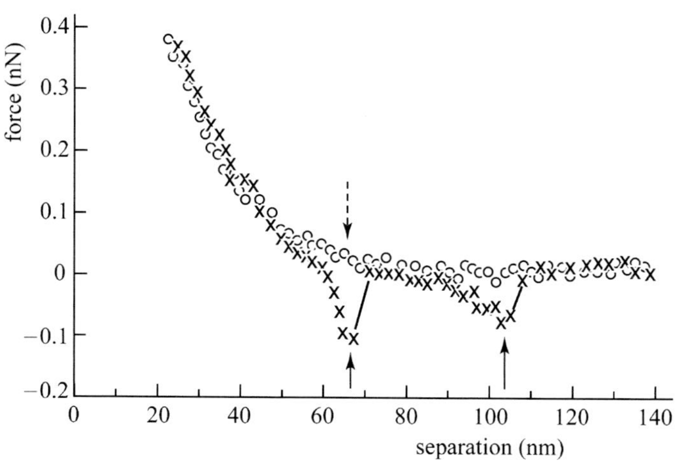

Fig. 7.21 Force profile obtained on a PDMA brush (molar mass of the chains 30600 g mol^{-1}; brush density $\sigma = 0.17\ nm^{-1}$), showing both the compression (○) and extension (×) of the brush. The repulsive force increases monotonically as the brush is compressed due to excluded volume effects. The dashed arrow indicates onset of repulsion corresponding to the equilibrium thickness of the brush. The solid arrows depict rupture events. (Results of Goodman et al. [151], redrawn with permission from the American Chemical Society).

curve. These data demonstrate that PEO brushes can provide appropriate physical barriers against adhesion of bacteria. It does not become clear, however, how general this conclusion is for various kinds of bacteria. Brushes prepared by adsorbing a diblock copolymer were studied by Kelley et al. [146], both by AFM (using a tip rather than a spherical particle) and by SFA. One case was a PS brush in toluene, the other a PSS brush in aqueous electrolyte. Qualitatively, the (repulsive) forces measured with the two methods were similar, but the forces measured in the SFA increased more steeply upon approach than those found with the AFM tip. Obviously, the AFM tip may easily indent the brush layer. In addition, salts made the charged brush shrink, as expected. Butt et al. [147] also investigated PEO–poly(methacrylic acid) (PMAA) diblock copolymers. For these, the curves showed both repulsion and an erratic kind of attraction, ascribed to adhesive contact. A clear temperature effect could not be ascertained.

Giesbers et al. [153] investigated the interaction between two silica surfaces both covered with an adsorbed PEO homopolymer in aqueous solution at different pH values and electrolyte concentrations and as a function of polymer chain length. PEO adsorbed on silica retains some mobility: free and adsorbed chains can readily exchange and conformation changes occur rapidly [154]. It is therefore likely that bridges between approaching surfaces can form fairly easily, so that attraction will outweigh repulsion. Indeed, in none of the measurements was steric repulsion (or excluded volume interaction) observed, but an attractive force depending on pH and electrolyte concentration was found on approach. Clearly, this must be some sort of bridging force. The range of the attraction is in the order of twice the hydrodynamic thickness of the PEO layers. On retraction, a larger bridging force was measured, which increases with decreasing pH (at a constant salt concentration of 1 mM NaCl). This increase is apparently correlated with the adsorption energy per segment, which is known to increase with decreasing pH. At intermediate pH the adhesion depends strongly on the load force, i.e. the force with which the surfaces have been pressed together (Fig. 7.22). Measurements at pH 4 and 1 mM NaCl show a strong dependence of the adhesion force on the chain length of the polymer. A plot of the adhesion force as a function of adsorbed amount of polymer shows a linear relationship between these quantities (Fig. 7.23); it can also be seen that for attraction to show up, a minimum polymer chain length is needed. It is tempting to conclude that the force is proportional to the number of bridges and that this number, in turn, is proportional to $(\Gamma - \Gamma_t)$, where Γ_t is the amount adsorbed below which there are almost exclusively trains on the surface. These findings tie in fairly well with the general observation that in practice short polymers are better dispersants than long polymers and with experiments by Ilekti [155], who found that particles covered with long polymers tend to form irreversible aggregates when compressed by centrifugation.

Forces between two surfaces embedded in a polymer melt were studied by Sun et al. [156]. Two samples of polymer were used, one of low and one of high molar mass ($M_w = 4200$ and $18\,000$ g mol^{-1}, respectively). The low molar mass

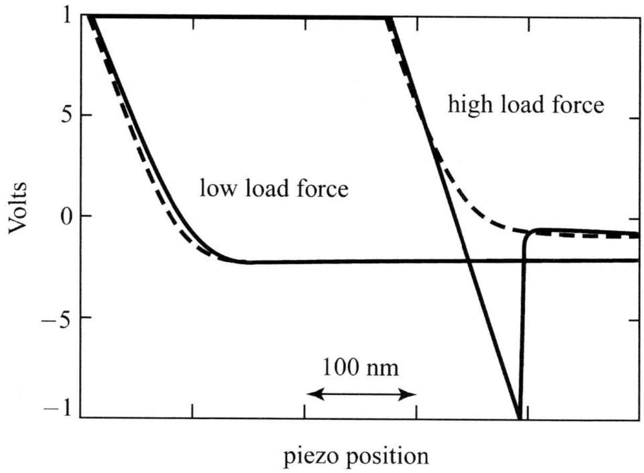

Fig. 7.22 Interaction between two PEO-covered silica surfaces at neutral pH ($\sim$6.5) at various load forces (raw data). Molar mass of PEO is 246 kg mol^{-1} and the NaCl concentration is 10^{-3} M. (Results of Giesbers et al. [153], redrawn with permission from Elsevier).

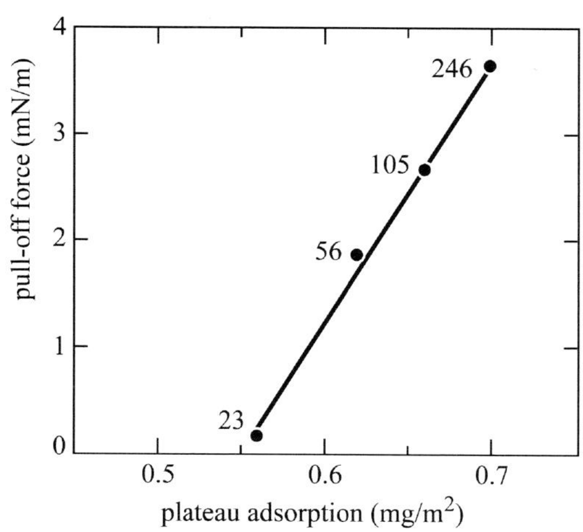

Fig. 7.23 Relation between the adhesion between two PEO-covered silica surfaces at pH 4 and 10^{-3} M NaCl and the plateau amount of PEO adsorbed for different chain lengths. (Data of Giesbers et al. [153], redrawn with permission from Elsevier).

led to attraction between the surfaces, as expected theoretically for a system in which thermodynamic equilibrium can be reached. A theoretical analysis on the basis of SCF calculations [157] shows that the attraction is due to the formation of bridges. The high molar mass polymer, however, gave rise to repulsion, in contrast to (equilibrium) theory. This repulsion can only be understood if it is assumed that at least a fraction of the chains that have contacts with the solid substrate is tied so strongly that it cannot leave on the time-scale of the experiment. These chains will then be compressed between the two surfaces and their distorted shapes lead to (entropic) repulsion.

7.3.6
Interactions in the Presence of Negatively Adsorbing Polymer

The presence of dissolved non-adsorbing polymers in a colloidal dispersion is known to give rise to an effect often called depletion attraction. Depletion attraction was first described and explained in 1954 by Asakura and Oosawa [158]. Near any hard surface, the concentration of non-adsorbing polymer coils is lower than the bulk concentration; this effect is called "depletion" and the zone of reduced concentration is called the depletion zone. It has a thickness δ which is given by the characteristic length in the polymer solution, i.e. the dimension of the polymer coils in dilute solution or the correlation length in semi-dilute solution. To be more precise, the polymer concentration near a non-adsorbing surface shows also a weak maximum in concentration just before it returns to the bulk value. This maximum is reminiscent of the oscillatory density profiles found for molecular (hard particle) systems; there is only a single detectable maximum, however, and this is also very weak because polymer coils are very "soft" particles that interpenetrate easily. As the distance between two opposing surfaces approaches twice the thickness of the depletion zone, overlap of density profiles occurs. In principle, one expects upon approach first a very weak repulsion due to the weak concentration maximum, but this is usually not observed. Once the depletion zones overlap the resulting imbalance in concentration and, hence, osmotic pressure drives the surfaces together. This results in a predominantly entropic force of range δ and a strength controlled by the osmotic pressure. In a colloidal system depletion attraction may lead to the formation of a dense phase (liquid, gel, flocs). Depletion interaction occurs also in dispersions of two non-interacting particulate phases with significant size or shape difference. For a detailed description and analysis of particle–polymer systems we refer to the book by Fleer et al. [145] and the review by Jenkins and Snowden [159]. Examples of practical systems in which depletion interaction occurs are latex–polymer mixtures in paints and protein–polysaccharide mixtures in food products. As a rule, the osmotic pressure in solutions of uncharged polymers is rather weak and so are the resulting depletion forces. Their measurement by means of colloidal probe AFM is a challenge. Much larger forces occur in polyelectrolyte solutions, where the counterions contribute strongly to the osmotic pressure. These latter forces are far more easily determined.

There have been only a few AFM studies to date that are devoted to depletion. Milling and Kendall [160] investigated the depletion interactions between silica surfaces in aqueous solutions containing polyacrylate, and Biggs and coworkers [66, 161] measured depletion forces between silica surfaces in solutions of PSS in water and investigated the dependence on the molecular weight of the polymer. In these systems, not only depletion but also electric double-layer interactions and van der Waals forces play a role. Biggs et al. found a secondary minimum in the interaction curves due to depletion attraction, the depth of which increased with both polymer concentration and molecular weight.

Depletion interactions in non-aqueous systems have been studied by Milling and Biggs [162] and Wijting et al. [163]. Both studies dealt with interactions between stearylated silica surfaces in cyclohexane with polydimethylsiloxane (PDMS) as the non-adsorbing polymer. In this system, electrostatic interactions are absent and van der Waals forces are very weak since the colloids have nearly the same refractive index as the medium. Milling and Biggs reported just one interaction curve for a given molar mass and a single concentration of PDMS. As expected from available data on the phase behavior of similarly coated silica particles, there is (weak) attraction in this system. Wijting et al. made a more systematic study, for various molar masses and concentrations, and found that the range of interaction decreased with increasing polymer concentration, in agreement with theory (see, e.g., Fig. 7.24). However, the strength of the depletion interaction was found to be much weaker than expected for a hard-wall type of interaction between polymer segments and the surface. Comparison with Scheutjens–Fleer self-consistent field (SCF) calculations showed that the results can be explained by a non-zero affinity between polymer segments and surface. This affinity is still well

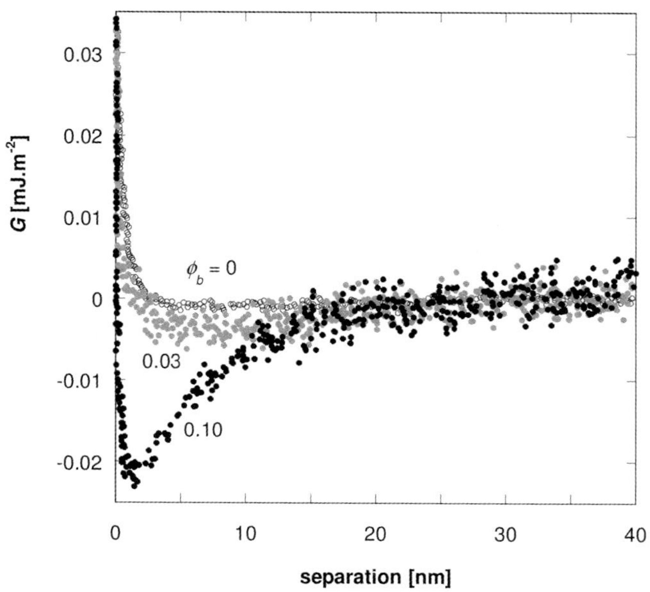

Fig. 7.24 Depletion interaction free energy between stearylated silica surfaces in solutions of PDMS in cyclohexane, for different PDMS bulk volume fractions ϕ_b as indicated. Molar mass of the polymer, 31 300 g mol^{-1}; radius of gyration, ca. 8 nm. (Data of Wijting et al. [163], redrawn with permission of the PCCP Owner Societies).

below the critical value for positive adsorption of the polymer coils, but reduces the strength of the depletion interaction to about one-sixth of its maximum value. This reduced magnitude of depletion interaction is in agreement with the additional experimental finding that more PDMS is needed to reach the colloidal gas–liquid binodal for stearylated silica particles in cyclohexane than predicted by the free volume theory for ideal polymers.

Since polyelectrolyte solutions have very pronounced density fluctuations, they provide a beautiful example of oscillatory ('structural') forces. Various methods of force measurement in the colloidal range have been used to detect them and these methods appear to be complementary. AFM can be used to measure the forces up to about two oscillations out from the surface (see, e.g., Fig. 7.25), provided the concentrations are not too low [66, 72]. SFA is more suitable for lower concentrations and the most sensitive methods (the thin-film balance [164, 165] and TIRM [66, 166]) detect forces up to several oscillations out from the surface. In salt-free polyelectrolyte solutions and over a wide range of concentrations c below the overlap concentration, the periodicity scales with $c^{-1/3}$ as expected for a closed-packed system of spheres. Above the overlap concentration the concentration dependence changes to $c^{-1/2}$ as expected on the basis of scaling theory for polyelectrolyte solutions. It is not entirely clear why AFM seems to be less

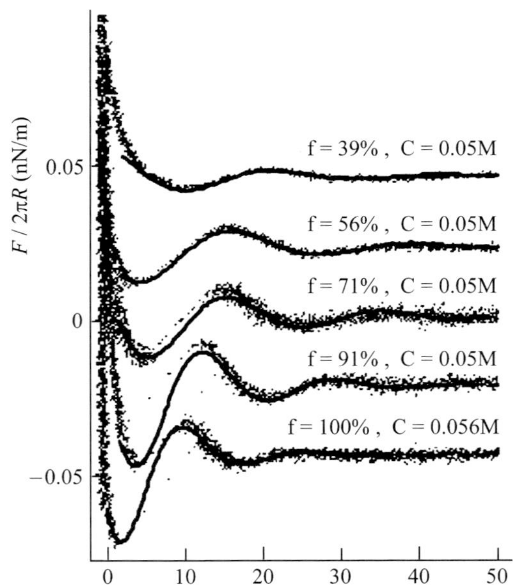

Fig. 7.25 Structural forces between a silicon wafer and a silica colloidal probe in an aqueous salt-free solution of partially sulfonated polystyrene (PSS) at different charge percentages f, measured with the AFM. PPS concentration, 0.05 M. For clarity, the curves are offset. (Results of Qu et al. [72], redrawn with permission from the American Chemical Society).

suitable than TIRM or the thin-film balance for measurement of structural forces: the sensitivity should be good enough. One reason may be that AFM is a "dynamic" method: the probe approaches at a finite velocity, possibly leading to perturbation of the fluid between the surfaces, especially if there is an attraction causing the probe to "snap" into the surface. This may wash out weak oscillatory force patterns. Better control over the rate of approach and retraction may improve the suitability of the AFM technique to analyze depletion (and other structural) interactions.

7.3.7 Single-molecule Force Spectroscopy

Many forces measured in the colloidal regime arise as the result of collective molecular effects: wetting, structural forces, double-layer forces, etc. In a chemical context, one usually attempts to interpret the (collective) forces in terms of molecular structures and interactions, which requires some sort of molecular model. The sensitivity of current equipment is such, however, that forces between individual molecules and intramolecular forces can often be measured directly. This is the challenge of single-molecule force spectroscopy (SMFS). One application is the measurement of the rupture force for specific interactions. The usual approach is to attach the separate partners of an interacting pair of molecules to tip and substrate, respectively, and then measure pull-off forces. We mentioned this type of measurements already in Section 7.3.2. The aim is to have at most a few molecules in the interaction area and this implies that the area should be kept small (sharp tip) and/or the density of active molecules low. In a successful experiment, one then observes pull-off forces that reflect sequential breaking of several molecular pairs. There is a typical stochastic variation around the mean pull-off force, which is represented by means of a histogram (number of events as a function of pull-off force). Such a histogram displays a periodicity related to the breaking force for a single pair. An example concerning the interaction between the biological ligand–receptor system of biotin and avidin was shown in Fig. 7.18. Typical single-molecule binding forces are in the range of tens to hundreds of piconewtons. Given typical molecular sizes (1 nm), this corresponds to interaction energies of order 10 kT and larger.

Another application of SMFS involves the stretching of polymers and other (bio)macromolecules to probe, e.g., the nanomechanical properties of single strands [167–169], the unfolding of proteins [170] and the unzipping of RNA and DNA chains. This rapidly developing field has been reviewed by Zhang and Zhang [171]. In stretching experiments, the macromolecule is bound with one part to a substrate and with the other to the AFM tip. The molecule is stretched until it detaches from one of the surfaces or breaks. A typical example of force versus extension curves is shown in Fig. 7.26, concerning the stretching of polyvinylamine (PVA) chains [168]. The polymer was covalently attached to both substrate and tip. At short separations the first force–distance profile is dominated by a strong adhesive force resulting from several events, such as stretch-

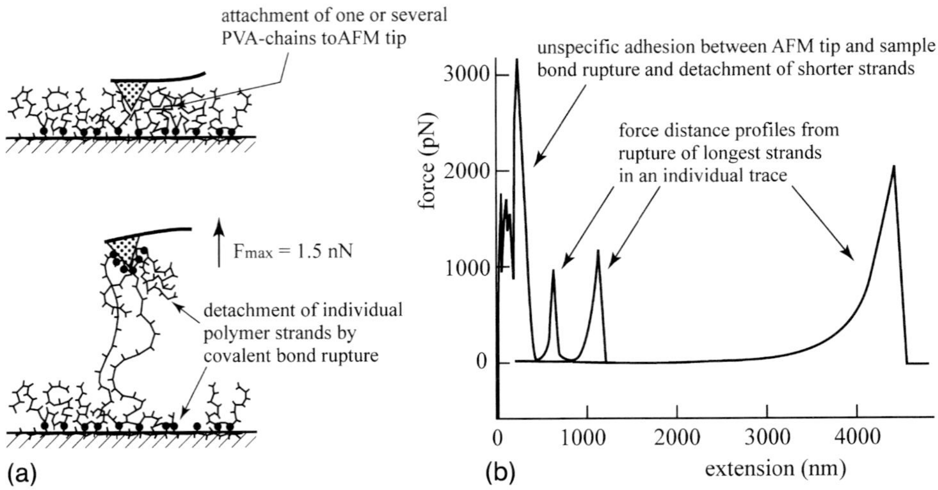

Fig. 7.26 Stretching of PVA chains covalently attached with one end to a glass substrate and with the other to an AFM tip. (a) Schematic diagram of the stretching experiment. (b) Force–extension curves for three individual chains of different contour length. Attraction is indicated as a positive force, which is commonly used in SMFS. (From Hugel et al. [168], redrawn with permission from the American Chemical Society).

ing of more than one chain, desorption of chains from substrate and/or cantilever and covalent bond rupture of short chains. In Fig. 7.26, for clarity this is shown for only one of the curves. In successive retraction–approaching cycles, avoiding contact between the tip and additional PVA chains at the substrate, the distance was increased until only one chain remained between substrate and tip. From fitting the results to the worm-like chain model, it was found that the bending rigidity of the PVA (as expressed by the persistence length) is considerably lower than predicted by theory.

SMFS is not limited to the use of an AFM and can also be performed with optical tweezers (see, e.g., [172, 173]).

7.4 Conclusion and Outlook

The AFM has very rapidly made its way into the field of colloid science as a tool to study the forces that act between colloidal objects or between surfaces representative for such objects. The field has attracted a large number of creative scientists who have contributed to extensions in many directions; it seems that nearly any system of colloidal relevance can be studied in this way. Because the AFM is such a versatile instrument, the amount of data on a plethora of systems is almost overwhelming. However, a quantitative interpretation of the data in terms of the molecules or materials making up the system is often still a long way ahead and complementary molecular modeling work should be carried out to bring understanding of the forces.

In order to help the scientific community to keep track of all the developments, reviews and specialist reports have been written to explain basic issues, highlight possibilities and describe good measurement practices [40]. It almost seems that alternative ways to study colloidal forces, such as the SFA, the thin-film balance or osmotic stress methods, are losing ground to the AFM (although the relatively new and very sensitive methods using TIRM and optical tweezers are attracting increasing attention). It should be borne in mind, however, that there are also drawbacks and specific disadvantages associated with the AFM. For example, in terms of precision and absolute force–distance data the SFA is still superior. Typical advantages of the AFM are that it is relatively easy to work with and can be applied to a large variety of surfaces, and this makes studies between two different surfaces (asymmetric systems), pertaining to hetero-interaction in mixed colloids, particularly attractive. This aspect is very relevant to problems in biological systems where one often wants to determine the interaction between two different colloidal objects.

In conclusion, the AFM has contributed enormously to the advancement of colloid science, certainly if one considers this as the field "where chemistry, physics, biology and technology come together ...". We believe it will continue to do so.

List of Symbols and Abbreviations

AFM	atomic force microscope/atomic force microscopy
c	concentration
CC	constant charge boundary conditions to solve the Poisson–Boltzmann equation
CFM	chemical force microscopy
CP	constant potential boundary conditions to solve the Poisson–Boltzmann equation
D	distance between interacting surfaces
DLVO	Derjaguin–Landau–Verwey–Overbeek (theory on the stability of lyophobic colloids)
DMT	Derjaguin–Muller–Toporov (theory of elasticity and adhesion)
E	elastic modulus/Young's modulus
e	unit charge
EFM	electric force microscopy
F	force
FFM	frictional force microscopy (also referred to as LFM, lateral force microscopy)
FMM	force modulation microscopy (also referred to as FMI, force modulation imaging)

JKR	Johnson–Kendall–Roberts (theory of elasticity and adhesion)
k	cantilever spring constant (N m^{-1})
k_B	Boltzmann's constant
l	length (of lever spring)
M	added mass to cantilever (for calibration of spring constant following the Cleveland method)
m	effective mass of cantilever (determines together with the spring constant the resonance frequency of an unloaded cantilever)
MASIF	measurement and analysis of surface interaction and forces (a surface force apparatus)
MFM	magnetic force microscopy
M_w	molar mass
n_0	bulk concentration of electrolyte ions
NSOM	near-field scanning optical microscopy (also referred to as SNOM)
PB	Poisson–Boltzmann
R	radius
S	spreading parameter
SCF	self-consistent field
SFA	surface force apparatus (developed by Israelachvili and coworkers)
SFM	scanning force microscope/scanning force microscopy (synonym for AFM)
SMFS	single-molecule force spectroscopy
SPM	scanning probe microscope/scanning probe microscopy (general term for STM, AFM and related techniques)
STM	scanning tunneling microscope/scanning tunneling microscopy
T	temperature (K)
t	thickness (of lever spring)
TIRM	total internal reflection microscopy
U	interaction energy per unit area
W	work of adhesion
w	width (of lever beam)
z	(i) coordinate; (ii) ion valency
z^*	displacement of the lever spring end in the z-direction as measured with the optical lever technique (in fact, the inclination dz/dx is measured, with x the coordinate along the longitudinal direction of the lever)
α	immersion angle (defined in Fig. 7.19)
γ	surface free energy/interfacial tension
Γ	adsorbed amount
ε	dielectric constant
ν	resonance frequency of cantilever
Π	disjoining pressure
θ	contact angle
θ_r	receding contact angle
σ	polymer brush density

σ^0	surface charge density
σ^d	diffuse double-layer charge density
ψ	potential
ψ^0	surface potential
ψ^d	diffuse double-layer potential; potential at the outer Helmholtz plane

References

1 Israelachvili, J. N. *Intermolecular and Surface Forces*, 2nd edn, Academic Press, London, **1992**.
2 Derjaguin, B. V., Titijevskaia, I. I., Abricossa, A. D. *Discuss. Faraday Soc.* **1954**, *18*, 24.
3 Lyklema, J. *Fundamentals of Interface and Colloid Science. Vol. I. Fundamentals*, Academic Press, London, **1990**.
4 Overbeek, J. T. G., Sparnaay, M. J. *J. Colloid Interface Sci.* **1952**, *7*, 343.
5 Overbeek, J. T. G., Sparnaay, M. J. *Discuss. Faraday Soc.* **1954**, *18*, 12.
6 Sheludko, A. *Adv. Colloid Interface Sci.* **1967**, *1*, 391.
7 van den Boomgaard, T., Lyklema, J. *Langmuir* **1989**, *5*, 245–249.
8 Manev, E. D., Pugh, R. *Langmuir* **1992**, *7*, 2253–2260.
9 Parsegian, V. A., Fuller, N., Rand, R. P. *Proc. Natl. Acad. Sci. USA* **1979**, 2750–2754.
10 Claesson, P. M., Ederth, T., Bergeron, V., Rutland, M. W. *Adv. Colloid Interface Sci.* **1996**, *67*, 119–183.
11 Raspaud, E. *Eur. Biophys. J.* **2003**, *32*, 402–404.
12 Tabor, D., Winterton, R. H. S. *Nature* **1968**, *219*, 1120.
13 Tabor, D., Winterton, R. H. S. *Proc. R. Soc. London* **1969**, *A312*, 435.
14 Israelachvili, J. N. *J. Colloid Interface Sci.* **1973**, *44*, 255.
15 Israelachvili, J. N., Adams, G. E. *J. Chem. Soc., Faraday Trans. I* **1978**, *74*, 975.
16 Parker, J. L. *Langmuir* **1992**, *8*, 551–556.
17 Stewart, A. M. *Meas. Sci. Technol.* **1995**, *6*, 114–123.
18 Prieve, D. C., Freij, N. A. *Langmuir* **1990**, *6*, 396–403.
19 Prieve, D. C. *Adv. Colloid Interface Sci.* **1999**, *82*, 93–125.
20 Clapp, A. R., Dickinson, R. B. *Langmuir* **1999**, 2182.
21 Behrens, S. H. *Eur. Phys. J. E* **2003**, *10*, 115–121.
22 Dogariu, A. C., Rajagopalan, R. *Langmuir* **2000**, *16*, 2770–2778.
23 Grier, D. G. *Nature* **2003**, *424*, 810–816.
24 Hansen, P. M., Dreyer, J. K., Ferkinghoff-Borg, J., Oddershede, L. *J. Colloid Interface Sci.* **2005**, *287*, 561–571.
25 Rohrbach, A., Tischer, C., Neumayer, D., Florin, E. L., Stelzer, E. H. K. *Rev. Sci. Instrum.* **2004**, *75*, 2197–2210.
26 Binnig, G., Quate, C. F., Gerber, C. *Phys. Rev. Lett.* **1986**, *56*, 930–933.
27 Binnig, G., Rohrer, H. *Helv. Phys. Acta* **1982**, *55*, 726.
28 Weisenhorn, A. L., Hansma, P. K., Albrecht, T. R., Quate, C. F. *Appl. Phys. Lett.* **1989**, *54*, 2651–2653.
29 Burnham, N. A., Dominguez, D. D., Mowery, R. L., Coltob, R. J. *Phys. Rev. Lett.* **1990**, *64*, 1931–1934.
30 Ducker, W. A., Cook, R. F., Clarke, D. R. *J. Appl. Phys.* **1990**, *67*, 4045–4052.
31 Hoh, J. H., Cleveland, J. P., Prater, C. B., Revel, J. P., Hansma, P. K. *J. Am. Chem. Soc.* **1992**, *114*, 4917–4918.
32 Ducker, W. A., Senden, T. J., Pashley, R. M. *Nature* **1991**, *353*, 239–241.
33 Butt, H. J. *Biophys. J.* **1991**, *60*, 1438–1444.
34 Dai, H., Hafner, J. H., Rinzler, A. G., Colbert, D. T., Smalley, R. E. *Nature* **1996**, *384*, 147–148.
35 Wong, S. S., Joselevich, E., Woolley, A. T., Cheung, C. L., Lieber, C. M. *Nature* **1998**, *394*, 52–55.

36 Arnaudov, L. N., de Vries, R., Ippel, H., van Mierlo, C. P. M. *Biomacromolecules* **2003**, *4*, 1614–1622.

37 Li, S., Palmer, A. F. *Macromolecules* **2005**, *38*, 5686–5698.

38 Richter, R., Mukhopadhyay, A., Brisson, A. *Biophys. J.* *85*, 3035–3047.

39 Hayes, R. A., Ralston, J. In *Adhesion Promotion Techniques*, Mittal, K. L., Pizzi, A. (Eds.), Marcel Dekker, New York, 1999, pp 121–138.

40 Ralston, J., Larson, I., Rutland, M. W., Feiler, A. A., Kleijn, J. M. *Pure Appl. Chem.* **2005**, *77*, 2149–2170.

41 Dufrene, Y. F., Lee, G. U. *Biochim. Biophys. Acta* **2000**, *1509*, 14–41.

42 Dufrene, Y. F. *Micron* **2001**, *32*, 153–165.

43 Jena, B. P., Hörber, J. K. H. *Atomic Force Microscopy in Cell Biology*, Academic Press, San Diego, **2002**.

44 Derjaguin, B. V. *Kolloid-Z.* **1934**, *69*, 155–164.

45 Notley, S. M., Biggs, S., Craig, V. S. J. *Macromolecules* **2003**, *36*, 2903–2906.

46 Meagher, L., Franks, G. V., Gee, M. L., Scales, P. J. *Colloids Surf. A* **1999**, *146*, 123–137.

47 (a) van Zessen, E., Heijne, B., de Keizer, A., *Closing the Sulfur Cycle – Applications of Biosulfur*, Internal Report, Wageningen University, Wageningen, **2004**; (b) Razatos, A., Ong, Y. L., Boulay, F., Elbert, D. L., Hubbell, J. A., Sharma, M. M., Georgiu, G. *Langmuir* **2000**, *16*, 9155–9158.

48 Giesbers, M., *Thesis*, Wageningen University, Wageningen, **2001**.

49 Toikka, G., Hayes, R. A. *J. Colloid Interface Sci.* **1997**, *191*, 102–109.

50 Larson, I., Drummond, C. J., Chan, D. Y. C., Grieser, F. *J. Phys. Chem.* **1995**, *99*, 2114–2118.

51 Muster, T. H., Toikka, G., Hayes, R. A., Prestidge, C. A., Ralston, J. *Colloids Surf. A* **1996**, *106*, 203–211.

52 Toikka, G., Hayes, R. A., Ralston, J. *Langmuir* **1996**, *12*, 3783–3788.

53 Dubreuil, F., Shchukin, D. G., Sukhorukov, G. B., Fery, A. *Macromol. Rapid Commun.* **2004**, *25*, 1078–1081.

54 Jaschke, M., Butt, H. J. *Rev. Sci. Instrum.* **1995**, *66*, 1258.

55 Albrecht, T. R., Akamine, S., Carver, T. E., Quate, C. F. *J. Vac. Sci. Technol.* **1990**, *8*, 3386–3396.

56 Sader, J. E., White, L. R. *J. Appl. Phys.* **1993**, *74*, 1.

57 Senden, T. J., Ducker, W. A. *Langmuir* **1994**, *10*, 1003–1004.

58 Sader, J. E., Larson, I., Mulvaney, P., White, L. R. *Rev. Sci. Instrum.* **1995**, *66*, 3789–3798.

59 Cleveland, J. P., Manne, S., Bocek, D., Hansma, P. K. *Rev. Sci. Instrum.* **1993**, *64*, 403–405.

60 Hutter, J. L., Bechhoefer, J. *Rev. Sci. Instrum.* **1993**, *64*, 1868–1873.

61 Butt, H. J., Jaschke, M. *Nanotechnology* **1995**, *6*, 1–7.

62 Lévy, R., Maaloum, M. *Nanotechnology* **2002**, *13*, 33–37.

63 Stark, R. W., Drobek, T., Heckl, W. L. *Ultramicroscopy* **2001**, *86*, 207–215.

64 Roters, A., Johannsmann, D. *J. Phys.: Condens. Matter* **1996**, *8*, 7561–7577.

65 Butt, H. J., Cappella, B., Kappl, M. *Surf. Sci. Rep.* **2005**, *59*, 1–152.

66 Biggs, S., Prieve, D. C., Dagastine, R. R. *Langmuir* **2005**, *21*, 5421–5428.

67 Piech, M., Walz, J. Y. *J. Phys. Chem. B* **2004**, *108*, 9177–9188.

68 Stuart, J. K., Hlady, V. *Biophys. J.* **1999**, *76*, 500–508.

69 Huntington, S. T., Hartley, P. G., Katsifolis, J. *Ultramicroscopy* **2003**, *94*, 283–291.

70 Considine, R. F., Drummond, C. M. *Langmuir* **2001**, *17*, 7777–7783.

71 Neto, C., Craig, V. S. J. *Langmuir* **2001**, *17*, 2097–2099.

72 Qu, D., Baigl, D., Williams, C. E., Möhwald, H., Fery, A. *Macromolecules* **2003**, *36*, 6878–6883.

73 Peschel, G., Belouschek, P., Müller, M. M., König, R. *Colloid Polym. Sci.* **1982**, *260*, 444.

74 Horn, R. G., Smith, D. T., Haller, W. *Chem. Phys. Lett.* **1989**, 404–408.

75 Ducker, W. A., Senden, T. J., Pashley, R. M. *Langmuir* **1992**, *8*, 1831–1836.

76 Hartley, P. G., Larson, I., Scales, P. J. *Langmuir* **1997**, *13*, 2207–2214.

77 Feiler, A. A., Jenkins, P., Ralston, J. *Phys. Chem. Chem. Phys.* **2000**, *2*, 5678–5683.

78 Duval, J., Lyklema, J., Kleijn, J.M., van Leeuwen, H.P. *Langmuir* **2001**, *17*, 7573–7581.
79 Lyklema, J., Duval, J.F.L. *Adv. Colloid Interface Sci.* **2005**, *114/115*, 27–45.
80 Duval, J., Leermakers, F.A.M., van Leeuwen, H.P. *Langmuir* **2004**, *20*, 5052–5065.
81 Giesbers, M., Kleijn, J.M., Cohen Stuart, M.A. *J. Colloid Interface Sci.* **2002**, *248*, 88–95.
82 Minor, M., van Leeuwen, H.P. In *Fundamentals of Interface and Colloid Science. Vol. IV. Particulate Colloids*, Lyklema, J. (Ed.), Elsevier, Amsterdam, **2005**, pp. 4.1–4.53.
83 Lyklema, J. *Fundamentals of Interface and Colloid Science, Vol. II, Liquid–Solid Interfaces*, Academic Press, London, **1995**.
84 Derjaguin, B.V., Landau, L. *Acta Physicochim. URSS* **1941**, *14*, 633–662.
85 Verwey, E.W.J., Overbeek, J.T.G. Theory of the stability of lyophobic colloids. The interaction of sol particles having an electric double layer. *Elsevier, New York*, **1948**.
86 Lyklema, J., van Leeuwen, H.P., Minor, M. *Adv. Colloid Interface Sci.* **1999**, *83*, 33–69.
87 Chan, D.Y.C., Pashley, R.M., White, L.R. *J. Colloid Interface Sci.* **1980**, *77*, 283–285.
88 Chan, D.Y.C. *J. Colloid Interface Sci.* **2002**, *245*, 307–310.
89 McCormack, D., Carnie, S.L., Chan, D.Y.C. *J. Colloid Interface Sci.* **1995**, *169*, 177–196.
90 Hillier, A.C., Kim, S., Bard, A.J. *J. Phys. Chem.* **1996**, *100*, 18808–18817.
91 Barten, D., Kleijn, J.M., Duval, J., van Leeuwen, H.P., Lyklema, J., Cohen Stuart, M.A. *Langmuir* **2003**, *19*, 1133–1139.
92 Wiese, G.R., James, R.O., Healy, T.W. *Discuss. Faraday Soc.* **1971**, *52*, 302–311.
93 Biesheuvel, P.M. *J. Colloid Interface Sci.* **2004**, *275*, 514–522.
94 Grabbe, A., Horn, R.G. *J. Colloid Interface Sci.* **1993**, *157*, 375–383.
95 Pericet-Camara, R., Papastavrou, G., Behrens, S.H., Borkovec, M. *J. Phys. Chem. B* **2004**, *108*, 19467–19475.
96 Toikka, G., Hayes, R.A., Ralston, J. *J. Chem. Soc., Faraday Trans.* **1997**, *93*, 3523–3528.
97 Pashley, R.M. *J. Colloid Interface Sci.* **1981**, *83*, 531.
98 Raiteri, R., Preuss, M., Grattarola, M., Butt, H.J. *Colloids Surf. A* **1998**, *136*, 191–197.
99 Biggs, S., Mulvaney, P., Zukoski, C.F., Grieser, F. *J. Am. Chem. Soc.* **1994**, *116*, 9150.
100 Rabinovich, Y.I., Churaev, N.V. *Russ. J. Chem. Phys.* **1990**, *52*, 256.
101 Grabbe, A. *Langmuir* **1993**, *9*, 797–801.
102 Chapel, J.P. *J. Colloid Interface Sci.* **1994**, *162*, 517–519.
103 Wang, J., Bard, A.J. *J. Phys. Chem. B* **2001**, *105*, 5217–5222.
104 Hu, K., Fan, F.R.F., Bard, A.J., Hillier, A.C. *J. Phys. Chem. B* **1997**, *101*, 8298–8303.
105 Frens, G. *Kolloid Z. Polym.* **1972**, *250*, 736–741.
106 Claesson, P.M., Christenson, H.K. *J. Phys. Chem.* **1988**, *92*, 1650–1655.
107 Craig, V.S.J., Ninham, B., Pashley, R.M. *Langmuir* **1998**, *14*, 3326–3332.
108 Senden, T.J. *Curr. Opin. Colloid Interface Sci.* **2001**, *6*, 95–101.
109 Kleijn, J.M., Barten, D., Cohen Stuart M.A. *Langmuir* **2004**, *20*, 9703–9713.
110 Mishchuk, N., Ralston, J., Fornasiero, D. *J. Phys. Chem. A* **2002**, *106*, 689–696.
111 Cappella, B., Dietler, G. *Surf. Sci. Rep.* **1999**, *34*, 1.
112 Pierce, M., Stuart, J., Pungor, A., Dryden, P., Hlady, V. *Langmuir* **1994**, *10*, 3217–3221.
113 Giesbers, M., Kleijn, J.M., Cohen Stuart, M.A. *J. Colloid Interface Sci.* **2002**, *252*, 183–184.
114 Kappl, M., Butt, H.J. *Part. Part. Syst. Charact.* **2002**, *19*, 129–143.
115 Johnson, K.L., Kendall, K., Roberts, A.D. *Proc. R. Soc. London A* **1971**, *324*, 301–313.
116 Derjaguin, B.V., Muller, V.M., Toporov, V.J. *Colloids Surf.* **1975**, *53*, 314–326.
117 Maugis, D.J. *J. Colloid Interface Sci.* **1992**, *150*, 243–269.
118 Leite, F.L., Herrmann, P.S.P. *J. Adhesion Sci. Technol.* **2005**, *19*, 365–405.

119 Dupré, A. *Theory Mechanique de la Chaleur,* Gauthiers-Villars, Paris, **1869**.

120 Drelich, J., Tormoen, G.W., Beach, E.R. *J. Colloid Interface Sci.* **2004**, *280*, 484–497.

121 Vakarelski, I.U., Ishimura, K., Higashitani, K. *J. Colloid Interface Sci.* **2000**, *227*, 111–118.

122 Vakarelski, I.U., Higashitani, K. *J. Colloid Interface Sci.* **2001**, *242*, 110–120.

123 Vakarelski, I.U., Toritani, A., Nakayama, M., Higashitani, K. *Langmuir* **2001**, *17*, 4739–4745.

124 Vakarelski, I.U., Toritani, A., Nakayama, M., Higashitani, K. *Langmuir* **2003**, *19*, 110–117.

125 Maugis, D.J., Pollock, H.M. *Acta Metall.* **1984**, *32*, 1323–1334.

126 Fery, A., Dubreuil, F., Möhwald, H. *New J. Phys.* **2004**, *6*, 18.

127 Berger, C.E.H., van der Werf, K.O., Kooyman, R.P.H., de Grooth, B.G., Greve, J. *Langmuir* **1995**, *11*, 4188–4192.

128 Heinz, W.F., Hoh, J.H. *Nanotechnology* **1999**, *17*, 143–150.

129 Williams, J.M., Han, T., Beebe, T.P. *Langmuir* **1996**, *12*, 1291–1295.

130 Lo, Y.S., Huefner, N.D., Chan, W.S., Stevens, F., Harris, J.M., Beebe, T.P. *Langmuir* **1999**, *15*, 1373–1382.

131 Butt, H.J. *J. Colloid Interface Sci.* **1994**, *166*, 109–117.

132 Ducker, W.A., Xu, Z.G., Israelachvili, J.N. *Langmuir* **1994**, *10*, 3279–3289.

133 Fielden, M.L., Hayes, R.A., Ralston, J. *Langmuir* **1996**, *12*, 3721–3727.

134 Mulvaney, P., Perera, J.M., Biggs, S., Grieser, F., Stevens, G.W. *J. Colloid Interface Sci.* **1996**, *183*, 614–616.

135 Hartley, P.G., Grieser, F., Mulvaney, P., Stevens, G.W. *Langmuir* **1999**, *15*, 7282–7289.

136 Nguyen, A.V., Nalaskowski, J., Miller, J.D. *Miner. Eng.* **2003**, *16*, 1173–1181.

137 Gillies, G., Prestidge, C.A. *Adv. Colloid Interface Sci.* **2004**, *108/109*, 197–205.

138 Bachmann, D.J., Miklavcic, S.J. *Langmuir* **1996**, *12*, 4197–4204.

139 Chan, D.Y.C., Dagastine, R.R., White, L.R. *J. Colloid Interface Sci.* **2001**, *236*, 141–154.

140 Dagastine, R.R., White, L.R. *J. Colloid Interface Sci.* **2002**, *247*, 310–320.

141 Dagastine, R.R., Prieve, D.C., White, L.R. *J. Colloid Interface Sci.* **2004**, *269*, 84–96.

142 Lyklema, J. *Fundamentals of Interface and Colloid Science. Vol. III. Liquid–Fluid Interfaces,* Academic Press, London, **2000**.

143 Preuss, M., Butt, H.J. *J. Colloid Interface Sci.* **1998**, *208*, 468–477.

144 Dagastine, R.R., Stevens, G.W., Chan, D.Y.C., Grieser, F. *J. Colloid Interface Sci.* **2004**, *273*, 339–342.

145 Fleer, G.J., Cohen Stuart, M.A., Scheutjens, J.M.H.M., Cosgrove, T., Vincent, B. *Polymers at Interfaces,* Chapman and Hall, London, **1993**.

146 Kelley, T.W., Schorr, P.A., Johnson, K.D., Tirrell, M., Frisbie, C.D. *Macromolecules* **1998**, *31*, 4297–4300.

147 Butt, H.J., Kappl, M., Mueller, H., Raiteri, R. *Langmuir* **1999**, *15*, 2559–2565.

148 Yamamoto, S., Ejaz, M., Tsujii, Y., Matsumoto, M., Fukuda, T. *Macromolecules* **2000**, *33*, 5608–5612.

149 Yamamoto, S., Ejaz, M., Tsujii, Y., Matsumoto, M., Fukuda, T. *Macromolecules* **2000**, *33*, 5602–5607.

150 Goodman, D., Kizhakkedathu, J.N., Brooks, D.E. *Langmuir* **2004**, *20*, 2333–2340.

151 Goodman, D., Kizhakkedathu, J.N., Brooks, D.E. *Langmuir* **2004**, *20*, 6238–6245.

152 Currie, E.P.K., Norde, W., Cohen Stuart, M.A. *Adv. Colloid Interface Sci.* **2003**, *100–102*, 205–265.

153 Giesbers, M., Kleijn, J.M., Fleer, G.J., Cohen Stuart, M.A. *Colloids Surf. A* **1998**, *142*, 343–353.

154 Dijt, J.C., Cohen Stuart, M.A., Fleer, G.J. *Macromolecules* **1994**, *27*, 3219–3228.

155 Ilekti, P., *Thesis,* Université Paris, **2000**.

156 Sun, G., Kappl, M., Pakula, T., Kremer, K., Butt, H.J. *Langmuir* **2004**, *20*, 8030–8034.

157 Leermakers, F.A.M., Butt, H.J. *Phys. Rev. E* **2005**, *72*, 021807/1–10.

158 Asakura, S., Oosawa, F. *J. Chem. Phys.* **1954**, *22*, 1255–1256.

159 Jenkins, P., Snowden, M. *Adv. Colloid Interface Sci.* **1996**, *68*, 57–96.

160 Milling, A. J., Kendall, K. *Langmuir* **2000**, *16*, 5106–5115.

161 Biggs, S., Burns, J. L., Yan, Y., Jameson, G. J., Jenkins, P. *Langmuir* **2000**, *16*, 9242–9248.

162 Milling, A., Biggs, S. *J. Colloid Interface Sci.* **1995**, *170*, 604–606.

163 Wijting, W. K., Knoben, W., Besselink, N. A. M., Leermakers, F. A. M., Cohen Stuart, M. A. *Phys. Chem. Chem. Phys.* **2004**, *6*, 4432–4439.

164 Kolaric, B., Jaeger, W., von Klitzing, R. *J. Phys. Chem. B* **2000**, *104*, 5096–5101.

165 von Klitzing, R., Kolaric, B., Jaeger, W., Brandt, A. *Phys. Chem. Chem. Phys.* **2002**, *4*, 1907–1914.

166 Biggs, S., Dagastine, R. R., Prieve, D. C. *J. Phys. Chem. B* **2002**, *106*, 11557–11564.

167 Rief, M., Oesterhelt, F., Heymann, B., Gaub, H. E. *Science* **1997**, *275*, 1295–1297.

168 Hugel, T., Grosholz, M., Clausen-Schaumann, H., Pfau, A., Gaub, H., Seitz, M. *Macromolecules* **2001**, *34*, 1039–1047.

169 Zou, S., Schönherr, H., Vansco, G. J. *Angew. Chem. Int. Ed.* **2005**, *44*, 956–959.

170 Rief, M., Gautel, M., Oesterhelt, F., Fernandez, J. M., Gaub, H. E. *Science* **1997**, *276*, 1109–1112.

171 Zhang, W., Zhang, X. *Prog. Polym. Sci.* **2003**, *28*, 1271–1295.

172 Liphardt, J., Dumont, S., Smith, S. B., Tinoco, I., Bustamante, C. *Science* **2001**, *292*, 733–737.

173 Bennink, M. L., Leuba, S. H., Leno, G. H., Zlatanova, J., de Grooth, B. G., Greve, J. *Nat. Struct. Biol.* **2001**, *8*, 606–610.

8
Ion Specificity in Colloidal Systems

Johannes Lyklema and Frans Leermakers

Abstract

The properties of electrified interfaces, as they typically occur in aqueous solutions, are partly well understood. The Poisson–Boltzmann (PB) theory describes the generic part of the properties accurately; however, as excluded-volume and other specific effects of ions are not accounted for, a more quantitative description is still missing. According to their size and their specificity with respect to various double-layer properties, ions may be ranked in *lyotropic series,* also known as *Hofmeister series.* These can have a direct or reversed trend. A critical survey of experimental (literature) data is given with emphasis on solid/liquid interfaces and association colloids. In this chapter, we argue that it is essential not only to consider the size of the ions as the leading property to rationalize these series, but also one rather should consider the pair interactions of ion–surface, ion–water and ion–ion as being responsible for the specificity. To first order, one can model such pair interactions in a lattice version of the PB theory. Such an extended PB approach can on a mean field level account for the size of ions, part of the specific surface adsorption, the solvation parameters and the polarization of the ions (as well as the surface). Focusing on the solid/liquid interface, we consider various scenarios of how parameters can vary with the sizes of the ions and we generate cases with lyotropic series that are direct or are reversed. The key reason why relatively small changes in parameters can reverse the Hofmeister series is that several of the important parameters in the problem have the natural tendency largely to compensate each other's effects. Small changes in parameters will then allow one tendency to dominate over the other. An interesting scenario is found for the case when there is a large disparity between the solvation of the cation and the anion. In this system, a cooperative (jump-like) change of the surface charge was found with a small change in surface potential. Such jumpy behavior is strongly influenced by the size of the ions (and other parameters the same) and there exists a very pronounced lyotropic effect.

Colloids and Interface Science Series, Vol. 3
Colloid Stability and Application in Pharmacy. Edited by Tharwat F. Tadros

ISBN: 978-3-527-31463-8

8.1 Introduction

In science in general and in interface and colloid science in particular, one can always make the distinction between *generic* phenomena or features and *specific* ones. To the first category belong laws that are universal and do not depend on the nature of the material. Examples are the ideal gas law, $pV = NkT$, valid for any gas at sufficient dilution, or the generally valid relation $F = -kT \ln Q$ between the Helmholtz energy and the canonical partition function in a system of given N, V and T. Specific features depend on the nature of the system. For instance, the constants a and b in the van der Waals equation of state are specific: they have values that differ between different systems.

This chapter refers to *ionic specificity*, implying that we shall consider charged systems. In charged systems one can directly recognize generic features, such as Coulomb's law for the interaction force between two charges which is $z^2e^2/\varepsilon_0\varepsilon r$ for two equal charges ze, a distance r apart, and which is independent of the sizes of the size carriers (ions) as opposed to specific features, as for instance in the non-electrostatic binding of, say, a proton or an Na^+ ion to silica surfaces.

A typical aspect of ionic specificities is that they come together in, or on top of, generic electrostatic features. In practice, only the sum of two is measured. Hence the electrostatic contribution has to be subtracted in some way to arrive at the specific part. This subtraction involves the challenge that the electrical part itself may be generated by preferential adsorption or desorption of other ions, so that some care must be exercised in the unraveling.

Let us make this concrete with ion binding to charged surfaces. This is a recurrent process, underlying several of the illustrations to follow below. Let the total Gibbs energy of adsorption of an ion i at given p and T be $\Delta_{\text{ads}}g_i$. It consists of an electric and a 'chemical' part, where chemical stands for all contributions except for those purely Coulombic. Formally,

$$\Delta_{\text{ads}}g_i = \Delta_{\text{ads}}g_i i(\text{el}) + \Delta_{\text{ads}}g_i(\text{chem}) \tag{1}$$

Simple as it may seem, some basic assumptions underlie this equation and its application. The first is whether the two terms are really additive. According to our definition of the chemical part they are, but in practice this is not obvious. Let the potential at the place x, where ion i is to adsorb, be $\psi(x)$, then

$$\Delta_{\text{ads}}g_i = z_i e\psi(x) + \Delta_{\text{ads}}g_i(\text{chem}) = u_i(\text{el}) + \Delta_{\text{ads}}g_i(\text{chem}) \tag{2}$$

Then the problem is simple if $\psi(x)$ stands for the mean potential. In reality, however, $\Delta_{\text{ads}}g_i(\text{el})$ is the potential mean force: it includes all changes in the ionic distribution resulting from taking it from its reference position (the bulk) to x. The difference between these two potentials depends on the double-layer capacitance, which in turn depends on charge and potential. The trend is that these two potentials differ very little at low $\psi(x)$ and at very high $\psi(x)$, with a maximum in between. As in practice usually only a mean potential is available and accounted for, the error in estimating $\Delta_{ads}g_i(\text{el})$ is transferred to $\Delta_{ads}g_i(\text{chem})$. However, in this chapter we shall mostly ignore this problem.

Another point to be aware of is that $\Delta_{ads}g_i(\text{el})$ can basically be defined by transporting a volumeless charge, whereas $\Delta_{ads}g_i(\text{chem})$ accounts for the non-zero volume of i. The Gibbs energy of this process involves four components because ion transport from bulk to x requires liquid (water) molecules to move the other way, which has to be accounted for. In this respect, $\Delta_{ads}g_i(\text{chem})$ is similar to the Flory–Huggins χ parameter. The implication is that this quantity involves liquid structure-mediated interactions; hence it is partly entropically determined. It is not merely an internal energy, but definitely a Gibbs energy.

In this chapter, we shall emphasize ion specificities of the order of a few kT, mostly restricting ourselves to monovalent ions. The rationales for this choice are (1) that much larger affinities mostly involve charge-determining mechanisms (such as H^+ ions on to oxides) with a strong chemical bond nature, which is not the aim of this chapter, and (2) that for monovalent (counter)ions $\Delta_{ads}g_i(\text{chem})$ is relatively more important than for multivalent ions. In addition, for tri- and tetravalent ions hydroxy complexes may form and adsorb, depending on pH. Hence differences between (counter)ions of different radii show up most clearly with monovalent ions. For the same reason, we shall restrict ourselves to relatively simple systems such as solid surfaces in aqueous media, monolayers and micelles, avoiding proteins with their abundant idiosyncrasies.

Regarding the order of magnitude, one unit of kT equals the electrical energy involved in transporting a unit charge from a reference position to a position where the potential is about 25 mV higher or lower.

8.2
Lyotropic or Hofmeister Series

The occurrence of ion specificity is often inferred from the observation that certain measured quantities depend systematically on the radius of an ion that participates in the physical phenomenon under study. Mostly this is the counterion. Typical illustrations include critical coagulation concentrations of hydrophobic sols, critical micellization concentrations (c.m.c.'s) and surface pressures. Other things being equal, c.m.c.'s decrease in the presence of stronger specifically adsorbing counterions (because the electric double layer around the micelle, which opposes micellization, is better screened), and so do the critical coagulation concentration and the surface pressure of ionized monolayers.

Sequences of such data, arranged by the size of the specifically adsorbing ion, are known as *lyotropic series* or *Hofmeister series*. The terms are equivalent. Historically, the former term is associated with the stability of hydrophobic colloids and the latter with salting-out of proteins, with some emphasis on anions. We shall use the first term. As a trend, the specificity between anions tends to be

Table 8.1 C.c.c. values in mM of monovalent nitrates for negatively charged (p*I* ~4.70) AgI-sols (5 mM). (After Kruyt and Klompé [2], and Basiński [3]).

Electrolyte	Common sol	Repeptized sol	Data of Basiński [3]
$LiNO_3$	165	135	186
$NaNO_3$	140	130	129
KNO_3	136	124	102
$RbNO_3$	126	108	
$CsNO_3$	92	86	
$TlNO_3$	7.5	9	

somewhat more pronounced than that between cations. The reason is that the usual anions are bigger, so that the water structure component plays a relatively larger role than for the smaller cations. Mostly lyotropic features are not small; differences between monovalent counterions in the coagulation concentration, surface pressure or c.m.c. may amount to over 10%.

Regarding the sequence, a distinction can be made between quantities that *increase* with the (counter)ion radius and those that *decrease* with it. Here, we shall call these *direct* and *inverse*, respectively. It depends on the nature of the system and on the property being measured which type is found. For instance, for the classical silver iodide sol, if negatively charged, the surface charge at given surface potential increases in the direct direction $\sigma^0(\mathrm{Rb}^+) > \sigma^0(\mathrm{K}^+) > \sigma^0(\mathrm{Li}^+)$, whereas the critical coagulation concentration (c.c.c.) is the other way around. This apparent paradox has a simple explanation: stronger specific adsorption of a given ion leads at the same time to a higher σ^0 (because of better screening) and to a lower c.c.c. (because there is relatively less charge in the diffuse part of the double layer). As to the effect of the nature of the system, on some negatively charged oxides the sequence is direct, on others it is indirect. Oxides can even be surface-treated (e.g. by calcinations), leading to sequence inversion (see Table 8.1).

The conclusion is that lyotropic (or Hofmeister) sequences are not purely ionic properties but properties of ion + substrate pairs. This principle recurs in the following discussion of some studies from the literature (Section 8.3) and new results from modeling (Section 8.4).

8.3 Some Literature Examples

The present book is dedicated to Professors D. Platikanov and D. Exerowa, who deserve to be acknowledged for their work on thin liquid films. One of their most recent contributions was an appreciated chapter in a book edited by one of us [1]. Ion specificities are also found in such films and some illustrations can be found in the references given in that chapter. In this domain there is a dramatic difference between aqueous films stabilized by sodium and lithium dodecyl sulfate in that with the latter no Newton-black films can be formed (these are very thin films, a little over just a surfactant bilayer thick). This qualitative difference in behavior between two similar cations is probably caused by differences in the height of the maximum in the disjoining pressure thickness, $\Pi(h)$ isotherm. As passing this barrier (see [1], Fig. 6.33) is an essential step for Newton-black film formation, it acts as an activating energy so that small differences lead to exponential differences in rate of passing.

8.3.1 The Silver Iodide System

Historically, this system has played an important role as a model colloid. Sols of it are readily made (unless homodispersity is required) and they are relatively stable and (almost) insensitive to light. The solubility product is low, about 10^{-16} M^2; Ag^+ and I^- are the charge-determining ions. The surface potential ψ^0 obeys Nernst's law as a function of p*Ag*; a range of over 400 mV can be covered. Electrophoresis experiments are relatively easy and were carried out long before modern instrumentation was developed.

Although ion specificities have been investigated since the late 19th century, the AgI system was the first for which this was done systematically. In the early part of the 20th century, one of the obstacles was that the charge-forming processes were hardly understood. For some systems, HNO_3, as an added electrolyte, appeared to behave very different from $NaNO_3$ or $Ca(NO_3)_2$, whereas for others there was no substantial difference. The distinction between charge-determining and indifferent ions had to wait until Verwey pointed this out in his PhD thesis in 1933. For negative AgI, H^+ is indifferent but Ag^+ is charge determining, whereas for many oxides it is the other way around. Only after the distinction between charge-determining and indifferent ions became obvious did systematic studies on the binding of counterions start, originally particularly with respect to their valency (in connection with the Schulze–Hardy rule), but later also with respect to lyotropic sequences. The existence of small but significant differences with respect to their affinities showed that few ions are fully indifferent.

The most complete set of coagulation concentrations for negative AgI sols was obtained by Klompé, whose work was published during the War with her supervisor Kruyt [2]. This 70-page long paper is historically interesting because it gives a good view of the state of thinking in one of the leading colloid research centers during the Second World War. Because of the severed relations with the Soviet Union at that time, the Dutch School knew nothing of the progress made by Derjaguin's group, and vice versa. On the Dutch side, various attempts towards stability theory (by Hamaker, Verwey and others) were published in the Dutch *Chemisch Weekblad*, which was scarcely read abroad. At the time of Klompé's work, it was not yet established whether critical coagulation concentrations (c.c.c.'s) are system specific, nor was it known to what extent the method of preparation of the sol, the sol concentration and the concentration of charge-determining ions played their roles. The main task set forth by Kruyt and Klompé [2] was to study these variables systematically and, in doing so, include the influence of the nature of a variety of electrolytes. Results are collected in Tables 8.1 and 8.2.

The three columns exhibit clear differences, reflecting methodical and system-specific variations. The main difference between common sols and repeptized sols is that the latter group have coarser particles (absence of very small particles in the very heterodisperse sols). The c.c.c. values were obtained by the "coagulation series" method: a series of particularly carefully cleaned test-tubes containing sols with increasing concentrations of the electrolyte. The c.c.c. was identified as that salt concentration at which just a transparent top layer is observable after a given observation time and, usually, once turning the tubes end-over-end. This last step was found to increase the reproducibility. Most likely some mixture of perikinetic and orthokinetic coagulation took place and steps such as these certainly led to the difference in c.c.c. values between different sols and different investigators. Basiński [3] used different sols, of different sol concentration and took a shorter time of observation. He chose chlorides instead of nitrates. Kruyt and Klompé challenged this choice, arguing that Cl^- ions are charge determining. However, Basiński's data are not that different. The reason is that at $pI \approx 4.7$, $pAg \approx 11$; under this condition the solubility product of AgCl is by no means reached in the relatively dilute solutions of the chlorides. Hence, under the conditions chosen, the chlorides behave as indifferent.

A discussion of the absolute reliability of the data in Tables 8.1 and 8.2 is in order. Nowadays the "coagulation series" method is considered obsolete; data are rather obtained from the rate of coagulation method [4, 5], where the break in $\log W - \log c$ plots (W is a measure of the capture efficiency of colliding particles) is identified as (log) c.c.c. Results of such c.c.c.-determinations are systematically about 20–30% higher than those recorded in Tables 8.1 and 8.2, ba-

Table 8.2 As Table 8.1, but for bivalent nitrates.

Electrolyte	Common sol	Repeptized sol	Data of Basiński [3]
$Mg(NO_3)_2$	2.52	2.60	3.25
$Cd(NO_3)_2$	2.50	2.60	3.55
$Zn(NO_3)_2$	2.50	2.55	2.81
$Cu(NO_3)_2$	2.45	2.55	2.37
$Pb(NO_3)_2$	2.43	2.40	
$Co(NO_3)_2$	2.40	2.55	2.82
$Ni(NO_3)_2$	2.40	2.55	2.77
$Ca(NO_3)_2$	2.38	2.40	2.66
$Mn(NO_3)_2$	2.35	2.50	3.05
$Sr(NO_3)_2$	2.33	2.38	2.54
$Ba(NO_3)_2$	2.20	2.26	1.95

sically because in the rate method the electrolyte is used less efficiently. However, lyotropic sequences are the same. Similar series are also observed from the coagulation of *in statu nascendi* sols [5]. Hence it may be concluded that these series are well established. For the sake of argument, let us reconsider the results in the tables. The following can be concluded.

1. The specificity is more pronounced for the alkali metal ions than for the alkaline earth metal ions (Kruyt and Klompé call this the "ion scattering"). As this trend has also been observed for other systems, in this chapter we shall emphasize monovalent ions.
2. According to our definition in Section 8.2 the sequence is *indirect*.
3. Kruyt and Klompé, not having DLVO theory at their disposal, let alone its extensions, nevertheless made the correct suggestion that upon double-layer overlap counterions will move from the diffuse part to the Stern layer. It is historically interesting that this insight into regulation was lost over the following decades, until it emerged again fairly recently [7].
4. The inference is that the specific binding of monovalent cations to AgI increases from Li^+ to Cs^+, i.e. it is *direct*.
5. Kruyt and Klompé did not have a clear picture of the underlying specific binding mechanism. They just stated that probably polarizability, hydration and ion radius played a role and that these phenomena were correlated. Hence they did not realize that reaction *pair* interactions have to be considered. In this respect, they were not to blame as nowadays many authors still think that lyotropic sequences can be interpreted in terms of *only one type* of ion [8]. Neither were they aware of the role played by the structure of water and hence of the entropic contribution to the Gibbs energy of adsorption.

The insight that the specific adsorption of alkali metal ions increases with increasing ion radius was correct, but at the same time it is begging the question of whether this can be measured directly. In fact, several such attempts were made during the first half of the 20th century, for instance by using radionuclides. In addition, attempts were made to measure changes in salt concentration and/or composition upon coagulation of sols. Most of these experiments gave results that were poorly reproducible.

It took another decade before it was realized that direct measurement of the surface charge σ^0 as a function of the surface potential ψ^0 offered better prospects. Basically, the amount of charge sitting on a surface at given ψ^0 will increase with increasing screening. Screening increases with increasing salt concentration, but at given concentration it will increase more strongly for counterions with higher Gibbs energy of adsorption. Two questions arise: (i) how can $\sigma^0(\psi^0)$ be measured and (ii) how can $(d\sigma^0/dc_{salt})_{\psi^0}$ be related to the specifically adsorbed amount? We briefly discuss these developments for the AgI system, but first make a few remarks on the mercury system.

During the 1940s, double-layer measurements on mercury were started and refined by Grahame; basically, he measured the differential capacitance at various applied potentials [9]. Integration of these curves yields the required $\sigma^0(\psi^0)$

dependence. As a model system, mercury has at least four advantages over silver iodide: (1) the surface is renewable, so adsorption of impurities is easily circumvented (Grahame used a dropping mercury electrode), (2) the surface area is well defined, (3) the surface is molecularly flat and (4) the interface is, except at extremely high potentials, perfectly *polarizable*, meaning that there are no Faradaic currents across the interface; the implication is that $d\psi^0$ is known except for a constant. However, this system also has a drawback, viz. that colloidal mercury is virtually non-existent, so coupling between lyotropic sequences in double layers and stability is not possible. In a later paper [10], Grahame found the affinity sequence $Li^+ < Na^+ < K^+ < Rb^+$, i.e. the sequence is *direct*, just as on silver iodide. However, quantitatively the specificity is very small: the difference between Li^+ and Rb^+ is about 10 times lower than it is on AgI. In fact, this specificity is so minor that in stability measurements, had they been available, it would hardly be detectable. A possible reason for this subdued specificity is that mercury does not possess surface sites, hence the interaction is not sitewise. In effect, this is also an argument in favor of the pairwise nature of the counterion binding, observed for silver iodide, oxides and other systems, to be discussed below.

For silver iodide, $\sigma^0(\psi^0)$ can be obtained by titration of AgI suspensions with KI or $AgNO_3$, measuring the surface concentration difference between Γ_{KI} and Γ_{AgNO_3} potentiometrically. The surface charge then follows as

$$\sigma^0 \equiv F(\Gamma_{AgNO_3} - \Gamma_{KI}) \tag{3}$$

In fact, by titration Γ_{AgNO_3} and Γ_{KI} cannot be individually determined; only their difference is accessible. This means that a reference point for σ^0 has to be established. The obvious choice is the point of zero charge (p.z.c.), which has to be measured independently. Another problem to be solved is that of the specific surface area. For a review addressing all these problems, see [11]. Seminal titration results by van Laar and Mackor were reproduced by Overbeek in [12]. In the present chapter, in the figures we give more recent results of our own measurements, as reviewed in [11]. As it has been established that the surface potential behaves according to Nernst's law, each p*Ag* unit corresponds to 58 mV at 20 °C.

Figure 8.1 exhibits various characteristic features. The typical sigmoid $\sigma^0(\psi^0)$ behavior for purely diffuse double layers is observed only at low c_{KNO_3} and low ψ^0. Beyond that the differential capacitance

$$C = \left(\frac{\partial \sigma^0}{\partial \psi^0}\right)_{c_{salt}} \tag{4}$$

is, at least to the negative side, fairly constant because there the capacitance is mainly determined by its inner or Stern layer C^i, which over this range is not very sensitive to σ^0 or to c_{salt}. The shift of the p.z.c. to the right at high KNO_3 concentrations (1 M) indicates specific adsorption of NO_3^- ions on pristine AgI surfaces. The vertical distance between curves at given pAg, but different c_{salt} reflects the extent of screening by electrolyte. Alternatively, the change of pAg with c_{salt} at fixed σ^0 can be chosen for that. This latter quantity, defined as

$$\beta \equiv \left(\frac{\partial \text{pAg}}{\partial \log a_S}\right)_{\sigma^0} \tag{5}$$

is known as the *Esin–Markov coefficient*. It is a measurable quantity and can by thermodynamic means be related to the ionic components of charge σ_+ and σ_-, i.e. the contributions of cations and anions in compensating σ^0:

$$\beta = 1 + 2\left(\frac{\partial \sigma_+}{\partial \sigma^0}\right)_{a_S} = -1 - 2\left(\frac{\partial \sigma_-}{\partial \sigma^0}\right)_{a_S} \tag{6}$$

Hence, by integration using the p.z.c. as the reference,

$$\sigma_+ - \sigma_-(\text{p.z.c.}) = -\frac{\sigma^0}{2} + \frac{1}{2}\int_0^{\sigma^0} \beta \mathrm{d}\sigma^0 \tag{7}$$

or

$$\sigma_- - \sigma_-(\text{p.z.c.}) = -\frac{\sigma^0}{2} - \frac{1}{2}\int_0^{\sigma^0} \beta \mathrm{d}\sigma^0 \tag{8}$$

Infinitely close to the p.z.c., in the absence of specific adsorption, where $\beta = 0$ (see Fig. 8.1, at KNO_3 concentrations up to 10^{-2} M), $\sigma_- = \sigma_+ = -\sigma^0/2$, meaning that any incremental change σ^0 is 50% compensated for by adsorption of counterions and 50% by negative adsorption (expulsion) of co-ions. This phenomenon is also known from diffuse double-layer theory, but the above analysis proves it to be more general, since the thermodynamic derivation is free of any model assumption.

One might ask how it is possible that by thermodynamic reasoning single ionic charge contributions (*ionic components of charge*) can be determined, since

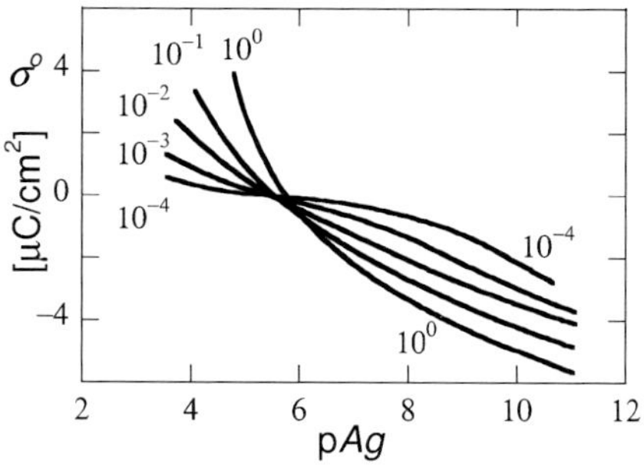

Fig. 8.1 The electrical double layer on silver iodide in the presence of different molar concentrations of KNO_3. $T = 25\,°C$.

double layers as a whole are always electroneutral: one can adsorb only electroneutral combinations of ions. The answer is according to the definitions

$$\sigma_+ = \sigma_{K^+} \equiv F(\Gamma_{KI} + \Gamma_{KNO_3}) \tag{9}$$

$$\sigma_- = \sigma_{NO_3^-} \equiv -F(\Gamma_{AgNO_3} + \Gamma_{KNO_3}) \tag{10}$$

$$\sigma^0 = F(\Gamma_{Ag^+} - \Gamma_{I^-}) \equiv F(\Gamma_{AgNO_3} - \Gamma_{KI}) \tag{11}$$

where σ_+ is the sum of two electroneutral contributions, the positive adsorption of KI (each adsorbing I^- brings along a compensating K^+ ion) and the negative adsorption of KNO_3 (the *Donnan expulsion*). In passing, Eq. (11) confirms that σ^0 (with respect to the p.z.c.) is obtainable by titration although Γ_{AgNO_3} and Γ_{KI} cannot be individually measured.

Generally, σ_+ and σ_- consist of two contributions, one from the diffuse part of the double layer and the other from the Stern layer, which is the seat of specific adsorption. By thermodynamic means there is no way of telling which part is diffuse, so that there is no thermodynamic procedure for subtracting this part in order to obtain the required Stern part σ^i_+or σ^i_-. Around 1947, when Grahame published his review [9], he resolved this issue by realizing that with increasingly negative σ^0, σ_+ keeps increasing, whereas the negative adsorption σ_- attains a plateau, given by [13].

$$\sigma_- = \sigma^d_- = (2\varepsilon_0\varepsilon RTc)^{1/2} \tag{12}$$

The equality $\sigma_- = \sigma^d_-$ stems from the fact that in this limit no Stern charge remains and the fact that a potential-independent limit is attained is also obvious because the co-ion concentration near to the surface can never reach a value lower than zero. Grahame found on negative mercury long stretches of constant σ_- values, interpreting these limiting values as diffuse contributions, obtaining the Stern part under non-limiting conditions. For silver iodide, this procedure does not work so well because the potential range that can be investigated is much shorter (about one-third of that for mercury). However, for this system a better option is to combine titration data with stability data.

To that end, the data in Table 8.1 must be compared with the part of the titration results in Fig. 8.2 to the right-hand side of the graph, where pAg has about the same value as that at which the coagulation studies were carried out. Note that making the comparison with the 10^{-1} M nitrates is good enough. (Strictly, c_{salt} ought to be exactly at the c.c.c. (Table 8.1), which is about 10–30% higher, but the concentration dependence of σ^0 is not that strong – roughly it increases with $\sqrt{c_{salt}}$, so the difference is less than 15%.) It is of historical interest that at the time these curves were measured (about 6 years after the Verwey–Overbeek book appeared and only 2 years after the publication of Kruyt's *Colloid Science* [12]) the belief in the universality of diffuse double-layer theory was so strong that several repetitions (by one of the present authors) were needed to convince

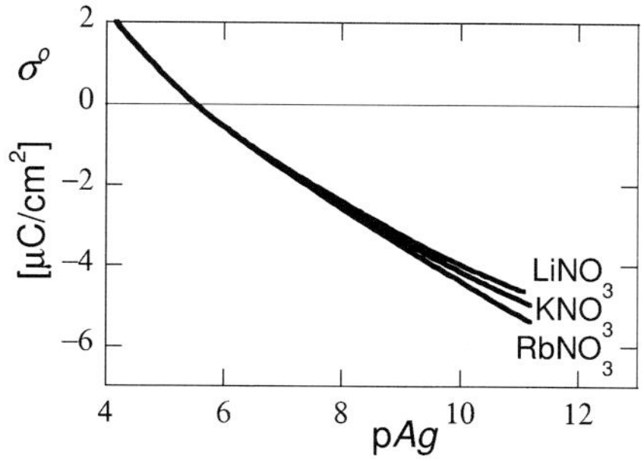

Fig. 8.2 Double-layer charge on silver iodide in decimolar solutions of some alkali metal salts. $T = 25\,°C$.

friend and foe that *higher* surface charges (in $RbNO_3$) corresponded to *lower* stability against coagulation. As stated, the basis of this insight is that alkali metal ions adsorb specifically on silver iodide in the direct sequence; the stronger a cation adsorbs (1) the more negative the surface charge is at given pAg (because of better screening) and (2) the lower the c.c.c. is (because ψ^d is lower). Before 1963 [14], this insight did not reach the literature.

The usual procedure of estimating $\Delta_{ads}g_i(\text{chem})$ in Eqs. (1) and (2) consists of the following steps:

1. DLVO theory is used to obtain ψ^d from the c.c.c.
2. Diffuse theory is used to compute σ^d from ψ^d.
3. From σ^0 and σ^d, the Stern charge σ^i is found.
4. An adsorption isotherm equation is set up for $\sigma^i(\sigma^0)$ with $\Delta_{ads}g_i(\text{chem})$ as the main parameter. Mostly, a mean-field isotherm in the Bragg–Williams approximation is employed; this leads to isotherms of the Frumkin–Fowler–Guggenheim (FFG) type. There are several things that can be said against this procedure, at least some parameters are not exactly known, but the trend is well established and to give some feeling for the order of magnitude two estimates are given in Table 8.3.

From the absolute magnitude, it may be inferred that the origin of the specificity is related to the water structure, rather than to complete dehydration: ionic hydration enthalpies are higher by a factor of 100 but hydration entropies ($T\Delta_{hydr}S_i$) compare better with the data in Table 8.3 and these reflect solvent structure rearrangements. We will come back to this in Section 8.3.2, but note

Table 8.3 Specific ionic Gibbs energy of adsorption on negative silver iodide from surface charge and stability measurements in units of kT.

	Low estimate [15]	High estimate [11]
Li^+	2.1	3.6
K^+	2.5	4.0
R^+	2.9	4.2

in passing an idiosyncrasy of the AgI system: above about 65 °C the specificity in σ^0 and c.c.c. suddenly disappears [16]. This has been interpreted as being caused by the "melting" of adjacent water, which at low temperature is expected to exist in a rather structured form and which has been held responsible for the good nucleation propensity for AgI as a cloud seeder.

8.3.2
Oxidic Surfaces

These may be considered as the counterparts of silver iodide because they are quite hydrophilic. The H^+ and OH^- ions are charge determining and for many oxides abundant titration, coagulation and electrokinetic data are available. Stern capacitances are much higher than on silver iodide and mercury, so that σ^0 (pH) curves behave more "diffuse-layer like". Research in this field includes surface characterization, identification of the extent to which Nernst's law applies for H^+ ions, determination of the p.z.c. and model studies which mostly involve site binding interpretations of σ^0 (pH) curves. The latter type of analyses automatically involve pK values or standard ionic Gibbs energies of adsorption, for protons, hydroxyl ions and counterions. The last-mentioned category is relevant for our present purposes.

Various conditions have to be met before reliable $\Delta_{ads}g_i^0$ values can be established. Some are of a practical nature and refer to the pristinity of the surface (spurious adsorbed species are often betrayed by shifts of the p.z.c.), non-inertness of the surface (non-reversibility of titration isotherms) or improper exclusion of CO_2 from the air during the titrations. Provided that all required experimental precautions have been properly addressed, isotherms become available, which, comparable to Fig. 8.2, indicate the required sequence for cations and anions (on negative and positive oxides, respectively). When additional experiments (coagulation and electrokinetic data shifts in the p.z.c.) support this trend, we may accept the observed sequence, i.e. the direct or inverse order, as established. As for the AgI and mercury systems, obtaining the sequences does not require a double-layer model, but assigning values to $\Delta_{ads}g_i^0$ does.

In Table 8.4, a number of well-established lyotropic sequences are collected. As before, we restrict ourselves to the alkali metal ions. Interesting trends emerge. The first is that the sequence depends on the nature of the oxide. As all these oxides are hydrophilic, it means that the general rule that the sequence is direct on the hydrophobic materials (AgI and Hg) and indirect on hydrophilic substances in terms of its Gibbs energy of hydration, (Li^+ is the most hydrophilic alkali metal ion) does not work. From the observation that changing the surface structure (by calcinations of rutile) also leads to sequence inversion, it may be concluded that interpretation in terms of interactions between the ions and the bulk of the solid (as in London–van der Waals forces) is not appropriate either. This was already concluded for the AgI system, where upon temperature increase the specificity suddenly disappears. We infer the same as for AgI, viz., that the interpretation of $\Delta_{ads}g_i^0$ must be sought in terms of pair interactions. What kinds of pairs and why would they be different between oxides?

Table 8.4 Lyotropic sequences for alkali metal ions on some negatively charged oxides.

Oxide	Sequence	Ref.
SiO_2 (BOH)	$Rb^+ > K^+ > Na^+ > Li^+$	17
TiO_2 (rutile)	$Li^+ > Na^+ > Cs^+$	18, 19
TiO_2 (low p.z.c.) [a]	$Cs^+ > K^+ > Na^+ > Li^+$	19
TiO_2 (rutile and anatase)	$Li^+ > Na^+ > K^+ > Cs^+$	20
α-Fe_2O_3	$Li^+ > Na^+ > K^+ > Cs^+$	21, 22
α-FeOOH	$Na^+ > K^+ > Cs^+$	23
γ-Al_2O_3	$Li^+ > Na^+ > K^+ > Cs^+$	24

a) Calcined sample.

The required correlation is found in the value of the p.z.c. Silica and calcined rutile have low p.z.c.'s of about 2 and 3, respectively, rutile and anatase have p.z.c.'s between about 5.5 and 6.5, depending on temperature and the origin of the sample, and the p.z.c.'s of hematite, goethite and the various aluminas are around 8.7, 7.5 and 8–9, respectively. A low p.z.c. means that the surface has more affinity for OH^- than for H^+, whereas for high p.z.c.'s it is the other way around. The correlation is obvious: surfaces with sites preferring H^+ over OH^- also prefer the smaller ion (Li^+) over the larger one (Cs^+) and the other way around.

All of this can be put on a thermodynamic footing. One of the most basic relationships is that between $\Delta_{ads}H_{H^+}$ and the p.z.c., as illustrated in Fig. 8.3 [33]. As with the titration data, a number of steps had to be made to obtain the adsorption enthalpies in this figure. One of these steps is interpretative in that ex-

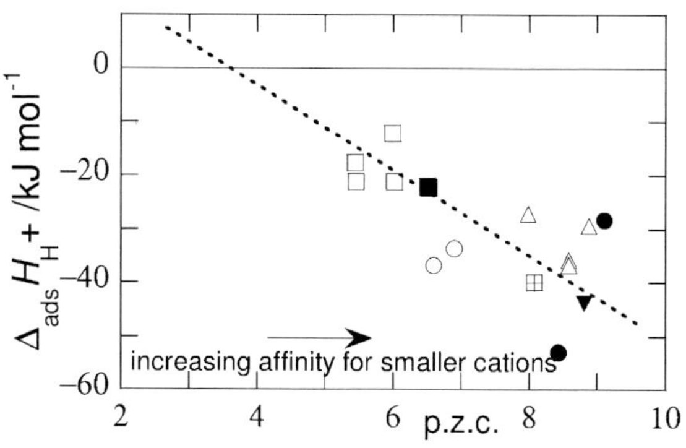

Fig. 8.3 Proton adsorption enthalpy, related to the point of zero charge of the oxide. (Taken from J. Lyklema, *Electric Double Layers in Fundamentals of Interface and Colloid Science*, Vol. II, Academic Press, London, 1995, Chap. 3, p. 3.157, after modification). Open squares [25–27] = rutile (TiO_2); open circle [28, 29] = magnetite (Fe_3O_4); open triangles [26, 27] = hematite (α-Fe_2O_3); square with cross [30] = goethite (α-FeOOH); closed triangle [31] = corundum (α-Al_2O_3); closed circle [27, 28] = alumina (α-Al_2O_3); closed square [32] = zirconia (ZrO_2). The dotted line is just a guide for the eye.

perimentally one cannot discriminate between adsorption of H^+ and desorption of OH^-. For a few systems (rutile and hematite), the values obtained have been obtained by two entirely independent methods: by direct microcalorimetry and from the temperature dependence of the p.z.c., using the appropriate Gibbs–Helmholtz relation [33], with gratifying agreement. Reliable data for oxides with p.z.c.'s below 5 are not available but extrapolation strongly suggests that, say on silica, the enthalpy of proton binding is positive. As double layers on silica form spontaneously, there must be compensation, perhaps from entropic contributions in the structure of adjacent water. It is well known that around the p.z.c. silica is strongly hydrated.

Absolute values for $\Delta_{ads}g_i^0$ for alkali metal ions on oxidic surfaces are not easy to come by. To our knowledge, the procedure of combining σ^0 and c.c.c. data, used for AgI, has not been applied. On the other hand, there are abundant sets of analyses of σ^0 (pH) c.c.c. in terms of site adsorption models. They have the disadvantage that these curves are relatively easily "explained" with rather many parameters; sometimes physically unrealistic fitting procedures have to be postulated, such as letting the thickness of the stagnant layer in electrokinetics vary with the electrolyte concentration [20]. Accepting the limited value of the obtained values for $\Delta_{ads}g_i^0$ one finds in the literature data for this parameter or rather for $-kT\log K_i$, that are in line with our trend, but with absolute values that may differ substantially between different authors and different models. For instance, the values reported in [24] are definitely lower than those in the review by Sverjensky [34].

The general conclusion of all the reported studies is two-fold: (1) specific adsorption must be interpreted in terms of pair interactions (i.e. pairs between sites on the surface and the alkali metal ion) and (2) the sequence depends on the nature of the sites. As a first step towards an interpretation of these sequences, it is useful to look at the water structure. Thermodynamically this means that the hydration enthalpy and hydration entropy of the pair formation should be considered. As a first step, in Table 8.5 these are reported for four ions [35]. It is immediately realized that the order of magnitude in the $T\Delta s$ term is more in line with the pair interaction Gibbs energy than the corresponding enthalpies and this suggests a water structure effect. It is also noted that, for both anions and cations, the hydration enthalpy decreases with increasing radius, whereas for the entropy term it is the other way around. Most entropies are positive, meaning that taking the ion

Table 8.5 Standard molar hydration enthalpies and entropies of four characteristic ions, $T = 300$ K. Units of kT per ion.

Ion	Radius (nm)	$\Delta_{hydr}h_i^0$	$T\Delta_{hydr}s_i^0$
Li^+	0.060	–211	–1.1
Cs^+	0.169	–111	13.4
OH^-	0.140	–214	1.38
I^-	0.216	–121	4.06

from its reference in the vapor phase involves the creation of more disorder or at least creates more accessible configurations.

A further step can be made in terms of a Gurney-type of binding [36], which was already invoked by Gierst et al. [37] to interpret these sequences. Connections can be made to the activity coefficients $f_{\pm}$ in electrolyte solutions. For cation–anion pairs in which both radii (a_+ and a_-) are large or when both are small, $f_{\pm}$ is lower than for pairs of different radii (large + small). The lower is $f_{\pm}$ (i.e. the further it is below unity), the stronger is the Gibbs energy for pair interaction in solution via

$$g_{\pm} = -kT \ln f_{\pm} \tag{13}$$

For instance, in 1 M molar solutions $f_{\pm}$ equals 0.544 for CsI and 0.910 for LiI [38]; in 3 M solutions, these activities become 0.434 and 1.715, respectively. Between 1 and 3 M solutions this would imply a difference between the two of $0.51kT$ and $1.36kT$, respectively, i.e. of the same order as found for the pair interactions. We do not have corresponding data for LiOH and CsOH.

The ensuing rule of thumb is simple: "like seeks like" for entropic reasons. This rule immediately accounts for the direct order on silver iodide (the adsorption sites are large I^- ions on the surface), in contrast to the indirect order on most oxides (small ions are the sites). Extension of this reasoning to SiO_2 and the lower p.z.c. rutile modification suggests that on those surfaces adsorption does not take place on surface OH^- groups but on larger entities, such as groups involving Si and O or Ti and O atoms.

8.3.3 Anionic Monolayers and Anionic Micelles

When ionic surfactants associate from aqueous solution to form a compact monolayer, spherical or cylindrical micelles or lamellar layers, an electric double layer is formed. Given the present theme, we consider anionic surfactants with monovalent cations as the counterions. The associates are kept together by hydrophobic bonding of the hydrocarbon tails; at the same time, this bonding forces the ionic head groups into close proximity, against their electrostatic repulsion. Cation binding to these head groups screens this repulsion and therefore promotes the association. The more electrolyte in the solution, the more effective is the screening. Experimentally, this trend can be seen in the lowering of the surface pressure and the critical micellization concentration, c.m.c., with increasing electrolyte concentration.

Specificity enters via the difference in screening power of different counterions. The better the affinity, i.e. the more negative $\Delta_{ads}g_i^0$, the stronger the screening. In this way, lyotropic sequences can, and have been, established experimentally.

In line with the trends, obtained in Sections 8.3.1 and 8.3.2, one should expect the sequence to be direct for large and inverse for small head groups. Look-

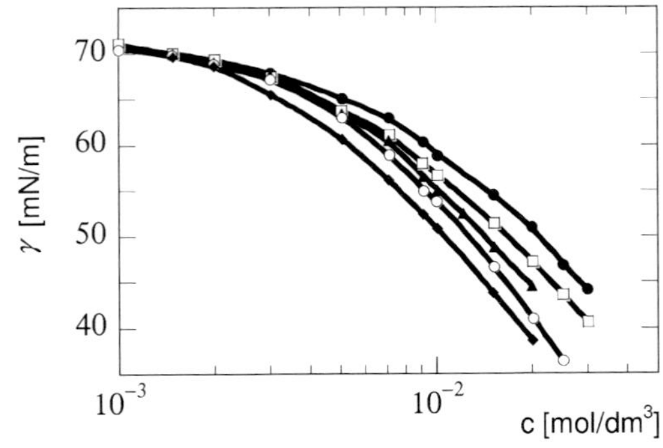

Fig. 8.4 Dependence of the surface tension of LiDS (●), NaDS (□), NH_4DS (▲), KDS (○) and CsDS (◆) on their concentration in solution. Lines represent fits of a theoretical model. (Redrawn from Warszyñski et al. [39]).

ing into the literature, it appears that there are abundant data for surfactants with large anionic head groups, as compared with the scarcity of those for small head groups. Hence only part of the predicted sequence can be checked experimentally, in particular for sulfate and sulfonic head groups. For this group, it appears that our rule generally applies.

The first illustration, Fig. 8.4, gives $\gamma(\log c)$ curves for various metal *n*-decyl sulfates (MDS). The slopes of these curves give the surface concentration Γ_i via Gibbs' law and it is concluded that Γ_i increases in the direction

$$\mathrm{Li^+ < Na^+ < NH_4^+ < K^+ < Cs^+} \tag{14}$$

These results compare well with those for the more extensively studied *n*-dodecyl sulfates, of which Fig. 8.5 gives an illustration. The data apply to 33 °C, but the authors also reported data at other temperatures. Figure 8.5 not only shows $\gamma(\log c)$ graphs before the c.m.c., but also gives c.m.c. values (see Table 8.6). The data at 25 °C are very closely in line with previous measurements [41–44]. One of the additional elements is that the temperature dependence is also included. The advantage of this is that, by application of the appropriate Gibbs–Helmholtz relationship, the Gibbs energy of micellization can be split up into an enthalpic and an entropic part.

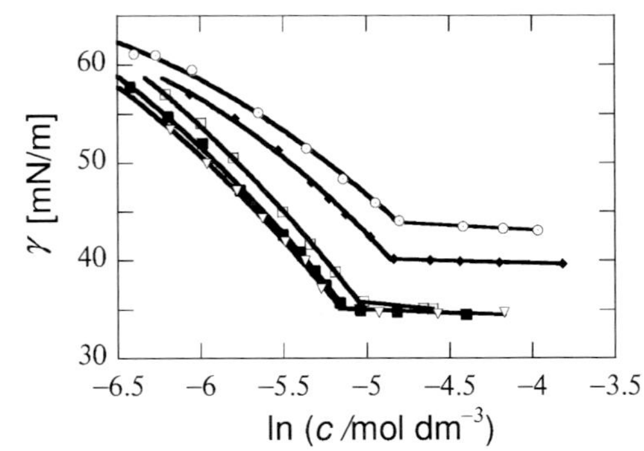

Fig. 8.5 Surface tension–ln(concentration) plots for lithium (spheres), sodium (closed diamonds), potassium (open squares), rubidium (closed squares) and cesium (open triangles) *n*-dodecyl sulfate in water at 33 °C. (Redrawn from Lu et al. [40]).

Table 8.6 C.m.c. values for alkali metal *n*-dodecyl sulfates (in mM) at different temperatures. Data taken from Lu et al. [40].

Temperature (°C)	Counterion				
	Li^+	Na^+	K^+	Rb^+	Cs^+
25	8.68	8.15			
33	8.46	8.10	6.71	5.90	5.90
40	8.25	8.05	6.90	6.01	6.10

The first observation from Table 8.6 is that the sequence is direct, as in Eq. (14), in line with our prediction (stronger counterion binding corresponds to a lower c.m.c. and a steeper decay of $\gamma(\log c)$ before the c.m.c.). The sequence does not vary with the temperature. In fact, the temperature dependence is a minor phenomenon, much less conspicuous than that of the ion specificity. At closer inspection it is seen that for Li^+ $\mathrm{d(c.m.c.)}/\mathrm{d}T < 0$, whereas for the other counterions it is the other way around. The thermodynamic characteristic adsorption parameters are represented in Table 8.7. Because of the scarcity of data, the inherent error margin is fairly large, about $4kT$ for $T\Delta_{\mathrm{ads}}s_i^0$ and $\Delta_{\mathrm{ads}}h_i^0$; as a corollary, entropy and enthalpy do not add up exactly to the tabulated $\Delta_{\mathrm{ads}}g_i^0$ value, which is more accurately known (from the c.m.c. and the surface pressure just below the c.m.c.). Accepting this uncertainty, nevertheless the different entropy–enthalpy balance for Li^+ is significant. For all other alkali metal ions the Gibbs adsorption energy is to a large extent enthalpically determined, with the sequence not very clearly expressed in the enthalpy and entropy. For the Li^+ ion, which "refuses" most strongly to bind specifically to the sulfate group, the entropy term is very high and positive; so that $-T\Delta_{\mathrm{ads}}s_{\mathrm{Li}}^0$ dominates $\Delta_{\mathrm{ads}}g_{\mathrm{Li}}^0$. The suggestion forces itself on us that some randomization of the adsorbate has to take place in order to accommodate the Li^+ ion electrically; this randomization also leads to a repulsive enthalpy. Obviously, the order of magnitude of $\Delta_{\mathrm{ads}}g_i^0$ in Table 8.7 compared with those obtained for the binding of those ions on AgI or oxides is much higher because here we are considering adsorption of whole

Table 8.7 Thermodynamic parameters (in units of kT) at 33 °C for the adsorption of alkali metal *n*-dodecyl sulfates at the aqueous solution/air interface just below the c.m.c. (Data taken from Lu et al. [40]).

	$\Delta_{\mathrm{ads}}g_i^0$	$\Delta_{\mathrm{ads}}h_i^0$	$T\Delta_{\mathrm{ads}}s_i^0$
Li^+	−13.2	6.0	18.3
Na^+	−13.7	−10.1	4.7
K^+	−14.1	−12.1	2.4
Rb^+	−13.6	−8	6.7
Cs^+	−14.1	−8	5.5

surfactant molecules; the binding Gibbs energy between the ion and the monolayer is only a small part of that. Lu et al. [40] carried out some neutron reflection experiments but were unable to find significant differences in the layer thicknesses as a function of the nature of the cation.

The ensuing question may be asked whether the procedure described to obtain the data in Table 8.3 for silver iodide may be modified to find the part played by ion–head group interactions. The answer is: no. It is feasible to establish the fraction of counterions that is bound. The number of free counterions per micelle can be estimated from electrophoresis, from the structure factor in light scattering, conductivity, osmotic pressure, etc. As the number of surfactant monomers per micelle is measurable (static light scattering), the fraction of bound counterions can be established. For NaDS, from data by Stigter [45], one can calculate this fraction to be 0.77 in the absence of NaCl, slightly *decreasing* with increasing c_{NaCl} until about 0.68 at 0.1 M electrolyte. This relatively minor salt effect is different from the rather major *increase* found for silver iodide and oxides. The main reason for this difference is that, unlike a solid, a micelle grows upon adding electrolyte, in the present example by as much as 40%. Moreover, there is no independent way of establishing the electric potential at the location where the counterion finds itself. Hence, even if one would find some procedure for establishing $\Delta_{\mathrm{ads}}g_i^0$, there is no way of sequestering the electrostatic part, demanded according to Eq. (2) to obtain the required chemical part.

As it transpires that for monolayers, micelles and more complicated association colloids no unambiguous procedures are available to assess this chemical part, other ways have to be found to interpret the ionic specificity. One of these is modeling, which will now be addressed. In the modeling section we shall focus on the charged solid/liquid interface and keep the applications to "soft" interfaces for future work.

8.4 Modeling Ion Specificity at the Electrified Solid/Liquid Interface

8.4.1 Introduction

The generic properties of charged interfaces in aqueous solutions are fairly accurately represented by the Poisson–Boltzmann (PB) equation. From this perspective, it is natural to use this formalism as the start of our modeling efforts and add on top of this various aspects that may lead to ion specificity. We will use a lattice implementation of the PB equation to account for various effects that have to do with the volume of the ions. In such models, ions occupy an integral number of lattice sites and are roughly spherical. The solvency and specific surface interactions are accounted for in the Bragg–Williams approximation and parameterized by the Flory–Huggins parameters. A polarization term is used in the segment potential that is used in combination with a local volume-fraction-averaged dielectric permittivity. The surface charge is taken to be controlled by the pH in solution and by taking only acidic groups we consider only negatively charged surfaces as an example.

In the PB line of thinking, ions are point charges that behave ideally. Similarly to the ideal gas law, where the particles are point-like, the ions lack specificity simply because they are volumeless. The ideal gas law is known to be accurate provided that the molecular concentration remains sufficiently low. Similarly, it is known that the PB equation remains correct in the sense that the volume of the ions can be ignored, provided that the ion concentrations remain low. However, in practical situations, where the surfaces have a significant surface charge density, the PB equation leads very quickly to high values for the local ion concentration, especially in close proximity to a charged surface where volume effects are very often important. In this context, it makes sense to search for a correction to the PB equation, which was found in assuming the presence of a Stern layer.

The Stern layer introduces a charge-free zone next to the surface with a dimension given by the (hydrated) ion size. Together with some adjustable value for the capacitance of the Stern layer one can also account for some of the variability in electric double layer phenomena. The elegance of the approach rests in the fact that for the diffuse part the PB-equation remains analytically solvable, while using this special boundary condition.

Below we will introduce a numerical route to solve more complex models that are in the spirit of the PB approach, but go beyond this. As is commonly done, we will not account for the spatial organization of the charges on the surface and assume that the charge density is laterally homogeneous. As a result, the PB equation becomes effectively one-dimensional and all quantities are typically expressed per unit area. In such a form the PB equation is a member of the so-called self-consistent field (SCF) theories, also known as mean-field methods. We use the implementation of the SCF equations of Scheutjens and Fleer (SF) [46–49], which allows us elegantly to account for the fact that the ions (and also other species) have a non-zero volume. The central idea is that a set of coordinates is used on which the molecules can be placed. These coordinates are arranged in layers parallel to the electrified interface and in each layer we shall assume that the fluctuations in the molecular composition away from the mean distribution can be ignored. We will therefore elaborate a lattice version of the extended PB theory. A lattice model is the simplest way to account for the excluded volume of the ions. The lattice model naturally incorporates the Stern layer concept, but goes beyond it [50–54]. More specifically, we shall invoke a compressibility relation to guarantee that the local volume is limited and that we cannot pack arbitrarily many molecules in any place. We also mention at this point that we shall consider ions that occupy more than one lattice site. We can, however, introduce most of the ideas without taking this into account.

Any model that describes an experimental system realistically will undoubtedly be very complex and loaded with parameters. The results of such a model will be exceedingly difficult to rationalize because of the many effects that add up to the total picture. As we are not close to this situation, it is better to construct simple models and make systematic variations in the parameters and try to understand the response and correlate this with experimental facts. The interesting point as far as computations are concerned is that the parameters used in the computations can also deviate strongly from experimentally realistic values. We will accept this only if this exercise will teach us something extra regarding the physics of the system.

8.4.2
Outline of the Method and Procedures of Elaboration

In this chapter, we shall not discuss all the details of the discretization scheme of Scheutjens and Fleer, or go into all the details on how the Poisson equation is implemented on a discrete set of coordinates. Instead, we will mention the main characteristics, go into the details of the molecular input of the models and refer to the literature for fuller details [46–54]. It should be understood from the start that by the use of a lattice model one accepts some shortcomings. By far the most relevant shortcoming is the fact that the smallest length scale in the analysis is given by the lattice length l, for which we will take a value of 0.3 nm. If one becomes interested in phenomena at an even shorter length scale, one has to abandon the lattice model. On the lengths that are larger than this lattice length, the discretization is rigorous (in the context of the model).

Within a classical Stern layer approach the dielectric constant of the Stern layer is an adjustable parameter. Below we will not follow this classical approach. More specifically, we will assume that the local dielectric permittivity is a unique function of the local molecular compositions and therefore is no longer a freely adjustable parameter. In such a model for the electric double layer, it is necessary to allow for gradients in the dielectric permittivity, for example because the solid phase typically has a dielectric constant significantly different from that of water. Below we therefore implement the idea that the dielectric permittivity in the system is assumed to be some function of the (local) molecular distributions. In this case the (continuous) Poisson equation reads

$$\varepsilon_0 \frac{\partial}{\partial \xi} \varepsilon(\xi) \frac{\partial}{\partial \xi} \psi(\xi) = -q(\xi) \tag{15}$$

where the ξ-coordinate is the distance from the surface and the dielectric permittivity ε, the electrostatic potential ψ (averaged along the surface) and the spatial charge density by q as well as the surface charge density σ (here assumed to be homogeneously distributed *along* the surface) are a function of it. Only when ε does not depend on the spatial coordinate is it allowed to bring it to the right-hand-side of Eq. (15). As the gradient in the potential is related to the electric field $E(\xi) = -\partial\psi(\xi)/\partial\xi$, the Poisson equation expresses the fact that a local charge density q can only exist if it is accompanied by a gradient in $D = \varepsilon E$, where D is the dielectric displacement.

In the SF–SCF approach there exists a z-coordinate which assumes discrete values $z=1, 2, 3, \ldots, M$, representing distances from the surface $\xi=l, 2l, 3l, \ldots, Ml$. Below we will use the discrete variable and trust that the relation to the continuous equations is evident. The machinery of the SF–SCF model is such that the so-called self-consistent solution of the equations guarantees that the Poisson equation is obeyed at each coordinate z in the system. At this stage it is necessary to pay attention to the boundary conditions. At $z=0$ we will have a surface which will carry some surface charge induced by a specified charge-determining21 mechanism. We will specify this below in detail. At $z=M$ we will assume that there exists a reflecting boundary. In other words, at the system boundary the electric field is assumed to be zero. When M is very large, we thus consider the electric double layer of a single surface in an electrolyte solution. When M is small we are dealing with a pair of interfaces that are a distance $H=2M$ apart. The latter is important for colloidal stability issues [55, 56], but in the following we will take the system size as being large enough that the electrostatic potential has vanished to a very good approximation. The reflecting boundary condition guarantees that the overall charge of the system, that is including the surface charge, is exactly zero. Near the boundary at $z=M$ there are no gradients in molecular species and we will refer to the concentrations of all components in this region as the bulk concentrations. We will use a superscript b to refer to these. The bulk serves as the reference phase for the electrostatic potential and the ion concentrations. In this region the local charge density is zero.

Within the *classical* PB equation the Poisson equation is combined with the Boltzmann equation. In our discrete coordinate system we can write this as

$$\varphi_x(z) = \varphi_x^{\mathrm{b}} \exp \frac{-u_x(z)}{kT} \tag{16}$$

where $\varphi_x(z)$ is the volume fraction of segments of type x, which is a dimensionless concentration. It is computed as the fraction of lattice sites filled at coordinate z by segments (monomers) of type x. When the number of monomers is given by $n_x(z)$ and the number of lattice sites per layer given by L, we compute the volume fraction as $\varphi_x(z) = n_x(z)/L$. Volume fractions are converted to molar concentrations by $c_x(z) = \varphi_x(z)/(1000\, l^3 N_{\mathrm{av}})$, where N_{av} is Avogadro's number. Within the *classical* PB model, the self-consistent potential $u_x(z)$ is given by

$$u_x(z) = z_x e\psi(z) \tag{17}$$

where z_x is the valence (including the sign) of the species. The valence is zero for uncharged segments (water), 1 for a monovalent cation and, –1 for the anion. In general, the charge density (number of charges per layer) is

$$q(z) = \sum_x z_x e\varphi_x(z) \tag{18}$$

where the sum is over all charged species in the system. Equation (18) is, of course, potentially confusing. The use of volume fractions is strictly not applicable for point charges. In the case of point charges, however, one should interpret $\varphi_x(z)$ as the number of point charges of type x per (volume of) L lattice (in the continuous version this quantity has a dimension and in the discrete version it is made dimensionless). The coordinate $z=0$ is not accessible for either ions or water molecules. Typically, the distribution of the segments that form the solid phase is fixed, having a volume fraction of unity, and there should be charged groups at the surface, i.e. at $z=0$. One way to vary the surface charge density in the calculations is to consider the surface layer to be composed of two types of segments, one type which is uncharged, named S, and another that has, e.g., a segment named Z with a unit negative charge, i.e. $z_Z = -1$. The surface charge is then controlled by the ratio S/Z or, for that matter, by the volume fraction of charged groups, i.e. $\varphi_Z(0)$. For a surface with a fixed surface charge the value of $\varphi_Z(0)$ is an input parameter and the electrostatic potential in the surface layer, $\psi^0 = \psi(0)$, is (with respect to a reference potential) the observable. For a given charge density profile, the electrostatic potential can be evaluated exactly once the dielectric permittivity is specified throughout the system.

Within a lattice approach, it is natural to compute a value for the dielectric permittivity at each position. The dielectric permittivity should depend on the local molecular composition. The simplest implementation of this is to evaluate this quantity per layer as the volume fraction-weighted average

$$\varepsilon(z) = \varepsilon_0 \sum_x \varepsilon_x \varphi_x(z) \tag{19}$$

where ε_x is the relative dielectric permittivity of a bulk phase composed of units of type x. In the case of point charges the sum over x in Eq. (19) may be replaced by the term for water only and $\varepsilon(z) = \varepsilon_0 \varepsilon_{\text{water}}$ for all z values (where the solvent can go). At this stage we also realize that it is impossible to generate a bulk phase of pure ions and that the relative dielectric constant for the ion species is not directly measurable, but ions can be ranked according to their polarizability. Differences in the polarizability between various ions may well be one of the ways according to which ion specific effects develop. We will return to this effect in more detail below.

In the *classical* PB approach, the solvent is assumed to have a constant dielectric permittivity and the above equations can routinely be solved (in the continuous limit often even analytically). In passing we note that at equilibrium the chemical potentials of all mobile components, i.e. water, cation and anion, do not depend on the local coordinate. This means that

$$\mu_x = \mu_x^0 + kT \ln a_x(z) + z_x e \psi(z) = \mu_x^0 + kT \ln a_x b \tag{20}$$

where a_x is the activity of the component x (which depends on local compositions) and μ_x^0 is the standard-state chemical potential.

Let us now consider the excluded volume of all species in the system in the SCF approach. The idea is that the previous equations remain valid and we only add various features on top of it. In general the start of such extended analyses is to formulate a mean-field partition function that contains the appropriate contributions for the system in which all components have a specified volume. This partition function leads to the (mean field) Gibbs energy of this system. This Gibbs energy can be written as a function of both the volume fractions and the segment potentials of all the components. Minimizing this free energy function, taking a compressibility constraint into account, leads to the self-consistent segment potentials and hence to the volume fraction profiles. From this optimization, we know that the potentials follow from the differentiation of the (interaction part of the) Gibbs energy to the volume fraction of a component. Van Male [49] has worked out all these details and found for the segment potential

$$u_x(z) = z_x e\psi(z) + u_x^{\mathrm{vol}}(z) \tag{21}$$

$$u_x^{\mathrm{vol}}(z) = u'(z) + \sum_y \chi_{xy}\left[\left\langle \varphi_y(z) \right\rangle - \varphi_y^{\mathrm{b}}\right] - \frac{1}{2}\varepsilon_0(\varepsilon_x - 1)E^2(z) \tag{22}$$

Each of the three new contributions in Eq. (22) represents one aspect related to the finite volume of the molecular species, which strictly are contributions to g_i(chem) as used in Eq. (2). The first term is the result of the compressibility constraint. In a lattice model it is customary to fill up all lattice sites with molecular components – this is known as the incompressibility condition – that is that at each coordinate z

$$\sum_x \varphi_x(z) = 1 \tag{23}$$

The value for $u'(z)$ is chosen such that this constraint is obeyed. For example, when in our numerical procedure in which we search for the appropriate solution of the equations, it appears that the total volume fraction tends to be less than unity, we will lower $u'(z)$ and this will attract more segments to this position. Conversely, when the volume fraction has the tendency to be larger than unity, we choose a higher value for $u'(z)$ so that segments are pushed away from the local coordinate. At the self-consistent solutions, to be elaborated on below, the $u'(z)$ value is well defined and unique; its value is such that Eq. (23) is strictly obeyed. Physically, we can link $u'(z)$ to the work needed to generate space (to vacate one lattice site) that makes it possible to introduce the component x. This contribution is linked to the incompressibility constraint and therefore it may be referred to as the Lagrange field or the compressibility contribution to the segment potential.

The second contribution accounts for non-ideal contacts between the molecular components. These contributions are accounted for here using the Bragg–Williams approximation. This means that the interactions are weighted by the local volume fractions. The angular brackets in Eq. (22) represent a three-layer average:

$$\langle \varphi(z) \rangle = \frac{1}{6}\varphi(z-1) + \frac{4}{6}\varphi(z) + \frac{1}{6}\varphi(z+1) \approx \varphi(\xi) + \frac{l}{6}\frac{\partial^2 \varphi(\xi)}{\partial \xi} \tag{24}$$

where on the right-hand side we switched to the continuous coordinate considering that the local averaging implements non-linear contributions to the interaction term. Of course, in the calculations the discrete version is implemented. The discrete version is non-local as in the segment potentials at coordinate z the volume fractions at coordinates $z-1$ *and* $z+1$ occur.

As we have a lattice model in which all lattice sites are filled, it is possible to work with effective Flory–Huggins interaction parameters [57]:

$$\chi_{xy} = \frac{Z}{2kT}(2g_{xy} - g_{xx} - g_{yy}) \tag{25}$$

where Z is the lattice coordination number (for our case of a simple cubic lattice $Z=6$) and g is the relevant contact Gibbs energy, which can have various origins. Any molecular model to evaluate these quantities from first principles should account for all energetic and entropic effects that are relevant for such a contact. For example, when water molecules are involved, one should thus need to account for the alterations in the structure of water around the contact, the (high-frequency) dispersion forces, etc. Here we will treat this issue phenomenologically and make (educated) guesses for the resulting Flory–Huggins parameters.

When the interaction (Gibbs) energy of an x–y contact is higher than the average of an x–x and a y–y contact, the x–y units effectively repel each other and the FH parameter is positive. When the FH parameter is zero, there is no difference between an unlike contact and the average between the pair of like contacts. This is the athermal case. In practice, such cases are rare. In this chapter, we cannot ignore this contribution because some non-ideal mixing (with water) and some specific surface interactions may be the cause of ion specificity.

It is instructive to distinguish two sets of FH parameters. The first set deals with the interactions between the components in the bulk and the second set are the interactions where the surface is involved. Typically in the systems discussed we have three components in the bulk solution, viz., the cation, the anion and the water molecules, hence there are three independent interactions to worry about, i.e. the water–cation interaction parameter χ_{W+}, the water–anion interaction parameter χ_{W-} and the interaction between cations and anions χ_{+-}. Note that in the last interaction the direct electrostatic contribution is not included because it is separately accounted for. As in all our systems we do not expect situations where the concentration of cations and anions are both high at some specific coordinate we expect that the χ_{+-} parameter is of little importance for the present analysis. For this reason, we will keep this parameter to the athermal value in all our calculations, but mention that non-zero values for this parameter will influence, e.g., the solubility limit of a particular pair of ions. The other two parameters are important and relevant. A high positive value for these parameters may lead to problems with the solubility of the salt and we can anticipate that a low solubility product will correspond to a strong binding as a pair to a surface. Although the two parameters are, of course, independent, below we will vary these parameters in concert. The parameters will be varied in a correlated way in which we will take identical values for the co- and counterions, i.e. we choose $\chi_{W-} = \chi_{W+} \equiv \chi_{WI}$. In this case we have to be aware of the presence of a solubility gap if a high positive value is chosen. The parameters may be varied in an anti-correlated way, where we assume that the sum of the two parameters remains zero, i.e. $\chi_{W+} = -\chi_{W-}$. This means that one of the ions is assumed to be hydrophobic and the other hydrophilic. As in a precipitant both ions should occur in stoichiometric amounts, we do not have to worry about the solubility problem mentioned in the correlated approach. The hydrophilic ion will "solubilize" the other one (as in surfactants).

The second set of parameters deals with the interactions of the three components (water, cation and anion) with the substrate, leading to three "adsorption" parameters χ_{SW}, χ_{S+} and χ_{S-}. These parameters will determine to a large extent whether or not specific adsorption of one or more of the species will occur. Qualitatively these effects were discussed earlier where they were collected in $\Delta_{ads}g_i(\text{chem})$. As the substrate has a fixed distribution and because in an incompressible lattice model all sites are filled by molecular species, the adsorption process actually involves the exchange of species next to the surface. As a result, only the *difference* between the two exchanging species is relevant. In other words, we can select, e.g., the solvent as our reference component and consider the exchange parameters $\Delta\chi_{Sx} = \chi_{Sx} - \chi_{SW}$. Now, if $\Delta\chi_{Sx} < 0$, this means that the species x is adsorbed preferentially over water and vice versa. In the literature, one typically finds that one writes $\chi_{Sx} = \Delta\chi_{Sx}$ and state that it is assumed that $\chi_{SW} = 0$. Although this is somewhat sloppy, we will do this also here.

The final contribution in Eq. (22) is directly linked to the volume fraction-weighted average of the dielectric permittivity equation given by Eq. (19). If Eq. (19) is implemented, the third term in Eq. (22) is needed in order to keep the extended PB theory thermodynamically consistent. If one leaves this third term out of the segment potential, one will violate the Gibbs adsorption equation, Eq. (24). It is a matter of choice whether in this third term one writes $(\varepsilon_x - 1)$ or just ε_x, because all segment type independent contributions to the segment potential can be grouped into the Lagrange term. This means that we need to explain the origin of the $\frac{1}{2}\varepsilon_x E^2(z)$ term. Here we note that because the molecular entities in the system have volume, it will in general be true that if such component (whether it is charged or not) will change its energy when it is taken from the bulk (where the electric field is zero) to the region near the charged surface where the electric field is finite. It is favorable for this species to have an internal dipole oriented such that it is well aligned with the electric field. How much a specific species is doing this is proportional to the electric field. The energy gain of doing so is also proportional to minus the electric field, hence the term

$-\varepsilon_x E^2(z)$. The minus sign reflects the fact that there is a gain in energy. Orienting the dipole in the particular direction is entropically unfavorable. We therefore have to correct for the entropy loss. A detailed analysis shows that exactly half the energy gain is lost by the entropic contribution. As a result the Gibbs energy gain is given by $-\frac{1}{2}\varepsilon_x E^2(z)$.

In order to have a way to manipulate the size of the molecular species within a lattice model, we will introduce molecular species that are larger than the lattice site, such that one ion or solvent (water) molecule occupies multiple sites. Within a cubic lattice one particular lattice site has six neighbors. The idea is to build a compact "star" with a central segment (named C) connected to a fixed number of neighbors, i.e. the central segment has a given number of arms (hence the label A). Let us call the number of arms N. If all arms have unit length, the maximum number is $N=6$. To calculate the volume fraction distribution of these star-like objects, it is necessary to account correctly for the possible directions in which the N-arms can point. Here we will use a Markov approximation which does not obey all excluded-volume correlations of the arms, but which does allow for a very efficient algorithm to compute the relevant distributions and which is easily extended. Only the central segment will carry the charge (either plus unity or minus unity). The surrounding A units are uncharged but have the same solvency and dielectric permittivity parameters as the central unit. For example, for water we may have a positively charged central segment, it may be negatively charged or it is neutral (the equilibrium between these species is discussed below). Calculations for all the star-like species proceed by a simple recipe in which the following abbreviations are used: water is referred to as $i=1$, the cation as $i=2$ and the anion as $i=3$.

The start of the recipe [48] is to introduce two Boltzmann weights for species i: $G_{iC}(z) = \exp -u_{iC}(z)/kT$ and $G_{iA}(z) = \exp -u_{iA}(z)/kT$. These Boltzmann weights would represent the statistical weight of the central or one arm unit, respectively, if these were to have been detached from the others. As the central segment is connected to the arms, there are entropic consequences (especially if $N<6$) and we have to account for the connectivity. The way to do this is well documented. The overall volume fraction of the star-like molecule is composed of the contribution of the central segment and N arms:

$$\varphi_i(z) = \varphi_{iC}^{b}[\Phi_{iC}(z) + N\Phi_{iA}(z)] \tag{26}$$

where φ_{iC}^{b} is the bulk volume fraction of just the central segment of molecule number i. The unnormalized volume fractions Φ are found by a combination of the Boltzmann weights. The central segment is connected to N arms and therefore we can write

$$\Phi_{iC}(z) = G_{iC}(z)\langle G_{iA}(z)\rangle^N \tag{27}$$

In this equation the angular brackets are the three-layer averaging similar to that given in Eq. (24). Physically the angular brackets give the appropriate averaging, leading to the statistical weight of all possible positions of the arm with respect to the central unit. The computations of the distribution of the arms are slightly more complicated. The idea is that the arms must be connected to the central unit. The statistical weight is computed in two steps going from each of the $N-1$ arms (first step), "through" the center (second step) to the remaining arm. All steps are implemented by the site-averaging procedure (angular brackets):

$$\Phi_{iA}(z) = \left\langle \langle G_{iA}(z) \rangle^{N-1} G_{iC}(z) \right\rangle G_{iA}(z) \tag{28}$$

The molecular species of the star-type have some internal flexibility especially if $N<6$. As in the bulk (which is the reference state) the flexibility is the same, there is little effect of this internal flexibility unless near the surface the flexibility is employed to bring the central charged segment to some preferred location with respect to the surface. The deformation of the star gives a conformational (entropic) penalty that counteracts this tendency. Although this particular molecular mechanism is not necessarily realistic, we believe that it has some intrinsic merits. The main advantage is that the (uncharged) arms provide some packing and conformational constraints for the ions. Such a packing effect may differ from one ion to another and the Ansatz is that the change in N is one of the reasonable ways to mimic this. Here we note that below we fix the molecular model for the water molecules at $N=6$ arms.

Finally, we should detail how the surface is charged. The most simplistic way to do this is to assign the surface groups Z to have a unit negative charge and vary the overall surface charge by changing the ratio S/Z, where S refers to the uncharged sites. At the next level of sophistication, one may invoke a model by which the charge develops on the surface as a response to the solution properties. Again the mechanism by which the surface charge is generated is a potential source for ion specificity that one may investigate in more detail. Below we will show results for a specific model in which acidic surface groups may lose a proton depending on the local pH:

$$ZH + H_2O \rightleftharpoons Z^- + H_3O^+ \tag{29}$$

which is controlled by a dimensionless equilibrium constant K_{ZH}. It is straightforward to extend the analysis to account for amphoteric surfaces that both can have negative and positive surface charges. However, in line with the results presented in Section 8.3, we shall limit ourselves here to negative surfaces. Equation (29) should be complemented by the proton-generating mechanism given by the auto-dissociation of water:

$$2H_2O \rightleftharpoons OH^- + H_3O^+ \tag{30}$$

This equilibrium is controlled by the equilibrium constant K_W. In passing we note that within this approach, a site-binding competition scheme may be developed for the counterions to go into competition with the protons for the neutralization of surface groups, i.e.

$$ZM + H_2O \rightleftharpoons Z^- + M^+ \tag{31}$$

with corresponding equilibrium constant. This site-binding model is complementary to the Bragg–Williams approach and we decided not to elaborate on this below. Below only Eqs. (29) and (30) are used to generate the surface charge and we will not go into further details on variations in surface charge-generating mechanisms and corresponding possible ion-specific implications.

The implementation of weak, i.e. pH-dependent, charges has been worked out within the SF–SCF approach by Israëls et al. [52]. It rests on the idea that segments can have internal states. One or more of the states are charged and therefore have an electrostatic contribution to the segment potential, whereas the neutral states do not have this contribution. A consistent way to include this in the SCF formalism was shown by Linse and Bjorling [58]. Here we shall not go into further details and refer to the original studies mentioning that Eqs. (29) and (30) are implemented locally, taking into account that the local electrostatic potential influences the local concentration of the ionic species. We also note that in the calculations of the ionic strength the speciation (protons or hydroxyl ions) of the water molecules is included.

In Eq. (26) we assumed that the bulk concentration of each species i is known. However, not all of these bulk volume fractions can be chosen independently. The Gibbs–Duhem relation shows that not all components can be chosen independently and when there are charges the electroneutrality adds a second constraint to our variables. It is instructive to detail these constraints within the present model. As the bulk should be electroneutral, we can freely choose the bulk concentration of all charged species but one. The last type of ion, which we typically refer to as the neutralizer, has a bulk concentration such that the space charge density in the bulk vanishes. Similarly to Eq. (23), there exists also a compressibility constraint in the bulk. This implies that the bulk volume fractions are coupled, i.e. $\sum_i \varphi_i^b = 1$. Effectively, this constraint is enforced by assigning one component, which we will refer to as the solvent (here called molecule j), for which the bulk concentration in the bulk follows from the bulk concentrations of all the others, $\varphi_j^b = 1 - \sum_{i \neq j} \varphi_i^b$, such that the incompressibility constraint is obeyed.

Numerical solutions of the governing equations are routinely generated with an accuracy of seven significant digits [59]. For these solutions the volume fractions are consistent with the potentials. This means that the volume fractions that are needed to compute the electrostatic potentials (via the evaluations of the local charge density and the local dielectric permittivity and the use of the Poisson equation) are also generated using the Markov procedure for the star-like species. In addition to this self-consistency requirement the total charge of the system (including the surface) is zero and at each local coordinate the sum of the volume fractions adds up to unity.

8.4.3
Choice of Parameters

As mentioned above a molecularly realistic model that takes a number of ion-specific features into account invariably requires a large number of parameters. Most of those which we shall use below have already been discussed above. Here we will repeat these and mention (to simplify the presentation of the results) the default parameter setting.

If not mentioned otherwise, all molecular species have a volume of seven lattice sites, i.e. they have $N=6$ arms. Below we will vary the size of the ions such that the co-ions and counterions always have the same volume. This is not a serious limitation because typically the volume of the co-ion is rarely important. The number of arms is varied from $N=0$ to 6. To compare one system with another, it is appropriate to fix the concentration of the ions. We do this by fixing the volume fraction of the central unit of the ions. Below we will use, e.g., the volume fraction of counterions in the bulk and we mean with this the volume fraction of the central unit in the bulk, i.e. we use $\varphi_{iC}^{\mathrm{b}}$ and note that we will typically omit the C (for simplicity). The uncharged segments around the central charged segment will add to the volume fraction of the salt, as given by the overall volume fraction $\varphi_i^{\mathrm{b}} = \varphi_{iC}^{\mathrm{b}} + \varphi_{iA}^{\mathrm{b}}$, which will be a function of the value of N used. Also for clarity we will use the convention that the bulk volume fraction of the counterions is specified as the concentration of salt (index s is used): $\varphi_s^{\mathrm{b}} = \varphi_{2C}^{\mathrm{b}}$. Below the default value for the bulk volume fraction of the counterions is given by $\varphi_s^{\mathrm{b}} = 10^{-2}$. This corresponds to approximately $c = \varphi_s^{\mathrm{b}}/1000\, l^3 N_{\mathrm{av}} = 10^{-5}/l^3 N_{\mathrm{av}} \approx 0.614$ mol L^{-1}, for the lattice-site dimension $l=0.3$ nm.

The reference value of all interaction parameters is the athermal case, i.e. $\chi_{xy} = 0$. When surface interactions are taken to deviate from zero, we will report values for the interaction parameters for the choice that the parameter with respect to the water component is zero, i.e. $\chi_{\mathrm{SW}} \equiv 0$.

If not mentioned otherwise, the relative dielectric constant, also for the solid substrate, is $\varepsilon_x=80$ and the temperature is $T=298.15$ K.

To control the dissociation of water we have set $\mathrm{p}K_{\mathrm{W}}=16.5$, which is the value needed in the lattice model. This value corresponds to the more familiar quantity $\mathrm{p}\tilde{K}_{\mathrm{W}}=14$, where the tilde refers to the fact that the dissociation constant is divided by the square of the molar concentration of water. pH is an input parameter in the calculations. Below we present the surface charge not as a function of pH, but as a function of the surface potential, but it is understood that experimentally such curves are measured as a function of the pH. The acid groups at the surface has been assigned a low value for the dissociation constant such that we mimic, e.g., silica: $\mathrm{p}K_{\mathrm{ZH}}=2$. This means that a neutral surface occurs only at extremely low pH. We recall that in this limit the ionic strength is necessarily rather high (remember that the protons also contribute to the ionic strength), which explains the need for the rather high default value for the counterion volume fraction $\varphi_s^{\mathrm{b}} = 10^{-2}$. The fraction of sites at the surface that can be charged is fixed to 0.5. The surface charge σ^0 that will be reported below is expressed in terms of the fraction of sites of the surface that is charged, where we include the minus sign to remind ourselves that we have a negative charge at the surface. Thus the maximum value of the dimensionless surface charge is $\sigma^0 = -0.5$, which corresponds to a surface charge of approximately $-1.8\ \mathrm{C\,m^{-2}}$, where we used again $l = 0.3$ nm, which is of similar magnitude to those found for, say, oxides. It will be clear that it is difficult to charge the surface to its maximum values; indeed, one has to go to extremely high pH to achieve this. In this case, the surface potential will have an extremely negative value.

8.4.4
Results and Discussion

Above we have argued that a lattice model is complementary to the continuous one provided that one is interested in length scales larger than the discretization length. To prove this, we should confront the lattice predictions with the analytical results that are available for the continuous PB equation. The exact result can conveniently be expressed by

$$\gamma(z) = \gamma^0 e^{-\kappa z} \tag{32}$$

where $\gamma = \tanh(e\psi/4kT)$ and $\kappa = (2e^2 \times 10^{-18}/\varepsilon kTl^3)^{\frac{1}{2}}\sqrt{\varphi_s^b} \approx 25.56\sqrt{\varphi_s^b}$ is the inverse Debye length in units of $\mathrm{nm^{-1}}$ and z is in units of nm. In Fig. 8.6 a, typical results are presented (continuous lines) for a fixed surface potential of $\psi^0 = -200$ mV for three values of the ionic strength in semi-logarithmic coordinates. The linear decay that starts some distance away from the surface, that is, where the electrostatic potential is below approximately 40 mV, clearly shows that the electrostatic potential profile decays exponentially. This range is usually referred to as the Debye–Hückel (DH) regime. Near the surface there are some deviations from the DH behavior, as expected. The corresponding lattice predictions are also given in Fig. 8.6 a by the symbols. Note that in our results the ions have a non-zero volume ($N = 6$) and therefore we expect some small deviations from the PB results just near the surface. However, in this example the numerical results copy the analytical results to a very good approximation, both in the DH and in the high-potential regimes. The good correspondence shows that discretization has few adverse effects. As the numerical lattice model has the advantage of accounting for various ion-specific effects, it makes sense to analyze its behavior further.

In Fig. 8.6 b we present the dimensionless surface charge on the surface as a function of the electrostatic surface potential for the same salt concentrations as in Fig. 8.6 a. Again, this type of result is very well known. For high ionic strength the double-layer capacitance is roughly constant, whereas this is clearly not the case for low ionic strength. Close inspection shows that near the point of zero charge (p.z.c.), the charge–potential curve for the low ionic strength case merges with the curves for higher ionic strengths before the p.z.c. is reached. Recall that the distance between curves at different ionic strengths is deter-

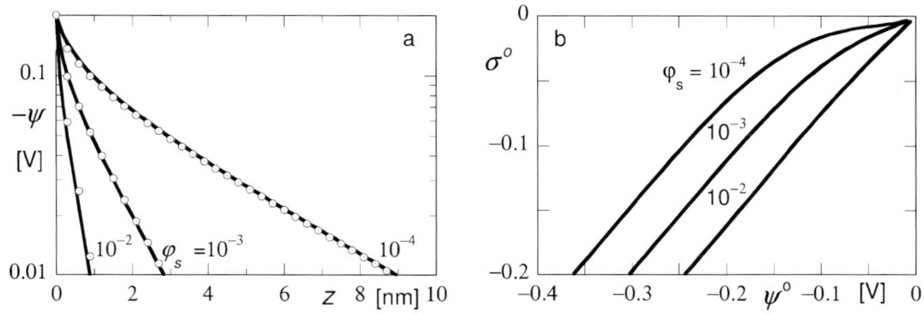

Fig. 8.6 (a) The electrostatic potential as a function of the distance from the surface for three ionic strengths as indicated by the volume fraction of counterions in the bulk (only the C-unit) in log–linear coordinates. The surface potential is fixed to $\psi^0 =$ −200 mV. The continuous lines are the 'exact' PB result and the symbols are the discrete SCF results. (b) The dimensionless surface charge (in units of e/l^2) as a function of the surface potential for three values of the ionic strength (in dimensionless units) as indicted. All three curves end at $\sigma^0 = -0.5$ (not shown) at some surface potential $\psi^0(\varphi_s^b)$ that depends on the ionic strength.

mined by the screening and measured in terms of the Esin–Markov coefficient, Eq. (5). The merging of the curves can therefore be attributed to an increase in screening. As for the total ionic strength the protons are also taken into account, it becomes true that the protons dominate the ionic strength, especially for the low added ionic strength cases. Below we shall restrict ourselves to some high ionic strength and therefore we do not pay further attention to this flaw and note that the same occurs at very high pH where the surface becomes fully charged (the maximum charge on the surface is $\sigma^0 = -0.5e/l^2$). In all further results we keep the volume fraction of salt (i.e. the volume fraction of the central ion of the star-like molecules) to $\varphi_s^b \equiv \varphi_{cs}^b = 10^{-2}$.

Let us start by analyzing the response of $\sigma^0(\psi^0)$ to some changes in the governing parameters. Figure 8.7 shows how the surface charge versus surface potential curves vary when the ion size is changed (Fig. 8.7 a) and when the polarizability of the ions is varied (Fig. 8.7 b). In comparison with the results shown in Fig. 8.6 a, we have extended the graphs to higher surface potentials and higher surface charges. In both Fig. 8.7 a and b the sensitivity towards the changed parameter shows up at high surface charge and thus at high surface potential. Above we argued that the size effects of ions would only occur when the ions are crowded near the surface. In the absence of specific adsorption, this occurs only when the surface charge is sufficiently high. From this point of view, it is easily explained why for a given surface potential the surface charge remains lowest when the ions are the largest (cf. Fig. 8.7 a). The larger ions cannot pack as easily just next to the surface and therefore the large ions cannot screen the surface charge as well as the smaller ions. This trend is classically interpreted as increasing the thickness (lowering the capacitance) of the Stern layer.

The explanation of why the ions with the highest polarizability can for given surface potential stabilize a highest surface charge is also fairly straightforward.

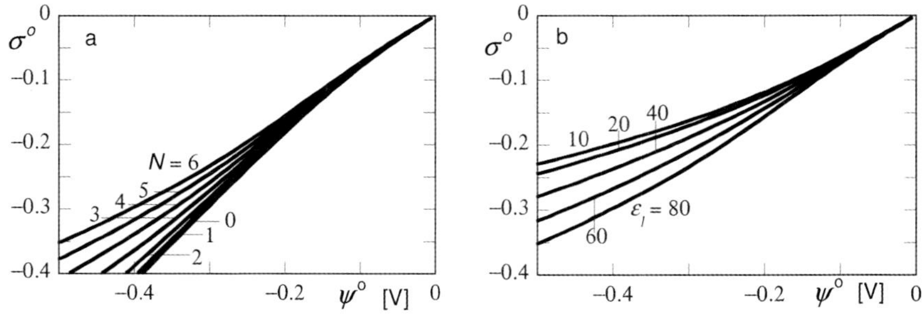

Fig. 8.7 (a) Dimensionless surface charge versus surface potential for ionic strength $\varphi_s^b = 10^{-2}$ for different values of the size of the ions as indicated by the number of arms N. (b) Surface charge versus surface potential for various values of the value of the dielectric constant for the ions (both ions have the same polarizability).

Near the surface the electric field is high. Ions with a high polarizability are drawn more strongly into these high electric field regions. As a result, the layers just next to the surface have a higher dielectric permittivity and as a result the surface charge can be higher (the Stern layer capacitance is high). When due to the presence of surface charges ions with a low polarizability are drawn to be near the surface, the region next to the surface becomes a region with a low dielectric permittivity. This results in a relatively low Stern layer capacitance and, hence, for a given potential the charge is relatively low.

Figure 8.8 shows the effect of the dielectric permittivity of the substrate (=solid) on the charge–potential curves. To understand how the surface charge can have an effect, one needs to know in some detail how the discretization is implemented in the lattice model. We assume that the charge of the ions that are placed on the lattice sits exactly in the center of a lattice site. In layer $z=0$ only surface units are present. In layer $z=1$ the first layer with solvent and mobile ions is found. Physically the boundary between the solid and the solvent thus is positioned at $z=1/2$ and therefore the charge at the surface is buried exactly half a lattice layer in the substrate. The dielectric permittivity is modeled to be homogeneous within each layer. As a result, there is a jump in the dielectric constant exactly half way between the layers. In the region $0 < z < 1/2$ the

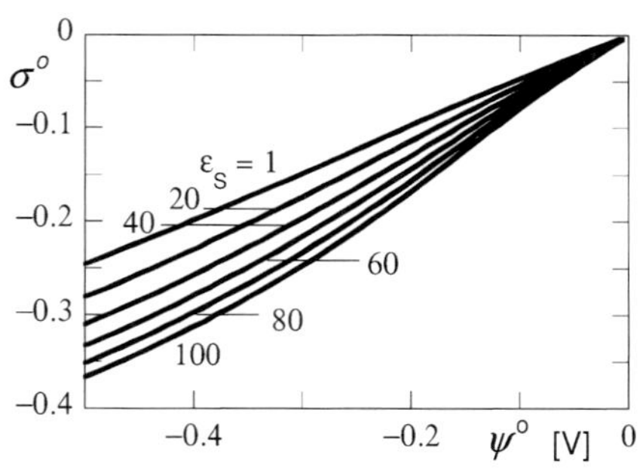

Fig. 8.8 Surface charge versus surface potential for different values of the dielectric constant of the solid surface (as indicated).

dielectric constant of the surface is found whereas from $1/2 < z < 3/2$ another dielectric constant is found. To integrate the Poisson equation from the surface charges to the charges in the first layer, one has to account for the jump in dielectric constant. From this, it is clear that the dielectric constant of the substrate does effect the electrostatic potential profile and thus the surface charge–surface potential curve. In practice, it may also be the case that the charges are buried somewhat in the solid. The extent to which this happens will correlate with the extent to which ε_s plays a role in the double-layer capacitance. We may add that Sverjensky [34] makes a point of the role of ε_s in site-binding interpretations, although in a rather different way.

As can be seen in Fig. 8.8, the lowest value of the dielectric constant of the surface suppresses the surface charge most. This may be interpreted as some sort of low average dielectric constant for the Stern-layer. Such a low value gives a low capacitance and this suppresses the charging-up process. The effect of the surface is clearly seen at high surface charge, but is also present in the regime where the surface is only weakly charged. The reason is that the double-layer capacitance is already affected by the substrate even when the molecular composition in the first lattice layers is still close to the bulk values.

A much more complicated behavior is found when on top of the electrostatic interactions the ions are drawn to the surface by non-electrostatic interactions. In Fig. 8.9 we quantify the effect that can be expected when the counterions have non-coulombic interactions with the substrate (Fig. 8.9 a) or when the counterions have such interactions (Fig. 8.9 b). When specific interactions operate, one finds significant effects already when the surface charge is very low. This is the most important quantitative feature that distinguishes the specific adsorption from the results presented above. It is easily seen that when our model would had accounted for amphoteric surfaces, we would have found that the p.z.c. would have shifted as a result of the specific adsorption. In the present model, the surface is only negative and formally the p.z.c. is found as a limiting situation.

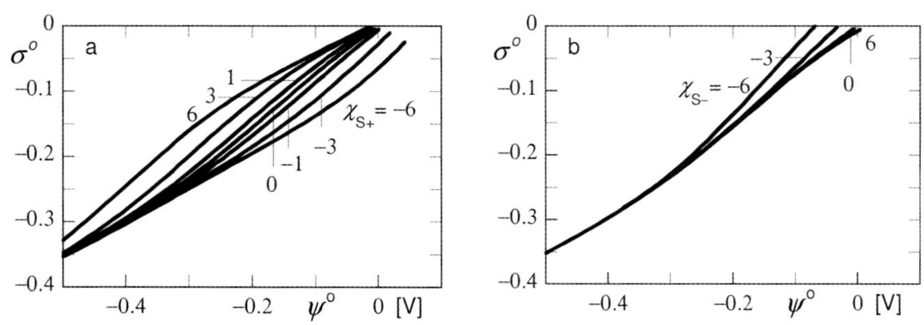

Fig. 8.9 (a) The dimensionless surface charge as a function of the surface potential in volts for various values for the adsorption energy of the counterion to the surface as indicated. (b) The same as (a) but now for the co-ion. In both graphs $N = 6$.

When the counterion is attracted preferentially to the surface, the effects are significantly larger than when the co-ions interact with the substrate. For this reason, we will first discuss the relevant features of Fig. 8.9 a. Recall that a negative value for χ_{S+} implies that the ion is specifically attracted more strongly to the surface than the solvent molecules and when this value is positive the specific interaction is repulsive. As the ions and water have the same molecular volume, we do not have to worry about complications resulting from size disparities. From Fig. 8.9 a, it is seen that the surface potential can be positive while the surface remains negatively charged. What happens effectively is that adsorption of counterions generates a local positive electrostatic potential which is accompanied by the induction of some negative charge on the surface. The electrostatic potential at the surface remains relatively low for given surface charge. For high pH and therefore at large surface charge and high surface potential, the specific effect of the extra non-electrostatic adsorption affinity gradually diminishes. In this regime, the excluded volume of the ions is of importance (cf. Fig. 8.7 a), which counteracts the specific adsorption. When the affinity of the ion is changed gradually from attractive to repulsive, the changes in the charge–potential curves in the region of small surface charges become relatively small. This should be contrasted with the effect of repulsive interactions in the regime of high surface charges. Again, in this region the excluded volume of the ions is important and repulsive contributions add to this effect in such a way that a relevant effect remains up to very high surface charges when χ_{S+} is sufficiently positive. Increasing the repulsion of ions to the surface has the same effect as increasing the size of the ions.

As co-ions are effectively repelled by the surface as soon as the surface is charged, one should not expect many effects of specific interactions of the co-ions. This is exactly what is found, as is seen in Fig. 8.9 b. Only when the surface is very weakly charged does the adsorption or depletion of the co-ions have some effect on the charge–potential curves. A strong attraction of the co-ions sweeps up the electrostatic potential even for the case when the proton charge on the surface remains essentially zero. The reverse effect, when the co-ion is repelled from the surface, has little effect, for obvious reasons. Again, if we had modeled amphoteric surfaces we would have found a richer behavior, as strong adsorption of co-ions could have led to change of the sign of the surface charge upon which they would become counterions.

Let us shift our attention to the bulk solvation effects of the ions. It is well known that ions in an aqueous solution are more strongly hydrated the smaller the ion is. Usually this is attributed to the fact that a small ion has a higher field around it such that the water dipoles and quadrupoles orient more strongly around the smaller than around the larger ion. As this effect is not included in the classical PB theory, or in the present extended version, it is of interest to mimic it within a Bragg–Williams approximation. Here we implement in a somewhat *ad hoc* way that ions interact differently with the solvent and that this interaction can be parameterized by short-range nearest-neighbor FH parameters. At this stage it should be understood that a finite value for the interaction parameter of an ion with the solvent also contributes to the effective adsorption energy if upon adsorption the degree of solvation changes. The key idea is that when the ion adsorbs on the surface, it exchanges a contact with the solvent for a contact with the surface. The loss of the contact with the solvent is the reason why the adsorption becomes sensitive to the way in which the ion is solvated. The better the ion remains solvated, the less likely it is to adsorb on the surface and vice versa.

In the parameter section we mentioned that we can change the interaction parameters for the co- and counterions in concert in the same direction or such that the sum of the FH parameters cancel. The result of these scenarios is given in Fig. 8.10a and b, respectively. In both scenarios the largest effects of the changes in the solvation energy are seen in the region of highly charged surfaces, albeit that there are some intricacies.

From Fig. 8.10a, where both the co- and counterions have identical interactions with the solvent, we see that there are very small effects in the region of low charge densities on the surface. As both the co-ion and the counterion contribute to the effective adsorption energy due to a change in hydration, there is no difference and therefore there is no charge-generating mechanism. This explains why for surfaces that are very weakly charged there is little effect of the value of χ_{WI}. The rather large effects seen at highly charged surfaces is linked to the fact that the ions have a significant volume. As soon as the surface is highly charged and the counterions next to it become densely packed, they have to expel water from the surroundings and therefore the solvation parameter is important. Attraction between the solvent and the ions has the same effect as increasing the size of the ion, whereas repulsion between solvent and ions has the same effect as decreasing the size of the ion. The parameter χ_{WI} is not increased above the value 3, because above this value the two ions have a solubility gap. From the Flory–Huggins theory [57], it is known that the critical interaction parameter obeys the equation $\chi_{WI}^{cr}(N+1)=2$. Inserting $N=6$ gives

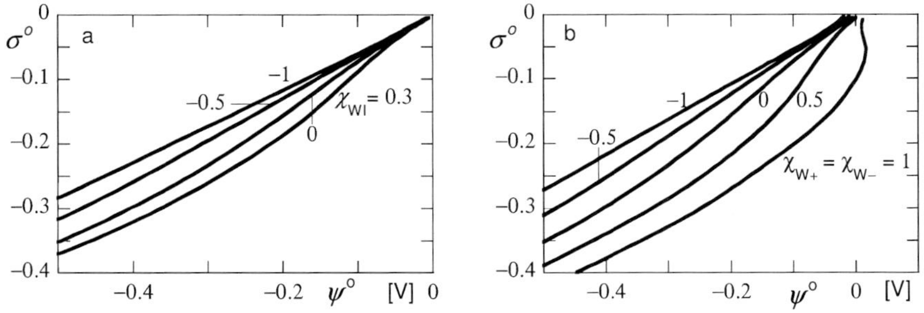

Fig. 8.10 (a) Surface charge as a function of potential for various values of the interaction between water and (both) ions as indicated. (b) Surface charge versus surface potential curves for systems in which the two ions have opposite values for the interaction with water (ions have balanced solvency parameters) as indicated.

$\chi_{\rm WI}^{\rm cr} \approx 0.286$ and the result of $\chi_{\rm WI} = 0.3$ is only possible because the ionic strength is kept below the binodal value.

In Fig. 8.10 b, the sum of the interactions with the solvent cancel and the hydrophilic ion solubilizes the hydrophobic ion. This means that macroscopic phase separation is not possible. Local phase separation is possible, however, as we will see. The results in Fig. 8.10 b can be understood when we consider the interaction parameter with the solvent for the counterion (cation) as our leading parameter. The value for the interaction parameter of the co-ion is of relatively low importance simply because next to the surface the co-ions are repelled by electrostatic forces. When $\chi_{\rm W+} < 0$, the cation is strongly hydrated and the result is very similar to the corresponding result in Fig. 8.10 a; for high surface charge densities, the ions like to be surrounded by water and therefore they appear relatively large. Until $\chi_{\rm W+} < 0.3$, the results in Fig. 8.10 b do not deviate strongly from those of Fig. 8.10 a. In contrast to what was impossible to do in Fig. 8.10 a, within the asymmetric variation of the solvency parameters it is possible to increase $\chi_{\rm W+} > 0.3$, as macroscopic phase separation will not occur. However, local phase segregation remains possible. The system can choose to put the cations, which want to leave the aqueous solution, next to the surface. The corresponding negative charge can be generated by dissociating protons from the acid groups. This can occur in a cooperative manner and this can for sufficiently high value of $\chi_{\rm W+}$ lead to a surface phase transition. We illustrate this by the graph $\chi_{\rm W+} = 1$. In this case, at slightly positive values for the electrostatic potential at the surface the surface charge jumps from a value near zero to a value slightly larger than –0.1. This jump in surface charge is accompanied by a jump in the number of counterions next to the surface, such that the surface charge remains roughly compensated, hence the electrostatic potential at the surface can remain close to zero. One has to employ an equal-area construction to locate the transition point exactly. We did not do this here.

It is of interest to elaborate briefly on the surface phase transition presented in Fig. 8.10 b. It is obvious that it is possible to shift this transition to higher values of the surface charge and correspondingly higher values for the surface potential if the co-ions are assigned a repulsion with the surface. This effect is illustrated in Fig. 8.11, where we can see that the jump-like change of the surface charge shifts to lower potentials with increasing repulsion of the co-ions to the surface. The higher the repulsion, the longer the condensation of the counterions on the surface is postponed. Interestingly, the size of the jump is not affected by the repulsive surface field. It is not known to the authors whether this phenomenon has already been found experimentally.

Above we have assumed that it is possible to vary several parameters independently. Of course, experimentally this is very hard (or impossible) to do. It is very likely that several of the effects discussed above work together or against each other. In combining effects we enter the field of scenarios. The first scenario is that the two effects discussed in Fig. 8.7 a and b occur at the same time. Here we assume that the dielectric permittivity of the layer will scale with the volume fraction of the ions, i.e. that it is proportional to our parameter N of the

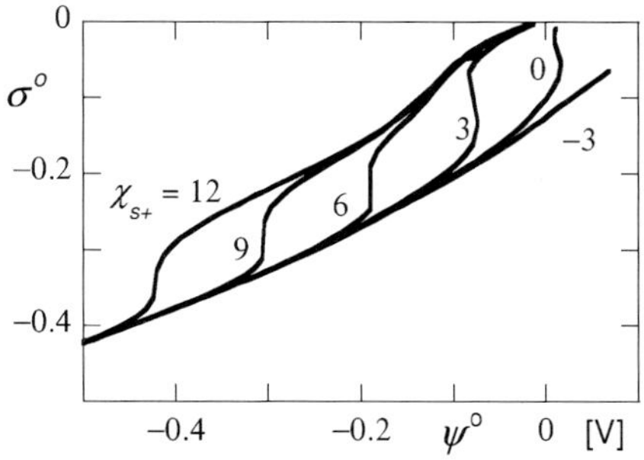

Fig. 8.11 Surface charge versus surface potential plots for systems wherein the ions have a balanced solvency parameter ($\chi_{W+} = -\chi_{W-} = 1$) for various values of the interaction of the positive ion with the surface is varied as indicated.

ion. Figure 8.12 shows what happens when two different dependences are superimposed. In Fig. 8.12 a we have assumed that $\varepsilon_I = 20 + 10N$, such that the smallest ion has a relative dielectric permittivity of $\varepsilon_I=20$ and the largest one $\varepsilon_I=80$, which is that of water. As the increase in the size pushes the ions away from the surface, the high polarizability attracts the ions to the surface and therefore we may expect some compensation. As can be seen in Fig. 8.12 a, full collapse of the curves does not occur: the smaller ions tend to allow a higher surface charge density at given surface potential. In Fig. 8.12 b, a slightly stronger dependence is assumed, i.e. $\varepsilon_I = 5 + 15N$, implying that the range of polarizabilities goes from $\varepsilon_I=5$ to 95. In this case there is more compensation. There is, of course, no physical reason why full compensation over the entire range must be expected. Here we focus on this case as such a situation can be used for systems in which even more parameters are varied.

The same two dependences are used in combination with a significantly lower value for the dielectric permittivity of the surface. Recall that a low ε_s has in the model the effect of a low dielectric permittivity in the "Stern" layer. As is shown in Fig. 8.13 a, the collapse of the surface charge versus surface potential

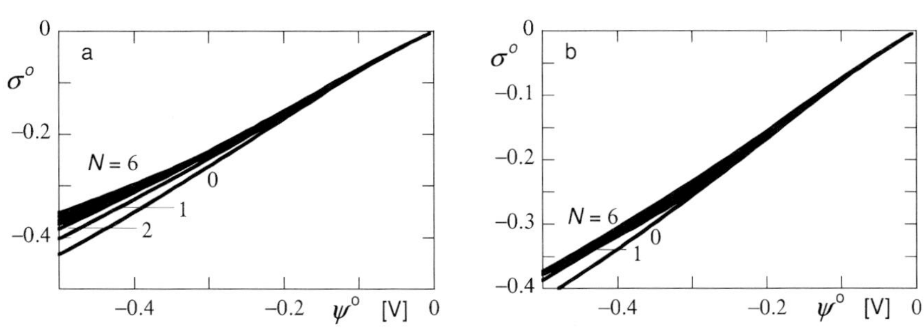

Fig. 8.12 (a) Surface charge versus surface potential for the systems in which both the size of the ions is varied by changing the number of segments N around the central ion and at the same time the dielectric permittivity of the ions using the *ad hoc* relation $\varepsilon_I = 20 + 10N$. (b) The same as in (a) but for a stronger dependence of the dielectric permittivity on the 'size' $\varepsilon_I = 5 + 15N$. In this figure, $\varepsilon_s=80$ (the default value).

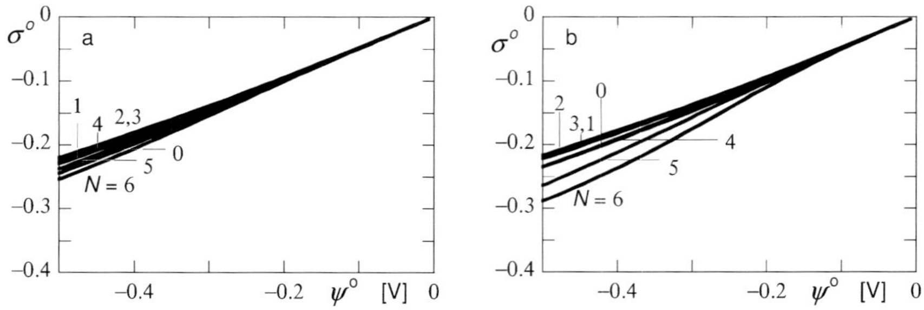

Fig. 8.13 (a) Surface charge versus the surface potential for a surface with a low dielectric permittivity $\varepsilon_s = 5$ assuming a weak size dependence for the dielectric permittivity of the ions $\varepsilon_I = 20 + 10N$, for various systems for which the ions differ in size as indicated. (b) The same but for the strong dependence $\varepsilon_I = 5 + 15N$.

curves is almost perfect when for $\varepsilon_s = 5$ the weak size dependence for the polarizability of the ions is used, i.e. for $\varepsilon_I = 20 + 10N$. In this case the stronger dependence $\varepsilon_I = 5 + 15N$ overcompensates the size effect. As can be seen in Fig. 8.13 b, the largest ion, i.e. the most polarizable one, clearly induces the highest surface charge for given surface potential. Close inspection of Fig. 8.13 show that in both graphs the N dependence is non-monotonic. In both cases an ion size of $N = 2$ gives the lowest surface charge for a given surface potential.

Summarizing, for a highly polarizable surface the smallest ions allow for the highest surface charges and the other way round. This implies that we have a first mechanism for an inversion of the Hofmeister series with the nature of the surface.

A second approach is to combine the size variation of the ions with the surface affinities. Above we have seen that specific adsorption has a significant effect on the charge–potential curves even when the size of the ions was not varied. In Fig. 8.14, the effect of a fixed adsorption energy parameter is combined with ion-size changes for a system for the case when there is a weak dependence of the ion polarizability $\varepsilon_I = 20 + 10N$ combined with a low polarizability of the surface (the case where the compensation was fairly good as shown in Fig. 8.13 a). When the ions are attracted to the surface (Fig. 8.14 a), we should look at changes in the titration curves especially in the region of low surface charge. As the overall adsorption energy increases, the more contact the star-like object can make with the surface, so we intuitively expect that the larger ions will shift the curves towards lower potentials. This is exactly what is found. This means that for a given surface potential the larger ions allow for more surface charge. In Fig. 8.14 a, the curves run almost parallel, implying that the double-layer capacitance is constant for these cases. This is expected because the layer next to the surface does not vary strongly in molecular composition in the surface charge range (surface potential range) probed here.

The opposite trend is found when for the same polarizability relation $\varepsilon_I = 20 + 10N$, the counterions are specifically repelled from the surface. Again,

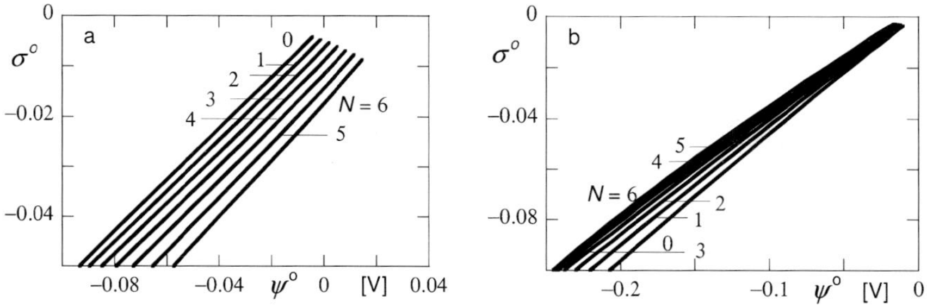

Fig. 8.14 (a) Surface charge versus surface potential for the case when the counterions are attracted to the surface with by a modest $\chi_{s+} = -3$ and a weak N dependence for the polarizability of the ions is implemented, namely $\varepsilon_I = 20 + 10N$. (b) The same as in (a) but now for a slight repulsion of the counterions for the surface $\chi_{s+} = 3$. In both graphs the surface has a low polarizability with $\varepsilon_s = 5$ (cf. Fig. 8.13 a).

in the repulsion mode the effects are more pronounced at higher surface charges, but the trends are clear. The largest ion now allows for the lowest charge density at the surface for a given surface potential. This implies a second mechanism to account for reversal of the Hofmeister series. Of interest is that at high surface charge the lines start to cross each other and a non-monotonic response with respect to the N variation is found, similarly as in Fig. 8.13. This points to the possible scenario that the Hofmeister series for a particular surface may change order simply by going to a higher charge density on the surface. Such switching behavior in the Hofmeister series may be expected in practice also because at low charge density the specific adsorption phenomena are important, whereas in the highly charged systems the size aspects of the ions dominate. These effects can therefore easily give rise to an inversion of the Hofmeister series but it is unknown by the authors whether this effect has already been seen experimentally.

Let us once more consider the effect of the ion-size variation in the Ansatz of a weak polarization dependence $\varepsilon_I = 20 + 10N$ for the case that the surface has a low dielectric permittivity ($\varepsilon_s = 5$) and compare this with the highly dielectric permittivity case ($\varepsilon_s = 80$). The results are summarized in Figs. 8.15 to 8.17. In these figures, various solvency conditions have been examined.

In Fig. 8.15 both ions are in a marginal solvent, $\chi_{\mathrm{WI}} = 0.3$. The situation for the not-so-good solvent condition was shown in Fig. 8.10 a for the case when the ions were all of the same size. In this graph it was shown that the curves shift to higher surface charge the better the solvent is. It should be realized that the effective solvent strength improves when the ions become smaller. This is because for a fixed χ_{WI} the product $\chi_{\mathrm{WI}}(N+1)$ becomes smaller with decreasing N. This implies that the system effectively goes deeper and deeper into the one-phase region (away from the critical point) when N decreases. We therefore would expect the titration curves to shift to the left with decreasing N. Inspection of the results in Fig. 8.15 a shows that this trend is roughly followed. On going from $N = 6$ to 2, we see the expected trend, but this reverses for the smal-

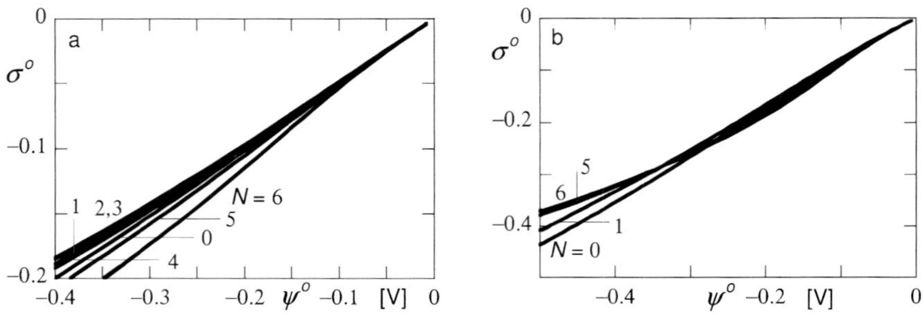

Fig. 8.15 (a) Plot of the dimensionless surface charge against the surface potential for different sizes of the ions as indicated by the number of arms N. The ions are in a not-so-good solvent situation $\chi_{WI} = 0.3$. The other parameters are as in Fig. 8.13 a: the weak size dependence for the relative dielectric constant of the ions $\varepsilon_I = 20 + 10N$, the surface has a low polarizability, $\varepsilon_s = 5$. (b) Same as in (a) but now for $\varepsilon_s = 80$. (Graphs for $N = 2$, 3, 4 lie between those for $N = 1$ and 5). Note that the range in surface charge is larger in (b) than in (a).

lest ions. The small rebound may be connected with the non-monotonicity seen already in Fig. 8.13 a and which was attributed to the fact that the size effects are not exactly balanced by the polarizability effects.

The results in Fig. 8.13 b illustrate (in comparison with Fig. 8.13 a) the effect of the dielectric permittivity of the surface. Figure 8.13 b shows the results for the higher ε_s case. Compared with Fig. 8.13 b, a surface with a higher ε_s is much more easily charged than a surface with a lower ε_s, as was explained above in detail. The trend at very highly surface potentials ($\psi^0 = -0.5$ V) is such that the larger ions tends to prevent further charging of the surfaces compared with the smaller ones. Around a surface potential of $\psi^0 = -0.2$ V, these trends are reversed (as in Fig. 8.13 a), but the curves remain very close to each other.

In Fig. 8.16, the opposite case is considered; now the solvent quality of the ions is improved to $\chi_{WI} = -0.3$. Recall that in our model with decreasing N the solvent quality goes in the direction of the athermal case. From this point of view, it is not unexpected that the trends reverse with respect to those presented in Fig. 8.15. In Fig. 8.16 a, similarly as in Fig. 8.15 a, a non-monotonic dependence on N is found, but now the smallest ion allows for the highest charge density on the surface. Compared with the reference system (Fig. 8.13 a), the curves have spread just a bit. Increasing ε_s gives a clear monotonic size dependence. The smallest ions allow for the highest surface charge (for given surface potential). This trend was expected because from the above results it is clear that at a surface with a high ε_s the surface charge is rather high and the size effects of the ions become relatively more important.

In Figs. 8.15 and 8.16 we chose the solvent conditions symmetric. In Fig. 8.17, the result is shown when the solvent quality of the cation is opposite to that of the anion, where the cation was chosen to be in the poor solvent condition ($\chi_{W+} = -\chi_{W-} = 1$). Recall that from Fig. 8.11 we know that for the case of $N=6$ this situation can give rise to a jump-like change in the adsorption of the cation. As the solvent quality improves when the ions become smaller, we expect that the transition does not occur for the smaller ions. Inspection of Fig. 8.17 shows that indeed the transition tends to give way to a more gradual (higher order) transition when the ions are small (obviously except in Fig. 8.17 b for $N=6$). Again, as in Figs. 8.15 and 8.16, the two graphs in Fig. 8.17 differ with respect to the polarizability of the surface. In both graphs there is a large size dependence for the charge as a function of the potential in the region of $\sigma^0 = -0.1$. To explain this extraordinarily large ion size effect, we have to point to the nearby surface phase transition. Similarly as in Figs. 8.15 b and 8.16 b, the surface with a higher dielectric permittivity allows for a higher surface charge and as the larger ions cannot support the high surface charge as well as the smaller ions, we see in Fig. 8.17 b that the lines come together and have a

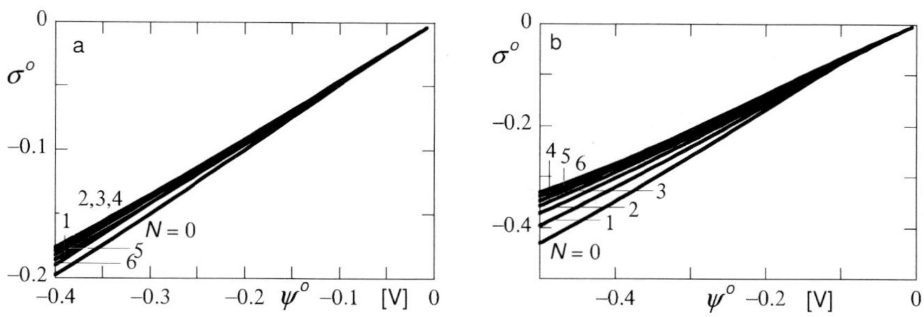

Fig. 8.16 (a) Plot of the dimensionless surface charge against the surface potential for different sizes of the ions as indicated by the number of arms N. Both ions are in a good solvent situation, $\chi_{Wl} = -0.3$. The other parameters are as in Fig. 8.13 a: the weak size dependence for the relative dielectric constant of the ions $\varepsilon_l = 20 + 10N$, the surface has a low dielectric permittivity $\varepsilon_s = 5$. (b) Same as in (a) but now for $\varepsilon_s = 80$.

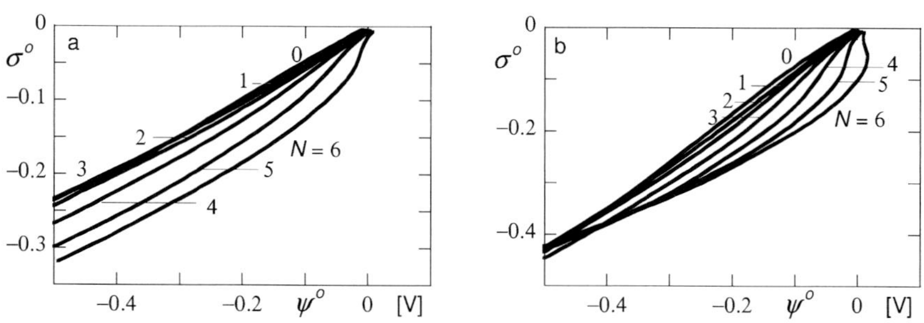

Fig. 8.17 (a) Plot of the dimensionless surface charge against the surface potential for different sizes of the ions as indicated by the number of arms N. The two ions are opposite solvency parameters $\chi_{W+} = -\chi_{W-} = 1$. The other parameters are as in Fig. 8.13 a: the weak size dependence for the relative dielectric constant of the ions $\varepsilon_l = 20 + 10N$, the surface has a low dielectric permittivity $\varepsilon_s = 5$. (b) Same as in (a) but now for $\varepsilon_s = 80$.

tendency to cross each other. The excluded-volume effect at high surface charge density compensates the cooperative accumulation of the larger ions at lower charge densities. However, it is repeated that in most real cases (amphoteric surfaces) the phenomena at low σ^0 are affected by the charge reversal.

8.4.5
Conclusions of the Modeling

From the above, it is clear that there are many possible scenarios to generate lyotropic or Hofmeister series. At this stage we are not going to speculate on which scenario is the more likely for one particular experimental system or to draw some firm conclusions for a particular case. We feel that not enough work has been done at this stage to justify this. In this context, it is important to note that it is expected that most (perhaps all) parameters needed for molecular modeling can in principle be extracted from measurements. Unfortunately, we are not yet near such a favorable situation. As a result, we can change parameters more or less at will and analyze the predictions. At this stage, it should be understood that to some extent the results that come out of the model are put in by the parameters (given the machinery that specifies the physics). Keeping this in mind, it is useful to summarize our findings.

In the SF–SCF model, it is possible to account not only for the generic properties of the electrified solid/liquid interfaces, but also for ion size, specific adsorption, ion solvation and ion polarizability effects. We can change these parameters systematically and independently such that the relative importance of these ion-specific effects varies. For reasonable parameter settings we found that several of the specific effects largely compensate each other. This observation is essential and we believe that it is not due to the particular choices of the parameters. Therefore, we expect that in experimental systems the same can happen. The compensation of specific effects explains why the classical PB equation, which completely ignores all specific properties of the ions (because all volume effects are ignored) remains successful. It also explains why it is possible to find lyotropic or Hofmeister series in the direct or reversed order, depending on relatively small changes in one or more parameters and concomitantly in the experimental case where one system is only slightly different from another.

The SF–SCF modeling is not limited to the special case presented above. More specifically, it is possible to consider amphoteric surfaces by assigning surfaces sites that accept a proton and other sites that donate a proton according to the pH [54]. Then both the positively and the negatively charged surfaces are treated in the same setting. For such a model, one can account for the fact that the p.z.c. may shift, e.g. due to the specific adsorption of the ions. It is also possible to extend the SF–SCF model to account for two gradients in the electrostatic potential and the charge density. It is then possible to account for inhomogeneous lateral surface charge distribution, chemical inhomogeneities and surface roughness [53]. All these extra phenomena can cause some specific ion-size effect. On the other hand, phenomena at a lower scale of discretization than l are difficult to account for, say the entropic liquid structure determined contributions to $\Delta_{ads}g_i$(chem).

The SF–SCF model is capable of accounting for flexible polymers at interfaces and has also been used for the molecular modeling of self-assembly [60–63]. In the case of ionic surfactants, we recently showed [60] that it is possible to predict the shift of the c.m.c., the size of the micelles, the sphere-to-rod transition and related properties as a function of the molecular properties of the monovalent ions. In this way the experimental trends in Section 8.3.3 can also be approached. Indeed, the micellization problem is even richer because of the strong coupling between the screening capabilities of the ions and the micellar degrees of freedom such as size distribution and shape. Also in other problems, such as the adsorption of (charged) surfactants or polyelectrolytes at (charged) interfaces, the specific effects of small ions can be investigated with the present SF–SCF approach [64, 65]. We expect that the SF–SCF model can be used to help explain many of the experimentally found non-generic ion size effects in these and related systems.

8.5 Outlook

Thermodynamic analysis, molecular modeling, computer simulations and experiments should always be used together to demystify ion-specific effects in colloidal systems. Ions have a finite volume and therefore interact pairwise with (sites on) the substrate (=solid), the solvent and other ions. They are (similarly to uncharged molecules) polarized in an electric field and affect the structure of the water in which they are embedded. These (and perhaps even other) effects lead to the lyotropic or Hofmeister series. This study shows how classical thermodynamic analysis on the basis of experimental evidence and modeling both contribute to the understanding of lyotropic or Hofmeister series. Each of these approaches has its intrinsic advantages and it is the combination of the two that offers most prospects.

References

1 D. Platikanov, D. Exerowa, Thin Liquid Films, in J. Lyklema (Ed.), *Fundamentals of Interface and Colloid Science*, Vol. V. Elsevier, Amsterdam, **2005**, Chapter 6.

2 H.R. Kruyt, M.A.M. Klompé, *Kolloid Beihefte 54* (**1943**) 484.

3 A. Basiński, *Roczn. Chem. 15* (**1935**) 430.

4 N. Fuchs, *Z. Phys. 89* (**1934**) 736.

5 H. Reerink, J.Th.G. Overbeek, *Discuss. Faraday Soc. 18* (**1954**) 74.

6 M. Mirnik, *Nature 190* (**1961**) 689.

7 J. Lyklema, J.F.L. Duval, *Adv. Colloid Interface Sci. 114/115* (**2005**) 27.

8 W. Kunz, P. Lo Nostro, B.W. Ninham (Eds.), Hofmeister Phenomena, *Curr. Opin. Colloid Interface Sci. 9* (**2004**).

9 D.C. Grahame, *Chem. Rev. 41* (**1947**) 441.

10 D.C. Grahame, *J. Electrochem. Soc. 98* (**1951**) 343.

11 B. H. Bijsterbosch, J. Lyklema, *Adv. Colloid Interface Sci. 9* (**1978**) 147.

12 J. Th. G. Overbeek, Electrochemistry of the Double Layer, in H. R. Kruyt (Ed.), *Colloid Science I*, Elsevier, Amsterdam, **1952**, Chapter 4.

13 J. Lyklema, *Fundamentals of Interface and Colloid Science*, Vol. II, Academic Press, London, **1995**, Section 3.5b.

14 J. Lyklema, *Trans. Faraday Soc. 59* (**1963**) 418.

15 J. Lyklema, *Mol. Phys. 100* (**2002**) 3177.

16 J. Lyklema, *Discuss. Faraday Soc. 42* (**1966**) 81.

17 Th. F. Tadros, J. Lyklema, *J. Electroanal. Chem. 17* (**1968**) 267.

18 Y. G. Bérubé, P. L. de Bruyn, *J. Colloid Interface Sci. 28* (**1968**) 92.

19 F. Dumont, J. Warlus, A. Watillon, *J. Colloid Interface Sci. 138* (**1990**) 543.

20 K. Bourikas, T. Hiemstra, W. van Riemsdijk, *Langmuir 17* (**2001**) 749.

21 A. Breeuwsma, J. Lyklema, *Discuss. Faraday Soc. 52* (**1971**) 324.

22 F. Dumont, A. Watillon, *Discuss. Faraday Soc. 52* (**1971**) 353.

23 R. Sprycha, *J. Colloid Interface Sci. 127* (**1989**) 1.

24 T. Hiemstra, R. Rahnemaie, W. H. van Riemsdijk, *Geochim. Cosmochim. Acta 69* (**2005**) A39.

25 Y. G. Bérubé, P. L. de Bruyn, *J. Colloid Interface Sci. 27* (**1968**) 305.

26 L. G. J. Fokkink, A. de Keizer, J. Lyklema, *J. Colloid Interface Sci. 127* (**1989**) 116.

27 M. Kosmulsi, J. Matysiak, J. Szczypa, *J. Colloid Interface Sci. 164* (**1994**) 280.

28 P. H. Tewari, A. W. McLean, *J. Colloid Interface Sci. 40* (**1972**) 267.

29 M. A. Blesa, N. M. Figliolia, A. J. G. Maroto, A. E. Regazzoni, *J. Colloid Interface Sci. 101* (**1984**) 410.

30 W. A. Zeltner, E. Yost, M. L. Macheski, M. I. Tejedor-Tejedor, M. A. Anderson, *ACS Symp. Ser. 323* (**1986**) 142.

31 D. A. Griffiths, D. W. Fuerstenau, *J. Colloid Interface Sci. 80* (**1981**) 271.

32 N. M. Figliolia, *PhD Thesis*, Universidad de Buenos Aires, **1987**.

33 J. Lyklema, *Fundamentals of Interface and Colloid Science*, Vol. II, Academic Press, London, **1995**, Section 3.10c.

34 D. A. Sverjensky, *Geochim. Cosmochim. Acta. 69* (**2005**) 225.

35 Y. Marcus, *Ion Solvation*, Wiley, New York, **1985**, Table 8.5.10.

36 R. W. Gurney, *Ionic Processes in Solution*, McGraw-Hill, New York, **1953**, p. 167.

37 L. Gierst, L. Vandenberghen, E. Nicolas, A. Fraboni, *J. Electrochem. Soc. 113* (**1966**) 1025.

38 R. A. Robinson, R. H. Stokes, *Electrolyte Solutions*, Butterworth, London, **1955**, pp. 476, 480.

39 P. Warszyñski, K. Lunkenheimer, G. Czichocki, *Langmuir 18* (**2002**) 2506.

40 J. R. Lu, A. Marrocco, T. J. Su, R. K. Thomas, J. Penfold, *J. Colloid Interface Sci. 158* (**1993**) 303.

41 P. Mukerjee, *Adv. Colloid Interface Sci. 1* (**1967**) 24.

42 P. H. Elworthy, K. J. Mysels, *J. Colloid Interface Sci. 21* (**1966**) 331.

43 M. J. Rosen, *J. Colloid Interface Sci. 79* (**1981**) 587.

44 P. Mukerjee, K. J. Mysels, *Critical Micelle Concentration of Aqueous Surfactant Systems*, National Bureau of Standards, Washington, DC, **1971**.

45 D. Stigter, *J. Phys. Chem. 68* (**1964**) 2603.

46 J. M. H. M. Scheutjens, G. J. Fleer, *J. Phys. Chem. 83* (**1979**) 1619–1635.

47 J. M. H. M. Scheutjens, G. J. Fleer, *J. Phys. Chem. 84* (**1980**) 178–190.

48 G. J. Fleer, M. A. Cohen Stuart, J. M. H. M. Scheutjens, T. Cosgrove, B. Vincent, *Polymers at Interfaces*, Chapman and Hall, London, **1993**.

49 J. van Male, *PhD Thesis*, Wageningen University, **2003**.

50 O. A. Evers, G. J. Fleer, J. M. H. M. Scheutjens, J. Lyklema, *J. Colloid Interface Sci. 111* (**1986**) 446–454.

51 M. A. Cohen Stuart, G. J. Fleer, J. Lyklema, W. Norde, J. M. H. M. Scheutjens, *Adv. Colloid Interface Sci. 34* (**1991**) 477–535.

52 R. Israëls, J. M. H. M. Scheutjens, G. J. Fleer, *Macromolecules 26* (**1993**) 5405–5413.

53 J. F. L. Duval, F. A. M. Leermakers, H. P. van Leeuwen, *Langmuir 20* (**2004**) 5052–5065.

54 A. W. P. Vermeer, F. A. M. Leermakers, L. K. Koopal, *Langmuir 13* (**1997**) 4413–4421.

55 J. M. H. M. Scheutjens, G. J. Fleer, *Macromolecules 18* (**1985**) 1882–1900.

56 W. J. Lokar, L. K. Koopal, F. A. M. Leermakers, W. A. Ducker, *J. Phys. Chem. B 108* (**2004**) 15033–15042.

57 P. J. Flory, *Principles of Polymer Chemistry*, Cornell University Press, Ithaca, NY, **1953**.

58 P. Linse, M. Bjorling, *Macromolecules 24* (**1991**) 6700.

59 O. A. Evers, J. M. H. M. Scheutjens, G. J. Fleer, *Macromolecules 23* (**1990**) 5221–5232.

60 F. A. M. Leermakers, J. C. Eriksson, J. Lyklema, Association Colloids and Their Equilibrium Modelling, in J. Lyklema (Ed.), *Fundamentals of Interface and Colloid Science*, Vol. V: Soft Colloids, Elsevier, Amsterdam, **2005**, Chapter 4, pp. 4.1–4.123.

61 F. A. M. Leermakers, A. L. Rabinovich, N. K. Balabaev, *Phys. Rev. E 67* (**2003**) 011910/1–17.

62 M. M. A. E. Claessens, B. F. van Oort, F. A. M. Leermakers, F. A. Hoekstra, M. A. Cohen Stuart, *Biophys. J. 87* (**2004**) 3882–3893.

63 W. J. Lokar, L. K. Koopal, F. A. M. Leermakers, W. A. Ducker, *J. Phys. Chem. B 108* (**2004**) 15033–15042.

64 J. Klein Wolterink, F. A. M. Leermakers, G. J. Fleer, L. K. Koopal, E. B. Zhulina, O. V. Borisov, *Macromolecules 32* (**1999**) 2365–2377.

65 F. A. M. Leermakers, J. M. P. van den Oever, E. B. Zhulina, *J. Chem. Phys. 118* (**2003**) 969–980.

9
Stabilization of Thin Films, Foams, Emulsions and Bifluid Gels with Surface-active Solid Particles

Krassimir P. Velikov and Orlin D. Velev

Abstract

This chapter summarizes the current advances in stabilization of thin films, foams, emulsions and bifluid gels by surface-active particles. Small particles of different shape and morphology with partial wettability can adsorb strongly at liquid/liquid and gas/liquid interfaces and are able to stabilize foams and emulsions. Factors such as particles size, shape and hydrophobicity and their effect on the stabilization mechanism are discussed in detail. The energy of attachment of a surface-active particle to the interface depends on the contact angle and interfacial tension and is proportional to the particle radius. Partially hydrophobic and shape anisotropic particles (e.g. rods) are more efficient than spheres in covering and protecting bubbles and emulsion droplets against coalescence. The mechanism of stabilization by surface-active particles is based on contributions from several effects. These particles stabilize dispersions of immiscible fluids by providing steric hindrance to their coalescence, by bridging the droplets or bubbles and by increasing the bending energy of the interface. Additional stability can be achieved by interconnecting the particles on the interface by chemical linking or physical entanglement. In addition, stabilization against gravity-induced separation (creaming and sedimentation) can be achieved by the formation of a network from dispersed phase or from excess particles in the continuous phase. Surface-active particles can be used to stabilize more complex structures such as double emulsions, which was proven to be very difficult with molecular surfactants. Finally, it was shown theoretically that surface-active particles with equal affinity for two fluids could create bicontinuous particle-stabilized fluid gels. Stabilization by surface-active particles has already found many industrial applications; however, a new level of stability and (structure) control in various dispersion systems and the possibility of deriving new materials need to be further explored.

Colloids and Interface Science Series, Vol. 3
Colloid Stability and Application in Pharmacy. Edited by Tharwat F. Tadros

ISBN: 978-3-527-31463-8

9.1 Introduction

Dispersions of two liquids (emulsions) or a liquid and a gas (foams, aerosols) are inheritably thermodynamically unstable systems. Given a long enough time, the droplets or bubbles will phase separate in order to minimize the interfacial and capillary energy of the dispersed system. However, they can be kinetically stabilized by various barriers against flocculation, coalescence and Ostwald ripening [1]. It is well established that, similarly to surface-active molecules (surfactants, polymers, proteins), small solid particles can absorb at fluid/fluid interfaces and stabilize emulsions and foams. The stabilization effect of small solid particles was recognized long ago following the work of Pickering [2]. Later, several groups have contributed to understanding the behavior and stability of particle-stabilized emulsions [3–7]. The interest in emulsions stabilized by surface-active solid particles has recently surged both in academia and in industry [4, 6–8]. This is partly a result of the current activity in colloidal and nanoparticle research, interest in novel functional materials synthesized by using particles at interfaces [9–17] and drug delivery [18].

Emulsion and foam systems containing solid particles can be found in many industrial formulations, including food, paint, pharmaceutical, personal care and agricultural products. Solid particles of organic or inorganic nature are often present in the components naturally or are added intentionally to improve or alter dispersion functionality (stability, texture, type, composition, appearance, etc.). Naturally derived particles from micrometer down to nanometer size, including, for example, casein micelles and globular proteins such as β-lactoglobulin, soy glycinin and ovalbumin, are surface active and impart additional stabilization. There is great fundamental interest in understanding the particle stabilization effect, controlling it and using it in the design of novel products and materials. In this chapter, the current understanding of stabilization of thin films, foams, emulsions and bifluid gels is reviewed.

There are several factors governing the stability of dispersions containing surface-active particles. The stabilization effect of particles is largely determined by their absorption behavior, which will be discussed in Section 9.2. The different stabilization mechanisms by spherical and non-spherical particles in films, foams and emulsions and the factors affecting them are discussed in Sections 9.3 and 9.4. Sections 9.5 and 9.6 are devoted to more complex particle-stabilized dispersions such as double emulsions and bifluid gels.

9.2 Absorption Behavior of Particles at Interfaces

Particles with partial wettability in two immiscible fluid phases can adsorb strongly at the interface between them. A transfer energy of $-\Delta E$ should be overcome to remove a spherical particle from the interface. This energy depends on the contact angle θ_{OW}, measured through the water phase, the particle radius r and the interfacial tension γ_{OW} [Eq. (1) and Fig. 9.1]

$$-\Delta E = \pi r^2 \gamma_{OW}(1 \pm \cos\theta_{OW})^2 \tag{1}$$

The sign in the parentheses in Eq. (1) is positive for transfer to the oil phase and negative for transfer to the aqueous phase. Equation (1) is valid for particles much smaller that the curvature of the interface (exact for plat interface) and when the gravity force is negligible. For larger particle sizes or smaller droplets, the effect of the curvature should be taken into account [19, 20]. An expression including the curvature effects, the transfer free energy $-\Delta E$ of a single spherical particle ($\theta_{OW} < 90°$) at a spherical curved drop interface (Fig. 9.1) is given by [21]

$$-\Delta E = 2\pi\gamma_{OW}[r^2(1 \pm \cos a)\cos\theta_{OW} - R_s{}^2(1 - \cos\beta)] \tag{2}$$

where $a = \arcsin(x/r)$, $\beta = \arcsin(x/R_s)$, R_s is the resulting size of the swollen droplet after adsorption of particles and x is the radius of the three-phase contact line around the particle at the interface given by

$$x = \frac{rR_s \sin\theta_{OW}}{\sqrt{r^2 + R_s{}^2 + 2rR_s\cos\theta_{OW}}} \tag{3}$$

Numerical estimates show that for particles as small as a few tens of nanometers, the energy change of particle absorption is higher than the kinetic energy of Brownian motion. Even particles of a few nanometers in size will require a few k_BT of energy to be removed from an interface, which makes the ad-

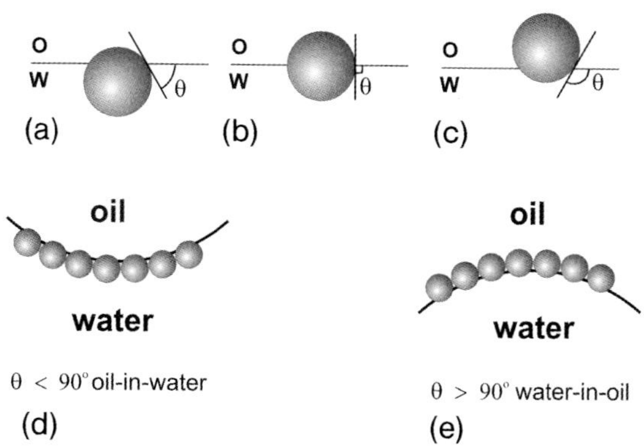

Fig. 9.1 (a–c) Position of a spherical surface-active particle at a planar oil–water interface at a contact angle (measured through the aqueous phase) of less than 90° (a), equal to 90° (b) and greater than 90° (c). (d and e) Corresponding configuration of particles at a curved fluid/fluid interface. For $\theta < 90°$, particle-stabilized o/w emulsions may form (d). For $\theta > 90°$, w/o emulsions are likely to form (e). (Modified after [8]).

sorption effectively irreversible and underlies the stabilization effects discussed here.

The wettability of the particles is the main factor determining their absorption behavior and, in the end, the type and stability of the emulsion formed. The wetting behavior of a particle attached to an interface is described by contact angles, which are related to the interfacial tensions by the Young equation (Fig. 9.1):

$$\gamma_{SO} - \gamma_{SW} = \gamma_{OW} \cos \theta_{OW} \tag{4}$$

where θ_{OW} is the contact angle measured through the water phase and γ_{SO}, γ_{SW} and γ_{OW} are surface tensions at the solid/oil, solid/water and oil/water interfaces, respectively. In order to predict values for the contact angle using Eq. (4), all three surface tensions are needed. The interfacial tension γ_{OW} can be measured accurately, but there are no precise methods for directly measuring γ_{SO} and γ_{SW}. Alternatively, the surface tensions at solid/oil and solid/water interfaces can be estimated by using combining rules and the interfacial tension can be approximated from the surface tension of the pure components. Only recently an advanced replica method was developed for direct measurement of the contact angle of micrometer-sized solid particles trapped at fluid/fluid interface [22].

Particle wettability can be changed by adding surface-active compounds that adsorb on the solid particles and/or water/oil interface. The adsorbed surfactant or protein molecules can modify the three-phase contact angle and hence alter the surface tension, surface charge and surface viscosity. Alternatively, coating with different material of grafting with polymers can also modify the particle surface properties. For example, stearic acid can be used to modify hydrophilic calcite particles [3] and silanol can be used for hydrophobization of silica particles [23]. Finally, particles having inhomogeneous surfaces [two surface regions of differing wettabilities ("Janus")] or surfaces with complex morphology can be more efficient than particles with homogeneous surface properties [24].

The rates of particle adsorption on liquid interfaces have been the subject of several recent studies [7, 25, 26]. Abdel-Fattah and El-Genk [27] studied the adsorption kinetics of hydrophobic, negatively charged, surfactant-free polystyrene latex particles at an air/water interface. They found that both the maximum particle coverage of the interface and the characteristic time of absorption initially increase with the increase in electrolyte concentration; however, beyond a critical concentration the maximum coverage starts to decrease while the characteristic time of absorption still increases. One should note that in they excluded the absorption of clusters of particles from the bulk and restricted their observations to single, non-aggregated particles only [27].

In contrast to these results, Paunov et al. showed theoretically that a higher concentration of background electrolyte should increase the amount of particles adsorbed on the interface [25]. In their analysis, they considered the particles hydrophobicity as determined by the contact angle, the free energy of the double layers around the particles upon adsorption, charge repulsion between the air/water interface and negatively charged particles and the free energy of mixing terms. Their theoretical calculations also indicate that the maximum particle coverage of the interface increases with the size of the particles [25]. The irreversible adsorption of surface-active particles on liquid/liquid interfaces was demonstrated experimentally by Aveyard et al. [26]. They showed that compression of layers of particles in a Langmuir trough apparatus beyond the collapse point caused folding of the interface, but the particles were not expelled from the surface.

The structure and stability of monolayers of surface-active silica particles at both horizontal and vertical air/water and silicone oil/water interfaces were studied in detail by Horozov et al. [28]. They showed that very long-range repulsion between particles absorbed on a liquid/liquid interface keeps them well separated and ordered at large distances. This repulsion can be sufficiently strong to oppose the gravity force acting on the particles at vertical fluid interfaces. In their work, the dipolar repulsion through the air (oil) due to charges at the particle/water interface [29] or dipoles at the particle/air (oil) interface do not play a significant role at interparticle distances studied [30]. Coulomb repulsion through the non-polar phase and the capillary attraction caused by the deformations of the fluid interface around the particles appear to be the major contributors to the pair interaction energy at large distances.

9.3 Mechanism of Stabilization of Foams and Emulsions by Spherical Particles

Solid particles can stabilize dispersions of immiscible fluids by steric hindrance to the contact of interfaces, by bridging the droplets, by creating a particle–particle, droplet–droplet or mixed network and by increasing the bending energy of the interface (Fig. 9.2). In the first mechanism (Fig. 9.2a), the particles adsorbed at the fluid interface form a dense layer around the dispersed droplet, inhibiting coalescence. At higher volume fractions of the dispersed phase and partial attraction between the emulsion droplets or between the particles, the emulsion can partially flocculate into a network with better stability against gravity-induced effects (Fig. 9.2b). At higher particle concentration, the excess of particles in the continuous phase may lead to the formation of a space-filling aggregate network (Fig. 9.2c), if the interactions between the particles are partially attractive. Another mechanism of stabilization is observed when particles at contact bridge the two interfaces and prevent further coalescence (Fig. 9.2d) [31, 32]. This "zipping" mechanism plays an important role and could explain the stabilization effect of particles at partial surface coverage only. In analogy with surfactants, the particle layer increases the viscoelasticity of the interface and the resistance to coalescence. Destabilization of the emulsion, which leads to coalescence, occurs when particles are displaced from the interface into one of the bulk phases or are laterally displaced away from the drop–drop contact region [3, 5].

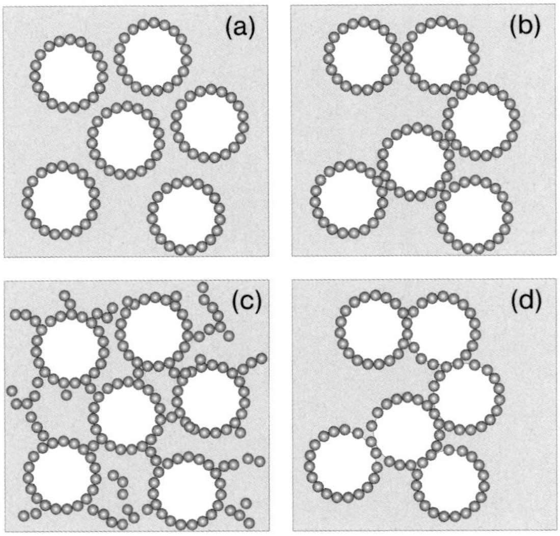

Fig. 9.2 Schematics of different possibilities for droplet stabilization by surface-active particles. (a) Steric hindrance to coalescence of emulsion droplets; (b) creating a droplet-droplet network; (c) creating a particle-particle network that bridge the droplets; (d) bridging of the emulsion droplets due to particle "zipping" of the interfaces.

9.3.1
Particles in Liquid Films

The first direct observation of the dynamics and mechanism of stabilization of surface-active particles was performed in model free liquid films [33, 34]. The microscopic observations were carried out in reflected monochromatic illumination using model free-standing foam films formed by sucking liquid out of a biconcave meniscus held in a capillary. This convenient experimental method has been proposed and developed by Scheludko and Exerowa [35] and has been the basis of the modern studies of thin-film properties and stability [36, 37]. The techniques have been used by others to study the stratification of micelles inside foam films [38–40], which also can be an important factor of stabilization of liquid films in foams and emulsions. Films containing nanometer-sized latex particles and proteins were been studied by a similar method [41–43].

The surface activity of the particles and their stabilization effect on the film could be controlled by the type of surface-active film stabilizer used. In films from solutions containing sodium dodecyl sulfate, a strong anionic surfactant, the latex particles have a similar charge to the film interface and were immediately expelled from the film area during the first few seconds of film thinning without any contribution to the film stabilization.

In experiments with films stabilized by a protein, bovine serum albumin (BSA), entrapment of a few particles inside the film area for short periods was possible due to the high interfacial viscoelasticity of the protein monolayers and

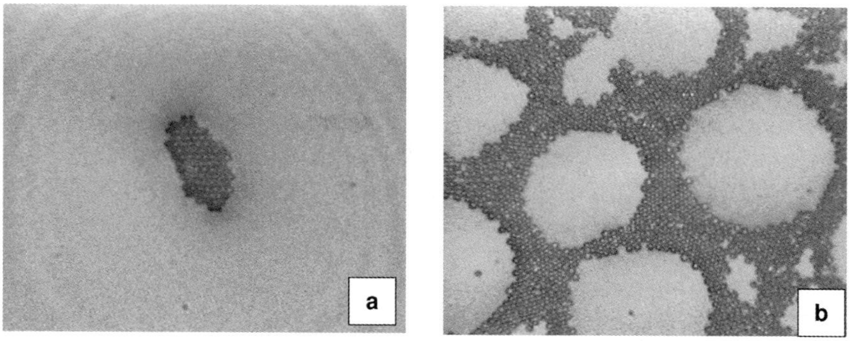

Fig. 9.3 Metastable 2D particle structures formed inside a protein (BSA)-stabilized foam film. (a) Formation of a 2D cluster due to the action of the long-range capillary forces between particles entrapped inside the film; (b) 2D foam-like particle structures formed at higher particle concentration. (Reproduced with permission from [33]).

maybe due to partial hydrophobization by absorption of BSA on the particle surface. The entrapped particles were typically collected inside the thicker "dimple" in the film center and formed small clusters of two-dimensional (2D) arrays (Fig. 9.3 a), attracting each other by the lateral capillary forces at distances as large as tens of micrometers. At higher latex coverage of the surfaces ($\sim$50% by area), the density of the particle clusters changed from single clusters to percolated 2D foam-like metastable clusters similar to those observed earlier on the surface of fluorinated oil [44] (Fig. 9.3 b). The entrapment of particle networks inside the film extends its lifetime. However, all of these metastable structures were eventually expelled from the film, which then thinned and eventually ruptured.

The particle dynamics changed dramatically in foam films stabilized by cationic surfactants [dodecyltrimethylammonium bromide (DTAB) or hexadecyltrimethylammonium bromide (HTAB)]. Cationic surfactants are charged oppositely to the polystyrene microspheres and modify their surfaces by coupling to the grafted negative charges. This resulted in partial hydrophobization of the spheres. The partially hydrophobic particles tend to adsorb strongly at the water/air interface because of the increased hydrophobic attraction. The basic stages of film evolution in the presence of these hydrophobized particles are shown in Fig. 9.4. During the first formation of the film, the particles were pushed out into the surrounding meniscus and the film area was devoid of particles. During the expansion of the first film, however, the particles in the ring closest to the periphery came into contact with both surfaces and stuck to them, linking together the lower and the upper surfaces (Fig. 9.4 a) [33, 34]. In this configuration, even at very low surface coverage, the particle could play a role in long-term film stabilization. After pulling away the surfaces by closing the film, these particles were sucked into the middle (Fig. 9.4 b and c). An ordered cluster of particles formed, binding the two film surfaces together. Each subsequent opening and closing of the film added new particles to those already stuck be-

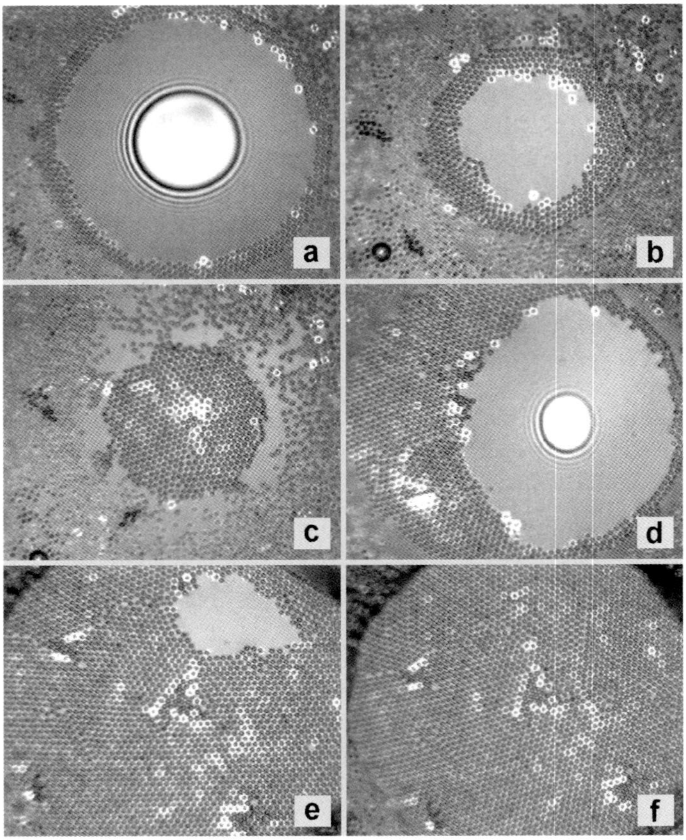

Fig. 9.4 Micrographs of the formation of 2D arrays of surface-active particles in a free-standing foam film in a Scheludko–Exerowa cell. Successive opening (a and d) and closing (c and f) of the film leads to particle structuring via the "zipping" mechanism – particles bridging the two liquid interfaces [33, 34]. (b, e) are stages observed before complete closing of the film. The microspheres are 7 μm in diameter. (Reprinted with permission from [33]).

tween the interfaces, leading to the formation of larger close-packed 2D arrays (Fig. 9.4 d–f) until the film area was completely filled by the crystal. This led to a progressive increase in the pressure needed to open a new film, as the stable dense particle layers were difficult to displace or break.

At high concentration of particles and low and intermediate DTAB concentration, adsorption of particles on both meniscus interfaces took place before the film was formed. Bringing the two surfaces together resulted in a bilayer particle structure, followed by transition to an ordered monolayer of microspheres. The film was extremely well stabilized. We were also able to observe momentarily the formation of a bilayer structure, followed by a transition to a monolayer in the case of films with high particle concentrations and stabilized by a mixture of protein

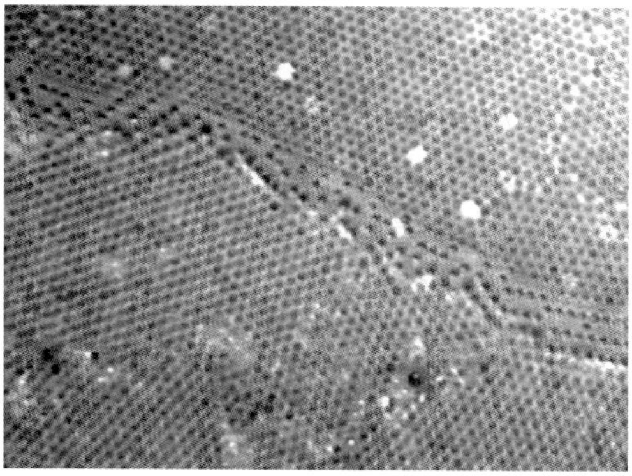

Fig. 9.5 Momentary coexistence of a metastable particle bilayer and a monolayer during the thinning of a film with a high concentration of particles. The transition region is seen between the top left side and bottom right side of the frame. The upper region is the monolayer. (Reproduced with permission from [33]).

(BSA) and cationic surfactant (HTAB). The bilayer was only a metastable phase and the transition to a monolayer was very rapid and uncontrollable. Both the transient bilayer and the monolayer were ordered in a hexagonal lattice and there was a sharp phase transition boundary between them (Fig. 9.5).

These observations revealed three major parameters controlling the particle stabilization: the particle–surface interactions, the capillary forces and the interfacial rheology. Both the capillary interactions between the particles and the particles and the meniscus play a major role in the formation of a 2D array. The capillary interaction between the particles is always attractive and could occur both above and below a film thickness equal to the particle diameter, although the menisci around the particles will have different curvatures. This attractive interaction energy can exceed the thermal energy k_BT by many orders of magnitude for distances as large as 100 times the particle diameters [33]. The second important component of the capillary interactions is the interaction between the particles and the curved menisci surrounding the film (e.g. formed due to deformation of the droplet/bubble, Fig. 9.6). The particle–meniscus interaction can be attractive or repulsive depending on the sign of the parameter $\sin\theta_1 \sin\theta_2$ (where θ_1 and θ_2 denote the slope angles at the cell (or deformed droplet) and particle menisci, respectively) [45–47]. The meniscus around the particles is convex ($\sin\theta_2<0$) for hydrophilic particles and concave ($\sin\theta_2>0$) for hydrophobic particles attached to the surfaces of a thicker film, whereas the film meniscus in a hydrophilic glass capillary or in a deformed droplet/bubble is always convex ($\sin\theta_1<0$). The 2D particle assemblies are expelled in thinning SDS or protein-stabilized films because $\sin\theta_1\sin\theta_2>0$, i.e. these particles will be attracted to the film periphery (Fig. 9.6a). Adversely, the interaction between the meniscus and hydrophobic particles will be repulsive ($\sin\theta_1\sin\theta_2<0$) and these particles will be pushed towards the film center (Fig. 9.6b) or equilibrate near the periphery (Fig. 9.6a). The particles bridging the surfaces are repelled from the film periphery and compressed in the center. In summary, intrinsically stable particle structures are obtained only in films with repulsive capillary forces between the

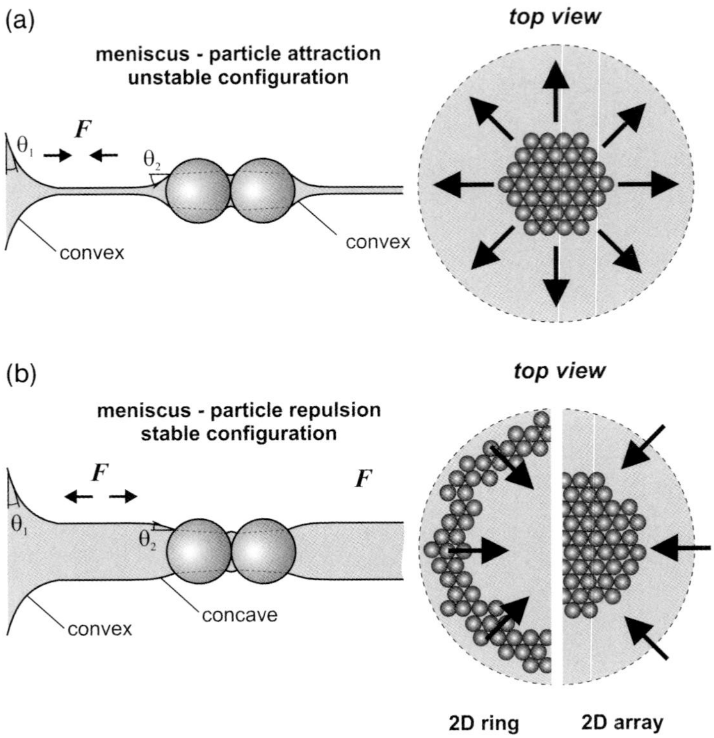

Fig. 9.6 Schematic representation of the capillary interactions acting between 2D particle arrays formed in the thin liquid film and the surrounding meniscus. The 2D structure within the film is compressed by the lateral capillary forces and assumes a stable 2D ring-like or close-packed array configuration due to the repulsion between the convex and concave menisci. The stable configuration forms only when the particles are partially hydrophobized.

particles and the meniscus. Organized particle layers play a major role in the rheological behavior and stability of thin films.

A similar zipping mechanism was later observed by Dimitrova et al. [48] in thinning films where protein aggregation led to cross-linking of the interfaces. The films cease to thin and become indefinitely stable. Stancik et al. observed that hydrophilic silica particles can bridge the interfaces of an oil–water–oil film, such that the oil droplets appear to become irreversibly stuck together [49]. A recent publication concerning the stabilization of octanol droplets by silica particles suggests that bridging by particles is common in many other systems [50]. The particle zipping mechanism in vertical emulsion films with particle monolayers at their surfaces was studied in detail by Horozov et al. [28]. They observed stable films only when the particle wettability fulfils the condition for stable particle bridges in agreement with the theory that hydrophilic particles can produce stable oil-in-water (o/w) emulsions, whereas hydrophobic ones give water-in-oil (w/o) emulsions. Hydrophilic particles in dilute disordered monolayers at the surfaces of water films are expelled from the film center toward its periphery, producing a dimple surrounded by a ring of particles bridging the film surfaces, as previously observed in foam thin films [33, 34]. In contrast, the thinning of octane films with dilute ordered monolayers at their surfaces ultimately leads to the spontaneous formation of close-packed crystalline arrays of hydrophobic particles bridging both surfaces at the center of the film. The dynamics of water and octane films with dense close-packed particle monolayers at their surfaces are very similar [28].

These studies clearly indicate that the formation of a bridging particle monolayer between the film surfaces is a common feature in particle-stabilized fluid–fluid dispersions. It occurs spontaneously both in foam and in emulsion films with particles absorbed at their surfaces. At higher particle concentrations, when the film surfaces are covered with close-packed particles, a bridging monolayer is formed as a result of bilayer to monolayer transition [28, 33, 34]. The rapid transition occurs at capillary pressure lower than the critical pressure for rupturing the film stabilized by a particle bilayer due to the limited ability of the monolayers in the meniscus to sustain the additional lateral stress generated in the bilayer at an elevated capillary pressure.

9.3.2 Emulsions

Beginning with the pioneering work of Pickering [2], shells of surface-active particles adsorbed around droplets have been used to as an alternative to surfactants to stabilize emulsions. The effects playing a role in these emulsion systems can be grouped in the following categories.

9.3.2.1 Effect of Particle Hydrophobicity

It is now well established that particles with contact angle $\theta < 90^\circ$ tend to stabilize o/w emulsions, whereas those with $\theta > 90^\circ$ stabilize w/o emulsions [3]. Particles that are completely wetted by the oil or water phase do not stabilize any type of emulsion. For particles of intermediate hydrophobicity ($\theta \approx 90^\circ$), the continuous phase of the preferred emulsion becomes that in which the particles are initially dispersed [6, 51]. In some systems a surfactant is needed to modify the particle contact angle to adsorb the particles adequately, analogously to the film studies described in the previous section. For example, several surfactants have been used to modify the wettability of hydrophobic fat crystals [4, 52–55].

Upon increasing the particle hydrophobicity, the average drop size passes through a minimum [23]. Accordingly, the stability to gravity-induced emulsion separation (creaming and sedimentation) passes through a maximum [23]. The effective average wettability of the particles at the drop interfaces can be gradually changed by using a mixture of hydrophobic and hydrophilic particles [56]. This can be achieved by addition of hydrophilic particles to w/o emulsions stabilized by hydrophobic particles (or by addition of hydrophobic particles to o/w emulsions stabilized by hydrophilic particles) or by varying the fraction of one of the particle types at constant particle concentration [56, 57]. However, a "symmetric" mixture of two types of particles, one that stabilize o/w and other that stabilize w/o emulsions, has a very weak or no stabilization effect [3].

The mechanisms of phase inversion in particle-stabilized emulsions are still not well understood [58]. Usually phase inversion occurs under conditions where the average contact angle at the interface passes through 90°. In the case of emulsions stabilized with a mixture of two types of particles, both types of emulsions containing particle mixtures are very stable to coalescence; however, the stability to creaming of o/w and sedimentation of w/o emulsions decreases when approaching the inversion point [56]. This coincides with a maximum in the median drop sizes [56]. This behavior is opposite to that in systems of one particle type (catastrophic inversion) by varying the hydrophobicity of the particles [23]. This is probably due to an ineffectiveness of the particle mixture to create a particle–particle, a droplet–droplet or mixed network in the continuous phase. Changing the volume fraction of one phase can also lead to phase inversion. For example, for $\theta < 90°$, w/o emulsions can be inverted to o/w if the fraction of oil is increased [57].

Partial attraction between the solid particles also seems to be a factor contributing to effective stabilization of emulsions [3]. The mechanical stability of the interfacial layer of particles and its rheological properties are strongly affected by the degree of interparticle interactions. Several studies have considered the effect of different colloidal forces on emulsion stability [20, 21, 59–61]. The attractive van der Waals forces are relatively short range, yet the displacement of a particle from the interface is more difficult when they are in contact and attract each other. Previous work has shown that good emulsion stabilization was achieved when the particles were weakly flocculated [62]. Coalescence can be enhanced by either deflocculation or complete flocculation of the particles. Complete deflocculation will destroy the network of particles and will increase their lateral mobility. On the other hand, complete flocculation will result in the formation of large aggregates, thereby reducing their ability to stabilize emulsions. Interparticle interactions are also responsible for non-Newtonian emulsion rheology due to gelation or glass transition of the continuous phase.

The particle roughness or, in general, complex surface morphology also plays a role in establishing the wetting contact angle. The effect of particle surface roughness has been elucidated by Vignati et al. [50]. They compared monodisperse, smooth-surfaced spherical colloids with particles of noticeable surface "roughness". Their results show that surface roughness appreciably lowers particle ability to stabilize emulsions. Specifically, surface roughness is responsible for time-dependent wetting and contact angle hysteresis (generally surface roughness is associated with large contact angle hysteresis). This is particularly important for contact angles close to 90°, where different type of emulsions can be formed depending on the phase in which the particles were initially dispersed. The continuous phase of the preferred emulsion becomes that in which the particles are first dispersed. As a result, the catastrophic phase inversion is dependent on the initial state of the system [63]. The hysteresis caused by particle surface roughness enhances this effect. Particles dispersed initially in the water have a higher receding angle than the advancing angle when they are initially dispersed in the oil. Maximum stability can be obtained for either emulsion type simply by changing the initial particle location. In contrast, emulsion type is independent of the initial location of surfactant due to the fast distribution of surfactant molecules between phases during emulsification.

Electrolyte concentration and pH of the aqueous phase are important factors governing the emulsion stability by changing the properties of the water/oil interface (due to ionic surfactant) or the particle/water interfaces. Recently, controlled coalescence was demonstrated using pH-sensitive particles [64]. Amalvy and coworkers employed pH-responsive, sterically protected polystyrene latex to stabilize o/w emulsions [64, 65]. The neutral weakly charged particles adsorb strongly at the oil/water interface, leading to the formation of stable emulsions. However, on lowering the pH of the aqueous phase from 5.6 to 2.2, the particle surface becomes fully protonated, increasingly hydrophilic and hence insoluble in the oil phase. This leads to progressive desorption of particles from the oil/water interface, which in turn destabilizes the emulsion. Coalescence of the oil droplets, followed by macroscopic phase separation, occurs. This effect is reversible with respect to pH. Weak flocculation can be achieved at pH near the isoelectric point of the particle. Completely flocculated particles form less stable emulsions.

The degree of flocculation can also be controlled by adding salt in the case of o/w emulsion or by surfactant in the case of w/o emulsion. For o/w emulsions, high ionic strength suppresses the electric double layer around the emulsion droplets and allows particle flocculation. The flocculation initially enhances the emulsion stability. However, further an increase in salt concentration decreases the emulsion stability due to the formation of large particle flocks. The particles aggregate at high electrolyte concentrations and these aggregates can no longer stabilize the emulsion and tend to sediment. Poorly stabilizing system (e.g. hydrophilic particles in pure water) can be made optimally stable with low levels of salt. A recent theoretical description of the adsorption of charged particles at charged liquid interfaces revealed that adsorption is favored at high ionic strength as particles become more hydrophobic [25]. This effect was observed experimentally by studying the dynamics of colloidal particles in thinning liquid films [31].

9.3.2.2 **Effect of Particle Size**

The size of the surface-active particles also plays an important role in defining the stabilization effectiveness because it controls the ability of the particles to be held at the curved and spherical interfaces. The basic rule is that the particles should be significantly smaller (by at least an order of magnitude) than the resulting emulsion droplets to provide effective stabilization. The effects of particle size on emulsion stability have been widely studied [3, 66], but a quantitative relation between particle size and equilibrium emulsion droplet diameter has not yet been established. Emulsion stability increases with decrease in particle size until a critical particle size is reached; however, this limit is not understood in detail [66]. The emulsion droplet size increases with increase in particle size, which leads to enhanced gravity-induced instabilities (sedimentation or creaming) [66]. During the emulsification process, large particles cannot stabilize the emulsion well because they cannot adsorb at the interface and remain in the continuous phase. Furthermore, large particles in contact with a droplet will detach at high shear [67].

9.3.2.3 **Effect of Particle Concentration**

The effectiveness of particle stabilization depends in part on the presence of a sufficiently dense layer of particles around the emulsion droplet. However, there is evidence that emulsion stabilization can also be achieved at partial coverage [50]. The particle concentration at the interface depends on the total particle concentration, volume fraction of the dispersed phase (number of droplets and their size distribution) and the adsorption–desorption equilibria. Accordingly, the total particle concentration will influence the number of droplets and their size distribution. Initially, all particles will be adsorbed on the interface, leading to incomplete coverage of the droplets. With an increase in the total particle concentration, the fraction of adsorbed particles decreases and reaches an asymptotic value [6]. For surfactant-stabilized emulsions, the droplet size decreases with concentration until the critical micelle concentration is reached, after which it remains constant. In the case of emulsions stabilized by surface-active particles, an increase in the particle volume fraction leads to a decrease in the size of the emulsion droplets and increased emulsion stability [60, 68, 69]. Tambe and Sharma demonstrated theoretically the effect of particle surface concentration on the drainage of thin liquid films [69]. An increase in the particle surface concentration leads to retardation of the film drainage and stabilizes the emulsion.

The change in particle concentration also changes the structures formed by the particles. Tarimala and Dai, using confocal microscopy, studied the 2D assembly of colloidal-sized polystyrene particles in polydimethylsiloxane-in-water solid-stabilized emulsions [70]. At low concentrations the monodisperse particles in the emulsions formed small patches with local "hexagonal" order, separated by other particle-free domains (Fig. 9.7). Simultaneous segregation of particles with different sizes and different wettability, including mixtures of hydrophobic and hydrophilic particles, was observed at the emulsion interface.

In high-resolution images from confocal microscopy, Vignati and Piazza observed that particles within the contact region can sometimes form a monolayer, suggesting simultaneous adhesion of a given particle to two droplets [50]. This "bridging" mechanism, as observed in both thin foam and emulsion films, may help to keep the droplets at a finite distance, hindering interdroplet film drainage [28, 33, 34, 49]. The particle-covered area undergoes noticeable morphology changes as the droplets drift apart: colloids on the small droplet are mostly lo-

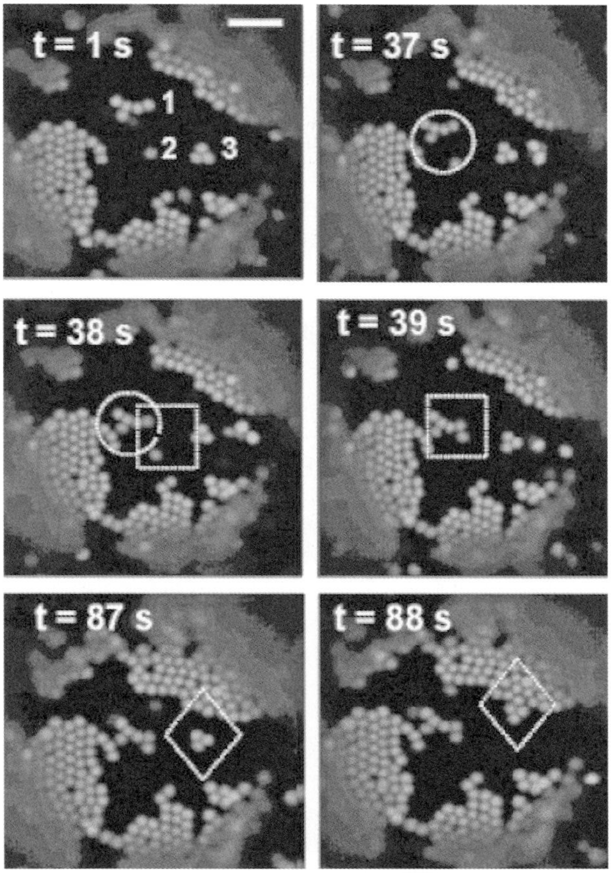

Fig. 9.7 Confocal fluorescence microscopy images showing the aggregation of particles at the interface of droplets in a particle-stabilized emulsion. The circles, squares and diamonds highlight the positions before and after particle attachment. The scale bar is 5 μm. (Reproduced with permission from [70]).

calized around the contact region when drops are touching but progressively spread out over the entire surface of the small droplet as the droplets separate. Their conclusion is that particle redistribution of the droplet surface may play a role in stabilizing against coalescence.

For high enough particle concentrations, not all particles could adsorb on the oil/water interface. The excess of particles in the continuous phase can lead to gelation (network formation) or an increase in the viscosity – retarding or preventing complete sedimentation of water droplets or creaming of oil droplets [57, 71]. In the case of strongly interacting or shape-anisotropic particles (see below), gelation can occur at much lower particle concentrations. Partial flocculation that leads to the formation of a network between the emulsion droplets also increases the emulsion stability [72].

9.3.2.4 **Interfacial and Bulk Rheology**

The stability of emulsions depends on the interfacial and bulk rheology of the continuous phase. The rheology of emulsions stabilized by solid particles is enhanced both by the layer of particles around the droplets and by aggregation of excess particles in the bulk. The viscosity of the continuous phase depends on the particle volume fraction and shape. The bulk liquid viscosity can easily be increased by increasing the concentration of particles or using thickening agents. Alternatively, stabilization can be achieved by creating an apparent yield stress in the continuous phase at low shear. An excess of solid particles leads to network formation or glass transition in the continuous phase. At high shear the system will display shear thinning and the system would be easily processable. The apparent yield stress can also be created by partial flocculation of the emulsion droplets (Fig. 9.2 b). In the latter case, the emulsion droplets assemble into a weak gel and in this way decrease or completely prevent the flocculation. Sintering between the particles can also affect the bulk rheology. Sintering has been observed in many w/o emulsions stabilized with solid fat particles [52, 58, 73, 74].

In analogy with bulk rheology, the droplet interfaces loaded with adsorbed particles can exhibit viscous, elastic and viscoelastic properties. The ability of emulsions to resist flocculation depends to a large extent on the properties of the interface. High surface viscosity or viscoelasticity will retard the thinning of the films between the droplets. Interactions between the particles and between particles and emulsifier are also important as they modify the interface and to a large extent dictate the emulsion stability. In the absence of any attraction between the particles, lateral particle displacement from the coalescence region will be easy and the stability will be significantly decreased. At high particle concentrations, monodisperse particles at contact form close-packed hexagonal arrays. Very high pressure or shear stress is required to break these structures apart. Tambe and Sharma modeled the rheological properties of particle layers on fluid/fluid interfaces [3]. Their findings indicate that viscosity effects are dominant at low particle concentrations, whereas viscoelastic effects are dominant at high particle concentrations. The viscoelastic properties of the interface can be modified by additional adsorption of proteins and/or surfactants [71, 75].

The shells of particles adsorbed around droplet "templates" can be bound together to make new types of materials and capsules. The first demonstration of the use of particle-stabilized emulsion for the fabrication of hollow superstructures of latex particles was reported by Velev and coworkers [9, 10, 12]. Latex microspheres were collected, ordered and fixed together on the surfaces of micrometer-sized octanol droplets (Fig. 9.8). Subsequently, the droplets were dissolved in the surrounding environment to obtain a suspension of microspheres in the form of ordered hollow "supraparticles" – capsules with surfaces that are composed of a close-packed layer of colloidal particles linked together to form a solid shell. A technique for making hollow, elastic capsules with sizes ranging from micrometers to millimeters by assembling particles around w/o emulsion droplets (Fig. 9.8) has been developed later by Weitz and coworkers [76]. The same group demonstrated the fabrication of capsules comprising a single-layer network of polymer adsorbed

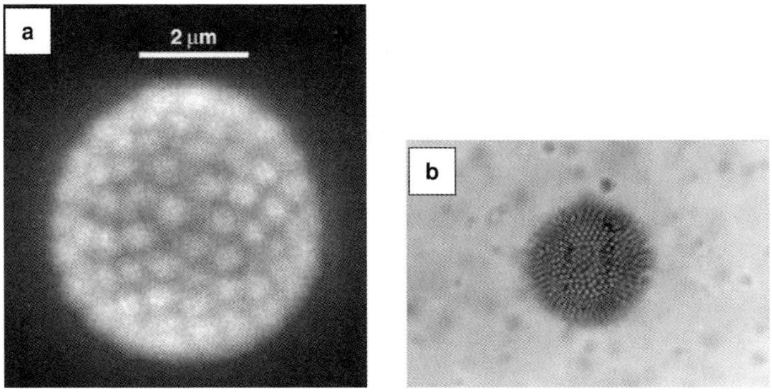

Fig. 9.8 Microscopy images of emulsion droplets covered by surface-active particles. (a) Projection of a three-dimensional confocal fluorescence microscope image of 0.7-µm diameter PMMA colloidal particles adsorbed on the surface of a water droplet in decalin. Reprinted with permission from [76]. (b) Optical microscopy image of an octanol droplet covered with 1-µm latex microspheres. (Reprinted with permission from [9]).

on colloidal particles [77]. These capsules are made using directed self-assembly of the polymer and particles at the interfaces of emulsion droplets.

Particle-stabilized emulsions formed the basis of many new materials and complex liquids. Recently, Croll and Stöver demonstrated that emulsions can be stabilized by soft, swellable microgel particles covalently linked at the oil-water interface to form "tectocapsules" [78, 79]. The particles are first allowed to self-assemble at the oil/water interfaces of a propyl acetate–water suspension and then are covalently fixed in place to form gel capsules. Emulsion stabilization by nanoparticle assembly and subsequent cross-linking on droplet interfaces was also demonstrated by Russell and coworkers [80, 81]. The ligand-stabilized nanoparticles are weakly attached to the fluid interface, which can be used to induce size-selective particle assembly. They showed that size-dependent adsorption and desorption from an interface gave rise to a 2D phase separation at the fluid/fluid interface. Later, the same group also demonstrated chemical cross-linking of the ligands attached to the nanoparticles. The composite organic–inorganic, nanometer thick membranes prevent convection but allow diffusion of small molecules across the interface and provide excellent stabilization. Russell et al. also used cowpea mosaic virus nanoparticles that self-assemble as a monolayer at perfluorodecalin/water interfaces. These close-packed shells were covalently cross-linked to form a robust membrane with interparticle connectivity [82]. Magnetic surface active particles have been used in a few other studies. For example, Melle et al. studied solid-stabilized emulsions containing magnetic particles [83]. Such emulsions show reversible macroscopic phase separation on application of external magnetic fields. Duan et al. used interfacial self-assembly of magnetite nanoparticles to demonstrate the formation of colloidosomes with shells predominantly composed of monolayers of liquid-like, close-packed nanoparticles [84]. Binks et al. demonstrated the formation of stable emulsions, of both single and multiple types, containing ionic liquids and stabilized solely by silica nanoparticles [85].

9.3.3 Foams

Spherical surface-active particles can promote or inhibit foam stabilization in ways very similar to the case of emulsions. The stability of foams has been found to depend on the particle size, shape, concentration and hydrophobicity, and also on the type of surfactant used [8, 86, 87]. Data and theoretical conclusions in the literature suggest that hydrophobic particles may initiate foam destruction by rupturing the foam films via a bridging–dewetting mechanism. In contrast, if the particles are hydrophilic, they could collect in the plateau borders, slowing the liquid drainage and kinetically increasing the foam stability. Very few studies of foams stabilized solely by solid particles have been reported [8], which include foams formed by a latex suspension close to its coagulation point or short-lived foams stabilized by sludge particles [88, 89].

The effect of surface-active particles on foam stability was recently reviewed by Murray and Ettelaie [7]. Similarly to emulsions, it has also been recognized that globular proteins or protein aggregates behave as surface-active nanoparticles. Some of the stronger protein films, such as those formed by β-lactoglobulin, soy glycinin and ovalbumin, collapse and form protein particles as the bubbles which they stabilize shrink in size [7, 90].

Dickinson et al. demonstrated that partially hydrophobic silica nanoparticles can form stable air bubbles in an aqueous phase [91]. As expected from thin liquid film experiments, no stable bubbles could be formed if the particles were too hydrophobic. They suggested that the bubbles in their system may be stable only when the silica particles in the bulk aqueous phase form a weak gel and/or that this bulk particle gel is continuous with aggregated particles adsorbed on the surfaces of the bubbles (see Fig. 9.2) [91]. They analyzed the shrinkage kinetics of a single bubble and showed that below a critical bubble size, the bubbles shrank too fast to be stabilized by the particles. The critical bubble size depends on the size of the particles or aggregates of particles in the dispersion, among other factors.

9.4 Stabilization by Non-spherical Particles

Shape-anisotropic particles such as plates or rods can also serve as effective foam and emulsion stabilizers. Non-spherical particles can orient in different ways at the interface and form jammed or intertwined networks, enhancing foam or emulsion stability. The orientation at the interface provides insight into possible mechanisms of emulsion (de)stabilization (e.g. partial flocculation; see Fig. 9.2).

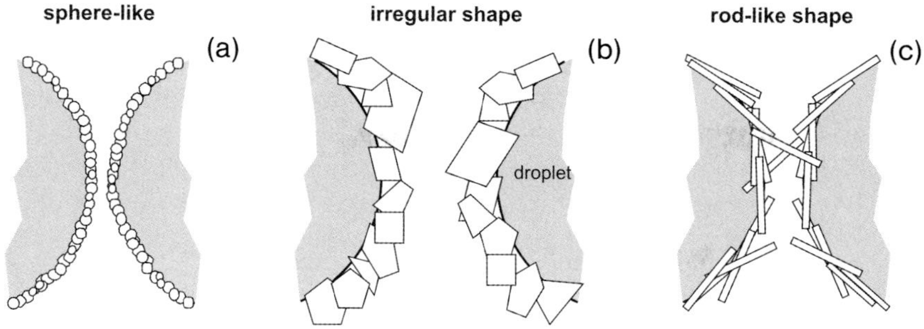

Fig. 9.9 Schematic illustration of the effect of particle morphology on droplet coverage and effectiveness of stabilization. (a) Small sphere-like particles that cover the droplet surface very uniformly; (b) large irregular particles that are not able to cover the droplet surface effectively; (c) anisotropic particles – coverage by rod-like particles that allows both physical entanglement and bridging (partial coalescence) between droplets due to piercing.

For example, platelets are likely to be much more efficient than spheres in covering and protecting emulsion droplets against coalescence (Fig. 9.9).

The viscosity of the continuous phase depends on the particle concentration and shape. Anisotropic particles such as rods and platelets can increase the viscosity more efficiently than spherical particles. An excess of solid particles leads to network formation or glass transition in the continuous phase. Both platelets and rods can create (weak) gels at low volume fractions in the continuous phase, stabilizing the emulsion against gravity-induced separation (creaming for o/w and sedimentation for w/o emulsions). The application of non-spherical particles in specific foam and emulsion stabilization will be discussed in detail below.

9.4.1 Foams

A large number of shape-anisotropic nano- and microparticles of various sizes have been synthesized recently following the thrust in nanoscience research, and possibly can be used as effective stabilizers with well-defined shape and size. Recently, we demonstrated the formation of extremely stable aqueous foams using polymer rod-like particles (Fig. 9.10) [92]. The process for synthesis of these new "micro-rods" by shearing of a polymer solution was developed in our group [93]. We found that the foams and single-foam films stabilized by these rods possess extraordinary long-term stability (Fig. 9.11).

Examination of the micro-rod foam under a microscope showed that it was made of small spherical bubbles (Fig. 9.10 b), each of which was covered by a dense shell of adsorbed entangled polymer rods extending into the water phase (Fig. 9.10 c). Observations of single-foam films formed between menisci in a capillary (similarly to a Scheludko–Exerowa cell) showed that the whole surface of films was densely covered with intertwined rods (Fig. 9.10 d). The rods entangle,

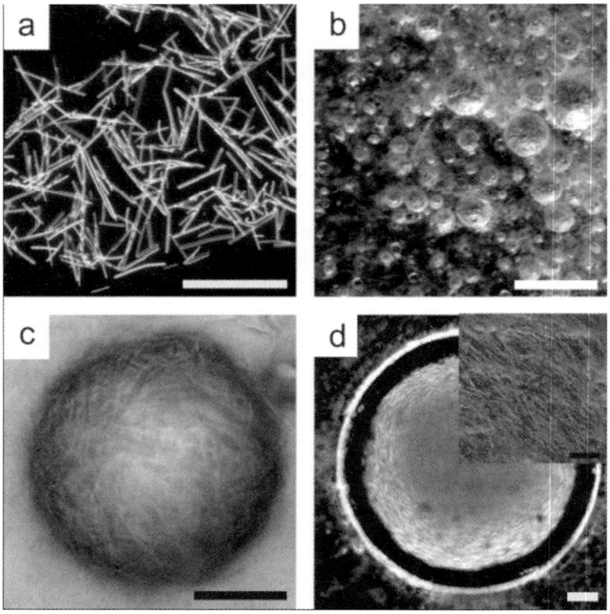

Fig. 9.10 (a) Optical micrograph of the SU-8 micro-rods used as foam superstabilizers; (b) foam bubbles stabilized by rods; (c) micrograph of a single "hairy" bubble covered by a layer of adsorbed rods; (d) single thin aqueous film formed from a suspension of SU-8 rods. The inset in (d) is a highly magnified area near the film center. The scale bar is 50 μm for (a) and (c) and the inset in (d) and 200 μm for (b) and (d). (Reproduced with permission from [117]).

overlap and sometimes form small liquid-crystal like domains. The film is thicker than at least two opposing layers of rods (ca. 1–2 μm). This is an overwhelmingly thick film compared with the equilibrium thickness of surfactant foam films stabilized by electrostatic repulsion [94, 95] or the thickness of common "black" films with steric repulsion between surfactant monolayers adsorbed on the opposing surfaces ($\sim$12 nm) [94, 95]. Hence the first factor in stability of the foam with polymer rods is the steric repulsion between the adsorbed particle layers, which keeps the films very thick, preventing breakage and suppressing gas diffusion.

The second unique factor of superstabilization is the mechanical rigidity of the continuous net of overlapping and entangled micro-rods at the film surface (cf. inset in Fig. 9.10 d). In contrast to spherical particles entrapped in thin foam films, the physically entangled micro-rods cannot rearrange and be expelled from the film [33, 34]. Similarly to the foam systems, single films with concentration above $\sim$1 wt% were infinitely stable and were not destabilized by drying, which leads to rapid breakdown of films containing molecular surfactants only. The films and membranes from intertwined rods could be dried and kept intact for more than 1 week.

One of the major reasons for the formation of thick and rigid films between the bubbles is the strong adsorption of the polymer rods at the air/water inter-

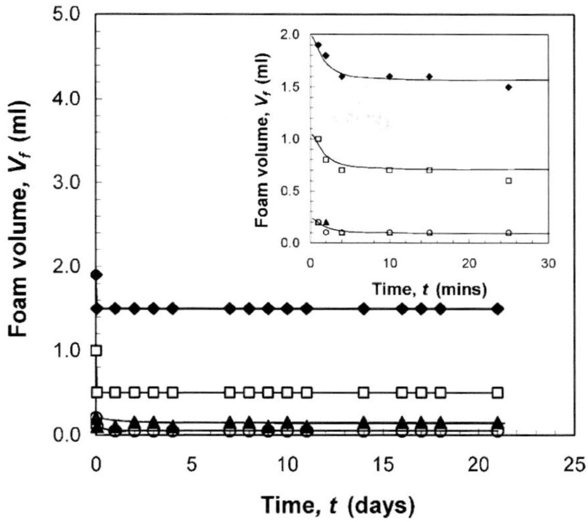

Fig. 9.11 Time dependence of the volume of bulk foams stabilized by micro-rods. The lines of experimental points (lack of any foam breakdown) correspond to different solid concentrations: diamonds, 2.18 wt%; squares, 1.09 wt%; triangles, 0.44 wt%; circles, 0.22 wt%. The inset shows the volume changes for the first 30 min. (Reproduced with permission from [117]).

face. The contact angle between rods and water was estimated to be between 80° and 90°, as measured by the gel trapping technique [96]. Thus, the adsorbed micro-rods were partially immersed in the water phase but did not aggregate spontaneously when dispersed in the bulk. As already mentioned above, partially hydrophobic particles are optimal stabilizers of various foam and emulsion films [33, 34, 58, 86, 87, 97]. The strong adhesion between the adjacent and overlapping rods in the films (responsible for the rigidity of the adsorbed layer) is probably augmented by their hydrophobicity and high friction. The rod entanglement appears to be the major difference between foams stabilized with spherical and rod-like particles [33, 34, 98]. The stabilization effect is lost when the rod surfaces are hydrophilized by addition of surfactant, causing their desorption from the interface (Fig. 9.12).

The bulk structure of foams stabilized with rod-like particles was also found to be very different from that of common surfactant-stabilized foams. The micro-rod foams consisted of approximately spherical air bubbles (Fig. 9.10b), not only for the very first stage of the foam formation but for the whole period of observation. Such a structure is typical of unstable wet transient foams, which contain large amounts of water and live for only a few seconds [99]. These bubbles did not deform to allow for the liquid from the films between them to drain and to form a "dry" foam. The reason for the lack of deformation is probably the rigidity of the dense "hairy" shells around the bubbles. The formation of such rigid shells from rods around emulsion droplets was recently observed by the authors too [100].

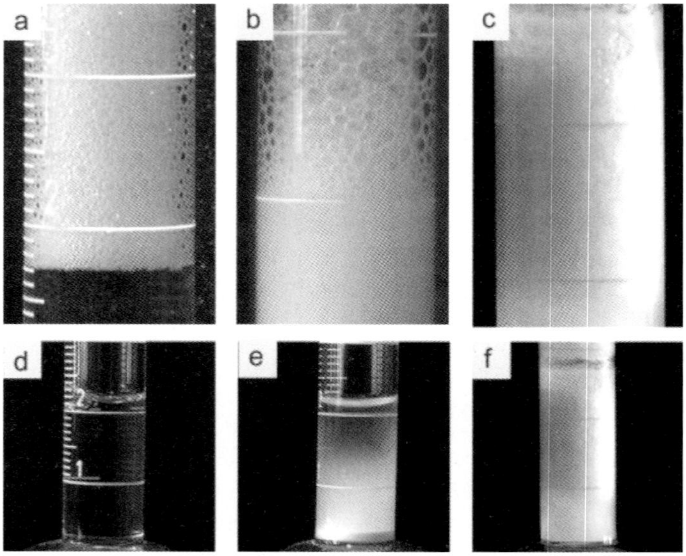

Figure 9.12 Comparison of the appearance and structure of foams stabilized by (a and d) 8.7 mM SDS, (b and e) a mixture of 8.7 mM SDS and 2.18 wt% SU-8 rods and (c and f) 2.18 wt% SU-8 rods. The top images (a–c) were obtained 10–15 min after the foam was formed. The bottom images (d and e) are of the top system after 2 days of storage. Note that SDS not only makes much less stable foams than the micro-rods but also effectively acts as a defoaming agent for the micro-rod foams. (Reproduced with permission from [117]).

The effects presented here differ significantly from literature data for foams stabilized with spherical particles. It has been reported that strongly hydrophobic particles decrease foam stability by breaking the lamellae by a "bridging dewetting" mechanism [8, 86]. Foam stabilization with slightly hydrophobic and with hydrophilic particles is usually a result of collecting the particles in the plateau borders, which slows the liquid drainage and therefore the thinning and rupture of the lamellae [86, 87]. In the case of polymer micro-rod foams, the stabilization mechanism includes the formation of a thick rigid net of entangled rods around the bubbles. The bubbles do not coalesce because of the strong steric hindrance between the shells. Diffusion of air between the bubbles is not likely to occur because the liquid films between them remain very thick. Another reason for the apparent lack of disproportionation is that the bubbles covered by a rigid shell cannot readily shrink or expand. The strong rod adsorption and entanglement, formation of rigid hairy shells and sustaining of thick films make possible the superstabilization effect. Due to their rigid structure and resistance to mechanical perturbations, the polymer rod-like foams can be used in applications where the common foam stabilizers are not effective (Fig. 9.12).

9.4.2 Emulsions

Rod- and plate-like particles have also been found effective in emulsion stabilization. Van Boekel and Walstra observed that globule- or needle-type crystals could stabilize o/w emulsions, whereas a mixture of the globule- and needle-type crystals could decrease the stability of the emulsions or even induce droplet coalescence [101]. Yan and Masliyah [102, 103] examined the effects of asphaltene-modified clay (plate-like particles) on the stability of a mineral o/w emulsion. They found that the amount of asphaltene coated on the clay has a significant impact on the contact angle of the clay, thereby affecting the stability of the emulsion, the size of the resultant oil droplets and the distribution of the clays in the continuous aqueous phase and at water/oil droplet interfaces.

The pH of the aqueous phase has been an important parameter in similar solid-stabilized emulsion systems [104]. Masliyah et al. investigated the role of suspension pH and added fine kaolinite clays in the emulsification of a toluene-diluted bitumen–water system [105]. Within a distinct pH range, mixing toluene-diluted bitumen with water resulted in a stable w/o emulsion. Outside this pH range, the mixture was completely emulsified to a homogeneous, gel-like creamed phase. However, a creamed layer of varying volume separated on the top of an aqueous phase when mixed outside these pH ranges. The creamed layer was found to consist mainly of an o/w emulsion. The removal of the original fine solids in the bitumen prior to emulsification resulted in a constant volume of o/w emulsion separated on the top of the aqueous phase over the pH range studied, 1.5–11.

Binks et al. investigated the rheological behavior of w/o emulsions prepared from dispersions of hydrophobic clay (bentonite) in oils where the particles either sediment or swell to form tactoids, which remain suspended [106]. Dispersions of hydrophobic clay in isopropyl myristate as an oil phase consist of large, micrometer-sized particles that sediment rapidly. The resulting o/w emulsions were partially flocculated, with the extent of flocculation increasing as the drop volume fraction increased. In contrast, the hydrophobic clay particles swell in toluene to form micrometer-sized tactoids that remain suspended in the oil due to the formation of networks. Emulsions of water-in-toluene stabilized by hydrophobic clay are highly viscous. The emulsion viscosity does not, however, increase with the drop volume fraction in the same way as for the water-in-isopropyl myristate–water emulsions. The elastic storage and viscous moduli of these emulsions are affected more by the particle concentration in the emulsions than by the drop volume fraction. It was proposed that some of the particles on the drop surfaces remain linked to the network in the continuous phase, as illustrated in Fig. 9.2 c, and thus an integrated network of drops and tactoids forms.

The polymer micro-rods used as foam superstabilizers also proved very effective in stabilizing emulsion droplets. They were used in the fabrication of w/o emulsions of melted agarose gels stabilized with a shell of polymeric micro-rods (Fig. 9.13) [100]. A hot aqueous solution of agarose was emulsified in oil in the presence of rod-like polymeric particles to produce a stable w/o emulsion and

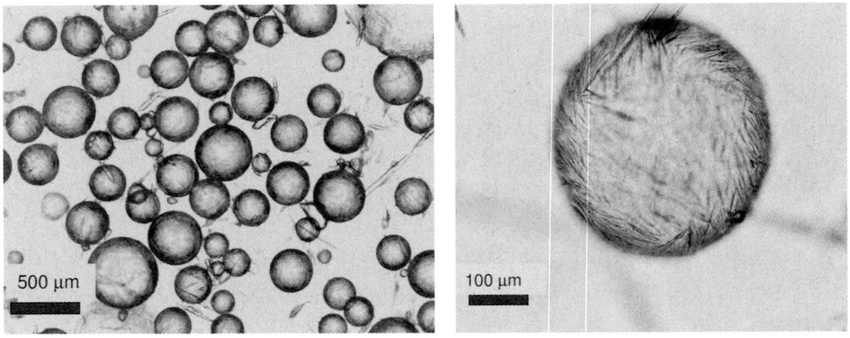

Fig. 9.13 Optical photograph of emulsion droplets from hot agarose solution in water dispersed in tricaprylin. The droplets are stabilized solely by SU-8 micro-rods [100].

the system was cooled to set the agarose gel. These emulsions were further used to fabricate rod–particle-stabilized capsules. The function of the gel cores is to support the particle shell and to give the microcapsules enough stiffness to be separated from the oil phase by centrifugation.

9.5 Multiple Emulsions Stabilized with Particles

It is well established that it is difficult to prepare stable multiple emulsions by using molecular surfactants as stabilizers. In contrast, due to high absorption energy, it is expected that particles could provide a means for stabilization of duplex emulsions. Attempts to prepare stable multiple emulsions by incorporating small solid particles as a mechanical barrier to coalescence in the surfactant formulations have been reported by Jahaniaval et al. [107], Oza and Frank [108], Garti et al. [109, 110] and Sekine et al. [111]. Recently, Binks et al. [112, 113] demonstrated fabrication of stable multiple emulsions using only a mixture of particles with different hydrophobicity as emulsifier (Fig. 9.14 a). Particle-stabilized multiple w/o/w emulsions are prepared in a two-stage process. In the first step, a w/o emulsion is formed by homogenizing water droplets in a dispersion of hydrophobic particles in oil. In the second step, the w/o emulsion is re-emulsified at low shear in an aqueous dispersion of hydrophilic particles. A minimal concentration of both particle types is required for the formation of multiple emulsions. Increasing the concentration of inner hydrophobic particles leads to an increase in the average globule diameter along with an increase in the number of inner water drops per globule. An increase in the outer hydrophilic particle concentration causes the globule diameter to decrease and the viscosity of the outer water phase to increase. In a similar manner, o/w/o emulsions can be prepared by first emulsifying oil in an aqueous dispersion of hydrophilic particles. In the next step, the o/w emulsion is gently re-emulsified in an oil dispersion of hydrophobic particles. Emulsions stable to both coalescence and sedi-

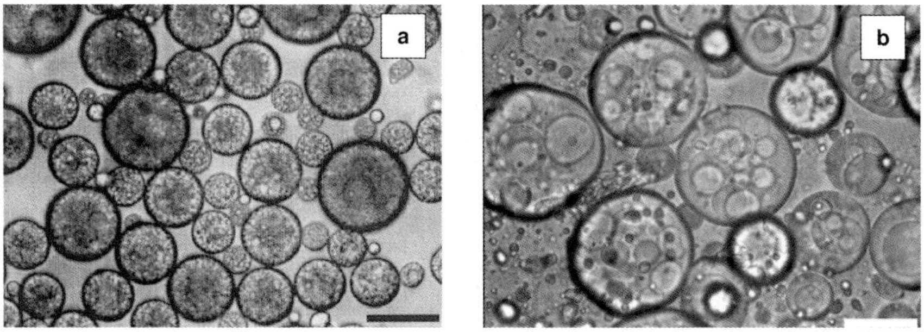

Fig. 9.14 Optical microscopy images of multiple emulsions stabilized with surface-active particles: (a) w/o/w multiple emulsion of triglyceride oil stabilized by two types of silica particles of different hydrophobicity. Image from Wacker-Chemie GmbH patent DE 10211313; (b) o/w/o multiple emulsion of Miglyol 810N-in-1,2-dimethyl-3-propylimidazoliumbis(trifluoro-methylsulfonyl)imide (DpimNT)-in-water multiple emulsion with inner interface stabilized with silica particles initially dispersed in the ionic liquid. The outer interfaces are stabilized with the same silica particles, which have been initially dispersed in water. Bar = 50 μm. (Reproduced with permission from [85]).

mentation can be prepared at high enough concentrations of the two particle types. Later, Binks et al. also demonstrated multiple emulsions from ionic liquids stabilized solely by silica nanoparticles of same type (Fig. 9.14 b) [85].

Double emulsions are an attractive system for many industrial applications, such as encapsulation and drug delivery. The mechanisms of particle stabilization in double emulsions have not been completely elucidated, but it is expected that these mechanisms are based on principles similar to those in single films and emulsions. Further research in this area is needed, as this important field remains largely unexplored.

9.6 Bifluid Particle-stabilized Gels

Colloidal particles with equal affinity for two fluids are known to adsorb irreversibly at the fluid/fluid interface. Therefore, theoretically, it is possible to create bicontinuous fluid gels stabilized with particles when the contact angle $\theta = 90°$. Such structures were expected to form near the phase inversion of particle-stabilized emulsions; however, they have not been observed so far. Cates et al. recently presented large-scale computer simulations of another approach for creating particle-stabilized bicontinuous gels – the demixing of a binary solvent containing such particles [114]. The newly formed interface sequesters the colloidal particles; as the interface coarsens, the particles are forced into close contact by capillarity. Coarsening is markedly curtailed and the jammed colloidal layer enters a glassy state, creating a multiply connected, solid-like film in three dimensions. The resulting gel contains percolating domains of both fluids (Fig. 9.15).

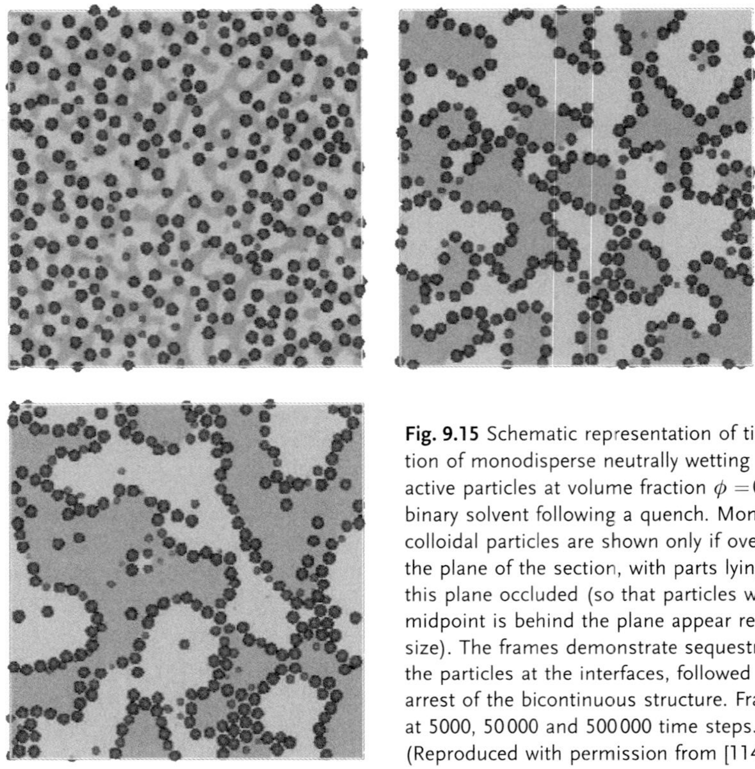

Fig. 9.15 Schematic representation of time evolution of monodisperse neutrally wetting surface-active particles at volume fraction $\phi = 0.2$ in a binary solvent following a quench. Monodisperse colloidal particles are shown only if overlapping the plane of the section, with parts lying behind this plane occluded (so that particles whose midpoint is behind the plane appear reduced in size). The frames demonstrate sequestration of the particles at the interfaces, followed by near-arrest of the bicontinuous structure. Frames are at 5000, 50 000 and 500 000 time steps. (Reproduced with permission from [114]).

Computer simulations have allowed the exploration of the kinetic pathway that leads to the creation of amorphous soft-solid materials. In such a bicontinuous emulsion gel, two interpenetrating fluid domains are frozen by a densely jammed monolayer of colloidal particles at the fluid/fluid interface. Such materials may have interesting properties, stemming directly from the arrested nature of the particle layers. They could be highly tunable in elasticity and pore size through the volume fraction and radius of the solid particles. The area of particle-stabilized bicontinuous (bifluid) gels is interesting and calls for further research and, importantly, experimental demonstration.

9.7 Conclusions and Outlook

Foam and emulsion systems stabilized by particles are interesting from both fundamental and applied perspectives. The "Pickering" emulsions have been known for more than 100 years [2], but are still an object of intense investigation. Small solid particles of different shape and morphology can adsorb strongly at the oil/water and air/water interfaces and stabilize emulsions and foams. Depending on their hydrophobicity, particles can stabilize o/w or w/o emulsions. Particles that are completely wetted by the two phases do not stabilize any emulsion. Using suitable combinations of particles of different hydrophobicity, stable multiple emulsions of w/o/w and o/w/o type can be formed. Particle wettability can be changed by adding surface-active compounds (e.g. surfactants, polymers, proteins), which adsorb both on the surface of the solid particles and water/oil interface, or by surface coating of grafting.

Excess of particles in the continuous emulsion phase leads to increased viscosity and gelation, which may improve the emulsion stability. Weak flocculation of the particles both provides lower mobility of the particles on the interface (preventing coalescence) and also contributes to the partial flocculation of the whole emulsion (preventing gravity-induced instabilities). This can be achieved by adjusting the pH and/or electrolyte concentration. Shape-anisotropic particles, e.g. platelets, are more efficient than spheres in covering and protecting the emulsion drops against coalescence. The continuous phase of the preferred emulsion becomes that in which the particles are first dispersed. For particles of intermediate hydrophobicity ($\theta \approx 90^\circ$), the continuous phase of the preferred emulsion becomes that in which the particles are first dispersed.

Surface-active particles can be used to stabilize more complex dispersions such as double emulsions, which is very difficult to achieve with common molecular surfactants. Finally, it has been shown theoretically that surface-active particles with equal affinity for two fluids could create novel bicontinuous particle-stabilized fluid gels; however, experimental demonstration of such a system is still awaited.

Stabilization of foams with particles is a less studied area and achieving long-term stability is difficult. Many of the factors of stabilization in foams are similar to those in emulsion systems, but some unique effects such as “plugging” of the three-phase regions between the films and preventing liquid drainage are also operational. The formation of superstable foams using rod-like particles has been demonstrated recently [92]. It can be expected that various new types of cylindrical, elongated and plate-like particles will allow the formation of very stable foam and emulsion systems in the future.

The development of effective particle stabilizers opens up great possibilities for the fabrication of surfactant-free emulsions and foams. Foams and emulsions stabilized solely by particles may be valuable for applications in which surfactants need to be avoided due to toxicity or price restrictions, for the formation of ultrastable foams and emulsions or making more complex systems such as double emulsions and bifluid gels. Using surface-active particles of different hydrophobicity for the fabrication of surfactant-free emulsions can have great potential for controlling their stability and rheology.

Important applications of particle-stabilized emulsions, which will not be discussed in detail here, can be found in the field of materials science. Particles assembled at liquid/air and liquid/liquid interfaces offer immense possibilities for the fabrication of new materials with special properties and controlled structure [13]. Some recent examples include ordered macroporous materials [15–17], colloidosomes [9, 10, 12, 115] and shape-anisotropic colloidal particles [116]. Colloidal assemblies such as colloidosomes can serve as capsules for delivery applications and basic units for the assembly of more complex materials with a hierarchical structure. A large number of nano- and microparticles of various sizes and shapes and surface patterns have been synthesized recently. These particles can bring forward a new level of stability and control in various emulsions, foams and new derivative materials. Particle-stabilized foams and emulsions constitute an old system, which has yet to expand and develop.

Acknowledgments

This study was supported by an NIRT grant from the National Science Foundation, USA. The authors dedicate this paper to Professors Dotchi Exerowa and Dimo Platikanov, two prominent scholars and teachers, who are among the founders of the excellent colloidal research school in their Alma Mater, the University of Sofia. K. P. Velikov acknowledges the management at Unilever Food and Health Research Institute for supporting this work.

References

1 A. Kabalnov, H. Wennerstrom, *Langmuir 12* (**1996**) 276.
2 S. U. Pickering, *J. Chem. Soc. 91* (**1907**) 2001.
3 D. Tambe, A. Sharma, *Adv. Colloid Interface Sci. 52* (**1994**) 1.
4 D. Rousseau, *Food Res. Int. 33* (**2000**) 3.
5 B. P. Binks, *Curr. Opin. Colloid Interface Sci. 7* (**2002**) 21.
6 R. Aveyard, B. P. Binks, J. H. Clint, *Adv. Colloid Interface Sci. 100* (**2003**) 503.
7 B. S. Murray, R. Ettelaie, *Curr. Opin. Colloid Interface Sci. 9* (**2004**) 314.
8 B. P. Binks, *Curr. Opin. Colloid Interface Sci. 7* (**2002**) 21.
9 O. D. Velev, K. Furusawa, K. Nagayama, *Langmuir 12* (**1996**) 2374.
10 O. D. Velev, K. Furusawa, K. Nagayama, *Langmuir 12* (**1996**) 2385.
11 O. D. Velev, *Adv. Biophys. 34* (**1996**) 139.
12 O. D. Velev, K. Nagayama, *Langmuir 13* (**1997**) 1856.
13 K. P. Velikov, O. D. Velev, Novel Materials Derived from Particles Assembled on Liquid Surfaces, in B. P. Binks, T. S. Horozov (Eds.), *Colloidal Particles at Liquid Interfaces*, Cambridge University Press, Cambridge (**2006**) pp. 225–297.
14 O. D. Velev, A. M. Lenhoff, E. W. Kaler, *Science 287* (**2000**) 2240.
15 B. P. Binks, *Adv. Mater. 14* (**2002**) 1824.
16 V. N. Manoharan, A. Imhof, J. D. Thorne, D. J. Pine, *Adv. Mater. 13* (**2001**) 447.
17 A. Imhof, D. J. Pine, *Nature 389* (**1997**) 948.
18 T. M. Allen, P. R. Cullis, *Science 303* (**2004**) 1818.
19 R. K. Niven, N. Khalili, D. B. Hibbert, *Chem. Eng. Sci. 55* (**2000**) 3013.
20 S. Levine, B. D. Bowen, *Colloids Surf. 59* (**1991**) 377.
21 R. Aveyard, J. H. Clint, T. S. Horozov, *Phys. Chem. Chem. Phys. 5* (**2003**) 2398.
22 V. N. Paunov, *Langmuir 19* (**2003**) 7970.
23 B. P. Binks, S. O. Lumsdon, *Langmuir 16* (**2000**) 8622.
24 B. P. Binks, P. D. I. Fletcher, *Langmuir 17* (**2001**) 4708.
25 V. N. Paunov, B. P. Binks, N. P. Ashby, *Langmuir 18* (**2002**) 6946.
26 R. Aveyard, J. H. Clint, D. Nees, V. N. Paunov, *Langmuir 16* (**2000**) 1969.
27 A. I. Abdel-Fattah, M. S. El Genk, *Adv. Colloid Interface Sci. 78* (**1998**) 237.
28 T. S. Horozov, R. Aveyard, B. P. Binks, J. H. Clint, *Langmuir 21* (**2005**) 7405.
29 P. Pieranski, *Phys. Rev. Lett. 45* (**1980**) 569.
30 G. Y. Onoda, *Phys. Rev. Lett. 55* (**1985**) 226.
31 K. P. Velikov, F. Durst, O. D. Velev, *Langmuir 14* (**1998**) 1148.

32 N. D. Denkov, I. B. Ivanov, P. A. Kralchevsky, D. T. Wasan, *J. Colloid Interface Sci.* (**1992**) 589.

33 K. P. Velikov, F. Durst, O. D. Velev, *Langmuir* (**1998**) 1148.

34 K. P. Velikov, O. D. Velev, in K. L. Mittal (Ed.), *Proceedings of ACS Symposium on Foams and Emulsions*, American Chemical Society, Washington, DC, **2000**, p. 84.

35 A. Scheludko, D. Exerowa, *Commun. Dept. Chem., Bulg. Acad. Sci. 7* (**1959**) 123.

36 D. Exerowa, D. Kashchiev, D. Platikanov, *Adv. Colloid Interface Sci. 30* (**1992**) 429.

37 D. Exerowa, P. M. Kruglyakov, *Foam and Foam Films*, Studies in Interface Science, Elsevier Science, Amsterdam, **1998**.

38 D. T. Wasan, A. D. Nikolov, P. A. Kralchevsky, I. B. Ivanov, *Colloids Surf. 67* (**1992**) 139.

39 A. D. Nikolov, P. A. Kralchevsky, I. B. Ivanov, D. T. Wasan, *J. Colloid Interface Sci. 133* (**1989**) 13.

40 V. Bergeron, C. J. Radke, *Langmuir 8* (**1992**) 3020.

41 A. D. Nikolov, D. T. Wasan, *Colloids Surf. A 128* (**1997**) 243.

42 K. Koczo, A. D. Nikolov, D. T. Wasan, R. P. Borwankar, A. Gonsalves, *J. Colloid Interface Sci. 178* (**1996**) 694.

43 N. D. Denkov, H. Yoshimura, K. Nagayama, T. Kouyama, *Phys. Rev. Lett. 76* (**1996**) 2354.

44 G. S. Lazarov, N. D. Denkov, O. D. Velev, P. A. Kralchevsky, K. Nagayama, *J. Chem. Soc., Faraday Trans. 90* (**1994**) 2077.

45 P. A. Kralchevsky, V. N. Paunov, I. B. Ivanov, K. Nagayama, *J. Colloid Interface Sci. 151* (**1992**) 79.

46 P. A. Kralchevsky, K. Nagayama, *Langmuir 10* (**1994**) 23.

47 V. N. Paunov, P. A. Kralchevsky, N. D. Denkov, K. Nagayama, *J. Colloid Interface Sci. 157* (**1993**) 100.

48 T. D. Dimitrova, F. Leal-Calderon, T. D. Gurkov, B. Campbell, *Adv. Colloid Interface Sci. 108/109* (**2004**) 73.

49 E. J. Stancik, M. Kouhkan, G. G. Fuller, *Langmuir 20* (**2004**) 90.

50 E. Vignati, R. Piazza, T. P. Lockhart, *Langmuir 19* (**2003**) 6650.

51 B. P. Binks, J. A. Rodrigues, *Langmuir 19* (**2003**) 4905.

52 D. Johansson, B. Bergenstahl, *J. Am. Oil Chem. Soc. 72* (**1995**) 205.

53 D. Johansson, B. Bergenstahl, E. Lundgren, *J. Am. Oil Chem. Soc. 72* (**1995**) 921.

54 D. Johansson, B. Bergenstahl, *J. Am. Oil Chem. Soc. 72* (**1995**) 933.

55 N. Garti, H. Binyamin, A. Aserin, *J. Am. Oil Chem. Soc. 75* (**1998**) 1825.

56 B. P. Binks, S. O. Lumsdon, *Langmuir 16* (**2000**) 3748.

57 B. P. Binks, S. O. Lumsdon, *Langmuir 16* (**2000**) 2539.

58 P. M. Kruglyakov, A. V. Nushtayeva, *Adv. Colloid Interface Sci. 108/109* (**2004**) 151.

59 S. Levine, B. D. Bowen, S. J. Partridge, *Colloids Surf. 38* (**1989**) 345.

60 S. Levine, B. D. Bowen, S. J. Partridge, *Colloids Surf. 38* (**1989**) 325.

61 S. Levine, B. D. Bowen, *Colloids Surf. 65* (**1992**) 273.

62 B. P. Binks, S. O. Lumsdon, *Phys. Chem. Chem. Phys. 1* (**1999**) 3007.

63 B. P. Binks, S. O. Lumsdon, *Phys. Chem. Chem. Phys. 2* (**2000**) 2959.

64 J. I. Amalvy, S. P. Armes, B. P. Binks, J. A. Rodrigues, G. F. Unali, *Chem. Commun.* (**2003**) 1826.

65 J. I. Amalvy, G. F. Unali, Y. Li, S. Granger-Bevan, S. P. Armes, B. P. Binks, J. A. Rodrigues, C. P. Whitby, *Langmuir 20* (**2004**) 4345.

66 B. P. Binks, S. O. Lumsdon, *Langmuir 17* (**2001**) 4540.

67 P. G. Smith, T. G. M. van Deven, *Colloids Surf. 15* (**1985**) 191.

68 A. Gelot, W. Friesen, H. A. Hamza, *Colloids Surf. 12* (**1984**) 271.

69 D. E. Tambe, M. M. Sharma, *J. Colloid Interface Sci. 162* (**1994**) 1.

70 S. Tarimala, L. L. Dai, *Langmuir 20* (**2004**) 3492.

71 C. L. A. Berli, D. Quemada, A. Parker, *Colloids Surf. A 215* (**2003**) 201.

72 E. Dickinson, *Curr. Opin. Colloid Interface Sci. 3* (**1998**) 633.

73 D. Johansson, B. Bergenstahl, *J. Am. Oil Chem. Soc. 72* (**1995**) 911.

74 D. Johansson, *J. Am. Oil Chem. Soc. 72* (**1995**) 1235.

75 B. S. Murray, *Curr. Opin. Colloid Interface Sci. 7* (**2002**) 426.

76 A. D. Dinsmore, M. F. Hsu, M. G. Nikolaides, M. Marquez, A. R. Bausch, D. A. Weitz, *Science 298* (**2002**) 1006.
77 V. D. Gordon, C. Xi, J. W. Hutchinson, A. R. Bausch, M. Marquez, D. A. Weitz, *J. Am. Chem. Soc. 126* (**2004**) 14117.
78 L. M. Croll, H. D. H. Stöver, *Langmuir 19* (**2003**) 5918.
79 L. M. Croll, H. D. H. Stöver, *Langmuir 19* (**2003**) 10077.
80 Y. Lin, H. Skaff, T. Emrick, A. D. Dinsmore, T. P. Russell, *Science 299* (**2003**) 226.
81 Y. Lin, H. Skaff, A. Boker, A. D. Dinsmore, T. Emrick, T. P. Russell, *J. Am. Chem. Soc. 125* (**2003**) 12690.
82 J. T. Russell, Y. Lin, A. Boker, L. Su, P. Carl, H. Zettl, J. B. He, K. Sill, R. Tangirala, T. Emrick, K. Littrell, P. Thiyagarajan, D. Cookson, A. Fery, Q. Wang, T. P. Russell, *Angew. Chem. Int. Ed. 44* (**2005**) 2420.
83 S. Melle, M. Lask, G. G. Fuller, *Langmuir 21* (**2005**) 2158.
84 H. W. Duan, D. Y. Wang, N. S. Sobal, M. Giersig, D. G. Kurth, H. Mohwald, *Nano Lett. 5* (**2005**) 949.
85 B. P. Binks, A. K. F. Dyab, P. D. I. Fletcher, *Chem. Commun.* (**2003**) 2540.
86 R. Averyard, B. P. Binks, P. D. I. Fletcher, T. G. Peck, C. E. Rutherford, *Adv. Colloid Interface Sci. 48* (**1994**) 93.
87 R. J. Pugh, *Adv. Colloid Interface Sci. 64* (**1996**) 67.
88 S. K. Bindal, A. D. Nikolov, D. T. Wasan, D. P. Lambert, D. C. Koopman, *Environ. Sci. Technol. 35* (**2001**) 3941.
89 S. K. Bindal, G. Sethumadhavan, A. D. Nikolov, D. T. Wasan, *AICHE Journal 48* (**2002**) 2307.
90 E. Dickinson, R. Ettelaie, B. S. Murray, Z. P. Du, *J. Colloid. Int. Sci. 252* (**2002**) 202.
91 E. Dickinson, R. Ettelaie, T. Kostakis, B. S. Murray, *Langmuir 20* (**2004**) 8517.
92 R. G. Alargova, D. S. Warhadpande, V. N. Paunov, O. D. Velev, *Langmuir 20* (**2004**) 10371.
93 R. G. Alargova, K. H. Bhatt, V. N. Paunov, O. D. Velev, *Adv. Mat. 16* (**2004**) 1653.
94 D. R. Exerowa, P. M. Krugliakov, Foam and Foam Films: Theory, Experiment, Application, Elsevier (**1998**).
95 I. B. Ivanov, *Thin Liquid Films*, Marcel Dekker (**1988**).
96 V. N. Paunov, *Langmuir 19* (**2003**) 7970.
97 P. M. Kruglyakov, A. Nushtayeva, N. G. Vilkova, *J. Colloid. Int. Sci. 276* (**2004**) 465.
98 Z. P. Du, M. P. Bilbao-Montoya, B. P. Binks, E. Dickinson, R. Ettelaie, B. S. Murray, *Langmuir 19* (**2003**) 3106.
99 K. Malysa, *Adv. Colloid. Int. Sci. 40* (**1992**) 37.
100 P. F. Noble, O. J. Cayre, R. G. Alargova, O. D. Velev, V. N. Paunov, *J. Am. Chem. Soc. 126* (**2004**) 8092.
101 M. A. J. S. van Boekel, P. Walstra, *Coll. Surf. 3* (**1981**) 109.
102 Y. H. Yan, J. H. Masliyah, *Coll. Surf. A – Phys. Chem. Eng. Asp. 75* (**1993**) 123.
103 N. X. Yan, J. H. Masliyah, *Coll. Surf. A – P hys. Chem. Eng. Asp. 96* (**1995**) 229.
104 N. X. Yan, J. H. Masliyah, *J. Colloid. Int. Sci. 181* (**1996**) 20.
105 G. X. Gu, Z. Zhou, Z. H. Xu, J. H. Masliyah, *Coll. Surf. A – Phys. Chem. Eng. Asp. 215* (**2003**) 141.
106 B. P. Binks, J. H. Clint, C. P. Whitby, *Langmuir 21* (**2005**) 5307.
107 F. Jahaniaval, Y. Kakuda, V. Abraham, *J. Am. Oil Chem. Soc. 80* (**2003**) 25.
108 K. P. Oza, S. G. Frank, *J. Disp. Sci. Tech. 10* (**1989**) 163.
109 N. Garti, A. Aserin, I. Tiunova, H. Binyamin, *J. Am. Oil Chem. Soc. 76* (**1999**) 383.
110 N. Garti, C. Bisperink, *Curr. Opinion Colloid Int. Sci. 3* (**1998**) 657.
111 T. Sekine, K. Yoshida, F. Matsuzaki, T. Yanaki, M. Yamaguchi, *J. Surf. Deterg. 2* (**1999**) 309.
112 B. P. Binks, A. K. F. Dyab, P. D. I. Fletcher, H. Barthel, Multiple emulsions. Wacker-Chemie GmbH (DE10211313), 14-3-2003, Germany.
113 B. P. Binks, M. Kirkland, *Phys. Chem. Chem. Phys. 4* (**2002**) 3727.
114 K. Stratford, R. Adhikari, I. Pagonabarraga, J. C. Desplat, M. E. Cates, *Science 309* (**2005**) 2198.
115 A. D. Dinsmore, M. F. Hsu, M. G. Nikolaides, M. Marquez, A. R. Bausch, D. A. Weitz, *Science 298* (**2002**) 1006.
116 V. N. Manoharan, M. T. Elsesser, D. J. Pine, *Science 301* (**2003**) 483.
117 R. G. Alargova, D. S. Warhadpande, V. N. Paunov, O. D. Velev, *Langmuir 20* (**2004**) 10371.

10
Forces in Aqueous Nanofilms Containing Polyelectrolytes

Dan Qu, Katarzyna Hänni-Ciunel, Daniel Rapoport, and Regine v. Klitzing

Abstract

This chapter deals with the structuring of polyelectrolytes within liquid nanofilms. The effect of the film core is separated from the influence of the interfaces. With respect to the influence of *interfaces*, films with different surfaces, such as foam films, wetting films and films between two solid interfaces, are compared and the effect of surface modification is considered. The response of the *film core* is studied by different parameters of the polymeric system, such as degree of polymer charge, molecular architecture and ionic strength. Methods for force measurements in thin films are critically compared with respect to their optimum application.

10.1
Introduction

Thin liquid films play an important role in many macroscopic systems which consist of a dispersed and a continuous phase. For instance, in an aqueous foam, air bubbles are dispersed in a surfactant solution. A thin surfactant film separates the air bubbles. In an emulsion, liquid droplets are dispersed in a second liquid. The most prominent emulsions are oil-in-water or water-in-oil emulsions. In all cases, the stability of the macroscopic system is dominated by the stability of the thin film separating the compartments of the dispersed phase. It is evident that the study of these thin films is of great interest for technical applications. For fundamental science, thin films are suitable model systems for studies of interfacial forces which dominate the properties of all colloidal systems. Hence the field is not restricted to foam and emulsion films, but also includes wetting films, pseudo-emulsion films, which occur in emulsion foams, and films between two solid interfaces as in suspensions. In order to study forces, *foam films* and *wetting films* are highly relevant, since their thickness can easily be varied by changing the outer pressure. This explains the great attraction of these films for studies in colloidal science since many decades [1].

Colloids and Interface Science Series, Vol. 3
Colloid Stability and Application in Pharmacy. Edited by Tharwat F. Tadros

ISBN: 978-3-527-31463-8

One of the first disjoining pressure isotherms was published by Scheludko and Exerowa in 1959 [2]. It was measured with a so-called Scheludko cell, where the foam film is formed on a glass ring and the film surfaces are made to approach each other by sucking out the liquid. In a later version the ring was replaced by a porous plate [3–5]. Here the outer pressure is varied during the measurement. Since then, many different types of apparatus were developed to measure forces in liquid films. Wetting films have mostly been studied by Platikanov and Exerowa with a captured bubble method, where an air bubble in a liquid is pressed against a solid plate [6–9]. Pressure changes are either applied by changing the pressure within the bubble, or the solution is sucked by a micrometric piston through an outlet. Overviews on liquid films are given in [10] and [11].

Free-standing films are interesting in two respects. As already mentioned, the film can be considered as the building block of a foam so that its properties affect the behavior of the whole macroscopic foam [12]. In this context, it is the molecular structure at and near the film surfaces rather than the structure of the film core which is important. Second, the free-standing film presents a cavity which allows the study of the effect of geometric confinement on the structuring of polymers.

This chapter focuses mainly on the second aspect, the confinement of polyelectrolytes in nanofilms. Two large fields of colloidal science which have been investigated for several decades are associated with the studies presented here: interactions between two opposing surfaces in aqueous films and the structuring of polyelectrolytes in aqueous solutions (e.g. [13–17]). With respect to the general miniaturization, the question arises of how this chain ordering is affected by geometric confinement. Here *confinement* means that the dimensions of the total system are of the same order of magnitude as the correlation length of the polymer system (e.g. Debye length, correlation length, persistence length, radius of gyration). In the present study, the polyelectrolytes are entrapped between the two interfaces of a horizontal single liquid film. The film thickness and therefore the confinement are changed by the outer pressure or force. The film thickness is between 5 and 120 nm. *The aim of this chapter is to separate interfacial effects from film core effects.* In order to study the *effect of the film interfaces,* different films such as foam films, wetting films and films between two solid interfaces are presented. Subsequently, the influence of surface modification is described. The structuring of the polyelectrolytes within the film core is compared with that in bulk solution as a function of molecular architecture and electrostatic parameters, e.g. ionic strength and polymer charge density.

Each type of film needs its own technique. These are discussed critically according to the questions which will be studied.

10.2 Forces in Films

10.2.1 DLVO Forces

The DLVO forces consist of electrostatic and van der Waals interaction [18, 19]. The electrostatic repulsion between two identically charged interfaces is described by an exponential decay:

$$\Pi_{el} = \Pi_0 \exp[-\kappa(h - 2h_0)] \tag{1}$$

with the Debye-Hückel screening length $1/\kappa$ and the thickness h_0 of the adsorbed interfacial layer. Π_0 is connected to the surface potential Ψ_0 by the relation [20]

$$\Pi(h) = 64kT\rho_\infty \gamma^2 \exp(-\kappa h) = (1.59 \times 10^8)[c_{el}]\gamma^2 \exp(-\kappa h) \quad [\mathrm{Pa}] \tag{2}$$

$$\gamma = \tanh(ze\Psi_0/4kT) \tag{3}$$

In a simple foam film, the thickness of the interface is similar to the length of a surfactant molecule. The thickness of the so-called common black film (CBF) is determined by the DLVO forces, and the thinner Newton black film (NBF) is stabilized by steric repulsion and does not contain any free solvent molecules. A transition from a CBF to an NBF can be induced by the addition of salt, leading to a screening of the surface potential. This confirms the electrostatic nature of the repulsive force stabilizing the CBF. The transition from a CBF to an NBF corresponds to an oscillation of the disjoining pressure because of attractive van der Waals forces. This attractive part of the isotherm is mechanically unstable, and it cannot be measured by a TFPB. However, a step in film thickness from the thicker CBF to a thinner NBF is detected.

10.2.2 Structural Forces

In addition to DLVO forces, so-called structural forces can occur between two interfaces, i.e. in the film, due to the layering of molecules. This is related to an oscillatory concentration profile with a decay in amplitude from the interface towards the film bulk. The relation between the oscillatory concentration profile and the oscillatory pressure is still under discussion. An ostensive image of the oscillatory pressure is the layer-by-layer expulsion of the molecules which induces attractive depletion forces. With respect to the class of molecules this kind of force is also called solvation or hydration forces. These structural forces has been observed for spherical molecules entrapped between two mica plates in a surface force apparatus (SFA) (e.g. [20]). The period of the oscillation is connected to the diameter of a spherical molecule. Oscillatory forces also occur in free-standing films containing liquid crystals, colloidal particles (e.g. [10] and references therein) or micelles. The micellar systems have been investigated both in an SFA between solid interfaces [21] and in a TFPB between liquid interfaces [22, 23]. The step size scales with the surfactant concentration c_s as ${c_s}^{-1/3}$ [24]. This can be simply explained by a homogeneous distribution of "hard spheres" in three dimensions where the distance between two centers of mass is determined by the concentration of spheres. To our knowledge, the stepwise thinning (i.e. stratification) of foam films containing micelles occurs if the surfactant is charged but it has been not published so far for non-ionic surfactants. This indicates that the electrostatic repulsion between the micelles could be responsible for the film stratification. The oscillatory forces are described by

$$\frac{F(h)}{2\pi} = A \exp\left(\frac{-h}{\lambda}\right) \cos\frac{2\pi h}{d} \tag{4}$$

where $F(h)$ is the force in dependence of the distance h between the surfaces. The three parameters which characterize the oscillation are the amplitude A, the decay length λ and the period d.

10.3 Methods for Force Measurements

In the following, techniques are described for measuring the force between two opposing surfaces. The description mainly focuses on aspects which are important for thin aqueous films containing polyelectrolytes.

10.3.1 Thin-film Pressure Balance (TFPB)

The thin-film pressure balance (TFPB) is mainly used to study foam films, but recently it has also been extended to study wetting films.

The disjoining pressure isotherms (disjoining pressure Π as a function of the film thickness h) can be measured with the porous-plate technique in a TFPB (Fig. 10.1), developed by Mysels and Jones [3] and Exerowa and coworkers [4, 5]. In this apparatus, the capillary pressure is balanced by the disjoining pressure in a horizontal free-standing film. The film is formed from an aqueous polyelectrolyte–surfactant solution over a hole (diameter about 1–2 mm) drilled through a porous glass plate. The plate allows the liquid to flow out of or into the film whenever the pressure is changed. On the other hand, small pores (diameter about 1 μm) of the fritted glass plate make it possible to apply a pressure of 10^4 Pa. The fritted glass plate is filled with the liquid of the film, and is connected with the external reference pressure P_r (atmospheric pressure) by a glass

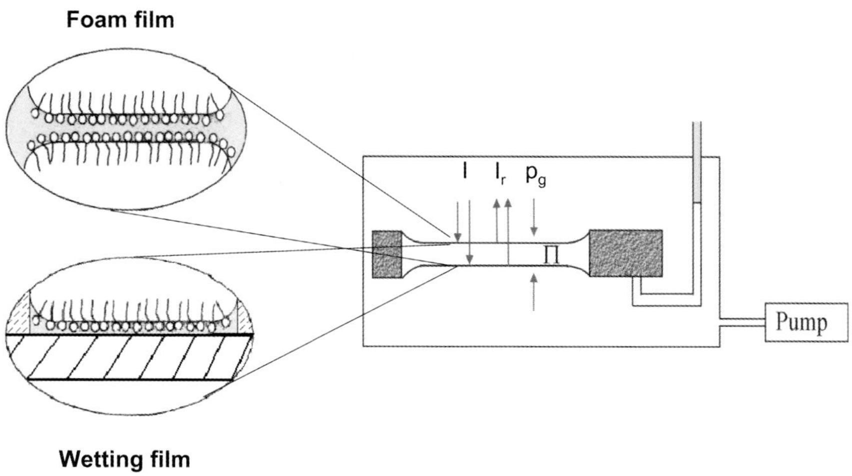

Fig. 10.1 Center piece of a thin-film pressure balance (TFPB). The inset shows the application for foam films and for wetting films.

tube. This film holder is enclosed in a metal cell, which allows the film to be pressurized using a piston pump. During the film drainage, the capillary pressure causes a sucking of film liquid into the plateau borders until the disjoining pressure begins to affect the dynamics. At equilibrium, the capillary pressure P_c and the disjoining pressure Π compensate each other.

$$P_c = \Pi = P_g - P_l = P_g - P_r - \Delta\rho g h_c + 2\gamma/r_c \qquad (5)$$

The hydrostatic pressure of the liquid column in the glass tube is given by $\Delta\rho g h_c$ ($\Delta\rho$=difference in density between the solution and the gas; g=gravitation constant; h_c=height of the liquid column in the glass tube above the film). The capillary pressure in the glass tube is determined by 2 γ/r_c (γ=surface tension of the liquid; r_c=radius of the capillary tube). The difference in pressure inside and outside the cell ($P_g - Pr$) is measured by a differential pressure transducer. The accuracy of the disjoining pressure is mostly limited by the difference pressure transducer, which usually has a specificity of 0.3% of its full range (i.e. 3–30 Pa). Above a pressure of 500 Pa, $P_g - P_r$ is the determining contribution to the disjoining pressure.

The film is illuminated by cold-filtered white light via a microscope through a quartz window on top of the cell and is monitored with video microscopy. The film acts like an interferometer, since the light which is reflected at the upper and the lower interface superposes. A maximum intensity occurs at a thickness of around 100–150 nm for visible light due to the phase shift of $\pi/2$ at one of the film surfaces. The intensity decreases with decreasing film thickness, which gives a change in "color" from white (at the maximum) to bright gray or dark gray and even black in the case of several nanometer thick films (e.g. NBF).

Parallel to video microscopy, the film thickness is determined by the interferometric method of Scheludko [25]. Therefore, the reflected intensity at one fixed wavelength ($\lambda = 550$ or 630 nm) is measured by a photomultiplier. The following equation is used to calculate the film thickness [25]:

$$h = \frac{\lambda}{2\pi n} \arcsin \sqrt{\frac{\Delta}{1 + \left[\frac{4R}{(1-R)^2}(1-\Delta)\right]}} \tag{6}$$

$$\Delta = \frac{I - I_{\min}}{I_{\max} - I_{\min}} \tag{7}$$

$$R = \frac{(n-1)^2}{(n+1)^2} \tag{8}$$

where n is the refractive index of the liquid. I is the actual intensity and $I_{\min}$ and $I_{\max}$ correspond to the last minimum and maximum intensity value. In the case of thin films (e.g. NBF, $h \approx 4$ nm) the different refractive indices of the different film regions have to be taken into account. In general, a sandwich structure of the two surfactant layers which are adsorbed at the film surfaces (n_{tail}, n_{head}, h_{tail}, h_{head}) separated by the film core (n_c, h_c) is used as a film model [26]. However, the determination of the refractive index and the thickness is rather imprecise, since the molecular density at the film interfaces is not directly accessible. The surface density at an air/water interface of a solution can be determined, for instance, by surface tension measurements, but probably these results cannot be transferred to the film interfaces. There is evidence for a higher surface density in an NBF than in the former CBF (e.g. [27]). Inside the cell, a reservoir of the film liquid is included which allows the air inside the cell to be saturated and avoids evaporation of the film liquid. It is assumed that the equilibrium film thickness is reached when the reflected intensity stays constant over a period of 20 min. The cell is temperature controlled and the quartz window within the top of the cell is heated to avoid condensation. Unless stated otherwise, the measurements are carried out under equilibrium conditions at room temperature.

Below the critical overlap concentration of polyelectrolytes c^*, the film drains in a continuous way. The isotherm shows a typical exponential decay, indicating electrostatic repulsion. The slope is directly related to the ionic strength. From c^* on, stepwise thinning of the film occurs. There is an abrupt change in the slope of the isotherm branches from a finite one below c^* to an infinite one above c^*, which is no longer related to the ionic strength [28, 29]. This phenomenon is explained in Section 10.4.1. With increasing polyelectrolyte concentration, the number of steps increases and the steps become smaller (e.g. [28–31]).

Figure 10.2 shows a typical $\Pi(h)$ isotherm of a polyelectrolyte–surfactant foam film above c^*. Note that the steps are irreversible. By decreasing the pressure no step backward to the former branch of the $\Pi(h)$ curve is possible, and the thick-

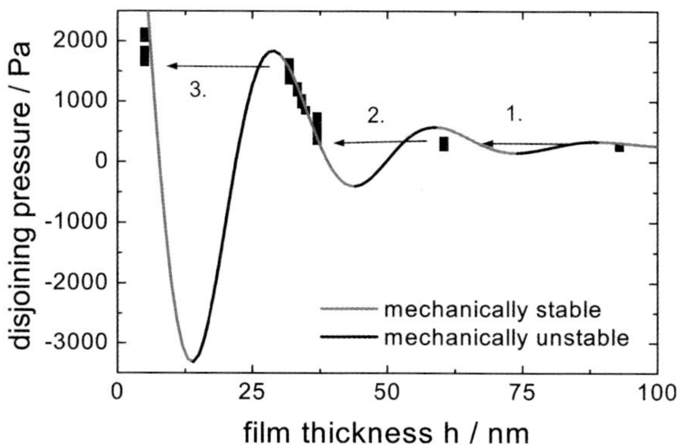

Fig. 10.2 Disjoining pressure Π as a function of the film thickness h of a $C_{12}G_2$–PDADMAC film and its simulation by a damped oscillatory function. In a TFPB experiment, only the mechanically stable parts are accessible whereas in AFM experiments, for suitable systems, the whole oscillation can be measured. (Adapted from [32]).

ness does not change in a significant way. The experimental data can be interpreted as parts of a damped oscillation, according to Eq. (4) for structural forces. The attenuation explains the irreversibility of the steps. If the film is at a mechanically unstable point after an increase in the pressure, it jumps to the next mechanically stable state, which is the next thinner isotherm branch. On the other hand, if the pressure is decreased, the system remains mechanically stable and therefore it does not jump back to the former thicker isotherm branch.

10.3.1.1 TFPB for Fluorescence Measurements (FTFPB)

For fluorescence measurements a usual TFPB is combined with a fluorescence microscope and a spectrometer [33]. In the case of pyrene studies, the fluorescence microscope has to be extended to the near-UV regime. Therefore, the condenser, filters and objective have to be exchanged as described in [34]. In order to have the possibility of switching between fluorescence and interferometric measurements, two light sources have to be mounted: an HBO lamp and a cold filtered lamp. In addition to the fluorescence filters, a filter block for white light has to be added. Since for interferometric measurements the same wavelength as the incident one is analyzed, there will be no crossover with fluorescence light, which is recorded at longer wavelengths. For the same reason, fluorescent dye labels do not disturb thickness measurements [33].

10.3.1.2 TFPB or Wetting Films

A TFPB can also be used to study wetting films [35]. Therefore, a solid substrate (typically quartz or silicon wafer) is approached from one side (see Fig. 10.1). It is attracted by the solution-filled porous plate via adhesion forces. For reasons of contrast, the application of a silicon wafer is advised. The film thickness is measured again by an interferometric method [36]:

$$h = \frac{\lambda}{2\pi n_2}\left\{\frac{2m+1}{2}\pi - \frac{\delta}{2} \pm \arcsin\sqrt{\frac{\Delta}{1+\left[\frac{4\sqrt{R_{12}R_{23}}}{\left(1-\sqrt{R_{12}R_{23}}\right)^2}(1-\Delta)\right]}}\right\} \tag{9}$$

$$\Delta = \frac{I - I_{\min}}{I_{\max} - I_{\min}} \tag{10}$$

$$R_{12} = \frac{(n_1 - n_2)^2}{(n_1 + n_2)^2} \tag{11}$$

$$R_{23} = \frac{(n_2 - n_3)^2}{(n_2 + n_3)^2} \tag{12}$$

where m is the order of the interference and $\delta = 0$ for the refractive indices $n_1 < n_2 < n_3$.

Figure 10.3 shows photographs of a foam film and a wetting film draining in a discontinuous way with a step in its thickness. Due to differences in phase shift at a solid and a fluid interface, the reflected intensities behave in the opposite way. In a foam film, the thinner film thickness occurs as dark spots which extend to a new homogeneous film thickness, and in a wetting film bright spots (thinner film) expand on a dark background (greater thickness). The origin of stratification can be a transition from a CBF to an NBF or structural forces, as mentioned in Section 10.2.

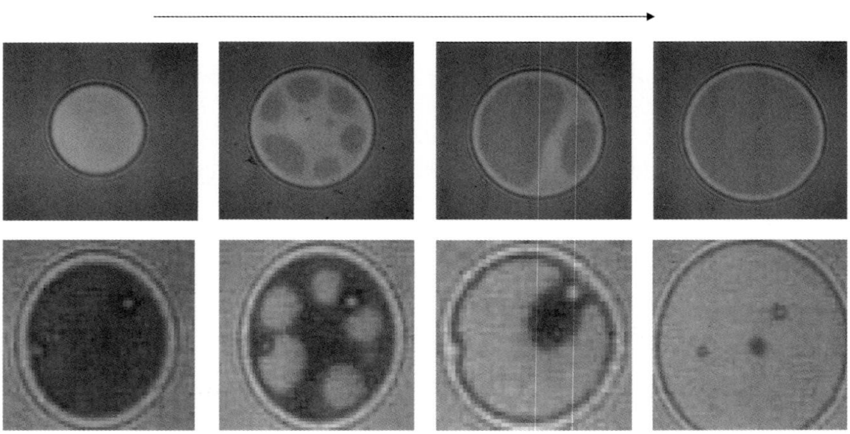

Fig. 10.3 Photographs of a transition from a thicker to a thinner film in a foam film (top row) and a wetting film (bottom row).

10.3.2
Colloidal Probe Atomic Force Microscope (CP-AFM)

The colloidal probe technique was first developed by Ducker et al. [37] and Butt [38]. Since for this type of experiment a low signal-to-noise ratio is required, an AFM is used which has to be designed for force spectroscopy. A colloidal probe (e.g. a silica particle) is glued to a tip-less cantilever. Usually, the particle has a radius in the order of micrometers. In many experiments the substrate used is a silicon wafer with a naturalized SiO_2 top layer. A drop of the polymer solution is placed on top of the substrate and the probe head is immersed within the solution. Force vs. distance $[F(h)]$ curves are measured. Since the distance between the colloidal probe and the substrate is much smaller than the diameter of the probe, by the Derjaguin approximation, the round silica particle can be considered as a flat surface and the interaction energy per unit area $E(h)$ can be acquired from the force $F(h)$ by

$$E(h) = F(h)/2R \tag{13}$$

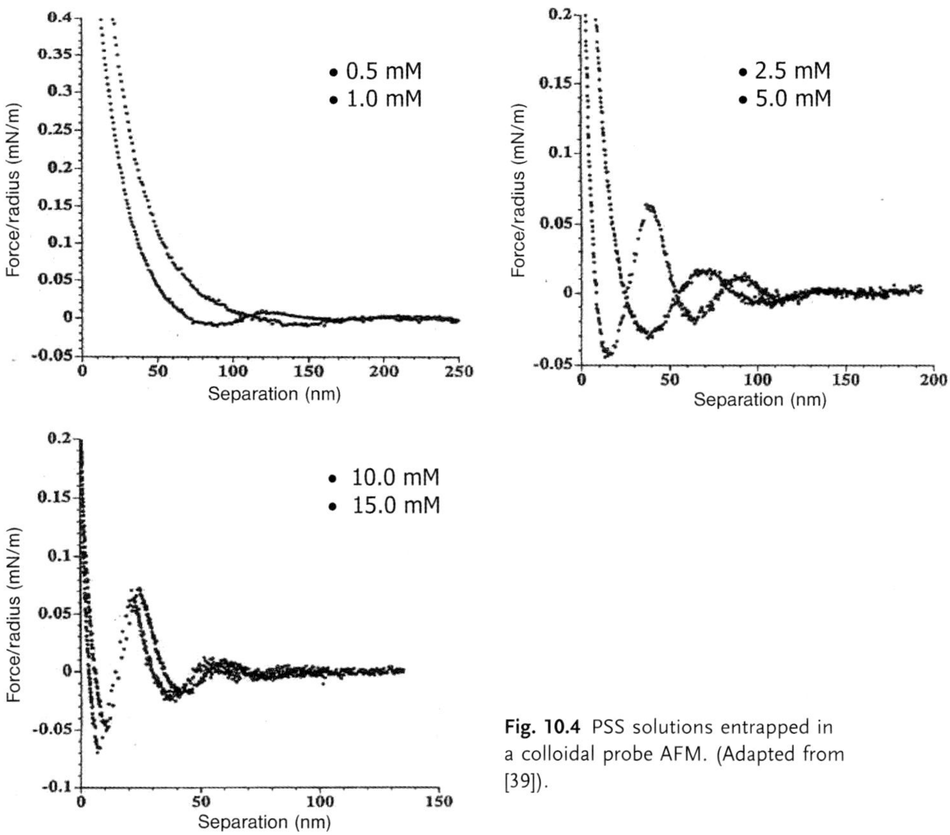

Fig. 10.4 PSS solutions entrapped in a colloidal probe AFM. (Adapted from [39]).

Structural forces of polyelectrolyte solutions confined between interfaces were observed for the first time by Milling and coworkers [39–41]. The experiments were carried out with a CP-AFM in aqueous solutions of different PSS concentrations. The force curves show oscillatory behavior with an exponential decay in forces towards thicker films. The period decreases with increasing PSS concentration. Figure 10.4 shows that the complete force curve can be measured with a CP-AFM.

10.3.3 Surface Force Apparatus (SFA)

The forces measured are those between two crossed cylindrical sheets of mica, glued to two optically polished curved glass discs. During an experiment, the apparatus is filled with a liquid so that the two mica surfaces are entirely immersed. The separation between the surfaces is measured by an interferometric technique. The incident white light normal to the surfaces passes through the lower surface and is multiply reflected between the cylinder surfaces. Due to constructive interference for certain wavelengths, interference fringes [fringes of equal chromatic order (FECO)] are transmitted through the upper surface. The FECO wavelengths depend on the thickness of the mica surfaces and the separation between the two mica surfaces. The surface separation can be determined by comparing the FECO wavelengths when the surfaces are in contact and apart. The forces measured between two crossed cylinders can be replaced in a mathematical model by forces between a sphere and a planar surface. The radius of the sphere is in the order of 1 cm. A detailed description is given in [20].

Whereas structural forces can be monitored for micellar systems and emulsions, in films of aqueous polyelectrolyte solutions no structural forces are detectable. This phenomenon is discussed in the following section.

10.3.4 Comparison Between the Different Methods: Advantages and Disadvantages, Fields of Application

The methods described above measure the force and the pressure between different types of interfaces, which leads to different fields of application. The method should be chosen according to the scientific question of interest.

The CP-AFM is the most suitable method to study (oscillatory) structural forces, since the complete oscillation (repulsive *and* attractive parts of the force curve) is accessible. This gives more precise fits than the disjoining pressure curves measured by a TFPB. With a TFPB, only the repulsive can be measured. Attractive interactions are not accessible, which leads to steps in film thickness and spots on the film during the transition from a thicker to a thinner film. In addition most of the polyelectrolytes which are investigated in aqueous nanofilms are hydrophilic and do not form stable free-standing films (i.e. foam films), and in general have to be stabilized by surfactants. The question is how the interaction between polyelectrolytes and surfactant affects the structural forces within the film. A CP-AFM gives the opportunity to study the pure polyelectrolyte system in water without any additives.

On the other hand, the effect of surface properties on film forces can be better investigated with a TFPB (see Section 10.6). The reason is that the properties of a *fluid* interface can be easily tuned by the adsorbed surfactant, e.g. by differently charged surfactants. Most of the *solid* interfaces are negatively charged. In order to change their charge, molecules such as polyelectrolytes or amides can be adsorbed. This brings two problems for CP-AFM studies: (1) in most cases, the adsorbed (mono)layer is not closed and leads to a lateral pattern at the surface. Since the patterns of both opposing surfaces are not the same, locally positively and negatively surface charges may attract each other. The surfaces might jump in contact, and it would be impossible to measure structural forces. (2) Polyelectrolytes can bridge from one surface to locally uncovered (i.e. oppositely charged) areas of the opposing surface, which leads again to a sudden approach of the interfaces.

The most precise measurements of structural forces are carried out with a CP-AFM between bare surfaces (see Sections 10.4.1 and 10.6). For the reasons mentioned above, adsorption of polyelectrolytes from the entrapped solution at the opposing interfaces has to be avoided. This restricts the choice of polyelectrolytes to negatively charged types.

Another advantage of TFPB is the better optical quality, important for visualization of the processes within the film. For instance, to visualize the film in a CP-AFM, a (transparent) glass slide has to be used instead of a silicon wafer. The reflection coefficient and therefore the contrast is much lower at the glass/water than at the air/water interface.

To summarize, TFPB and CP-AFM can be applied to study different questions and they are complementary to each other.

In an SFA, liquid films are also studied between two solid interfaces, but no oscillations are detectable. Reasons for this difference have not been clarified so far. The charge cannot be an explanation, since mica has a higher surface potential (about –150 mV) than silicon (about –80 mV), which should enhance the oscillation amplitude, as shown in Section 10.6.1.1. However, in all SFA measurements at aqueous polyelectrolyte films the oscillation is more or less completely suppressed. Surface viscoelasticity and slippage at the interface seem to play no role, since oscillations or parts of it are detectable between both solid (by CP-AFM) and fluid interfaces (by TFPB). Also, differences in curvature should not be important. The curvature of a planar foam film is zero (i.e. infinite radius of curvature). In a CP-AFM, the silica sphere has a radius of about 5 μm. The radius of curvature of the surfaces in an SFA is, at 1 cm, between these values. With respect to the Derjaguin approximation, the radius of curvature plays no role in all apparatus, since it is much larger than the distance. The TFPB is the only method where the data points of equilibrium thickness are measured. A typical CP-AFM measurement of an oscillatory force curve is recorded within 1 min. It is not possible to stop the measurement at a certain separation and to keep a certain data value constant (i.e. constant force at the respective surface separation) for times longer than several tens of seconds. There is a drift due to streaming in the apparatus.

A force–distance curve which is recorded at slow approach with a velocity of less than 10 nm/s no longer shows any oscillations [42]. This drift seems to be much stronger in an SFA. The measurement of the whole force–distance curve takes several minutes. It is not possible to drive the measurements faster, since the signal-to-noise ratio is much higher than in a CP-AFM. This could be caused by the fact that the SFA probe unit is immersed in the liquid, whereas in a CP-AFM only a drop is placed between the two surfaces and the probe head itself stays in still air. Therefore, fluctuations are much more pronounced in an SFA than in a CP-AFM. Subsequently, the larger volume of liquid in an SFA could induce a stronger streaming between the surfaces. For micellar or particle systems or emulsions, this streaming might induce an additional force enhancing the expulsion velocity of particles, micelles and droplets. However, the shape of the colloidal particles changes only marginally under streaming. This is different in films containing polyelectrolytes. The conformation changes drastically in a streaming field and the whole intermolecular structuring (i.e. the transient network, Section 10.4.1) might be destroyed. Hence SFA seems not to be a suitable method for studying forces of aqueous films containing polyelectrolytes. Therefore, only CP-AFM and TFPB measurements are considered.

10.4 Effect of Confinement

In the following, the influence of confinement on the structuring of polyelectrolytes within an aqueous solution, which is entrapped within a thin film, will be described. The effect can be separated with respect to two regions of the film, the *interfacial regions* h_s and the *film core* ($h_c = h - 2h_s$). The first effect is described in Section 10.6 and the latter is discussed in the present section. Section 4.1 focuses on the intermolecular interactions and the comparison between the characteristic lengths of the bulk and within the film. Section 4.2 deals mainly with intramolecular interactions which are monitored by fluorescence measurements in FTFPB.

10.4.1 Comparison of Polyelectrolyte Structuring Within Bulk and Film

The structuring of polyelectrolytes within the film core is related to structural forces. Therefore, in the following, results of CP-AFM measurements are presented. The fit of the oscillatory force–distance curves delivers the period d, the decay length λ and the amplitude A as shown in Fig. 10.5. The three parameters

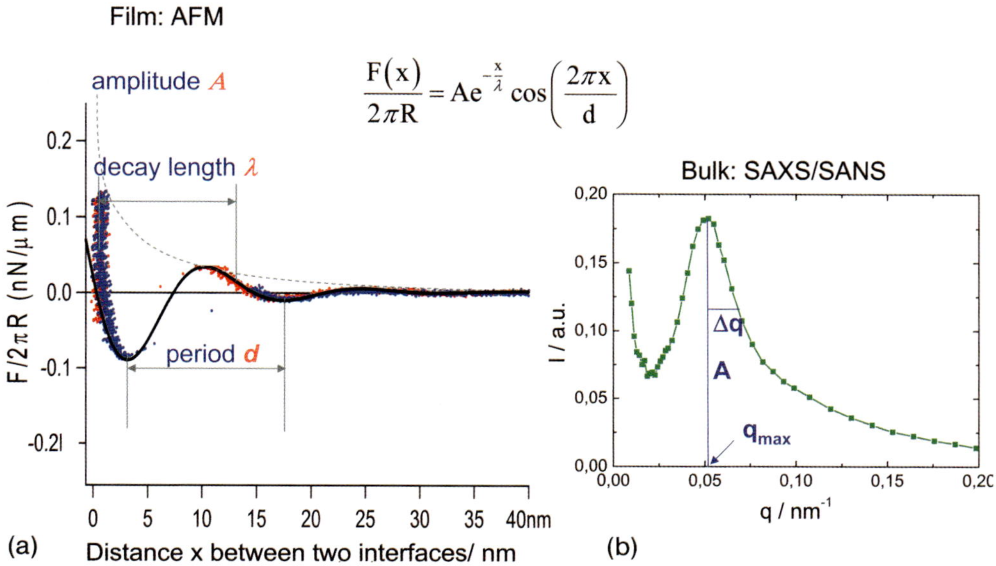

Fig. 10.5 (a) Oscillatory force curve measured with CP-AFM. The three fitting parameters d, λ and A are indicated. (b) The three fitting parameters q_{max}, Δq and A are indicated.

are compared to the characteristic lengths of the respective solution bulk, gained by small-angle scattering [small-angle X-ray scattering (SAXS) or small-angle neutron scattering (SANS)]. The scattering spectra show a broad peak, which clearly indicates interactions between the chains [13–17, 32, 43, 44]. With increasing polyelectrolyte concentration, the peak is shifted to larger q values (q=momentum transfer). Under the assumption that the Bragg relation is valid, a length ξ can be assigned to the peak position q_{max} via $\xi = 2\pi/q_{max}$. This length corresponds to the distance between polyelectrolyte chains and will be compared with the period d of the force oscillation occurring within the film. A measure of the correlation of this length is given by the decay length λ of the force oscillation in the film and the width Δq of the structure peak from bulk measurements. The latter can be converted into a length η via $\eta = 2/\Delta q$. The amplitude of the force oscillation and of the structure peak gives information about the strength of the potential, which is affected by both the intermolecular potential and the surface potential. The amplitude will be discussed in Section 10.6.

Figure 10.6 shows the comparison between the chain distance in the film and in the respective bulk solution of the same polyelectrolyte. The chain distance given by d and $2\pi/q_{max}$ is the same in the film and in the bulk. The same is valid for the correlation lengths λ and $2/\Delta q$. From this result, it is concluded that the polyelectrolyte concentration in the film and in the bulk is the same. Hence there is no effect of confinement on the characteristic lengths (i.e. chain dis-

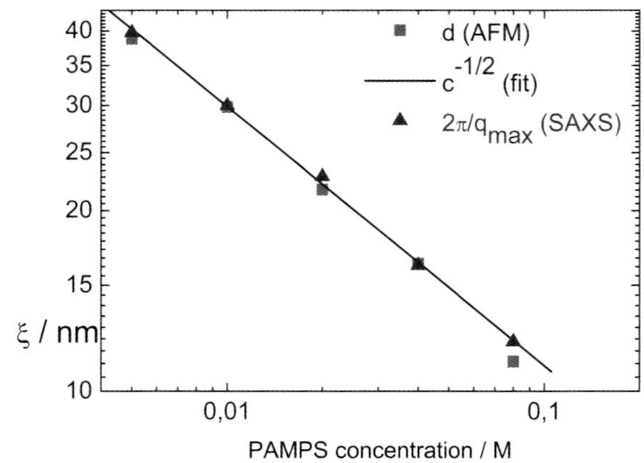

Fig. 10.6 Comparison between the period of force oscillation from film measurements and the chain distance in bulk solution. (Adapted from [45]).

tance and correlation length). The good agreement between the structuring in bulk and within film was shown for the first time in [43] and has since been described in other publications [32, 44].

In the following, the molecular model for the structuring of the polyelectrolyte chains will be discussed, which is strongly related to the chain distance. The correlation length will be in the center of focus in Section 10.5.2. In the measurements shown in Fig. 10.6, a linear polyelectrolyte was investigated. The chain distance scales with the polyelectrolyte concentration c_p as $c_p^{-1/2}$. This scaling law was also found for all other investigated linear polyelectrolytes in foam films, for both polyanions [30, 44, 46] and polycations [29, 32, 33]. There, not the full period but steps in film thickness are analyzed (see Section 10.3).

With respect to the stratification of films containing hard colloidal particles where the steps in film thickness are somehow related to the diameter of the particles, the question arises of which characteristic length corresponds to the period of the oscillatory curve. In analogy with the systems discussed in Section 10.2.2, one could assume that the period is related to the radius of gyration. If the radius of gyration were the step determining size, the step size would have to change with the degree of polymerization. However, the TFPB measurements show *no effect of the molecular weight* on the step size (see Fig. 10.7), which means that a layering of polyelectrolyte coils cannot be the right model. The similarity of the characteristic lengths in the film and in the bulk indicates that the interactions between the polyelectrolyte chains which induce structuring of the chains in the aqueous solution are the same as the interactions in the film which lead to film stratification.

For the bulk solution, models already exist for the structuring of polyelectrolyte chains. It is assumed that the structure peak is related to the mesh size of a transient network of overlapping polyelectrolyte chains [47–51]. Assuming that

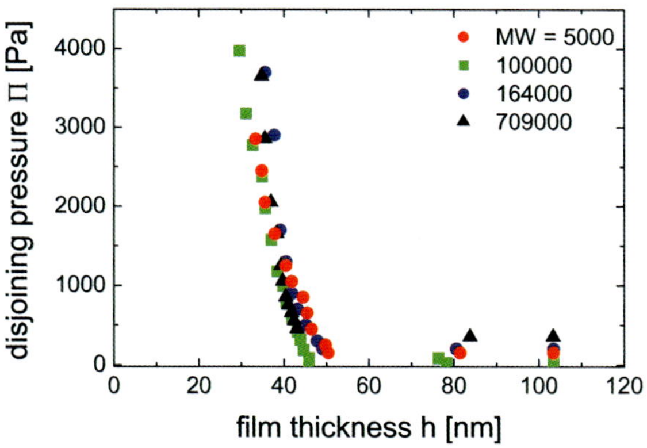

Fig. 10.7 Disjoining pressure isotherms of free-standing $C_{12}G_2$–PDADMAC films at different molecular weight of PDADMAC. The PDADMAC concentration was 5×10^{-3} monomol L^{-1}, and the $C_{12}G_2$ concentration was 10^{-5} mol L^{-1} [32].

the polyelectrolyte chains form a network-like structure in aqueous solution, this structure can obviously be transferred to the film bulk.

Furthermore, at low concentrations the surfactant has no effect on the step size, which means that the interactions between polyelectrolytes and surfactant are negligible. This has also been observed by SANS measurements in aqueous polyelectrolyte solutions, where the position of the structure peak is not influenced by the addition of surfactant [32, 44].

To summarize, neither the chain distance nor the correlation length within the polyelectrolyte solution changes under confinement from bulk to film. An explanation could be image charges at the outside of both film surfaces, which extends the network towards "infinity" and makes it isotropic, as it is in the bulk solution [52].

The question is whether there is a confinement effect on another scale. The occurrence of structural forces itself is a confinement effect. They are caused by the oscillatory concentration profile of polyelectrolytes [53, 54]. In terms of a transient network, that means a higher ordering of the network near the film interfaces with regions parallel to the interface of high concentration separated by regions of low concentration [55]. The meshes then form layers. Theoretical simulations showed a dominant orientation of chains parallel to the surfaces, although the mean chain distance remains the same [56]. Far from the interfaces the chains have a mean distance (mesh size), but the network is irregular as in bulk solution, which gives a flat concentration profile. Therefore, thick films, where the bulk properties dominate, do not show any oscillatory forces. When the film thins, the interfacial regions dominate and the force oscillation becomes more pronounced.

10.4.2
Local Polyelectrolyte Concentration

The previous section rises to two questions: (1) how does the concentration profile within the film look like? and (2) what happens to polyelectrolyte chains during force oscillation and the stepwise thinning? Are they expelled like spheres or does the network collapse mesh by mesh? For both studies, fluorescence measurements at foam films were carried out.

To address the second question, polyelectrolytes were labelled with fluorescein [33]. The fluorescence measurements were carried out in an FTFPB. Spectroscopic measurements show that the fluorescence intensity decreases abruptly during a step in film thickness. This indicates an expulsion of polyelectrolyte chains during each step. In comparison with layered sphere-like particles or molecules described in Section 10.2.2, the expulsion of polyelectrolyte chains is less insightful. However, as one polyelectrolyte chain may be involved in several meshes, it is not simply one chain that is expelled. The idea is that the network breaks down and is rebuilt much faster than the time resolution of the TFPB [32]. Up to a certain pressure, the network rebuilds itself with n meshes in the film, and above this pressure the network is only rebuilt with $n-1$ meshes. Hence, before the transition the total film thickness equals h_t, which is reduced to $h_t - \xi$ after one step. The total film thickness before and after a step is determined by

$$h_t = 2h_s + h_c = 2h_s + n\xi \tag{14}$$

$$h_t - \xi = 2h_s + (n-1)\xi \tag{15}$$

where h_c is the thickness of the film core and h_s is the thickness of one surface layer. The polyelectrolyte chains which are no longer involved in the network are pressed out of the film. Therefore, the polyelectrolyte concentration is reduced in comparison with the concentration in the surrounding meniscus, which induces (attractive) depletion forces between the film surfaces so that the film thins out stepwise.

In order to study the polyelectrolyte distribution within the film, pyrene-labeled polyelectrolytes were used. Pyrene tends to form excimers either between different chains (intermolecular interactions) at high polymer concentrations or along one chain (intramolecular interactions) in the case of chain coiling and high degree of labeling (typically more than 2% labeled monomers). The *mean* concentration remains constant during film thinning, since no change in the period of the force oscillation is detectable, as mentioned in the previous section. If the polyelectrolyte concentration profile were flat within the film, the excimer: monomer ratio should stay constant during film drainage. However, the excimer: monomer ratio increases with decreasing film thickness. In the case of charged polyelectrolytes, this happens stepwise directly related to steps in film thickness. Under the assumption of constant *mean* concentration, this can be only explained by changes in *local* concentration. It is assumed that the increase in excimer: monomer ratio reflects the oscillatory concentration profile. Regions with higher polyelectrolyte concentration gives a higher excimer: monomer ratio and they dominate the spectrum due to their higher integrated intensity. These measurements clearly show a confinement effect, which is related to an oscillatory concentration profile, leading in turn to oscillatory forces, as mentioned above.

10.5 Effect of Different Parameters on Film Forces

10.5.1 Molecular Architecture

10.5.1.1 Amphiphilic Diblock Copolymers

In addition to surfactants, low molecular weight amphiphilic triblock copolymers [57, 58] of the type ABA (A, hydrophilic; B, hydrophobic) or amphiphilic diblock copolymers [59–61] of the type AB stabilize free-standing films. From X-ray reflectometry and ellipsometry experiments, it is known that the hydrophobic part collapses at the air/water interface, and the hydrophilic part adopts a brush-like conformation directed towards the solution [62, 63]. Such a brush-like structure is also assumed at the surface of a PSSH–PEE film [29, 64]. If the two opposing brushes are pressed together in a TFPB experiment, the disjoining pressure increases exponentially (Fig. 10.8), which indicates that the brush layers are not in contact. Otherwise, an increase as $\Pi(h) \propto 1/h$ would be predicted [66]. For the example of PSSH–PEE, a 22.5-nm thick interfacial layer h_0

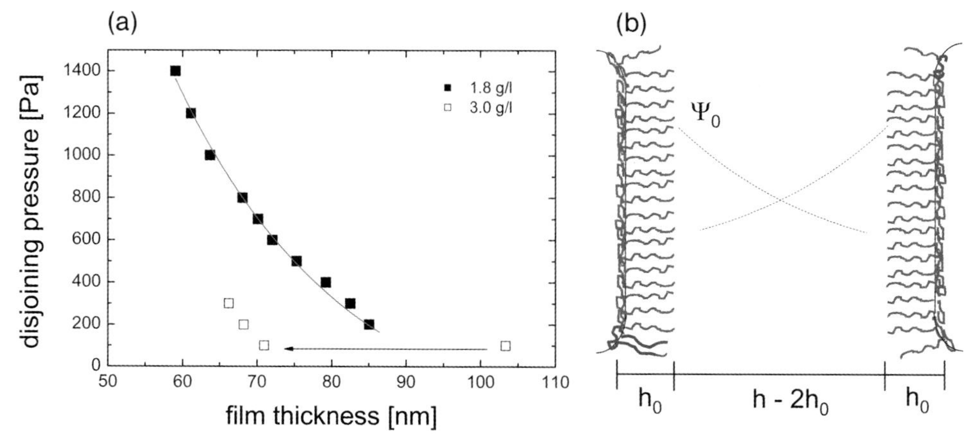

Fig. 10.8 (a) Disjoining pressure isotherms of PSSH–PEE at concentrations of 1.8 and 3 g L^{-1}. The solid line corresponds to an exponential fit of the experimental data according to Eq. (1). (b) Scheme of the opposing brushes. (Adapted from [65]).

is assumed [63], and the corresponding exponential fit results in a surface potential (at the PSSH tails) of about 22 mV. The fitted Debye length of 15 nm means that about 20% of the counterions are free and that most of the counterions are entrapped within the polymer brushes.

With increasing ionic strength, the film becomes thinner due to the screening of the electrostatic repulsion between the equally charged film surfaces (this is also observed for films of small surfactant molecules). In addition, the charged hydrophilic blocks coil with increasing ionic strength, thereby also contributing to the thinning of the film. The film thickness decreases with the salt concentration c_S as $c_S^{-0.18}$ instead of $c_S^{-1/3}$ as predicted by Pincus [67]. One explanation for the mismatch between experiment and theory could be that the theoretical model is based on volume-free counterions.

This class of block copolymers behaves like large surfactant molecules. Above a certain polymer concentration a stepwise thinning occurs (see Fig. 10.8), and the step size corresponds to the diameter of a micelle in the corresponding solution [64, 65]. These findings suggest the assumption that there is a layer of diblock copolymer micelles, entrapped within the film, being squeezed out. The structuring of the polymer micelles within the film seems to be weaker than the structuring of micelles consisting of low molecular weight surfactant, since the pressure which is needed to induce a step in film thickness is much lower.

10.5.1.2 **Branched Polyelectrolytes**

Free-standing films which are formed from aqueous surfactant solution containing branched PEI show stepwise thinning. With increasing polymer concentration, the step size decreases and the number of steps increases. Figure 10.9 shows the step size as a function of the corresponding concentration of monomer units for two molecular weights. The step size Δh scales with the polyelec-

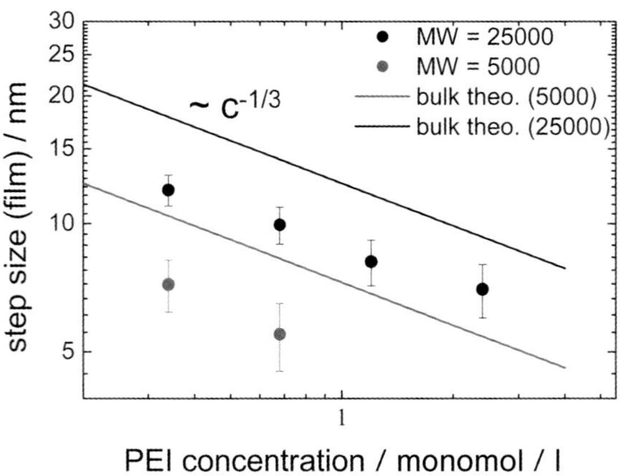

Fig. 10.9 Step size of film thickness as a function of PEI concentration. The solid lines are calculated distances in solution. (Data from [52]).

trolyte concentration c as $c^{-1/3}$ [52, 68]. In analogy with the spherical particles discussed above, a layer-by-layer expulsion of PEI molecules is assumed. The exponent –1/3 is explained by geometric arguments: with increasing concentration of the branched polyelectrolytes in a three-dimensional system the distance between two molecules decreases proportional to $c^{-1/3}$ in every dimension. At a fixed concentration of monomer units, the number of PEI molecules decreases with increasing molecular weight. The diameter is about 5 nm for MW = 25 000 [69] and 3 nm for MW = 5000. This leads to an increase in the intermolecular distance and therefore to an increase in the step size at higher molecular weight. PEI is a weak polybase. At a pH 9 (PEI in Milli-Q water) about 17% of the monomer units and at a pH 4 (adjusted with HCl) about 74% of the monomer units are charged (measured by streaming potential). With decreasing polymer charge density, the film stratification becomes less pronounced. The pressure which is needed to induce a step is lower and the film thickness between two steps cannot be stabilized any longer at pH 9. This leads to the conclusion that the layering of the PEI molecules is less ordered due to decreasing intermolecular electrostatic repulsion. This finding is also supported by drainage curves at these two pH values. The thinning velocity at pH 4 (i.e. higher polymer charge density) is much higher than that at pH 9. Due to electrostatic intermolecular repulsion, the expulsion of the polyelectrolytes from the film is much faster at pH 4 than at pH 9. Similar results are found for branched with respect to linear polyelectrolytes. For linear PEI, the disjoining pressure isotherm is less structured and the drainage process is slower than for branched PEI. This could be due to a lower charge density in the case of the linear PEI. The number of charged monomers at a certain pH is almost independent of the degree of branching, but the volume of a branched PEI molecule is much smaller than that of a swollen coil of linear PEI. In addition, the linear chains interdigitate as described in the following section, which could again slow down the drainage process.

These results support the assumption that the oscillatory forces are caused by electrostatic repulsion between the layered polyelectrolytes. On the other hand the $c^{-1/3}$ dependence indicates that the step size is determined by the geometry but not by electrostatics. A simple model (see Fig. 10.10) might illustrate how geometric arguments could be combined with electrostatic arguments. It is assumed that the step size Δh is

$$\Delta h = d + 2\kappa^{-1} \tag{16}$$

where d is the diameter of the PEI molecules. The diameter of one PEI molecule of molecular weight 5000 is about 3 nm. The measured step size at a PEI monomer concentration of 0.34 monomol L^{-1} is about 7 nm. This corresponds to $\kappa^{-1} \approx 2$ nm, which means an ionic strength of 0.025 mol L^{-1}. On the other hand, a PEI monomer concentration of 0.34 monomol L^{-1} (74% of the monomer units are charged at pH 4) corresponds to a concentration of monovalent counterions (i.e. ionic strength) of 0.25 mol L^{-1}. This difference in ionic

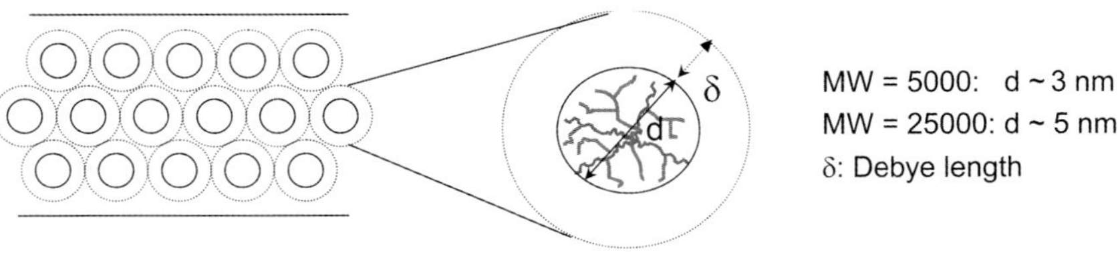

Fig. 10.10 Model for the layering of branched PEI molecules in a slit pore: combination of geometric and electrostatic arguments.

strength means that only 10% of the counterions are free and that the majority are entrapped within the branched molecules. This is in good agreement with theoretical predictions [70]. One has to take into account that the ratio between free and entrapped counterions may change with the polymer concentration.

10.5.1.3 **Rigidity of the Polymer Backbone**

The stratification behavior of liquid films containing polyelectrolytes is observed for many polyelectrolytes, not only for flexible and semi-flexible chains [71] but also for worm-like micelles of gemini surfactants [72], which can be regarded as "living" polymers. However, there seems to be a trend that the pressure which is needed to induce a step in film thickness is reduced with increasing persistence length. For instance, films containing the food stabilizer xanthan show stratification at a disjoining pressure of about 100 Pa whereas for films containing the much more flexible PSS a pressure of several hundred Pa is needed. A similar observation for the reduction in pressure amplitude was observed for double-stranded DNA [73]. In terms of oscillatory force measurements, that would mean that either the amplitude or the decay length is reduced. Since the ionic strength is the same, it is assumed that the amplitude decreases with increasing persistence length. The reasons for that have not been clarified so far. One could speculate that due to the greater stiffness the DNA and xanthan chains are more strongly oriented in parallel to the interface and that they lose their bulk structuring. In addition, the film becomes less stable with increasing persistence length. It is assumed that it is more difficult for the stiffer molecules to follow fluctuations of the interface.

10.5.2
Electrostatics

The structuring is strongly correlated with the ion distribution around the polyelectrolyte chains. The ion distribution is determined by both the counterion concentration, regulated by the polyelectrolyte charge density, and the distribution of external ions, determined by the addition of salt. In addition to the chain distance, the correlation length will also be discussed.

10.5.2.1 Polymer Charge Density

Figure 10.11 shows the results of the same system already presented in Fig. 10.6. Above a polymer charge density of 45%, the chain distance (i.e. the period of oscillation d and $2\pi/q_{max}$) is constant. This can be understood by the condensation of a certain ratio of counterions. The minimum distance between two identically charged free ions in a solvent, i.e. the Bjerrum length l_B, is determined by the compensation between electrostatic and thermal energy:

$$l_B = \frac{e^2}{4\pi\varepsilon_0\varepsilon_r kT} \tag{17}$$

The Bjerrum length is 7.1 Å in water at 298 K. The distance b between two neighboring charges along a strongly charged polyelectrolyte chain is shorter than the Bjerrum length. If $b < l_B$, counterion condensation takes place [74]. One concept of counterion condensation is that according to Manning [75, 76].

It is based on four assumptions:

- The interactions between different polyelectrolytes are negligible.
- The chain is locally stretched, i.e. the distance between two charges is an integer multiple of the monomer length.
- The free non-condensed counterions are treated by the Debye-Hückel theory.
- The dielectric constant is the one of the pure solvent and is uniform (mean field theory).

The Manning counterion condensation takes place up to an *effective* charge distance of the Bjerrum length. This means that the effective charge distance b_{eff} is constant above a polymer charge density of $1/l_B$. In other words, the charge density is *renormalized* to the value of $1/l_B$ by the counterion condensation.

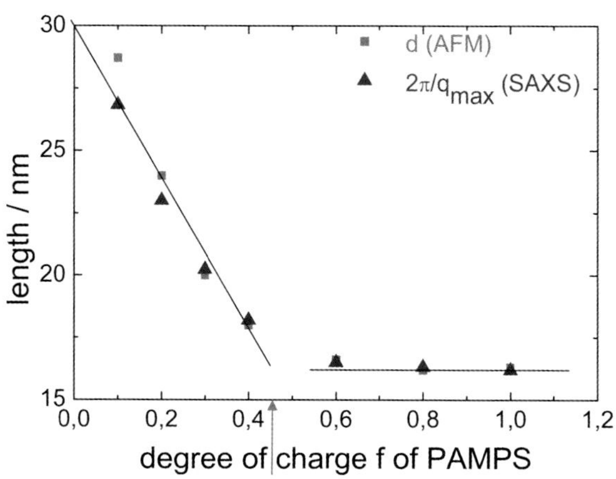

Fig. 10.11 The period of force oscillation from film measurements and the chain distance in bulk solution as a function of degree of PAMPS charge. The arrow indicates the Manning threshold. (Adapted from [45]).

Hence one can distinguish between condensed counterions and free counterions in the solution. Below a polymer charge density of $1/l_B$ the charge density corresponds to the nominal charge density determined by the chemical structure. At one end of the spectrum are the strongly charged polyelectrolytes ($b < l_B$) and at the other end the weakly charged polyelectrolytes ($b \gg l_B$). In the former case a single polyelectrolyte chain is assumed to be more swollen whereas in the latter case the chain is assumed to be partially collapsed [50]. The size of one blob is determined by the balance between electrostatic energy due to the repulsion between the charges and the thermal energy. (The term "blob" is used with different meanings in polymer science. Here, the blobs are located along *one* chain due to a partial coiling related to charge screening. In the literature, often the mesh of the transient polymer network is replaced by a "blob". This "blob" is a measure of the mean distance between *different* chains.) A similar behavior is predicted for chains in a poor solvent [77], where the balance between electrostatic energy and interfacial energy of the blobs against the solvent becomes important [78]. The transition from a swollen chain to a collapsed one, e.g. in a poor solvent, is also assumed to take place by the formation of an increasing number of blobs along the chain [79].

For the system presented in Fig. 10.11, the charge threshold, where Manning condensation starts, is about 45%. Below a charge density of about 45%, the mean chain distance increases with decreasing polyelectrolyte charge density due to stronger coiling as discussed.

As mentioned above, the Manning model is limited, since it does not take into account the interactions between different chains. It is clear that the interactions between different chains are not negligible. For instance, the free counterions act like a salt, which means that with increasing polyelectrolyte concentration the chains become more coiled. Also, the counterion condensation is affected by neighboring chains and is inhomogeneous along the chain [80, 81], which could induce the formation of blobs as discussed above.

For the present system, the Manning condensation holds on a scale of the chain distance, but what happens on a scale of the correlation length [i.e. decay length λ (film) and $2/\Delta q$ (bulk)]? Figure 10.12 shows a comparison between the chain distance and the correlation length. The chain distance follows, as already discussed, the Manning concept, which is indicated by coincident curves for a polymer charge of 60 and 100%. The decay length is smaller than the chain distance and differs for 60 and 100%. This means that the differently charged polymers attract the ions in a different way.

To summarize, within a cylinder of a radius d around the chain, the Manning condensation concept holds, which means that all "condensed ions" are located inside this cylinder. The charge regulation leads to the same electrostatic interactions on a scale of the chain distance. On a shorter length scale of the correlation length, the Manning concept is no longer valid, which means that not all "condensed ions" are within this distance around the chain, and that they are not condensed in the truest sense. Hence, condensed and free ions cannot be sharply distinguished. Otherwise, the Manning model has to hold on length

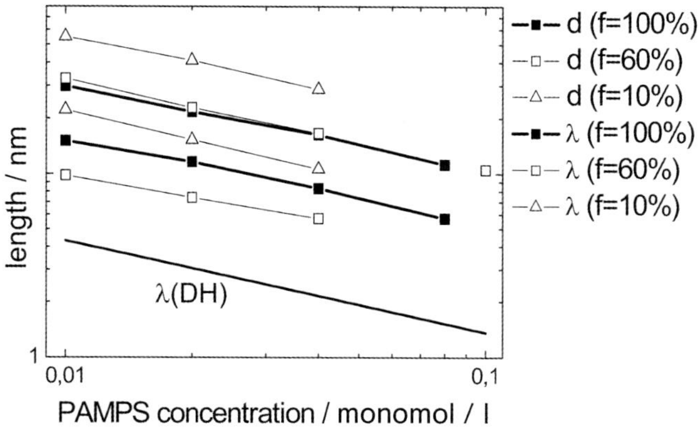

Fig. 10.12 Comparison between the chain distance and the correlation length at different degrees of charge as a function of PAMPS concentration. (Adapted from [45]).

scales even smaller than λ and $2/\Delta q$. The presented measurements give an insight into the ion distribution. Again, the ion distribution is similar in bulk and in the film, and no confinement effect can be detected.

Also in foam films the step size remains constant above the Manning threshold and increases with decreasing degree of polymer charge below the threshold. In addition, the pressure which is necessary to induce a step in the film thickness is reduced. This is due to the decreasing decay length, found in CP-AFM measurements. At low degree of charge the stratification vanishes [29]. This leads to the conclusion that the electrostatic repulsion between the polyelectrolyte chains is reduced and that the structuring becomes "softer" at a low polymer charge density.

10.5.2.2 **Ionic Strength**

Increasing the ionic strength has a similar effect on the correlation length as decreasing the polymer charge density. The correlation length decreases due to screening of the electrostatic repulsion between the polyelectrolyte chains [42]. Again in foam films the stratification of the film occurs at lower pressures with increasing ionic strength (Fig. 10.13) [29]. It is noteworthy that this effect is already observed at salt concentrations below the concentration of polyelectrolyte counterions. Usually, in solution, salt effects become important only at salt concentrations above the concentration of polyelectrolyte counterions. In contrast to the effect of polymer charge density, an increase in ionic strength up to the polyelectrolyte concentration does not influence the chain distance either in the bulk or in the film. This phenomenon has not yet been clarified. At higher salt concentrations, the structure peak cannot be resolved due to strong forward scattering (i.e. at low q values) related to an increasing osmotic compressibility and a decreasing osmotic pressure [32].

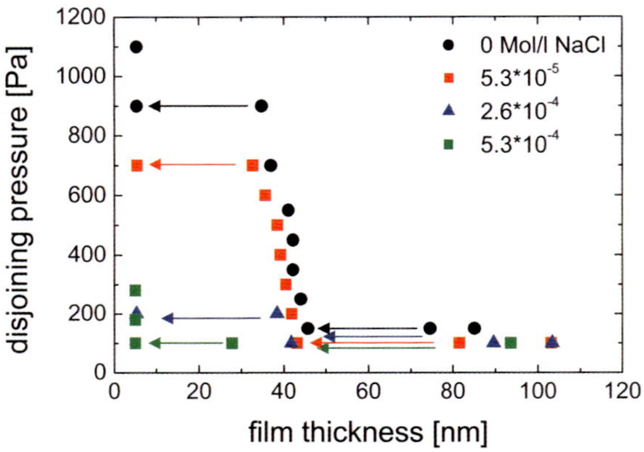

Fig. 10.13 Disjoining pressure isotherms of $C_{12}G_2$–PDADMAC films at different NaCl concentrations. (Adapted from [29]).

10.6
Effect of Film Interface Properties

So far, we have focused on the structuring of polyelectrolyte chains within the film core irrespective of the composition of interfaces. Now the influence of the properties of surfaces on the forces in nanofilms will be considered.

10.6.1
Charge of Interface

10.6.1.1 **Modification of Solid Interfaces**

The absence of oscillatory forces in an SFA raised the question of the influence of surface charge at the film interface on the structuring of polyelectrolytes within a thin liquid film. In an SFA, mica interfaces approach each other and in a CP-AFM usually a silica and a silicon surface approach each other. In order to compare the interfaces, the silicon wafer was coated with a mica sheet in a CP-AFM. The results in Fig. 10.14 show that the coating with mica gives a greater oscillation amplitude due to the higher surface potential of about –150 mV [82] against water in comparison of –80 mV in the case of silica [42]. The decay length is more or less the same. This shows clearly that the surface charge dominates the amplitude and the effect of an increase in counterion concentration, related to an increase in surface charge, is marginal.

10.6.1.2 **Variation of Polyelectrolyte–Surfactant Composition in Foam Films**

The choice of the surfactant has no detectable influence on the structuring of the polyelectrolyte chains within the *film core*, which leads to a fixed period of force oscillation at a certain polyelectrolyte concentration independent of the choice of the surfactant [28, 88]. On the other hand, the total film thickness is

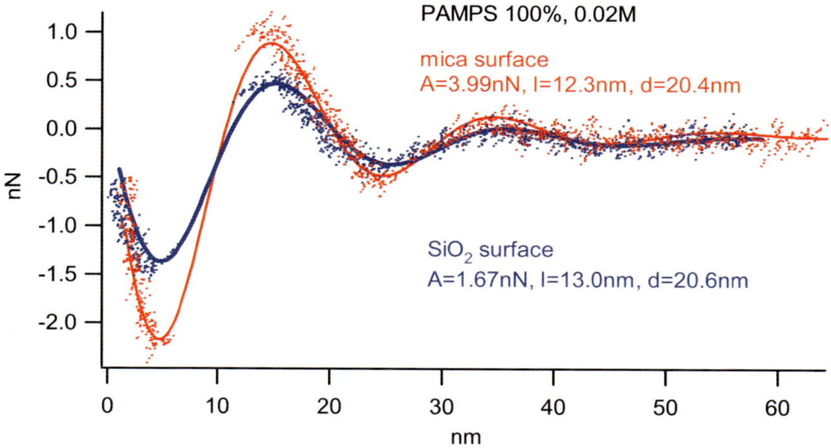

Fig. 10.14 CP-AFM curves of aqueous PAMPS film between a silica bead and either a silcon wafer or a mica sheet [42].

different for different surfactants, which leads to the assumption that the interactions between polyelectrolyte and surfactant at the *film interfaces* are not negligible. To investigate the interactions between both classes of compounds, different polyelectrolyte–surfactant combinations have been investigated [88]. The choice of the surfactant affects the total film thickness at higher pressure. Either an NBF with a thickness of about 4 nm or a CBF of about 30 nm occurs. The CBF is very stable and no transition to an NBF has been observed up to a pressure of several thousand Pa. In contrast, the NBF is not stable and it breaks after a few minutes. The results are summarized in Table 10.1.

Obviously, only the combination of a positively charged polyelectrolyte with a non-ionic surfactant results in an NBF. All other combinations result in a CBF.

The CBF is electrostatically stabilized. In films of ionic surfactants, pronounced repulsion between the surfaces takes place. In the case of equally charged polyelectrolytes and surfactants, both compounds repel each other,

Table 10.1 Type of the limiting free-standing film observed for specific combinations of polyelectrolyte and surfactant charge. Comparison with results obtained by other groups. (Adapted from [88, 89]).

	Non-ionic surface	Cationic surface	Anionic surface
Polyanion	CBF [28, 44, 71, 83, 84, 88]	CBF [28, 30, 46, 71, 83]	CBF [44, 84]
Polycation	NBF [29, 32, 33, 43]	CBF [32, 33, 68]	–
Non-ionic polymer	CBF [29]	–	CBF [85, 86]

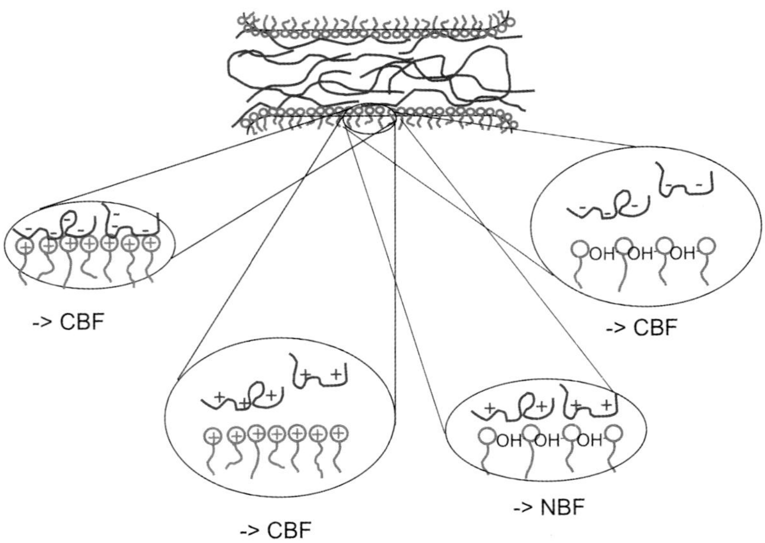

Fig. 10.15 Scheme of the structure of film interfaces for different surfactant–polyelectrolyte combinations.

which enhances the overall repulsion within the film. Surface tension measurements indicate strong complexation of oppositely charged surfactant and polyelectrolyte at the air/liquid interface [87, 90]. It is assumed that a charge reversal at the interface occurs, which again leads to electrostatic repulsion between the interface and the polyelectrolyte within the film core. Both situations lead to the formation of a CBF at high pressure. The only combination where the film interfaces and the polymers are oppositely charged is in films of non-ionic surfactant and positively charged polyelectrolytes. This can be understood by the fact that the free water interface is negatively charged [35] due to the preferred adsorption of OH^- ions [91]. The non-ionic surfactant partially replaces the OH^- ions, but the surface remains negatively charged at low surfactant concentrations [92]. The polycation is attracted by the interface, but no complexes are formed, as shown by surface tension measurements [88]. This could reduce the surface potential and therefore the electrostatic repulsion, and a transition to an NBF occurs. Schemes of different film structures are shown in Fig. 10.15. The formation of surface-active polyelectrolyte–surfactant complexes increases the film stability, since the pure surfactant films are not stable at these low concentrations [30, 46]. However, polyelectrolytes stabilize the foam films even in the absence of surface-active complexes, which is the case if both compounds are identically charged or if the surfactant is neutral. In this case, the stabilizing effect is due to oscillatory forces, which might hamper the film drainage [12].

10.6.2
Wetting Films Containing Polyelectrolyte Solutions

So far, only symmetrical films with either two fluid or two solid interfaces have been discussed. In the following section asymmetric films with one solid and one fluid interface are considered. PSS solutions were entrapped within the wetting film [93]. In Fig. 10.16, a disjoining pressure isotherm of a PSS film is shown.

In contrast to the observation in symmetrical films, only one step occurs irrespective of the polymer concentration. The step in film thickness occurs at very low pressure. This suppression of steps could be due to the asymmetric streaming profile in a wetting film which stretches the polyelectrolyte chains differently in different film positions perpendicular to the interfaces. The assumed correlation between the asymmetry in streaming profile and the suppression of steps has not been proven and understood so far, and will be a topic of future studies.

10.7
Summary and Conclusions

Different methods for force measurements in nanofilms have been presented. According to the particular center of focus, each method has its advantages. CP-AFM is the optimum technique to study structural forces, since the whole oscillation is available. A comparison with small-angle scattering shows that in the semi-dilute regime the chain distance and the correlation length remain the same as in the respective bulk solution. The response of the polymeric systems to certain inner parameters such as ionic strength, polymer charge degree and molecular architecture are the same within the film and in the bulk. From this

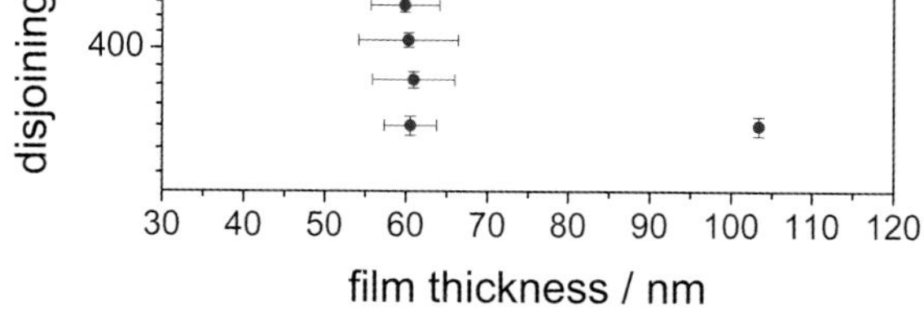

Fig. 10.16 Aqueous wetting film against silicon wafer [93].

point of view, a confinement effect is not detectable. Fluorescence measurements give a hint for an inhomogeneous (oscillatory) concentration profile of the polyelectrolytes within the film. This in turn leads to oscillatory forces which present the real confinement effect. Both for fluorescence measurements and in studies where the surface has to be modified, the thin-film pressure balance is the better method, since the surfaces can be easily varied. The final film thickness of a foam film can be tuned by the polyelectrolyte–surfactant combination while the structural forces are not affected by the choice of the surfactant.

The results show clearly that in liquid polyelectrolyte nanofilms interfacial effects can be separated from film core properties.

Acknowledgments

Tommy Nylander, Dominique Langevin, Jan Skov Pedersen are acknowledged. We thank the DFG (Schwerpunkt 1009, "Polyelektrolyte mit definierter Molekülarchitektur") for financial support.

References

1 B.V. Derjaguin, *Kolloid Z.* (**1934**) *69*, 155.
2 A. Scheludko, D. Exerowa, *Commun. Dept. Chem. Bulg. Acad. Sci.* (**1959**) *7*, 123.
3 K.J. Mysels, M.N. Jones, *Discuss. Faraday Soc.* (**1966**) *42*, 42.
4 D. Exerova, A. Scheludko, *Chim. Phys.* (**1971**) *24*, 47.
5 D. Exerova, T. Kolarov, K.H.R. Khristov, *Colloids Surf. A* (**1987**) *22*, 171.
6 B. Diakova, M. Kaisheva, D. Platikanov, *Colloids Surf. A* (**2001**) *190*, 61.
7 B. Diakova, C. Filiatre, D. Platikanov, A. Foissy, M. Kaisheva, *Adv. Colloid Interface Sci.* (**2002**) *96*, 193.
8 D. Exerowa, N.V. Churaev, T. Kolarov, N.E. Esipova, N. Panchev, Z.M. Zorin, *Adv. Colloid Interface Sci.* (**2003**) *104*, 1–24.
9 B. Diakova, D. Platikanov, R. Atanassov, M. Kaisheva, Adv. Colloid Interf. Sci 104 (**2003**) 25.
10 I.B. Ivanov, D.S. Dimitrov, in I.B. Ivanov (Ed.) *Thin Liquid Films*, Marcel Dekker, New York, (**1988**) p. 379.
11 D. Exerowa, K. Kruglyakov, *Foam and Foam Films*, Elsevier, Amsterdam, **1998**.
12 R. v. Klitzing, H.J. Müller, *Curr. Opin. Colloid Interface Sci.* (**2002**) *7*, 18.
13 M. Nierlich, C. Williams, F. Boué, J.P. Cotton, M. Daoud, B. Farnoux, G. Jannink, C. Picot, M. Moan, C. Wolff, M. Rinaudo, P.G. de Gennes, *J. Phys.* (**1979**) *40*, 701.
14 S. Förster, M. Schmidt M., *Adv. Polym. Sci.* (**1995**) *120*, 51.
15 K. Nishida, K. Kaji, T. Kanaya, *Macromolecules* (**1995**) *28*, 2472.
16 W. Essafi, F. Lafuma, C.E. Williams, *Eur. Phys. J. B* (**1999**) *9*, 261.
17 P.M. Claesson, M. Bergström, A. Dedinaite, M. Kjellin, J.F. Legrand, I. Grillo, *J. Phys. Chem. B* (**2000**) *104*, 11689.
18 B.V. Derjaguin, L. Landau, *Acta Physicochim. (URSS)* (**1941**) *14*, 42.
19 E.J.W. Verwey, J.T.G. Overbeek, *Theory of the Stability of Lyophobic Colloids*, Elsevier, Amsterdam, **1948**.
20 J. Israelachvili, in J. Israelachvili J (Ed.) *Intermolecular and Surface Forces*, Academic Press, New York, **1998**, p. 266.
21 P. Richetti, P. Kekicheff, *Phys. Rev. Lett.* (**1992**) *68*, 1951.

22 V. Bergeron, C. J. Radke, *Langmuir* (**1992**) *8*, 3020.
23 A. A. Sonin, D. Langevin D., *Europhys. Lett.* (**1993**) *22*, 271.
24 V. Bergeron, P. M. Claesson, *Adv. Colloid Interface Sci.* (**2002**) *96*, 1.
25 A. Scheludko, *Adv. Colloid Interface Sci.* (**1967**) *1*, 391.
26 S. P. Frankel, K. J. Mysels, *J. Appl. Phys.* (**1966**) *37*, 3725.
27 D. Exerowa, A. Nikolov, M. Zacharieva, *J. Colloid Interface Sci.* (**1981**) *81*, 419.
28 R. v. Klitzing, A. Espert, A. Asnacios, T. Hellweg, A. Colin, D. Langevin, *Colloids Surf. A* (**1999**) *149*, 131.
29 B. Kolaric, W. Jaeger, R. v. Klitzing, *J. Phys. Chem. B* (**2000**) *104*, 5096.
30 A. Asnacios, A. Espert, A. Colin, D. Langevin, *Phys. Rev. Lett.* (**1997**) *78*, 4974.
31 R. v. Klitzing, A. Espert, A. Colin, D. Langevin, *Colloids Surf. A* (**2001**) *176*, 109.
32 R. v. Klitzing, B. Kolaric, W. Jaeger, A. Brandt, *Phys. Chem. Chem. Phys.* (**2002**) *4*, 1907.
33 J. L. Toca-Herrera, R. v. Klitzing, *Macromolecules* (**2002**) *35*, 2861.
34 D. Rapoport, K. Hänni-Ciunel, E.-P. Resewitz, R. v. Klitzing, Soft materials, submitted.
35 K. Ciunel, M. Armelin, G. Findenegg, R. v. Klitzing, *Langmuir* (**2005**) *21*, 4790–4793.
36 A. Scheludko, D. Platikanov, *Kolloid Z.* (**1961**) 175, 150.
37 W. A. Ducker, T. J. Senden, R. M. Pashley, *Nature* (**1991**) *353*, 239.
38 H.-J. Butt, *Biophys. J.* (**1991**) *60*, 1438.
39 A. J. Milling, *J. Phys. Chem.* (**1996**) *100*, 8986.
40 A. J. Milling, B. Vincent, *J. Chem. Soc. Faraday Trans.* (**1997**) *93*, 3179.
41 A. J. Milling, K. Kendal, *Langmuir* (**2000**) *16*, 5106.
42 D. Qu, R. v. Klitzing, unpublished results.
43 R. v. Klitzing, *Tenside Surf. Deterg.* (**2000**) *37*, 338.
44 O. Theodoly, J. S. Tan, R. Ober, C. E. Williams, V. Bergeron, *Langmuir* (**2001**) *17*, 4910.
45 D. Qu, J. S. Pedersen, S. Garnier, A. Laschewsky, H. Möhwald, R. v. Klitzing, *Macromolecules* (**2006**) *39*, 7364.
46 V. Bergeron, D. Langevin, A. Asnacios, *Langmuir* (**1996**) *12*, 1550.
47 P. G. de Gennes, P. Pincus, R. M. Velasco, F. Brochard, *J. Phys. (Paris)* (**1976**) *37*, 37.
48 T. Odijk, *J. Polym. Sci. Polym. Phys. Ed.* (**1977**) *15*, 688.
49 J. Skolnick, M. Fixman, *Macromolecules* (**1977**) *10*, 944.
50 A. V. Dobrynin, R. V. Colby, M. Rubinstein, *Macromolecules* (**1995**) *28*, 1859.
51 J. L. Barrat, J. F. Joanny, *Adv. Chem. Phys.* (**1996**) *94*, 1.
52 R. v. Klitzing, B. Kolarić, *Prog. Colloid Polym. Sci.* (**2003**) *122*, 122.
53 M. Schmidt, H. Löwen, *Phys. Rev. E* (**1997**) *55*, 7228.
54 A. Yethiraj, *J. Chem. Phys.* (**1999**) *111*, 1797.
55 D. Rapoport, D. Anghel, G. Hedicke, H. Möhwald, R. v. Klitzing, *J. Phys. Chem. B*, in press.
56 J. de Joannis, K. Kremer, C. Holm, unpublished results.
57 R. Sedev, D. Exerowa, *Adv. Colloid Interface Sci.* (**1999**) *83*, 111.
58 K. Khristov, B. Jachimska, K. Malysa, D. Exerowa, *Colloids Surf. A* (**2001**) *186*, 93.
59 A. Saint-Jalmes, A. A. Sonin, M. Delsanti, P. Guenoun, J. Yang, J. W. Mays, D. Langevin, *Langmuir* (**2002**) *18*, 2103.
60 P. Guenoun, A. Scalchli, D. Sentenac, J. W. Mays, J. J. Benattar, *Phys. Rev. Lett.* (**1995**) *74*, 3628.
61 B. Rippner, K. Boschkova, P. M. Claesson, T. Arnebrant, *Langmuir* (**2002**) *18*, 5213.
62 K. Schillén, P. M. Claesson, M. Malmsten, P. Linse, C. Booth, *J. Phys. Chem.* (**1997**) *101*, 4238.
63 H. Ahrens, S. Förster, C. A. Helm, *Phys. Rev. Lett.* (**1997**) *81*, 4172.
64 S. Förster, S. Hermsdorf, W. Leube, H. Schnablegger, M. Regenbrecht, S. Akari, *J. Phys. Chem. B* (**1999**) *103*, 6657.
65 B. Kolarić, S. Förster, R. v. Klitzing, *Prog. Colloid Polym. Sci.* (**2001**) *117*, 195.

66 H. Ohshima, *Colloid. Polym. Sci.* (**1999**) *277*, 535.

67 P. Pincus, *Macromolecules* (**1991**) *24*, 2912.

68 R. v. Klitzing, B. Kolarić, *Tenside Surf. Det.* (**2002**) *39*, 247.

69 T. Hellweg, N. Henry–Toulme, M. Chambon, D. Roux, *Colloids Surf. A* (**2000**) *163*, 71.

70 A. Jusufi, C. N. Likos, H. Löwen, *Phys. Rev. Lett.* (**2002**) *88*, 018301.

71 R. v. Klitzing, A. Espert, A. Colin, D. Langevin, *Colloids Surf. A* (**2001**) *176*, 109.

72 A. Espert, R. v. Klitzing, P. Poulin, A. Colin, R. Zana, D. Langevin, *Langmuir* (**1998**) *14*, 4251.

73 D. Qu, D. Langevin, unpublished work.

74 F. Oosawa, *J. Polym. Sci.* (**1957**) *23*, 421.

75 G. S. Manning, *J. Chem. Phys.* (**1969**) *51*, 924.

76 J. Ray, G. S. Manning, *Macromolecules* (**1997**) *30*, 5739.

77 A. V. Dobrynin, M. Rubinstein, S. P. Obukhov, *Macromolecules* (**1996**) *29*, 2974.

78 A. R. Khoklov, *J. Phys. A* (**1980**) *13*, 979.

79 U. Micka, C. Holm, K. Kremer, *Langmuir* (**1999**) *15*, 4033.

80 G. V. Ramanathan, C. P. Woodbury, *J. Chem. Phys.* (**1982**) *77*, 4133.

81 S. A. Safran, P. Pincus, M. E. Cates, F. C. Mackintosh, *J. Phys.* (**1990**) *71*, 503.

82 A. M. Stewart, presented at the 27th Annual Condensed Matter and Materials Meeting, 4–7 February 2003, Wagga Wagga, Australia.

83 J. Toca, B. Kolarić, D. Anghel, R. v. Klitzing, unpublished results.

84 C. Marques, S. Guillot, D. Langevin, personal communication.

85 F. Bouchama, J.-M. di Meglio, *Colloid Polym. Sci.* (**2000**) *278*, 195.

86 O. Krichewsky, J. Stavans, *Phys. Rev. Lett.* (**1994**) *73*, 696.

87 A. Asnacios, R. v. Klitzing, D. Langevin, *Colloids Surf. A* (**2000**) *167*, 189.

88 B. Kolarić, W. Jaeger, G. Hedicke, R. v. Klitzing, *J. Phys. Chem. B* (**2003**) *107*, 8152.

89 R. v. Klitzing, *Adv. Colloid Interface Sci.* (**2005**) *114/115*, 253.

90 A. Asnacios, D. Langevin, J. F. Argillier, *Macromolecules* (**1996**) *29*, 7412.

91 K. G. Marinova, R. G. Alargova, N. D. Denkov, O. D. Velev, D. N. Petsev, I. B. Ivanov, R. P. Borwankar, *Langmuir* (**1996**) *12*, 2045.

92 C. Stubenrauch, J. Schlarmann, R. Strey, *Phys. Chem. Chem. Phys.* (**2002**) *4*, 4504.

93 K. Hänni-Ciunel, *PhD Thesis*, Technical University of Berlin, **2006**.

11
Association Between Polyelectrolytes and Oppositely Charged Surfactants in Bulk and at Solid/Liquid Interfaces

Per M. Claesson, Andra Dedinaite, Róbert Mészáros, and Imre Varga

Abstract

Mixtures of polyelectrolytes and oppositely charged surfactants find applications in many processes and products that are used in our daily life. Such systems also show many interesting features from a scientific point of view. Due to the combined technological relevance and scientific challenge, considerable research efforts have been made in this area in recent years. This has resulted in new theoretical approaches, the development of simulation methods and new experimental techniques and, of course, a large body of new findings. Together, these efforts have increased the understanding significantly, especially during the last 5 years. This chapter reviews some aspects of this topic, focusing on bulk association and association at solid/liquid interfaces. The review is focused on developments during the 21st century, even though in some cases a historical perspective is also offered.

11.1
Introduction

Many of the products we use in our daily life, such as household care and personal care products, utilize the combined properties of polymers and surfactants. It may even be fair to say that the striving to find optimum use of the interactions between polymers and surfactants, combined with increasing environmental and health concerns, are the main factors that have driven, and still drive, development. To increase the understanding of polymer–surfactant interactions is scientifically challenging since the structural changes in both the polymer and the surfactant affect their interaction and the properties of the mixed system. This is not all, however, as also the properties of the solvent (e.g. pH, ionic strength, presence of divalent ions) have a pronounced bearing on the final properties of the mixed system. In most applications the situation is even more complex since more than one type of surfactant is used in combination

Colloids and Interface Science Series, Vol. 3
Colloid Stability and Application in Pharmacy. Edited by Tharwat F. Tadros

ISBN: 978-3-527-31463-8

with different classes of polymers and other additives, such as perfume. In addition, the mixture should often fulfill its purpose at surfaces, which means that also the interactions between the surface and all components and the mixed aggregates of the system have to be optimized. Hence it is not surprising that even though significant progress in understanding has been made, much of the technological development is due to trial-and-error or, at best, educated guesses.

In this chapter, we focus on polyelectrolytes mixed with oppositely charged surfactants, a topic that in itself is very broad and any single review cannot cover all aspects or even a large fraction of all interesting aspects that deserve attention. Fortunately, several reviews on this and related topics have been published previously (e.g. [1–8]), and in these the interested reader can find various other aspects presented and discussed. Our approach is first to discuss association phenomena in bulk solution and then turn to the more complex issue of polyelectrolyte–surfactant association at solid/liquid interfaces.

11.2 Bulk Association

In this section we first focus on how to determine the surfactant binding isotherm. Next, we review the different thermodynamic approaches used for interpreting the binding isotherm and for drawing important conclusions from them. Recently, theoretical simulations of polyelectrolyte–oppositely charged surfactant systems have provided new understanding and this progress is discussed in the following section. A separate section is devoted to the stability of polyelectrolyte–surfactant aggregates and the important issue of trapped non-equilibrium states is stressed. The section on bulk association ends with some information about the internal structure of polyelectrolyte–oppositely charged surfactant aggregates.

11.2.1 Determination of Binding Isotherms

One of the major sources of information about macromolecule–surfactant interaction is provided by the surfactant binding isotherms (the amount of surfactant bound to unit mass of polymer, B), which can be obtained from the surfactant mass balance as follows:

$$B = \frac{c_0 - c_{\text{free}}}{c_{\text{pol}}} \tag{1}$$

where c_{pol} is the polymer concentration and $c_0 = n_0/V$ and $c_{\text{free}} = n_{\text{free}}/V$ denote the analytical (total) and the free monomer surfactant concentration, respectively (n_0 indicates the total amount of surfactant present in the system, n_{free} is the amount of surfactant present in monomer form and V is the volume of the polymer–surfactant system).

According to Eq. (1), binding isotherm measurements require the determination of the monomer surfactant concentration (c_{free}), which is usually based on either indirect or direct activity measurements. In a typical experimental setup (indirect activity measurements) equilibrium is established between the polymer–surfactant system (Cell P/S) and a polymer-free surfactant solution (Reference system, Cell Ref.). The thermodynamic condition of equilibrium is the constancy of the mean chemical potential of the surfactant, which reduces to the equality of the mean surfactant activity in the two cells ($a_{\pm,P/S} = a_{\pm,ref}$, where $a_{\pm} = \gamma_{\pm}(c_+c_-)^{1/2}$ for 1:1 electrolytes, assuming that the pressure difference is negligible between the two cells).

For the sake of simplicity, assuming that the binding measurements are carried out in the absence of inert electrolytes ($c_{+,ref} = c_{-,ref}$) and taking into account that the polymer-free reference system is a dilute electrolyte solution ($\gamma_{\pm,ref} \approx 1$), it is easy to demonstrate that by measuring the surfactant *concentration in Cell Ref.* ($c_{Surf,ref}$), the surfactant *activity in the polymer–surfactant system* can be determined [$c_{Surf,ref}$ ($= c_{-,ref}$ or $c_{+,ref}$) $\approx a_{\pm,ref} = a_{\pm,P/S}$]. It should be noted that the above reasoning holds only if $c_{Surf,ref} <$ c.m.c. (critical micelle concentration) because otherwise $c_{Surf,ref} = c_{free,ref} + c_{mic} \neq a_{\pm,ref}$.

From an experimental point of view, one may either develop the equilibrium of the polymer–surfactant solution with the polymer-free reference systems by equilibrium dialyses or separate a portion of the polymer-free supernatant of the polymer–surfactant system, e.g. by ultracentrifugation or ultrafiltration. As a convenient alternative to indirect activity measurements, the surfactant activity can be measured directly in the polymer–surfactant system by means of surfactant-selective electrodes. Since several excellent reviews [9–12] are available on the technical details and limitations of the experimental techniques, we will focus our attention on the theoretical aspects of the conversion of experimental activity data to monomer surfactant concentration.

A widespread assumption in the literature is that since the surfactant concentration is rather low in the investigated polymer–surfactant systems, the interactions of the surfactant ions can be neglected ($\gamma_{\pm,P/S} \approx 1$). Hence the measured mean surfactant activity is practically equal to the monomer surfactant concentration and it can be used directly in Eq. (1). However, Gilanyi et al. [13] showed that this assumption is valid only in a few special cases. In the presence of macroions (e.g. surfactant aggregates, polyelectrolytes), the distribution of small ions (surfactants) in the solution is strongly affected. The ions having the same sign of charge as the macroions are expelled from their close vicinity, while the ions having an opposite sign of charge are accumulated around them. Taking into account the electrostatic interaction of the macroions and the surfactant ions Gilanyi et al. derived the following formal expression between the free monomer surfactant concentration in the polymer–surfactant system (c_{free}) and the surfactant activity (surfactant concentration in the reference system):

$$c_{free} = \frac{1}{V}\int^{V} c_{Surf,ref}\, e^{y}\, dV' = c_{Surf,ref}\langle e^{y}\rangle = a_{\pm,Surf}\langle e^{y}\rangle \qquad (2)$$

where $y = e\Psi/kT$ is the reduced electrical potential with a reference potential chosen as $\Psi = 0$ in the polymer-free reference system and the angle brackets denote volume averaging. Using the above expression, the binding isotherm can be given in the following form:

$$B_{\text{pol}} = \frac{c_0 - a_{\pm,\text{Surf}} \langle e^y \rangle}{c_{\text{pol}}} = \frac{c_0 - c_{\text{Surf,ref}} \langle e^y \rangle}{c_{\text{pol}}} \tag{3}$$

In order to determine quantitatively correct binding isotherms, the term $\langle e^y \rangle$ in Eq. (3) must be determined, which can be done by independent measurements.

Recently, Gilanyi et al. suggested a method for the determination of $\langle e^y \rangle$ [14, 15]. This method is based on activity measurements of a probe electrolyte added to the system in a trace amount. The probe electrolyte should meet the following requirements:

- The counterion (A) of the surfactant (Surf-A) and one of the ions of the probe electrolyte (AB) must be identical.
- The probe electrolyte cannot bind specifically to the polymer.

If the above conditions are met, the binding isotherm can be given in the following form:

$$B_{\text{pol}} = \frac{c_0 - (a_{\pm,\text{Surf}}{}^2 / a_{\pm,\text{AB}}{}^2) c_{0,\text{AB}}}{c_{\text{pol}}} = \frac{c_0 - (c_{\text{Surf,ref}} / c_{\text{B,ref}}) c_{0,\text{AB}}}{c_{\text{pol}}} \tag{4}$$

where $a_{\pm,\text{AB}}$ and $c_{0,\text{AB}}$ are the mean activity and the analytical concentration of the probe electrolyte, respectively, and $c_{\text{B,ref}}$ is the equilibrium concentration of the co-ion of the surfactant (B) in the polymer free reference system. Since all quantities on the right-hand side of Eq. (4) can be experimentally determined, it provides an exact way for the determination of the surfactant binding isotherms from activity measurements.

A possible alternative to trace probe electrolyte measurements is to determine the surfactant activity in the presence of a large amount of inert electrolyte. In this case the $\langle e^y \rangle$ term in Eq. (3) becomes practically equal to one and the measured surfactant activities can be directly used for the binding isotherm calculations. However, it should be noted that it is well documented in the literature that the presence of inert electrolytes has a significant effect on the polymer–surfactant interaction itself, so in these cases the investigated systems are different from the salt-free systems.

11.2.2 Thermodynamics of Association

Surfactant binding isotherms provide essential information about the nature of the polymer–surfactant interaction. However, this information can be extracted only in terms of appropriate thermodynamic and physical interaction models.

Thermodynamic models provide the general framework for the description of polymer–surfactant interaction. They are formulated by identifying the most significant components of the system and assigning a standard free energy change to their formation. Thermodynamic models can provide the distribution of the different components of the system as a function of the thermodynamic variables. In order to gain a deeper insight into the details of the polymer–surfactant interaction, an in depth picture of the structure and geometry of the complex molecule must be developed and the physical interactions of the building blocks (e.g. surfactant molecules, polymers, aggregates) must be described explicitly to determine the contribution of each interaction to the overall standard free energy change of the complex formation. In recent decades, several different approaches have been developed for the description of the macromolecule–surfactant interaction, which have been summarized in a number of reviews [5, 6, 9–12]. The major aim of this section is to provide a brief overview of the different approaches, focusing on the underlying physical picture embedded in them.

The most general framework for the description of the polymer–surfactant interaction is provided by the multiple-binding equilibrium model. The underlying assumption of this approach is that the surfactant molecules (S) interact with (bind to) the polymer (P) one by one and each step can be characterized by an equilibrium constant (k_i):

$$\begin{array}{ll} \mathrm{P} + \mathrm{S} \leftrightarrow \mathrm{PS} & k_1 = [\mathrm{PS}]/[\mathrm{P}][\mathrm{S}] \\ \mathrm{PS} - \mathrm{S} \leftrightarrow \mathrm{PS}_2 & k_2 = [\mathrm{PS}]/[\mathrm{PS}][\mathrm{S}] \\ \mathrm{PS}_2 + \mathrm{S} \leftrightarrow \mathrm{PS}_3 & k_3 = [\mathrm{PS}_3]/[\mathrm{PS}_2][\mathrm{S}] \\ \vdots & \vdots \end{array} \tag{5a}$$

Within the framework of this approach, the macromolecule–surfactant interaction is characterized when the set of the equilibrium constants and the driving force of each binding step [$k_i = \exp(-\Delta G_i^\circ/RT)$] is determined. Using the above notation, the surfactant binding isotherm can be easily expressed in terms of the equilibrium constants:

$$\bar{B}([\mathrm{S}]) = \frac{\sum\limits_{i=1} i[\mathrm{PS}_i]}{[\mathrm{P}] + \sum\limits_{i=1} [\mathrm{PS}_i]} = \frac{\sum\limits_{i} iK_i[\mathrm{S}]^i}{1 + \sum\limits_{i} K_i[\mathrm{S}]^i} \tag{5b}$$

where

$$K_i = \prod_{j=1}^{i} k_j$$

In principle, if the binding isotherm is measured, Eq. (5 b) can be fitted to the experimental data and the equilibrium constants can be determined. However, this approach is limited to cases where the number of binding steps is small (<5, e.g. the ligand-binding reactions of certain proteins [16]). Nevertheless, in the case of polymer–surfactant and polyelectrolyte–surfactant association, the number of bound surfactant molecules and thus the number of parameters are much larger. Consequently, such a fitting procedure cannot be performed. Additional model assumptions must be introduced in order to overcome this difficulty.

The simplest model that can be derived from the multiple-binding equilibrium is based on the assumptions that a maximum number of surfactant molecules N can bind to a polymer chain and that the binding of the surfactant molecules is independent of each other ($\Delta G_i^0 = \Delta G^0$, thus $k_i = K_0$ and $K_i = K_0^i$). Using these assumptions, Eq. (5b) reduces to the well-known Langmuir isotherm:

$$\bar{B}([\mathrm{S}]) = \frac{K_0[\mathrm{S}]}{1 + K_0[\mathrm{S}]} N \qquad \text{or} \qquad \beta([\mathrm{S}]) = \frac{K_0[S]}{1 + K_0[\mathrm{S}]} \tag{6}$$

where β is the degree of binding ($\beta = B/N$). If the model assumptions hold, the equilibrium constant (K_0) must be independent of the degree of binding. However, in the case of polymer–surfactant and polyelectrolyte–surfactant systems, the equilibrium constant usually increases steeply as the amount of bound surfactant increases, when it is calculated by means of Eq. (6). This is an indication that bound surfactant molecules promote the binding of further surfactants and this phenomenon is usually referred to as cooperative binding.

A typical surfactant binding isotherm and a Langmuir isotherm are shown in Fig. 11.1. The surfactant binding starts from zero surfactant concentration in the case of monomer binding (the Langmuir case). However, in the case of co-

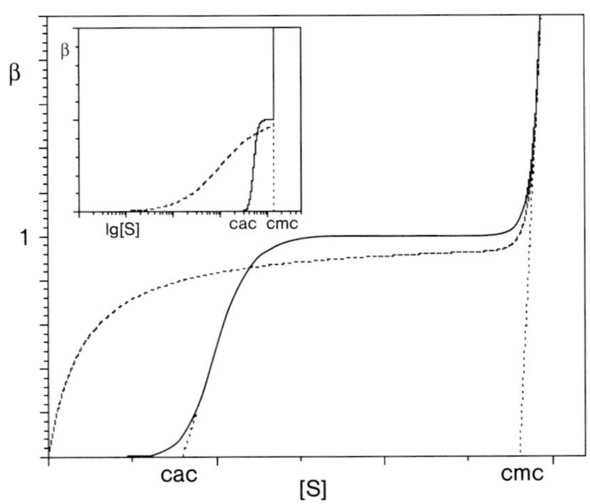

Fig. 11.1 Illustration of a Langmuir (dotted lines) and a cooperative (solid lines) binding isotherm. The inset depicts the same isotherms on a semi-logarithmic scale.

operative binding it starts at a relatively well-defined, finite surfactant concentration. This concentration is called the critical aggregation concentration (c.a.c.). The existence of the c.a.c. is usually regarded as a clear indication of cooperative binding. However, it should be noted that this is a sufficient condition only when the binding isotherm is plotted on a linear scale. When a semi-logarithmic scale is used (as is usually done for polyelectrolyte–oppositely charged surfactant association), a "c.a.c." appears even in the case of monomer binding (see the inset of Fig. 11.1). It should be emphasized that in this case the steepness of the binding isotherm can provide information about the cooperativity of the binding, as discussed below.

The cooperative binding of surfactants to macromolecules is one of the most striking and interesting features of macromolecule–surfactant interaction and considerable efforts have been made to gain an understanding of this phenomenon. On the other hand, the majority of the developed models belong to one of two approaches. The first family of models assumes that the surfactant molecules bind to specific binding sites on the macromolecules, whereas the other family treats the complex formation as micellization perturbed by the presence of the macromolecules.

11.2.2.1 Site Binding Models

The common physical picture in site binding models is that the polymer molecules are assumed to be linear chains composed of N segments, each of which can bind one surfactant molecule (or more generally a small ligand). This means that the macromolecules are considered to be independent of each other and they are treated simply as providing sites for the binding (adsorption) of the surfactant molecules. If an empty and an occupied adsorption site are denoted by the digits 0 and 1, respectively, the binding state of a macromolecule can be described by a sequence such as

$$01100011010001011110 \tag{7}$$

The cooperativity of the binding is reflected by the fact that the free energy of an adsorption state depends not merely on the number of adsorbed surfactant molecules but on the whole sequence. Equation (7) is a typical representation of the Ising model, which was originally developed for the description of ferromagnetic systems [17]. The different site binding approaches use different approximations developed for the treatment of the Ising model.

Bragg–Williams Approximation Hill [18] was first to treat the problem by using a statistical mechanical approach. In order to simplify the problem, he assumed that the nearest-neighbor interactions were dominant. Thus, the free energy of the adsorption state is determined by the number of nearest neighbors. To determine the binding isotherm, he derived the grand canonical partition function of the system within the framework of the Bragg–Williams approximation. This approximation assumes that the distribution of the molecules among the adsorption sites is random but takes into account the nearest-neighbor interactions. By means of the derived partition function, the average number of adsorbed molecules can be expressed as

$$\bar{B}([\mathrm{S}]) = \frac{\sum\limits_{i=1} i K_0^i \exp\left(-\frac{\varepsilon_{11}}{RT}\frac{i-1}{N}\right)^i [\mathrm{S}]^i}{1 + \sum\limits_{i=1} K_0^i \exp\left(-\frac{\varepsilon_{11}}{RT}\frac{i-1}{N}\right)^i [\mathrm{S}]^i} \tag{8}$$

where ε_{11} denotes the interaction energy between two adsorbed molecules on nearest-neighbor sites and K_0 is the equilibrium constant of binding to a site which does not have occupied neighbors. By comparing Eqs. (5 b) and (8), it is straightforward to show that this approximation predicts that the driving force of binding becomes proportional to the degree of binding:

$$k_{i+1} = \frac{K_{i+1}}{K_i} = K_0 \exp\left(-\frac{\varepsilon_{11}}{RT}\frac{2i}{N}\right) \tag{9}$$

where

$$K_i = K_0{}^i \exp\left(-\frac{\varepsilon_{11}}{RT}\frac{i-1}{N}\right)^i$$

It is interesting that although the applied formalism gave steep but smooth binding isotherms in the case of finite number of binding sites, the distribution function of the complex molecules [$p(\mathrm{PL}_i)$] exhibited two sharp maxima if the interaction energy between the nearest neighbors exceeded a critical value, which implied the occurrence of a phase transition. However, thorough analysis of the results clearly indicated that the presence of two maxima is the consequence of the applied approximation and a rigorous theory would give rise to a distribution function with a single broad maximum.

Zimm–Bragg Model Although the above treatment takes into account the nearest-neighbor interaction of the statistically formed pairs, it neglects the correlations introduced by the interaction of the adsorbed molecules. An alternative approach, which is suitable for the treatment of such a correlation, is provided by the matrix method [19]. This method has been widely used to describe different phenomena, which can be represented by an Ising model (e.g. titration curves of weak polyelectrolytes [20] or the helix–coil transition [21]). First, Schwarz [22] used this approach for the description of cooperative binding of small ligands to biopolymers by extending the Zimm–Bragg theory [21] for the coil–helix transition to this problem. Later, Satake and Yang [23] derived the same model independently, also using the Zimm–Bragg theory as a starting point for the description of cooperative binding of surfactants to polypeptides.

According to the model, the ligand binding to the macromolecule takes place in two competing steps, each of which can be characterized by an equilibrium constant:

$$00 + \mathrm{S} \leftrightarrow 01 \qquad K_{00} = [01]/[00][\mathrm{S}]$$
$$10 + \mathrm{S} \leftrightarrow 11 \qquad K_{10} = [11]/[10][\mathrm{S}]$$

The degree of cooperativity (u) is determined by the interaction of the nearest-neighbor ligands (ε_{11}), which can be expressed by the ratio of the two equilibrium constants:

$$u = \frac{K_{10}}{K_{00}} = \exp\left(-\frac{\varepsilon_{11}}{RT}\right) \tag{10}$$

By applying the matrix method, the fraction of occupied binding sites (β) and the average number of uninterrupted occupied sites (m) can be derived. In the case of infinite polymer chains, the following results are obtained:

$$\beta = \frac{1}{2}\left(1 - \frac{1 - K_{00}u[\mathrm{S}]}{\left\{(1 - K_{00}u[\mathrm{S}])^2 + 4K_{00}[\mathrm{S}]\right\}^{\frac{1}{2}}}\right) \text{ and } m = \frac{2\beta(u-1)}{[4\beta(1-\beta)(u-1)+1]^{\frac{1}{2}} - 1} \tag{11}$$

Furthermore, it can also be shown that the cooperativity of the binding (u) is characterized by the slope of the binding isotherm at half saturation in a semi-logarithmic plot:

$$\left(\frac{\mathrm{d}\beta}{\mathrm{d}\ln[\mathrm{S}]}\right)_{\beta=\frac{1}{2}} = \sqrt{\frac{u}{4}} \tag{12}$$

Many experimental binding isotherms can be reasonable well fitted with Eq. (11), but usually clear deviations appear at large degrees of binding (Fig. 11.2). As Shirahama pointed out [24], when charged ligands are bound to an oppositely charged macromolecule, the overall charge and thus the attractive interactions are decreased, giving rise to the observed decrease in the equilibrium constants. Satake and Yang formally indicated this effect in their model by expressing the equilibrium constant as

$$K_{00} = K_{00}^{*} \exp\left(-\frac{z_i e \Psi_0}{RT}\right) \tag{13}$$

However, they did not treat the dependence of the electric potential at the macromolecule surface (Ψ_0) on the degree of binding explicitly.

In order to take into account the reduction of the electrical potential with increasing binding, Delville [25] used the Poisson–Boltzmann theory in a cell model. He determined the local surfactant concentration at the surface of the

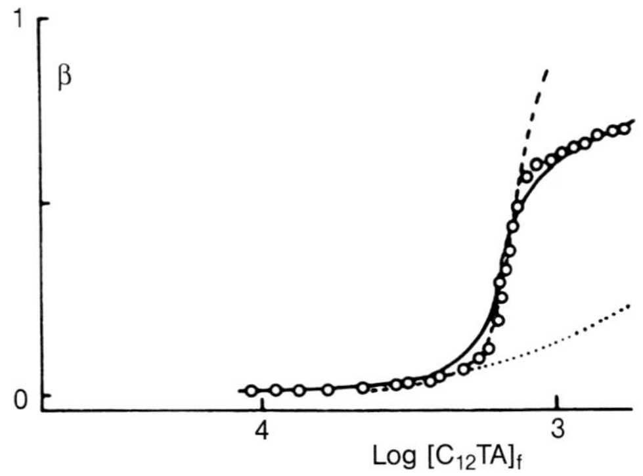

Fig. 11.2 Binding isotherm of dodecyltrimethylammonium bromide by DNA. Dashed line, cooperative binding; dotted line, Poisson–Boltzmann treatment; solid line, ion condensation and cooperative binding. (Reprinted from [25] with permission from Elsevier).

macromolecule (DNA) and assumed that the cooperative binding equilibrium is developed between the macromolecule and the locally accumulated surfactant molecules. By this approach, the agreement between the calculated and the experimental binding isotherms could be considerably improved (Fig. 11.2). Furthermore, the effect of ionic strength on the binding isotherms could also be interpreted as being the consequence of competitive accumulation of the surfactant and its co-ions from the added electrolyte present in the diffuse double layer around the polyelectrolyte.

The Nature of Surfactant Aggregates Since the site binding models are based on a linear Ising model, they do not provide information about the structure of the macromolecule–surfactant complex and the nature of the surfactant aggregates formed in the complex. Nevertheless, the uninterrupted sequence of bound surfactant molecules is usually considered as micelle-like surfactant aggregates in the literature. However, this interpretation bears an important contradiction to experimental results. As reflected by Eq. (11), with increasing degree of binding the average aggregate size increases steeply and eventually approaches the number of binding sites as the polymer becomes saturated. This should lead to the formation of a few large surfactant aggregates near saturation. Taking into account that the number of binding sites can be as large as a few thousand, surfactant aggregation numbers should be in the range of several hundred as saturation is approached. However, the results of several aggregation number measurements performed on flexible polymer (polyelectrolyte)–surfactant systems indicate that the surfactant aggregation number remains much smaller than the number of binding sites even near the saturation [26–31].

To conclude, the site binding models may provide a consistent and efficient method for the description of the binding of small ligands to stiff macromolecules (e.g. DNA or enzymes), which can provide well-defined binding sites (e.g. hydrophobic pockets for the binding of surfactants). However, in the case of flexible polymers and polyelectrolytes, alternative approaches should be used, which can resolve the qualitative disagreement between the predicted and experimental size of the polymer bound micelles.

11.2.2.2 Closed Aggregate Formation Models

This family of models is based on the explicit assumption that the polymer–surfactant complex formation involves the formation of micelle-like surfactant aggregates. In terms of the multiple-binding equilibrium model, this means that the set of equations in Eq. (5 a) represents the formation of a surfactant aggregate along a polymer region [P (it should be noted that in this notation P denotes a region of the polymer capable of accommodating a surfactant aggregate and a polymer molecule may contain several such regions)] and K_n is the intrinsic equilibrium constant for the formation of a surfactant aggregate from n surfactants:

$$K_n = \exp\left(-\frac{n\Delta\mu^0_{\mathrm{aggr},n}}{RT}\right) \tag{14}$$

where $\Delta\mu^0_{\mathrm{aggr},n}$ denotes the difference in the standard chemical potential per surfactant in an aggregate of n surfactants and the standard chemical potential of a surfactant molecule free in the solution. By specifying $\Delta\mu^0_{\mathrm{aggr},n}$ as a function of the aggregation number, the c.a.c., the size distribution of the surfactant aggregates and any aggregate size-dependent property of the polymer surfactant system (e.g. the binding isotherm) can be calculated.

Optimum Aggregate Size Models In order to simplify the problem, the surfactant aggregates are usually considered to be monodisperse and, instead of the whole distribution function of the surfactant aggregate size, the "optimum micelle size" is used. This approximation was originally developed for describing micelle formation [32] and is based on the pseudophase nature of the surfactant aggregates. The underlying assumption of the pseudophase model is that the surfactant aggregates can be treated as a new phase, since the chemical potential of the singly dispersed surfactant molecules becomes almost constant when the polymer–surfactant complex formation starts. Since in equilibrium the free energy of the system must be in a minimum, it is usually argued that the equilibrium aggregate must have the minimum standard free energy change per surfactant [33]. Hence the optimum aggregate size (n_{opt}) is represented by the condition:

$$\frac{\partial(\Delta\mu^0_{\mathrm{aggr},n})}{\partial n} = 0 \tag{15}$$

A further important result of the pseudophase approximation is that it allows the estimation of the standard free energy change of aggregate formation from the c.a.c.:

$$\ln(x_{\mathrm{c.a.c}}) = \frac{\Delta\mu^0_{\mathrm{aggr},n_{\mathrm{opt}}}}{RT} \tag{16}$$

Using the optimum micelle size approximation, Eq. (5 b) (the binding isotherm) reduces to the following form:

$$\beta([S]) = \frac{(K_0[S])^n}{1 + (K_0[S])^n} \quad \text{or} \quad \log\frac{\beta}{1-\beta} = n\log([S]) + n\log(K_0) \tag{17}$$

where n is the aggregation number of the optimum size surfactant aggregate. An important application of Eq. (17) is that it allows the estimation of the surfactant aggregation number either from the slope of the binding isotherm when it is given in a log–log plot or from the slope at half saturation if the binding isotherm is given in a semi-logarithmic plot $\{(d\beta/d\ln[S])_{\beta=1/2} = n/4\}$. It should be noted that Eq. (17) is formally identical with the well-known Hill equation [34], which was derived for the description of adsorption when the adsorbed species form clusters of size n.

The first comprehensive thermodynamic analysis of polymer–ionic surfactant complex formation based on the formation of optimum-sized surfactant aggregates was given by Gilanyi and Wolfram [35]. They took into account both polymer–surfactant interaction and micelle formation and allowed counterion (M) binding to the aggregates:

$$P + nS + mM \leftrightarrow \text{complex} \qquad K_{\text{complex}} = [\text{complex}]/[P][S]^n[M]^m \tag{18 a}$$

$$n'S + m'M \leftrightarrow \text{micelle} \qquad K_{\text{micelle}} = [\text{micelle}]/[S]^{n'}[M]^{m'} \tag{18 b}$$

where n and n' are the surfactant aggregation numbers of the complex and the micelle, respectively, and m and m' denote the number of apparently bound counterions. Using the mass balance equation of the surfactant and the counterion, which are given below, the equilibrium concentrations of the different species were calculated as a function of the total surfactant concentration (Fig. 11.3).

$$c_0 = [S] + n[\text{complex}] + n'[\text{micelle}] \tag{19 a}$$

$$[M] = [S] + an[\text{complex}] + a'n'[\text{micelle}] \tag{19 b}$$

where a and a' are the apparent degree of dissociation $[(n-m)/n]$. As indicated by Fig. 11.3, the model predicts that in the presence of the polymer three characteristic regions of the interaction can be distinguished. In the first region ($c_0 < c_{cr} \equiv$ c.a.c.), the amounts of the complex and the micelles are negligible and binding practically does not take place. In the second region ($c_{cr} < c_0 < c_M$), the complex concentration continuously increases but free surfactant micelles are not present in the solution. The average activity of the surfactant, $a_{\pm} = ([S][M])^{1/2}$, increases only slightly in this region. Finally, at $c_0 = c_M$, the surfactant activity reaches the c.m.c. and free micelles appear in the system. With further increase in surfactant concentration, the amount of free micelles increases rapidly, but the surfactant activity remains practically constant. As pointed out by the authors, at $c_0 = c_M$ the polymer does not have to be saturated with surfactant. In such a case, further saturation of the polymer slows considerably due to the almost constant surfactant activity (Fig. 11.3).

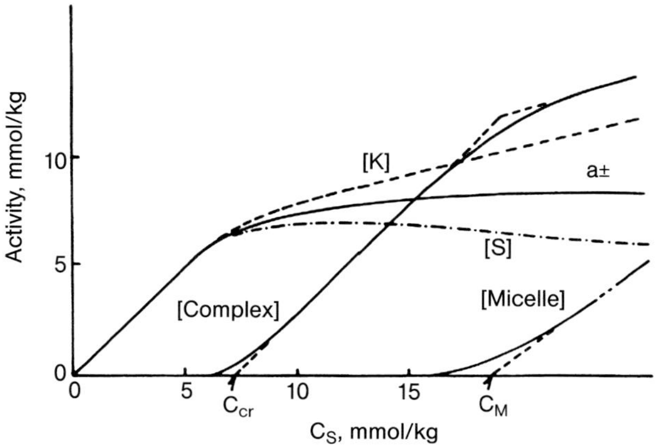

Fig. 11.3 Calculated activity of the different components (surfactant, counterion, complex and micelle) of the polymer–surfactant system as a function of the total surfactant concentration. (Reprinted from [35]; Copyright 1981, with permission from Elsevier).

Later, Nagarajan [36] reproduced the model of Gilanyi and Wolfram in a slightly simplified form by neglecting the possibility of counterion binding, but his analysis gave practically the same results. Furthermore, he also presented a molecular interaction model, which aimed at the derivation of the aggregation numbers (n and n') and also the equilibrium constants (K_{complex} and K_{micelle}) from molecular parameters within the framework of the pseudophase approximation [by means of Eqs. (15) and (16)]. His approach [33, 36, 37] was based on the same type of models that were developed previously for the description of micellization [38–40]. In the framework of these models, the standard free energy change of aggregate formation ($\Delta\mu^0_{\text{aggr},n}$) is estimated as a sum of different contributions accompanying the transfer of a surfactant molecule from the bulk phase to a surfactant aggregate. In the case of free micelles, $\Delta\mu^0_{\text{mic},n}$ is usually expressed as

$$\Delta\mu^0_{\text{mic},n} = \Delta\mu_{\text{w/hc}} + \Delta\mu_{\text{hc/mic}} + \Delta\mu_\sigma + \Delta\mu_{\text{steric}} + \Delta\mu_{\text{el}} \tag{20}$$

The first term in Eq. (20) takes into account the free energy change associated with the transfer of an alkyl chain from the water to the micelle core. The second term corresponds to the free energy change associated with the conformational restrictions of the surfactant tail in the micelle core. The third term describes the free energy requirement of the formation of the interface separating the micelle core and the surrounding water. The last two terms account for the steric and the electrostatic interaction of the surfactant headgroups accumulated at the micelle surface.

In his model, Nagarajan extended Eq. (20) in order to incorporate the effect of polymer on the free energy change of aggregate formation. He assumed that the polymer penetrates the interfacial region of the micelle, which has three main consequences: (1) the polymer shields a portion of the micelle core from water; (2) unfavorable polymer–water contacts are replaced by more favorable polymer–micelle core contacts; and (3) the presence of the polymer in the headgroup region increases the steric repulsion. These interactions were described by two parameters: the micelle core area shielded by the polymer and the hydrophobic interaction of the polymer with the micelle core. It should be noted that in this treatment the shielded area per surfactant served as an effective parameter, in which formally all the relevant polymer properties were included and which was independent of the surfactant aggregation number.

Ruckenstein et al. [41] used a different approach to take into account the effect of the polymer on the surfactant aggregate formation. They assumed that the presence of the polymer around the bound surfactant aggregates changes the microenvironment from pure water to an aqueous polymer solution, which is more hydrophobic. This change in the microenvironment gives rise to (1) a decrease in the interfacial free energy between the micellar core and its environment and (2) an increase in the interfacial free energy between the hydrophilic surfactant headgroups and their environment. Since the former effect favors whereas the latter opposes complex formation, their balance determines if polymer–surfactant complexation takes place. The model predicts that the complex formation is more favorable the smaller the surfactant headgroup is.

Nikas and Blankschtein [42] also presented a molecular interaction model in which they explicitly treated the polymer molecule. Their theoretical treatment takes into account the polymer hydrophobicity and flexibility and also the specific interaction between the polymer segments and the hydrophilic headgroups of the surfactant molecules. The model allowed the binding of a variable number of identical (optimum-sized) surfactant aggregates in an optimum arrangement (equidistantly) along the chain. Furthermore, the electrostatic repulsion of the neighboring bound surfactant aggregates and the entropic cost of stretching the polymer regions between the two interacting aggregates were treated. The model identified the interaction of the polymer segments and the micellar core and also the specific interaction of the segments and the surfactant headgroups as the major driving forces of the polymer–surfactant complexation, which can be further enhanced by poor solvent quality and small surfactant headgroup size. A central prediction of the model is that with increasing surfactant chemical potential, complexes containing one micelle form first. When all the available polymers have bound one surfactant aggregate, the binding of a second aggregate can start. The predicted binding mechanism should give rise to a stepwise binding isotherm (Fig. 11.4); however, as the authors indicate, such isotherms have not yet been observed. On the other hand, as they argue, the observation of the stepwise isotherms would require nearly monodisperse polymers.

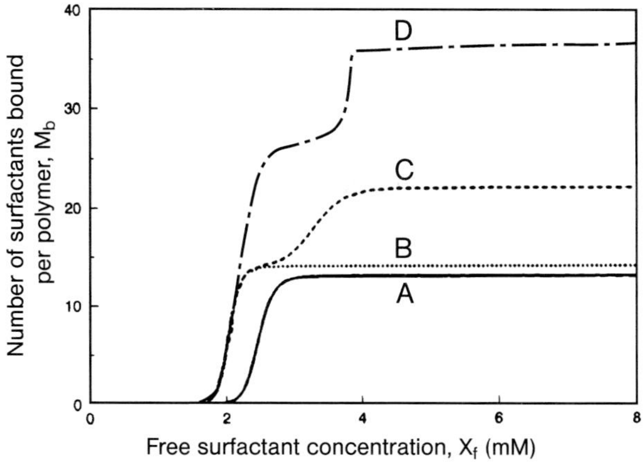

Fig. 11.4 Predicted average number of surfactant molecules bound to a polymer chain as a function of polymer molecular weight: $N=$ (A) 100, (B) 200, (C) 400 and (D) 800. (Reprinted with permission from [42]; Copyright 1994, American Chemical Society).

11.2.2.3 **Small Systems Thermodynamic Model**

The optimum-sized aggregate is a good approximation if the aggregate size distribution is narrow, symmetrical and constant in the concentration range of interest. These assumptions are excellently fulfilled in the case of micelles, where the aggregation number is large and the surfactant chemical potential becomes nearly constant above the c.m.c. However, in the case of polymer–surfactant aggregates, where the surfactant aggregation number is considerably smaller and the chemical potential can change in a relatively wide range (from the c.a.c. to the c.m.c.), the ensemble average of the aggregate size, which is relevant from a thermodynamic point of view, and the optimum size may differ considerably. In order to address this problem, Gilanyi developed a small system thermodynamic model of the polymer–surfactant complexation [43]. He treated the system at three levels of organization: (1) the individual surfactant aggregate combined with the polymer coil (subsystem): the subsystem was open to both surfactants and polymer segments, which means that their number can fluctuate around an average value; (2) the polymer–surfactant complex molecule (small system): the complex was considered as a collection of the subsystems and it was treated as open to the number of surfactant molecules and closed to the number of polymer segments; and (3) the macroscopic system. The author derived the thermodynamic functions of the macroscopic system. The surfactant binding isotherm was given as

$$\beta([\mathrm{S}]) = \frac{\exp(\Delta/kT)[\mathrm{S}]^{\bar{n}}}{1+\exp(\Delta/kT)[\mathrm{S}]^{\bar{n}}} \qquad \text{or} \qquad \ln\frac{\beta}{1-\beta} = \bar{n}\ln([\mathrm{S}]) - \Delta/kT \tag{21}$$

where $\Delta = \bar{G}^0_{\mathrm{sub}} - \bar{n}\mu_0$ is the standard free energy change of the mean subsystem formation and $\bar{n}$ is the size of an average aggregate. Although Eq. (21) is formally identical with Eq. (17), it should be noted that in this case both Δ and $\bar{n}$ are functions of the surfactant activity. Since the number of surfactant aggregates within a complex molecule is limited, the model predicts that the average aggregation number increases continuously as the surfactant activity increases. This prediction is in good agreement with the results of fluorescence probe measurements [44] and trace probe electrolyte measurements [14, 45], which indicates that the mean aggregation number indeed increases significantly with increasing surfactant activity. As a consequence, the correct thermodynamic analysis of the binding isotherms requires the determination of the surfactant aggregation number as a function of the surfactant activity by an independent method. In the absence of appropriate aggregation number data, one can determine only the standard free energy of aggregate formation per surfactant molecule at half-saturation:

$$(\Delta/\bar{n})_{\beta=\frac{1}{2}} = kT\ln\left([\mathrm{S}]_{\beta=\frac{1}{2}}\right) \tag{22}$$

and the surfactant aggregation number at the c.a.c. [from the initial slope of the binding isotherm on a log–log scale; see Eq. (22)].

Gilanyi et al. [45] also presented a molecular interaction model calculation based on the small system thermodynamic model for the interaction of non-ionic polymers and non-ionic surfactants. They derived a similar expression for the standard free energy change of aggregate formation that was introduced by Nagarajan, but instead of a constant effective area parameter they used the number of actually interacting polymer segments (s) as an independent variable. From the resulting expression they calculated the probability distribution function of the polymer–surfactant subsystem as a function of the surfactant aggregation number and the number of interacting polymer segments ($p_{n,s}$) at constant surfactant activity. Finally, the average aggregation number

$$\bar{n} = \sum_n \sum_s p_{n,s} n$$

the average number of interacting segments

$$\bar{s} = \sum_n \sum_s p_{n,s} s$$

and the average standard free energy change of aggregate formation

$$\bar{G}^0_{\mathrm{sub}} = \sum_n \sum_s p_{n,s} G^0_{\mathrm{sub}}(n,s) + kTp_{n,s}\ln(p_{n,s})$$

were determined at given surfactant activities. In agreement with previous findings, the model indicated that the major driving force of complex formation is the interaction of the polymer segments with the micellar core (shielding and hydrophobic interaction). However, the model also indicated that in the case of non-ionic surfactants, this driving force is not sufficient to counteract the increased steric repulsion, hence complex formation does not take place before the micelle formation. Furthermore, it was also found that there is a competition between the polymer segments and the surfactant molecules to take part in the aggregate, i.e. with increasing number of interacting segments the surfactant aggregation number decreases. In addition, both the increasing segment–surfactant interaction and segment size (shielded area) give rise to the decrease in aggregation number.

Oppositely Charged Polyelectrolytes and Surfactants The previously summarized aggregate formation models were developed for the description of non-ionic polymer–surfactant association. However, the charged nature of the macromolecule has a profound effect on the polymer–surfactant interaction. For instance, in the case of similarly charged surfactants and polyelectrolytes, polymer–surfactant association does not take place, whereas in the case of oppositely charged surfactants and polyelectrolytes, the association is known to be particularly strong, resulting in the typical phase behavior depicted in Fig. 11.5. As can be seen, when the surfactant concentration is sufficiently large, the polyelectrolyte–surfactant system separates into two phases, one concentrated in the polyion and the surfactant (a precipitate or a coacervate) and the other containing mostly the small ions and the excess of polyelectrolyte. In the one-phase region the experimental surfactant binding isotherms indicate strong cooperative binding and the c.a.c. is usually orders of magnitude smaller than in the case of non-ionic polymers. Another dramatic difference exhibited by non-ionic polymer–surfactant and polyelectrolyte–surfactant systems is the effect of an inert electrolyte on the c.a.c. In the former case the c.a.c. decreases with increasing

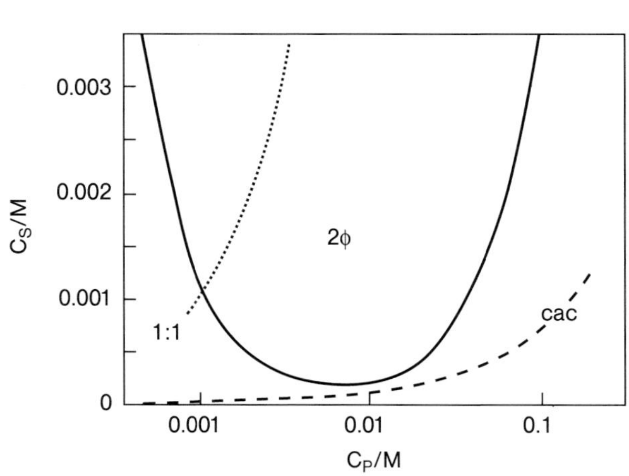

Fig. 11.5 Typical phase diagram of a dilute polyelectrolyte–oppositely charged surfactant system. The solid line indicates the onset of the phase separation. (Reprinted with permission from [46]; Copyright 2001, American Chemical Society).

ionic strength due to the shielding of the electrostatic repulsion of the surfactant headgroups in the surfactant aggregate. On the other hand, in the case of oppositely charged polyelectrolytes and surfactants, the c.a.c. increases when an inert electrolyte is added to the system, while the shape of the binding isotherm changes considerably. Finally, it should also be noted that while the c.a.c. is independent of the polymer concentration in non-ionic polymer–surfactant systems (since the segment concentration within the polymer coil is independent of the overall polymer concentration), the c.a.c. increases with increasing polyelectrolyte concentration in polyelectrolyte–surfactant systems [46].

In order to rationalize these observations Skerjanc et al. [47] and Löfroth et al. [48] presented simple calculations based on a Poisson–Boltzmann cell model. These calculations demonstrated that due to the high electrostatic potential in the close vicinity of the polyelectrolyte, the surfactant concentration could be several orders of magnitude larger in this region than in the bulk. These results suggest that the local concentration of surfactant can reach the c.m.c. in the neighborhood of the polyelectrolyte chain, even when the total concentration is orders of magnitude below the c.m.c., and, as a consequence, surfactant aggregates can form within the polyelectrolyte coil. However, it should be noted that from a thermodynamic point of view the c.m.c. is the mean activity of the ionic surfactant $[a_{\pm} = \gamma_{\pm}(c_+c_-)^{1/2}]$, which must be constant throughout the system in equilibrium. This is assured by the decrease in the surfactant counterion concentration accompanying the increase in the surfactant ion concentration in an external electric field. This means that in general micelle formation cannot take place in external electric fields before the c.m.c. is reached (all over the system), even if the surfactant ion concentration becomes much larger locally than it is at the c.m.c. in a homogeneous solution. The major difference between an external electric field in general and a polyelectrolyte molecule is that in the latter case the charged segments of the macromolecule can act as a counterion of a surfactant aggregate, leading to the formation of a polyelectrolyte–surfactant aggregate far below the c.m.c.

A consistent thermodynamic model, based on the above physical picture, was presented by Hansson and Almgren [49]. They assumed that the surfactant molecules can be present in the system in three different forms: (1) free in the bulk; (2) condensed (accumulated in the close vicinity of the polyelectrolyte due to the electrostatic interaction); (3) surfactant aggregates bound to the polyelectrolyte (formed from the condensed surfactants). Furthermore, they assumed that the binding process can be characterized by two equilibria. The first equilibrium describes the condensation of the surfactant ions as an ion-exchange reaction:

$$\mathrm{S} + \mathrm{C}_{\mathrm{cond}} \leftrightarrow \mathrm{S}_{\mathrm{cond}} + \mathrm{C} \qquad (K_{\mathrm{IE}})$$

where the subscript cond denotes the ions accumulated (condensed) in the electrostatic field of the polyelectrolyte and C denotes the simple counterion of the polyelectrolyte. The second equilibrium describes the aggregate formation from the condensed surfactant ions:

$$nS_{cond} \leftrightarrow S_n \qquad (K^n)$$

Introducing the notation $K_0 = K_{IE}/[C]$, the following binding isotherm could be derived:

$$\beta([S]) = \frac{K_0[S] + (K_0 K[S])^n}{1 + K_0[S] + (K_0 K[S])^n} \tag{23}$$

It should be noted that K_0 is a true constant only in excess salt, whereas at low salt concentrations the release of polyion counterions must be taken into account. The presented model correctly predicts the low c.a.c. values and also the effect of inert electrolytes on the c.a.c. and on the shape of the binding isotherm.

Later, Hansson presented a molecular interaction model for surfactant aggregate formation in the presence of oppositely charged polyelectrolytes [46]. As a starting point, he assumed that the polyelectrolyte–oppositely charged surfactant interaction can be considered as the formation of a "dressed" micelle [50]. In other words, he assumed the formation of a surfactant aggregate wrapped by m polyelectrolyte segments, which is accompanied by the release of k small ions previously bound in the electric field of the polyelectrolyte:

$$nS^+ + (P_m M_k)^{k-m} \rightarrow (S_n P_m)^{n-m} + kM^+ \tag{24}$$

In order to describe the polyelectrolyte–surfactant interaction, Hansson followed a similar approach to that used in the molecular interaction models of non-ionic polymer–surfactant association and derived the standard free energy change of the aggregate formation by the modification of the standard free energy expression derived for ionic micelles. Assuming that the interaction between the polyion and the surfactant aggregate is purely electrostatic, he divided the electrostatic contribution into three separate parts. The first and second parts describe the electrostatic free energy contribution of the polyion-dressed micelle due to its net charge and the electrostatic free energy of the free polyion. These contributions were calculated from the Poisson–Boltzmann cell model. The third part of the electrostatic contribution originates from the separation of the positive (surfactant) and negative (polyion) charges in the headgroup region of the aggregate.

The model indicates that the formation of the aggregate is basically driven by two factors: (1) there is a considerable entropic gain, since the loss of conformational entropy originating from the binding of a fraction of the polyelectrolyte chain is overcompensated by the entropic gain induced by the release of the bound counterions; (2) the reduction of the electrostatic energy of the system. The latter factor is due to the fact that the strong electrostatic repulsion of the neighboring charged groups along the polyelectrolyte chain is reduced more effectively in the double layer of the micelle than in the bulk solution phase. The model could explain the increase of the c.a.c. with increasing salt and polyelectrolyte concentration and the observation that monovalent and divalent surfactants with identical chain length have approximately the same c.a.c., and furthermore it predicted the experimentally found correlation between the c.a.c. and the surfactant aggregation number.

11.2.2.4 Monomer Binding to Polyelectrolytes

Although the cooperative nature of surfactant binding both to neutral linear polymers and to linear polyelectrolytes is well established in the literature, recently Mészáros et al. found that in the case of the hyperbranched polyethylenimine (PEI) non-cooperative monomer binding takes place [51]. Based on the measured binding isotherm, they distinguished two different binding mechanisms for the surfactant [sodium dodecyl sulfate (SDS)] to PEI. In a first binding step, the dodecyl sulfate ions bind in monomer form to the protonated amine groups (Eq. 25 a), which is accompanied by the re-establishment of the protonation equilibrium of the free amine groups (Eq. 25 b), leading to an increase in the solution pH:

$$\text{P-NH}^+ + \text{S}^- \leftrightarrow \text{P-NH}^+\text{S}^- \qquad K_\text{S} \tag{25 a}$$

$$\text{P-N} + \text{H}^+ \leftrightarrow \text{P-NH}^+ \qquad K_\text{H} \tag{25 b}$$

Above a critical amount of bound surfactant, the PEI–SDS complex molecules collapse and precipitate. After the collapse of the polyelectrolyte–surfactant molecules, the surfactant binding takes place as the adsorption of the surfactant molecules on the surface of the collapsed particles (causing a charge reversal). This means that at low surfactant concentrations, the system is a thermodynamically stable solution of the polymer–surfactant complex molecules. Above a critical concentration, the system is an unstable colloidal dispersion of the complex particles and, at even higher surfactant concentrations, the system may become a kinetically stable dispersion of the PEI–SDS particles, depending on the method of preparation. Based on the above physical picture, the authors could describe the measured binding isotherm and also the experimental pH curves measured as a function of the surfactant concentration at several initial pH values.

11.2.3
Simulations and Additional Theoretical Approaches

Recent years have seen significant progress in the theoretical treatment of polyelectrolyte–surfactant mixtures in bulk and also the development of modeling approaches. It is encouraging to note that predictions of theories and simulation results largely agree and that experimental results in many cases are in line with these predictions. Hence it is fair to say that a reasonably good understanding of the bulk association between polyelectrolytes and oppositely charged surfactants has been obtained.

11.2.3.1 One Polyelectrolyte and One Macroion

Monte Carlo simulations have been used for modeling the complexation between a single polyelectrolyte chain and a single macroion, i.e. a charged hard sphere that serves as a model for a charged micelle. The association is found to be driven by favorable electrostatic interactions and counterion release, whereas it is counteracted by bond energy strain and loss of conformational energy of the polyelectrolyte. The complexation process was concluded to be favorable both energetically and entropically [52] and more so with increasing charge density and flexibility of the polyelectrolyte chain. Complexation leads to a significant contraction of the polyelectrolyte chain, which, for a flexible polyelectrolyte, leads to a conformation that is more compact than a Gaussian chain [52]. It is also found that the segments accumulate closer to the surface of the charged macroion with increasing charge density and flexibility of the polyelectrolyte chain [53]. The critical aggregation concentration (c.a.c.) increases with chain stiffness, decreasing charge density of the polyelectrolyte [54] and decreasing size of the macroion [55]. The structure of the polyelectrolyte–macroion complex has been modeled in detail and found to depend crucially on polyelectrolyte charge density, chain flexibility and charge ratio between polyelectrolyte and macroion [56]. In many situations the macroion binds preferentially to the middle of the polyelectrolyte [57], but preferential binding at the tail of the macroion is also observed for some systems. Some typical snapshots from the simulations are provided in Fig. 11.6.

The effects of the surfactant chain length on the complexation between polyelectrolytes and micelles have been investigated using Monte Carlo simulations.

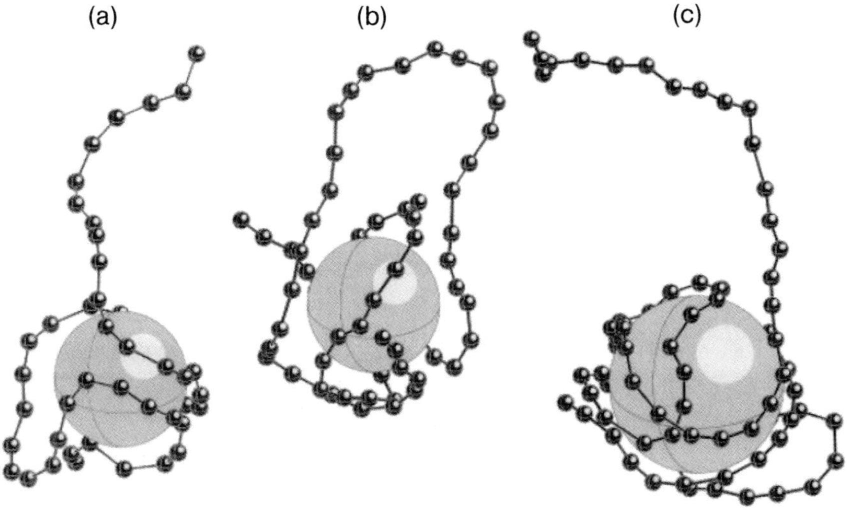

Fig. 11.6 Typical configuration of polyelectrolyte–macroion conformations. The size of the macroion corresponds to a surfactant aggregation number of (a) 30, (b) 41 and (c) 53. (Reprinted with permission from [55]; Copyright 1997, American Chemical Society).

One main finding is that the c.a.c. to c.m.c. ratio decreases as the surfactant chain length increases [55], a result that is in accord with experimental observations. It was suggested to be a consequence of the increasing size of the macroion (micelle) with increasing surfactant chain length, which allows for the formation of a tight layer of polyelectrolyte segments on the macroion surface with reduced conformational penalty.

Self-consistent mean field lattice models have also found use for investigating polyelectrolyte–surfactant complexation [58]. The results obtained were largely consistent with Monte Carlo simulations, but it was additionally deduced that hydrophobic polyelectrolytes are able to penetrate into micelles. However, the theoretical modeling failed to reproduce the increase in c.a.c. to c.m.c. ratio with increasing salt concentration that is observed experimentally.

11.2.3.2 Several Polyelectrolytes and One Macroion

The complexation between one macroion and several polyelectrolyte chains has been investigated by employing Monte Carlo simulations [59]. Wallin and Linse considered the situation when the magnitude of the charge of the polyelectrolyte and the macroion is the same. They found a strong preference for formation of a 1:1 complex when the polyelectrolyte chain was flexible and with a high charge density. On the other hand, when the charge density of the polyelectrolyte was reduced, complexes containing one macroion and two polyelectrolytes were favored since it becomes increasingly difficult for a single polyelectrolyte to cover the macroion surface fully.

11.2.3.3 One Polyelectrolyte and Several Macroions

The reverse situation, one polyelectrolyte and several macroions, has also been considered in Monte Carlo simulations. It was found that the binding of macroions to the polyelectrolyte resembles a Langmuir isotherm. The initial binding results in compaction of the polyelectrolyte chain and, for flexible polyelectrolytes, the charge of the segments next to the macroion overcompensates the macroion charge. At the adsorption plateau, the charges of the macroions present in the complex overcompensate the charge of the polyelectrolyte and the polyelectrolyte conformation becomes more stretched again. In this situation, the charge of the polyelectrolyte segments close to each macroion undercompensates the charge of the macroion [57]. The degree of overcompensation of the polyelectrolyte charge increases with the chain stiffness, which is a consequence of the fact that fewer segments are in close contact with each macroion [60]. It is worth noting that the first macroion binds more favorably to a flexible polyelectrolyte chain whereas a stiffer chain favors binding of an additional macroion to a charge-neutral polyelectrolyte–macroion complex. One important reason for this latter finding is that the more extended chain reduces the electrostatic repulsion between the excess charges of the macroions. The structure of the complex at saturation is of the bead-and-necklace type, where the macro-

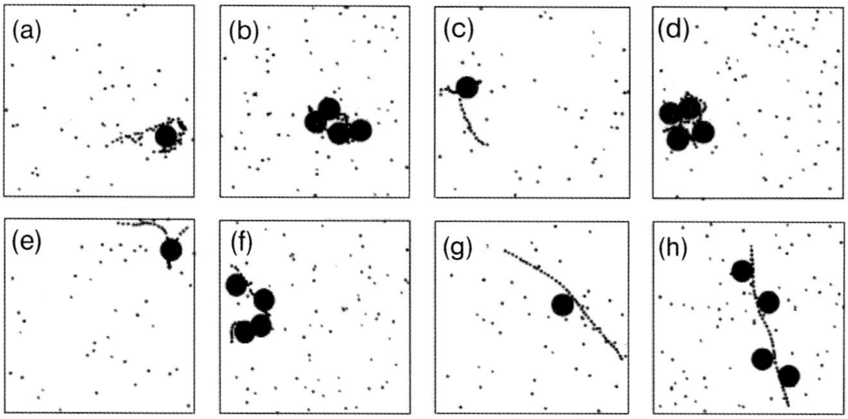

Fig. 11.7 Typical configurations of polyelectrolyte-macroion complexes. The figure shows complexes with one macroion (left column) and four macroions (right column). The polyelectrolyte stiffness increases for each row. The bare persistence length is 7 Å (row 1), 42 Å (row 2), 164 Å (row 3) and 1480 Å (row 4). (Reproduced from [60] with permission from Marie Jonsson and Per Linse; Copyright 2001, American Institute of Physics).

ions are closer together for more flexible polyelectrolyte chains. Some typical snapshots from the simulations are shown in Fig. 11.7.

In their theoretical considerations of complexation between polyelectrolytes and macroions, Nguyen and Shklovski reached similar conclusions to those described above. In addition, they emphasized that electrostatic correlation forces provide the driving force for the overcharging phenomena [61]. Similar results were also obtained by Schiessel et al., using scaling arguments [62]. It should be noted that the theoretical approaches and the simulations are valid for slightly different conditions. That nevertheless the conclusions reached are similar provides a strong indication that the results discussed above have a broad validity.

11.2.3.4 **Several Polyelectrolytes and Several Macroions**

The stability of polyelectrolyte–macroion solutions has been investigated by means of Monte Carlo simulations [63]. The general sequence of events observed when polyelectrolyte chains were added to a macroion solution was found to be formation of small repelling clusters, phase separation and redissolution when excess polyelectrolyte charges were present. This sequence of events is also often observed experimentally. The theory of Nguyen and Shklovski also predicts phase separation for close to charge-neutral clusters [61]. They also noted that the instability range increases as the electrostatic interaction is screened. In the Monte Carlo simulations, it was demonstrated that the ability to form extended clusters was increased when the polyelectrolyte charge density and length were increased [63]. More surprisingly, it was noted that polyelectrolytes with semi-flexible chains, with a persistence length similar to the size of the macroion, had the smallest tendency to form large clusters. This was suggested to be related to the observation that this situation resulted in the most compact layer of polyelectrolyte segments at the macroion surface, which reduced the driving force for formation of polyelectrolyte bridges between macroions.

Finally, we note that real polyelectrolyte–surfactant mixtures are more complex than the polyelectrolyte–macroion systems considered theoretically. First, the surfactant aggregates associating with the polyelectrolyte have the freedom to change aggregation number and shape as the association proceeds, whereas macroions with a fixed charge and shape have been considered in the theoretical work mentioned in Section 11.2.3 and in the simulations. Second, real polyelectrolyte–surfactant systems are prone to be trapped in persistent non-equilibrium states [64], which is difficult to capture theoretically and by simulation methods. Despite these differences, it is clear that the theoretical development and the simulation results provide a good basis for understanding solutions containing surfactants and oppositely charged polyelectrolytes; in particular, the trends associated with changes in polyelectrolyte charge, length and stiffness have been elucidated.

11.2.4 Stability of Polyelectrolyte–Surfactant Aggregates

Experimentally, the easiest way to obtain an understanding of the bulk properties of polyelectrolyte–surfactant aggregates is to measure the turbidity of the solution and follow the mobility of the aggregates formed [1, 51, 65, 66]. The general behavior emanating from such studies is that at high polyelectrolyte to surfactant ratios the solution is clear and the complexes have the same sign of charge as the polyelectrolyte (for an example see Fig. 11.8). As the surfactant concentration is increased, the solution becomes turbid and the magnitude of the mobility decreases. At even higher surfactant concentration a charge reversal of the aggregates is observed, as predicted by theory [61, 62] and simulations [57]. At this point, the turbidity of the solution decreases again.

We note that the turbidity maximum in each case is observed close to the zero mobility point and that the high-turbidity region becomes larger as the polyelectrolyte charge density increases. Light scattering experiments have demonstrated that the size of the polyelectrolyte chain initially contracts as surfactants associate with the chain and that the aggregate size increases dramatically close to the charge neutralization concentration [64, 67–69].

The charged aggregates formed away from the charge stoichiometry point display a remarkably small polydispersity, as first demonstrated by Langevin and coworkers [67, 68] and later confirmed by others [64, 69]. Away from charge stoichiometry, the aggregates formed are colloidally stable [51, 65] and the stability is largely due to interaggregate electrostatic repulsion. This conclusion is drawn from the observation that the region of polyelectrolyte to surfactant ratio over which colloidal instability is observed is increased by addition of salt [51, 64, 69].

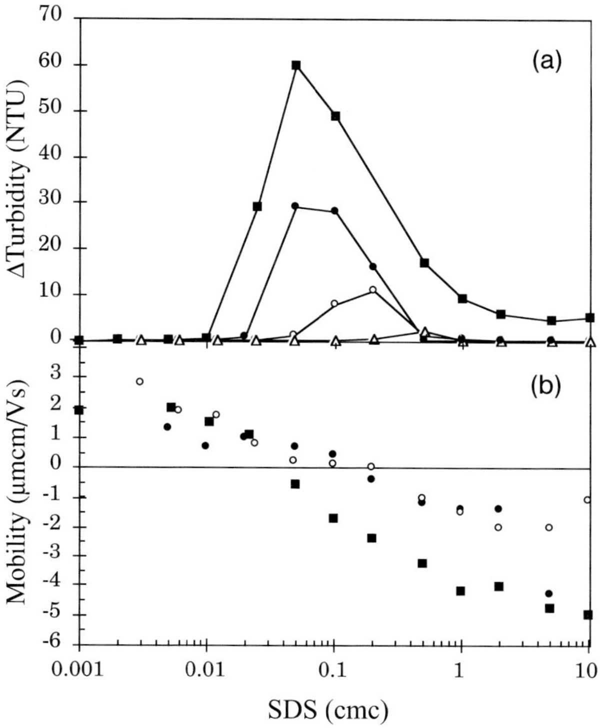

Fig. 11.8 (a) Turbidity difference between polyelectrolyte-free SDS solutions and solutions containing 20 ppm polyelectrolyte as a function of SDS concentration. Data are presented for PCMA, 100% charged (■), AM-MAPTAC-30, 30% charged (●), AM-MAPTAC-10, 10% charged (○), and AM-MAPTAC-1, 1% charged (△). (b) Electrophoretic mobility of polyelectrolyte–SDS aggregates. (Adapted from [1] by permission; Copyright 2001, Routledge/Taylor & Francis Group, LLC).

A more remarkable observation is that the size of the aggregates formed depends strongly on how the solution is mixed and that the initial difference persists over very long periods [64, 69]. Thus, mixtures of highly charged polyelectrolytes and oppositely charged surfactants are prone to be trapped in long-lived non-equilibrium states. A consequence is that great care should be taken to control the mixing process in both fundamental research and in technical applications utilizing such mixtures.

11.2.5
Structure of Polyelectrolyte–Surfactant Complexes

The internal structure of the polyelectrolyte–surfactant complexes can be investigated by, e.g., fluorescence quenching experiments that allow aggregation numbers to be determined and to identify the presence of non-polar domains [29]. A more detailed picture can be obtained by small-angle neutron scattering (SANS) and small-angle X-ray scattering (SAXS) experiments. Such studies have shown that the internal organization can vary substantially and depends on the polyelectrolyte architecture and the surfactant chain length. Cubic, hexagonal, lamellar and other phases have been reported (e.g. [70–79]). A trend is that more ordered structures are formed with more highly charged polyelectrolytes [76, 78, 79] and that the water content within the aggregates is rather small. For the charge-regulating and branched polyelectrolyte PEI mixed with SDS, an ordered lamellar structure is obtained at low pH, i.e. when the polyelectrolyte is highly charged. As the pH is increased to 10 and the charge density of PEI is reduced, the lamellar phase disappears but a stacking of individual polyelectrolyte–surfactant complexes in the bulk solution is still noticeable [78]. Interestingly, and broadly consistent with the above-mentioned result, neutron reflectivity studies recently demonstrated that multilayers of PEI and SDS formed at the air/water interface at pH 10 in a certain SDS concentration interval [80].

An example of a SAXS spectrum obtained for a mixture of a cationic, highly charged polyelectrolyte mixed with SDS is shown in Fig. 11.9. The relative positions of the peaks demonstrate that in this case a hexagonal phase has formed. It is suggested that cylindrical SDS micelles are wrapped by the polyelectrolyte and pack into a hexagonal arrangement.

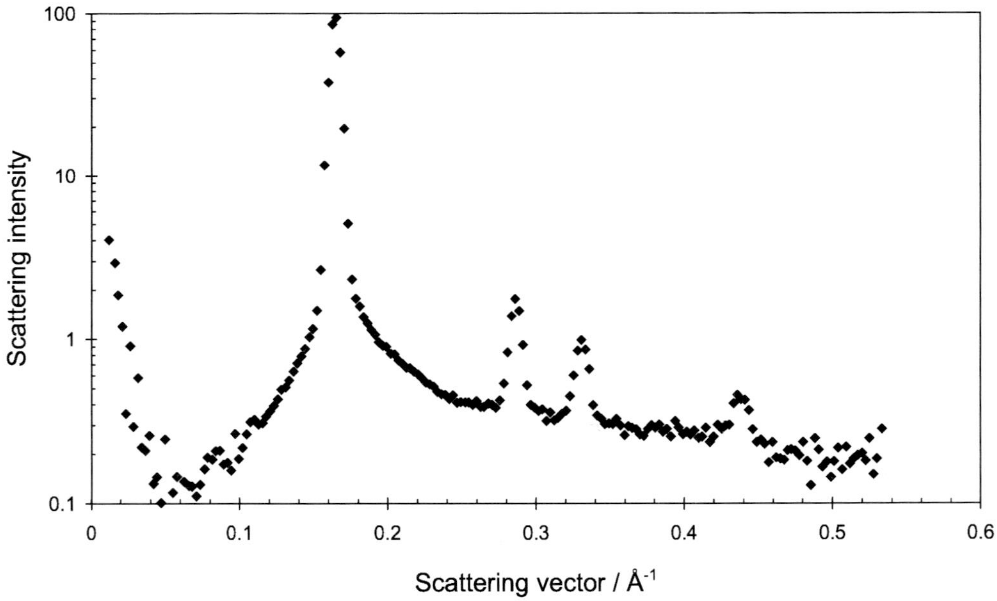

Fig. 11.9 Normalized small-angle X-ray scattering intensity as a function of scattering vector q for charge-neutralized mixture of SDS and PCMA (100% charge density). The concentration of PCMA was 1 wt%. (Adapted with permission from [79]; Copyright 2002, American Chemical Society).

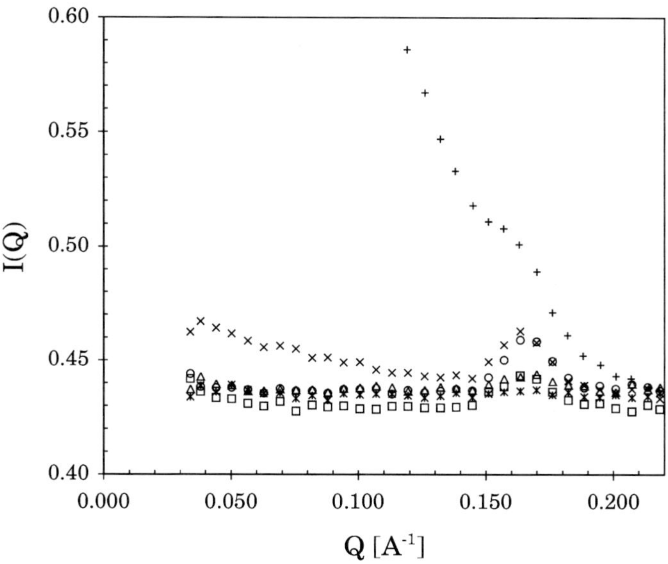

Fig. 11.10 Scattering intensity as a function of scattering vector at high q values for a 0.1 wt% solution of PCMA, a cationic polyelectrolyte carrying one permanent charge on each segment, in an H_2O–D_2O (80:20) mixture that contrast matches the polyelectrolyte. The deuterated SDS (d-SDS) concentration was 0.02 wt% (r=0.12, ◇), 0.05 wt% (r=0.3, △), 0.1 wt% (r=0.6, □), 0.2 wt% (r=1.2, ○), 0.5 wt% (r=3, ×) and 2 wt% (r=12, +), where r is the ratio of d-SDS to charged polyelectrolyte segments. The Bragg-like peak was also observed when the surfactant was contrast matched. (Adapted with permission from [75]; Copyright 2000, American Chemical Society).

This organization within the aggregates is not very sensitive to the polyelectrolyte-to-surfactant charge ratio, as demonstrated by SANS measurements [75]. The hexagonal packing exists for positively charged, charge-neutral and negatively charged aggregates, as demonstrated by the Bragg-like peak observed at large q values. The position of this peak remains unaltered, but it changes in intensity as the polyelectrolyte-to-surfactant charge ratio changes and the net charge of the aggregates changes sign (Fig. 11.10). This, combined with electrophoretic mobility measurements for the same system (Fig. 11.8), provides a very strong indication that the excess charges of the complex are located at the aggregate surface whereas the interior is close to uncharged.

Piculell and coworkers determined the complete phase diagram for mixtures of polyelectrolytes and oppositely charged surfactants. They noted that the true ternary phase diagram could be obtained by using mixtures of the complex salt of polyelectrolyte–surfactant, the surfactant and water [81]. The ternary phase diagrams for the complex salt cetyltrimethylammonium–polyacrylic acid (CTA-PA), the surfactant cetyltrimethylammonium bromide (CTABr) and water displayed a micellar phase, a small cubic phase and a large hexagonal phase and a number of two- and three-phase regions, as illustrated in Fig. 11.11.

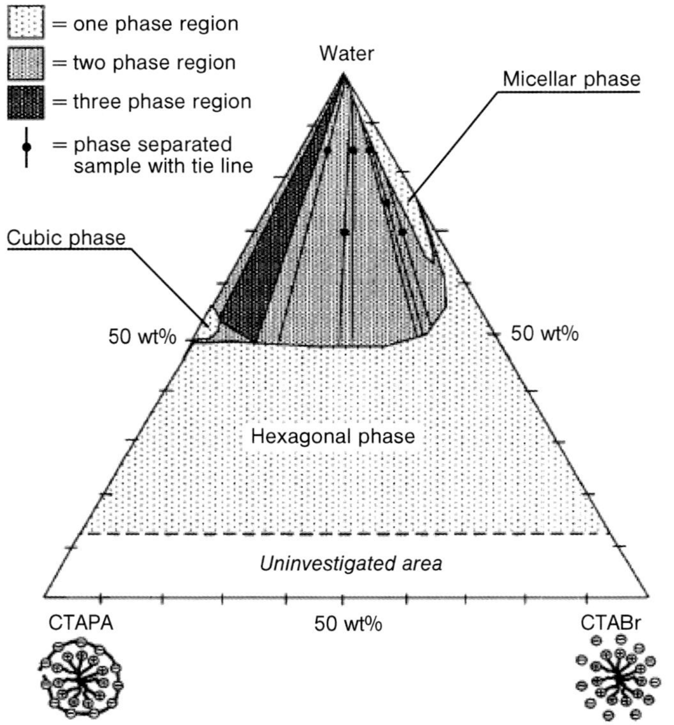

Fig. 11.11 Phase diagram of the complex salt CTAPA, the surfactant CTABr and water at 40 °C. (Reprinted with permission from [81]; Copyright 2002, American Chemical Society).

11.3 Interfacial Association

To achieve an understanding of the interfacial association between polyelectrolytes and oppositely charged surfactants is more complex than to understand the corresponding bulk association. One has to consider the interactions between the polyelectrolyte and the surfactant, but also the interaction between each component and the solid surface. Clearly, one can envisage a range of different situations. Braem et al. [82] suggested a simple and useful classification scheme to distinguish between the major different interaction possibilities. In their terminology, in a Class (I) system the surfactant has an affinity for the polymer and the surface is selective, i.e. only the surfactant or the polymer, but not both, has an affinity for the surface and it is suggested that this can be distinguished by a further subdivision into Class (Ip) (the surface is selective for the polymer) and Class (Is) (the surface is selective for the surfactant). A Class (II) system is defined as one where the surfactant binds to the polymer and the surface is non-selective. In a Class (III) system the surfactant has no affinity for the polymer and the surface is selective. This situation also gives rise to the division into Class (IIIp) and Class (IIIs). A Class (IV) system constitutes a system where the surfactant does not bind to the polymer and the interface is non-selective. Mixtures of polyelectrolytes and oppositely charged surfactants are obviously strongly associating and they will belong to either Class (I) or (II) depending on the nature of the solid surface.

From an experimental point of view, it is also important to distinguish experiments where the components are adsorbed sequentially, most often first the polyelectrolyte and then the surfactant, from experiments dealing with adsorption of preformed aggregates. The majority of the data reported in the literature relate to sequential adsorption experiments, whereas adsorption of preformed aggregates has, until recently, attracted less attention. This is largely due to experimental difficulties in investigating adsorption from turbid solutions, whereas in technical applications this situation is highly relevant.

In the following we discuss some experimental techniques that are useful for shedding light on polyelectrolyte–oppositely charged surfactant systems at interfaces and highlight a few of the investigations in the area with emphasis on results obtained in the 21st century.

11.3.1
Ellipsometry

Ellipsometry is an optical technique that is suitable for the determination of the optical properties of thin films on different surfaces. The underlying physical principle of the technique is the detection of the changes in the polarization state of a light beam when it is reflected from a surface. In this case, the electric field vector of the light can be resolved into two components, one lying in the plane of incidence (p component) and the other being perpendicular to this plane (s component). Since the two components generally exhibit different reflective properties and phase shifts during reflection, a change in the polarization occurs. As is customary in ellipsometry, the so-called ellipsometric angles (ψ and Δ) are measured from the orientation of the polarizer, compensator and analyzer. These ellipsometric angles are directly connected to the complex reflectance ratio (r) according to the following equations:

$$r = \frac{R_p}{R_s} = \tan\psi \exp(i\Delta) \tag{26}$$

$$\psi = \arctan\left|\frac{R_p}{R_s}\right|; \ \Delta = \phi_p - \phi_s \tag{27}$$

where R_p and R_s are the complex reflection coefficients and ϕ_p and ϕ_s are the phases of the parallel and perpendicular components of the polarized light.

Assuming an appropriate dielectric surface structure, for instance homogeneous, isotropic surface layers, the refractive index and thickness of the adsorbed layer (n_{ads} and d_{ads}, respectively) can be determined numerically from minimizing the difference between the experimental and simulated ellipsometric angles. However, the solutions are model dependent and extreme care must be taken to find the global minimum via sophisticated algorithms. Detailed descriptions of this procedure and the ellipsometric technique can be found elsewhere [83–87].

The measurements are not very sensitive to the separate determination of n_{ads} and d_{ads} when the adsorbed layer thickness is significantly smaller than the wavelength of the incident light beam, especially if the measurements are carried out at a single wavelength and angle of incidence. However, $(n_{ads} - n_b)d_{ads}$, which is proportional to the adsorbed amount, remains accurate due to the covariant nature of the standard deviation in n_{ads} and d_{ads} [88] (n_b denotes the refractive index of the bulk solution). A significant increase in the sensitivity towards the adsorbed layer thickness can be achieved by appropriate tuning of the substrate structure (for instance, varying the thickness of the silica layer on the top of the bulk silicon) and by sophisticated measurement and evaluation procedures [89]. A further improvement can be achieved by performing the measurements at multiple angles of incidence or at different wavelengths because of the gain in additional pairs of ellipsometric angles, which allows the determination of more unknown parameters.

The adsorbed amount (Γ) is only well defined in the case of a two-component solution [88]:

$$\Gamma = \frac{(n_{ads} - n_b)d_{ads}}{dn/dc} \tag{28}$$

where dn/dc is the refractive index increment of the solute which adsorbs on the surface.

One should keep in mind that the measured refractive index profile represents the inner, dense part of the adsorbed polymer layer [90]. Hence the layer thickness obtained from ellipsometry is generally smaller than the corresponding hydrodynamic thickness obtained by dynamic light scattering or the range of the steric force evaluated from surface force measurements. Because of this and the previously mentioned uncertainty in the adsorbed layer thickness determination, it is better to focus on the trends in the changes of d_{ads} and handle with care the absolute values of this quantity.

11.3.1.1 **Polyelectrolyte Adsorption**

The main thermodynamic information of the adsorption from dilute, binary solutions is provided by the experimental adsorption isotherm, e.g. in the equilibrium adsorbed amount versus equilibrium activity curves of the adsorbing component. The adsorption isotherms can be simply generalized according to the following expression [90]:

$$\frac{\Gamma}{\Gamma_0 - \Gamma} = K(\Gamma)a \tag{29}$$

where Γ and Γ_0 are the adsorbed amount and the saturation adsorbed amount, respectively, a is the activity of the adsorbing species and $K(\Gamma)$ is an apparent equilibrium constant, which is a function of the adsorbed amount and incorporates all interactions giving rise to deviations from ideal behavior. In adsorption studies of small molecules (surfactants), the adsorption isotherms are usually directly accessible experimentally. Therefore, via careful analysis of the experimental isotherms, the main parameters [Γ_0 and $K(\Gamma)$] governing the adsorption can be determined and understood. For instance, in [90], by analyzing the isotherms of alkyl sulfate homologues, a successful separation of $K(\Gamma)$ into the electrostatic and hydrophobic driving force of ionic surfactant adsorption was achieved.

On the other hand, in the case of polyelectrolyte adsorption, the situation is far more complicated. Due to the high number of surface–polymer segment contacts, the driving force of adsorption [$K(\Gamma)$] could be orders of magnitude larger for polymers than the values typical for small-molecule adsorption, especially at low polymer concentrations. This is pronouncedly true for the case of oppositely charged polyelectrolytes and surfaces and/or polyelectrolytes with non-electrostatic affinity towards the surfaces. According to Eq. (29), this results in the well-known high-affinity character of the adsorption isotherms of macromolecules, i.e. the beginning of these isotherms is so steep that saturation adsorption is attained at extremely low polymer concentrations. Various numerical approaches, based on extended versions of self-consistent mean field (SCF) theories of polyelectrolyte adsorption [91–93], also predicts that above 10^{-4} volume fraction, ~ 100 ppm polymer concentration, there is a negligible effect of the polymer concentration on the adsorbed amount.

In practice, this means that the adsorption isotherm of polyelectrolytes cannot be measured with appropriate accuracy and usually the adsorbed amount at one fixed polyelectrolyte concentration is measured and discussed in the majority of ellipsometric investigations under different conditions. This also means that the main thermodynamic information of adsorption, e.g. the driving force of adsorption [$K(\Gamma)$], is inaccessible experimentally. If the different experimental parameters investigated do not change the high-affinity character of the polyelectrolyte adsorption isotherm, then the effect of these parameters on the adsorbed amount at or close to saturation are usually measured [analogous to Γ_0 in Eq. (29)]. This quantity is related mainly to the effective size (surface area per molecule) of the adsorbed macromolecule, which is influenced by the chain length, stiffness, charge density and non-electrostatic affinity to the surface, the charge density of the surface, the ionic strength, pH and solvent quality of the medium.

The majority of ellipsometric investigations are concerned with oppositely charged macromolecules and surfaces since the adsorption of macromolecules having the same sign of charge as the surface or on neutral surfaces is electrostatically hindered, especially at low ionic strength and high charge density values of the polyelectrolytes and/or of the surface [94, 95]. Ellipsometric studies of adsorption of polyelectrolytes on oppositely charged surfaces reveal the effect of polymer solvency and charge density and also the ionic strength on the adsorbed amount. The results compare favorably with predictions of classical SCF theories [83, 94, 96]. Increasing the ionic strength can either increase [97, 98] or decrease [99] the adsorbed amount, depending on whether there is a considerable non-electrostatic affinity of the polymer segments or not towards the surface. On the other hand, the variation of the polymer segment charge may result in a maximum of the adsorbed amount at a low charge density value of the polymer [100]. The effect of the hydrophobic character (solvent quality) of the polyelectrolyte has a complex effect on the adsorbed amount. For cationic modified polyacrylamides with additional hydrophobic groups, the adsorbed amount on silica was found to decrease and the layer thickness to increase with increasing density of hydrophobic groups [95]. This finding is connected with the changes in the adsorbed molecular cross-section at the surface and conformation due to the hydrophobic side-chains. On the other hand, if the solvent quality decreases significantly it can induce a dramatic increase in the adsorbed amount since not single macromolecules but larger aggregates adsorb on the surface, which can act as nuclei for phase separation [100, 101].

As discussed above, the interpretation of the adsorption data is usually based on the equilibrium adsorption theories of polyelectrolytes. However, the kinetics of adsorption and desorption should also be carefully considered since the observed plateau value of the adsorbed amount as a function of time might be below the equilibrium amount adsorbed since the electrostatic barrier against further adsorption can shift the equilibration time far out of the practical laboratory time range [102, 103]. Finally, it should be noted that the desorption of polyelectrolytes is usually hindered due to the large number and strength of the segment–surface contacts [103, 104]. This feature and the fact that polyelectrolyte adsorption on an oppositely charged surface often results in overcompensation of the surface charge led to the recent technology of alternate polyelectrolyte multilayer deposition [105, 106].

11.3.1.2 Polyelectrolyte–Surfactant Systems

As mentioned above, the adsorbed amount is well defined only in the case of adsorption from binary solutions. If there are two adsorbing components (polyelectrolyte, surfactant) in the surface layer (in addition to the medium), then only the total optical surface excess, $(n_{\mathrm{ads}} - n_{\mathrm{b}})d_{\mathrm{ads}}$, is well defined. If the refractive index increment of the surfactant and that of the polymer are roughly equal, then the sum of their adsorbed amount might be calculated according to Eq. (28). However, the determination of the individual amounts of the polyelectrolyte and surfactant adsorbed can only be done by independent measurements of the adsorbed amount of one of the components or by making further assumptions. Consequently, ellipsometric measurements only provide information about the extent of adsorption from the polyelectrolyte–surfactant mixture. However, the adsorbed layer structure can be more efficiently investigated provided that the experimental setup gives sufficient sensitivity towards the separate determination of n_{ads} and d_{ads}.

Another important aspect of these systems is that due to the irreversible or, more correctly, extremely slowly reversible [103], nature of polyelectrolyte adsorption, significant order of addition effects and trapped non-equilibrium states in the adsorbed layers containing both polyelectrolytes and surfactant can be expected. Therefore, surfactant adsorption on preadsorbed polyelectrolyte layers must be distinguished from adsorption from polyelectrolyte–surfactant mixtures. The interfacial behavior of these mixed systems is often very complex and the results can be qualitatively interpreted on the basis of the individual adsorption properties of the surfactant and that of the polymer, in addition to the features of their bulk association. Recent studies, discussed in detail in the following, were chosen to represent a spectrum of the charged nature of the polyelectrolytes, surfactants and that of the surfaces.

Surfaces Selective for the Macromolecule: Polyelectrolytes with Oppositely Charged Surfactants and Surface, Class (Ip) Systems The majority of the recent and earlier studies belong to this group where the polymer is driven to adsorb on the surface whereas the individual surfactant adsorption is hindered electrostatically without the presence of the polyelectrolyte.

Branched Polyethylenimine–Sodium Dodecyl Sulfate–Silica PEI is a hyperbranched polyamine with high molecular weight containing primary, secondary and tertiary amine groups in a 1:2:1 ratio [107]. In a reflectometry study, the amount of PEI adsorbed on silica was found to increase with increasing ionic strength and pH, which can be interpreted qualitatively according to a significant non-electrostatic affinity of the PEI segments towards the silica surface [108]. This was also confirmed by streaming potential measurements, which revealed a considerable overcompensation of the silica surface charge, a result fully in line with direct force measurements [109]. This large overcompensation of the surface charge and the largely irreversible nature of the adsorption with respect to rinsing with the medium [104] are important reasons why PEI is often successfully used as the first layer in polyelectrolyte multilayer assemblies. Mészáros et al. also investigated the adsorption behavior of PEI on silica surfaces in the absence and presence of SDS by ellipsometry [107]. It was found that the adsorbed layer of PEI is very compact at low and moderate pH and at low ionic strength. At higher pH and ionic strength, the adsorbed layer is more extended but it can still be considered as fairly compact. These results indicate that the adsorbed branched PEI molecules contract significantly, in the direction perpendicular to the surface, to achieve an increasing number of surface–segment contacts.

In order to investigate the adsorption features of mixed PEI–SDS surface layers, measurements were performed where surfactant adsorption took place on silica precoated with PEI.

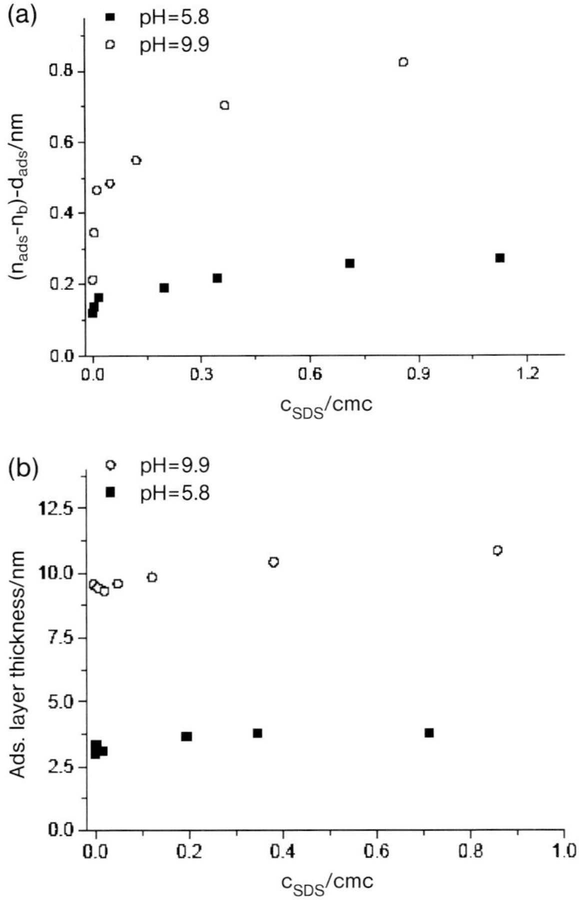

Fig. 11.12 (a) Optical, total adsorbed amount and (b) adsorbed layer thickness for SDS adsorption on preadsorbed PEI layers as a function of SDS concentration. The PEI was adsorbing on the silica wafer for 30 min, which was followed by a rinsing with the medium [0.01 M NaCl, pH (○) 9.9 and (■) 5.8]. After that the SDS solution was introduced at the same salt concentration and pH. $c_{PEI} = 50$ mg dm^{-3}. (Reprinted with permission from [107]; Copyright 2003, American Chemical Society).

In Fig. 11.12 a and b, the total optical surface excess and the adsorbed layer thickness of the preadsorbed PEI layers are shown, respectively, as a function of the added surfactant concentration (at pH 5.8 and 9.9 in 0.01 M NaCl). According to Fig. 11.12 a, the total adsorption increases with increasing surfactant concentration according to a saturation-type adsorption isotherm which correlates with the monomer binding mechanism of the DS^- ions to the PEI molecules in the bulk solution [51]. At pH 5.8, the amount of the surfactant adsorbed is considerably less than that at high pH. This observation can be explained by the lower amount of PEI adsorbed at this pH than at high pH, but it may also be connected with the very compact structure of PEI at the lower pH. This means that because of the vertical contraction of the adsorbed PEI molecules, the number of amine groups that are accessible to interact with the surfactant may be significantly reduced. This explanation is in line with the results presented in Fig. 11.12b, which reveal that the adsorbed layer thickness increases only slightly with SDS concentration. This finding suggests that the DS^- ion binds preferentially to the outer surface shell of the bound PEI molecules. Very similar results were found in the surface force study of Dedinaite et al. for the adsorption of SDS on preadsorbed PEI layers on mica surfaces, which indicated the formation of a surfactant layer adsorbed on top of the adsorbed PEI molecules [110]. According to this work, the association of PEI and SDS on mica shows a strong resemblance to the adsorption of ionic surfactants on oppositely charged surfaces. Namely, at low surfactant concentration a surfactant monolayer forms which neutralizes the excess charge of the polyelectrolyte and makes the surface hydrophobic. With increasing SDS concentration, a surfactant bilayer forms on the adsorbed PEI layer. This aspect of the SDS adsorption on preadsorbed PEI layers was recently utilized to build novel types of alternating multilayers containing PEI, anionic polyelectrolytes and SDS [111] and also PEI and anionic surfactant only [112].

The situation is different when the surfactant is allowed to adsorb simultaneously with PEI on silica surfaces. For this case, the total adsorption and the adsorbed layer thickness are shown as a function of the SDS concentration in Fig. 11.13. The surface was silica and the initial pH (pH^{in}) was 9.7.

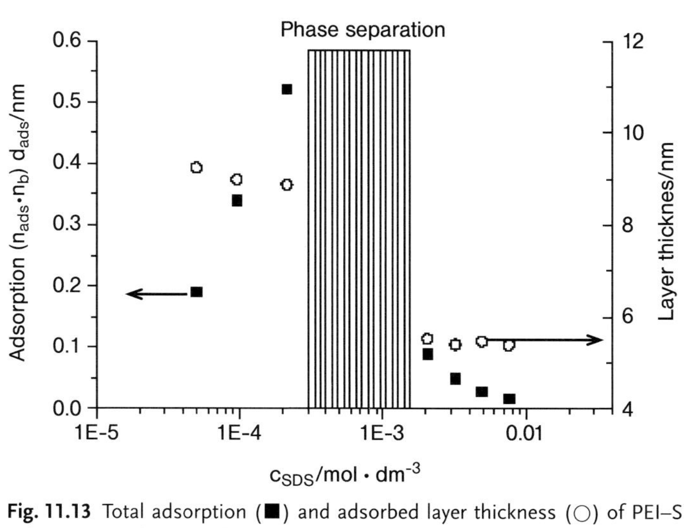

Fig. 11.13 Total adsorption (■) and adsorbed layer thickness (○) of PEI–SDS solutions on silica as a function of the total SDS concentration without added supporting electrolyte. $c_{PEI} = 50$ mg dm^{-3}; $pH^{in} = 9.7$. At this pH, the degree of protonation of PEI, $\Theta_0 \approx 0.11$. The pH was increased by more than one unit at the highest surfactant concentration (to pH 10.8). (Reprinted with permission from [107]; Copyright 2003, American Chemical Society).

A significant increase in the total adsorption takes place at low SDS concentrations, whereas above the precipitation concentration range the accumulation of the polymer and surfactant molecules in the adsorbed layer is small and decreases gradually with increasing surfactant concentration to nearly zero. These results can be interpreted by assuming that the PEI–SDS aggregates, formed in the bulk solution, adsorb on the silica surface. In the pre-precipitation concentration range, the initial positive charges of the PEI are gradually compensated with increasing SDS concentration and the polymer–surfactant complex shrinks due to the binding of the surfactant. Therefore, the total amount adsorbed may increase due to the decreasing extent of the segment–segment repulsion and the increasing hydrophobicity of the polymer–surfactant complex in addition to its decreasing size. On the other hand, in the post-precipitation domain the collapsed PEI–SDS particles possess a considerable negative surface charge density, which increases with increasing SDS concentration [51]. Taking into account that the silica surface is pronouncedly negatively charged at high pH, there is a significant energy barrier preventing adsorption of negatively charged PEI–SDS complexes. The adsorbed layer shows a roughly two-fold shrinking in the post-precipitation concentration range compared with the low surfactant concentration regime. This correlates well with the extent of the collapse of the PEI–SDS complex in bulk solution, which was recently detected by dynamic light scattering [51].

Cationic Amphiphilic Polyelectrolytes and Oppositely Charged Surfactants at the Silica/Aqueous Interface Samoshina et al. investigated the adsorption properties of copolymers of triethyl(vinylbenzyl)ammonium chloride and dimethyldodecyl(vinylbenzyl)ammonium chloride in the presence of SDS at silica surfaces [113]. The polymers denoted 0DT, 40DT and 80DT contained 0, 40 and 80 mol% monomers with dodecyl side-chains, respectively. The last two amphiphilic polyelectrolytes form intramolecular aggregates [114]. In the case of 40DT, essentially all dodecyl chains are packed into the hydrophobic core of the unimolecular aggregate. In the case of 80DT, some of the dodecyl chains remain outside the hydrophobic core of the intramolecular aggregate, which facilitates the formation of larger aggregates in bulk solution.

The surface excess of 0DT, 40DT and 80DT on silica was found, at low ionic strength, to be relatively low with an essentially flat adsorbed conformation of the polymers. As the ionic strength of the solution increased, an increase in the amount adsorbed and a more extended conformation away from the surface were observed, which is in line with the adsorption behavior of polyelectrolytes with a non-electrostatic affinity towards the surface. The adsorbed polyelectrolyte layers showed a high resistance towards rinsing, except 80DT. Due to its largely decreased solubility, the 80DT forms weakly adsorbed multilayers at the silica-water interface and upon rinsing a monolayer is left on the surface [114]. A tendency for weak multilayer formation has also been observed for 40DT on highly negatively charged mica surfaces [66].

The effect of SDS addition on the preadsorbed polyelectrolyte layers at low ionic strength is illustrated in Fig. 11.14. In this case, the polyelectrolyte was

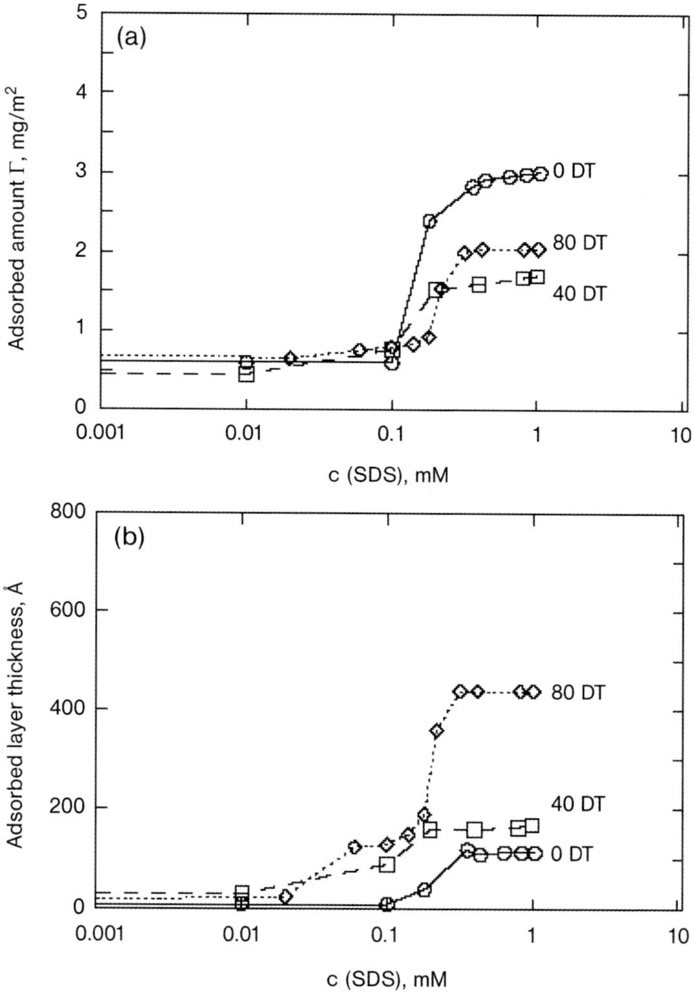

Fig. 11.14 Effect of SDS addition on the (a) amount adsorbed and (b) adsorbed layer thickness for the preadsorbed layers of 0DT, 40DT and 80DT at 1 mM NaCl. (Reprinted with permission from [113]; Copyright 2005, American Chemical Society).

not removed from the solution phase during SDS addition, which resulted in the co-adsorption of polyelectrolyte–surfactant complexes and individual surfactant molecules. Since the binding characteristics of the surfactants towards these polyelectrolytes are not known with sufficient accuracy, the plotted SDS concentrations are not the equilibrium monomer concentrations and a detailed interpretation of the data is difficult. However, in each case a large increase in the total amount adsorbed accompanied by swelling of the adsorbed layer occur as the SDS concentration approaches conditions where one would expect charge neutralization of the polymer with the surfactant; for the case of a 20 ppm 40DT solution, charge neutralized aggregates are formed when the SDS concentration reaches 0.3 mM [66]. Although no phase separation in the solution was detected under the conditions studied, the order of magnitude increase in the adsorbed layer thickness indicates multilayer formation, especially in the case of 40DT and 80DT. The total amount adsorbed increased continuously with SDS concentration for 40DT and 80DT, whereas it started to increase at a critical SDS concentration for 0DT. This was interpreted in terms of cooperative binding of SDS to the latter and non-cooperative binding of the surfactant to the hydrophobically modified polyelectrolytes. The adsorbed amount was observed to decrease only slightly on rinsing with SDS-free medium, suggesting that SDS promoted the incorporation of additional polyelectrolytes into the layer.

The authors also investigated the adsorption behavior of premixed 40DT–SDS solutions. A maximum in the amount adsorbed as a function of SDS concentration was found. These results can be similarly interpreted as the adsorption from PEI–SDS solution on silica reported in Fig. 11.13. At low SDS concentration, polyelectrolyte–surfactant complexes with net positive charge form, which increases the amount adsorbed compared with the adsorption of the polyelectrolyte alone. However, at high surfactant concentration, the overall charge of the 40DT–SDS complex reverses, which prevents adsorption on the negatively charged silica surface. The authors also reported that the adsorbed amount and layer thickness depend significantly on the method of polymer–surfactant solution preparation. These latter findings are in good agreement with the observed non-equilibrium aspects of the polyelectrolyte–surfactant complex formation [51, 64, 69].

Adsorption of Cationic Cellulose Derivatives–Anionic Surfactant Complexes on Silica Surfaces The effect of SDS on the adsorption behavior of cationic hydroxyethylcellulose derivatives (Polymer JR-125, JR-400 and JR-30M) and a hydrophobically modified cationic cellulose (Quatrisoft LM-200) at silica surfaces in the presence of 10 mM NaCl solution was investigated by Terada et al. [115]. The JR polymers have different molecular weights and almost the same cationic charge density. The cationic charge density of LM-200 is much smaller than those of the other cationic cellulose derivatives used in this study. The amount adsorbed and layer thickness of the hydrophobically modified LM-200 on silica were found to be considerably larger than for the other polymers investigated.

The effect of SDS was investigated under three different conditions: additions of SDS to preadsorbed polymer layers, co-adsorption of the polymer and SDS on a preadsorbed polymer layer and adsorption from premixed polymer–SDS solutions. As an illustration, some of the data for LM-200 are presented in Fig. 11.15 a and b. Here the adsorbed amount and layer thickness are shown on preadsorbed LM-200 for the cases of co-adsorption of surfactant and polymer, and also for adsorption of SDS alone. In absence of polymer in the bulk and at low SDS concentrations, the adsorbed amount increases slightly whereas the adsorbed layer compacts with increasing SDS concentration due to incorporation of the surfactant in the preadsorbed layer. However, above a critical concentra-

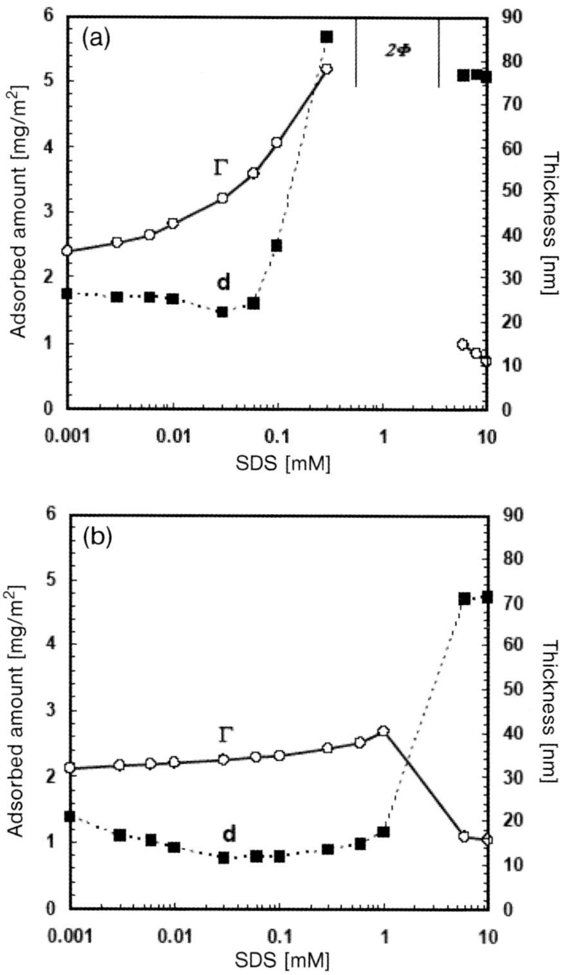

Fig. 11.15 (a) Effect on the amount adsorbed and the layer thickness of SDS addition to preadsorbed Quatrisoft LM-200 layers. The polymer concentration was fixed at 100 ppm and the NaCl concentration was 10 mM. The polymer was not removed from bulk solutions before SDS addition. (b) Effect on the amount adsorbed and the layer thickness of SDS addition to preadsorbed Quatrisoft LM-200 layers. The polymer was removed from bulk solutions before SDS addition. The NaCl concentration was kept at 10 mM. In both cases, open circles show the adsorbed amount and filled squares show the layer thickness. (Reprinted with permission from [115]; Copyright 2004, American Chemical Society).

tion, the adsorbed layer swells dramatically and the adsorbed amount decreases significantly. In this concentration range the surfactant interaction seems to compete effectively with the surface for the oppositely charged groups of the polymer, which results in the unbinding of a large number of segment–surface contacts. It should be noted that these observations are in contrast to the observed characteristics of adsorbed PEI layers, discussed previously, where no significant desorption and swelling of the adsorbed layer were detected, presumably due to the strong nature of PEI segment–silica surface contacts [107].

The co-adsorption of LM-200 and SDS on the preadsorbed LM-200 layers results in much larger amounts adsorbed at low surfactant concentration compared with when only SDS is added. This is attributed to further adsorption of both polyelectrolyte and surfactant molecules. At SDS concentrations close to the charge neutralization of the bulk polyelectrolyte–SDS complex, multilayer formation might also occur. On the other hand, in the post-precipitation region the adsorbed amount is found to be low and the adsorbed layer is swollen due to SDS incorporation and a reduction in the number of surface–segment contacts. Since the polyelectrolyte–surfactant complex has a negative net charge density in the post-precipitation region, no significant adsorption of aggregates is expected or observed. These findings are in line with previous studies of co-adsorption of SDS and a cationic cellulose derivative on preadsorbed layers of the polymer [116].

The adsorption behavior from premixed LM-200–SDS solutions [115] was found to be similar to the previously reviewed adsorption characteristics PEI–SDS [107] and 40DT–SDS solutions [113] on silica. In particular, a maximum in the adsorption as a function of SDS concentration was detected and attributed to the variation in the electrostatic driving force of the polyelectrolyte–surfactant complexes towards the silica surface. An interesting application of this study is connected with hair-care. Cationic cellulose derivatives are typical polymers used in hair shampoo formulations. These formulations also contain oppositely charged ionic surfactants. Therefore, diluting the polymer–surfactant solution in contact with the adsorbed layer mimics the situation during hair washing. It was found that at low surfactant concentration this type of rinsing decreased the adsorbed amount, possibly due to removal of the surfactant from the adsorbed layer. On the other hand, in the post-precipitation region a sharp increase in adsorbed amount was observed during the rinsing process. This is directly related to the precipitation of polymer–surfactant complexes on the surface upon rinsing.

Non-selective Surfaces: Polyelectrolytes with Oppositely Charged Surfactants on Hydrophobic (Neutral) Surfaces, Class (II) Systems There are only a few studies dealing with adsorption from mixtures of polyelectrolytes and oppositely charged surfactants on to non-selective surfaces. In most of these cases, the driving force for adsorption is hydrophobic interactions. Terada et al. investigated the effect of SDS on the adsorption behavior of cationic hydroxyethylcellulose (Polymer JR-400) and a hydrophobically modified cationic cellulose (Quatrisoft LM-200) at hydrophobized silica by null ellipsometry [117] and compared with the data for adsorption on hydrophilic silica surfaces [115] obtained under the same conditions. In all cases, the maximum amount adsorbed was obtained at an SDS concentration just before the phase separation region whereas the complex desorbed from the surface at high SDS concentrations.

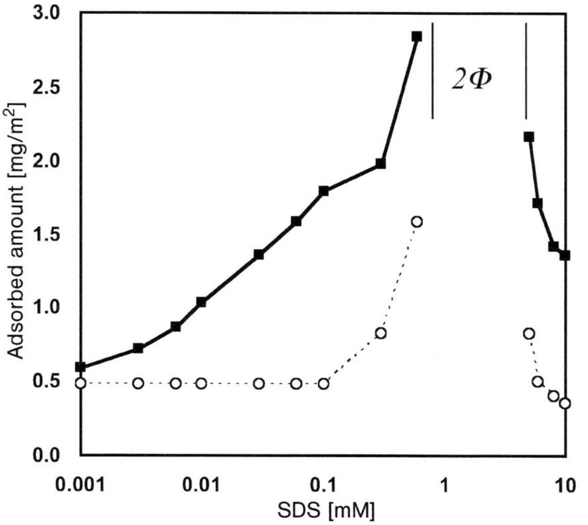

Fig. 11.16 Effect on the amount adsorbed of SDS addition on preadsorbed JR-400 layers on hydrophobized silica and on silica. The polymer concentration was fixed at 100 ppm and the NaCl concentration was 10 mM. The polymer was not removed from the bulk solution before SDS addition. Filled squares show the data for hydrophobized silica and open circles show the data for silica. (Reprinted with permission from [117]; Copyright 2004, American Chemical Society).

For the hydrophobically modified LM-200, no large difference in the adsorption behavior on the two surfaces was observed. On the other hand, for JR-400, the amount adsorbed on hydrophobized silica increased when a small amount of SDS was added. This effect was observed already at SDS concentrations far below the reported c.a.c. in bulk. On the other hand, on silica the adsorbed amount did not increase sharply until the SDS concentration was above the c.a.c. (Fig. 11.16). This difference is due to the fact that the SDS adsorbs on hydrophobized silica surfaces from very low surfactant concentrations whereas it does not adsorb on silica surfaces.

The authors also investigated the effect of dilution of the polymer–surfactant solutions in contact with the surface layer. It was found that when the polyelectrolyte–surfactant solution in the post-precipitation region was diluted by adding water, the adsorbed amount increased due to precipitation of the complexes. However, the kinetics of the adsorption (or desorption) during this rinsing process depended strongly on the hydrophobicity of the polymer and the state of the surface.

In another study, Cardenas et al. investigated the adsorption of anionic DNA, cationic cetyltrimethylammonium bromide ($C_{16}TAB$) and their mixtures at hydrophobized silica surfaces by means of ellipsometry [118]. In this work, interesting effects could be seen upon co-adsorption of the cationic surfactant and DNA on the preadsorbed DNA layer. A significant increase in the adsorbed amount was observed which was accompanied by a dramatic compaction of the adsorbed layer at low surfactant concentrations. This observation is correlated with the formation of DNA globules in bulk solution in the presence of the cationic surfactant. Rinsing with salt solution induced a decrease in the adsorbed amount, which was interpreted as being due to desorption of the cationic surfactant, leading to a denser DNA layer structure at the interface than formed by surfactant-free DNA. The final structure of the adsorbed layer was found to depend on the history of formation of the complexes, illuminating the importance of trapped non-equilibrium states and the length of the DNA.

Surfaces Selective for the Surfactant: Polyelectrolytes with Oppositely Charged Surfactants and Like Charged Surface, Class (Is) Systems In this case, the adsorption of the polyelectrolyte alone is electrostatically hindered whereas for the surfactant alone adsorption occurs. An interesting example of this situation was presented recently by Cardenas et al. [119], who investigated adsorption from aqueous mixtures of DNA and dodecyltrimethylammonium bromide ($C_{12}TAB$) on silica. The adsorbed amount as a function of surfactant concentration in the presence and absence of DNA is illustrated in Fig. 11.17. It should be noted that in the case of DNA–$C_{12}TAB$ mixtures the equilibrium concentration of the $C_{12}TAB$ is not equal to the total $C_{12}TAB$ concentration plotted in Fig. 11.17 due to the binding of the surfactant to the DNA. This makes it difficult to compare directly the adsorption for the two different cases. However, there is a very small amount adsorbed on the negatively charged silica surface at low surfactant concentration where the DNA–surfactant complex is still negatively charged and the adsorption of the free surfactant is also negligible. However, with increasing surfactant concentration, approaching the DNA charge neutralization, a very steep increase is observed. Since in this concentration range the adsorption of individual surfactant molecules is still negligible, the adsorption of aggregates of closely neutral DNA–surfactant complexes takes place. The adsorbed layer formed from these DNA–surfactant complexes could be easily removed by

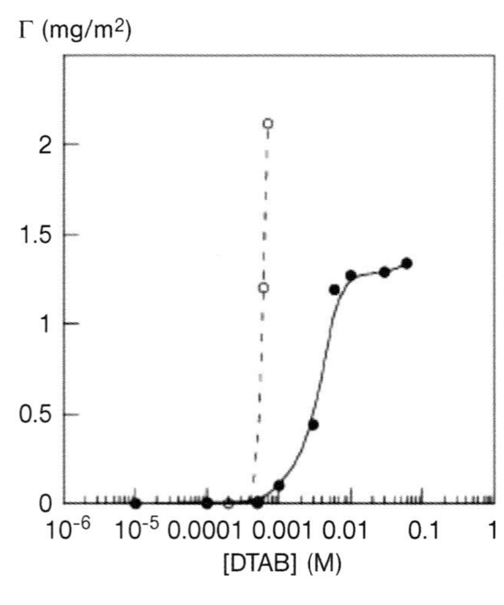

Fig. 11.17 Adsorption isotherm for $C_{12}TAB$ (filled circles) and DNA–$C_{12}TAB$ mixtures (open circles) as measured by ellipsometry. 0.06 mg mL^{-1} DNA was used for the mixed system. Lines are included as guides for the eye. (Reprinted with permission from [119]; Copyright 2004, American Chemical Society).

rinsing with the medium. This finding is in sharp contrast to the majority of cases discussed in previous sections where a significant driving force for polyelectrolyte adsorption towards the surface exists.

A further increase in $C_{12}TAB$ concentration leads to phase separation both in bulk and at the silica interfaces. An interesting result of this work was connected with the observed adsorption kinetics. During the adsorption process, significant decompaction of the surface layer takes place at high surface coverage, in spite of the fact that large aggregates of DNA–surfactant complexes adsorb on the surface. This observation is in line with the adsorption behavior of DNA–$C_{16}TAB$ complexes on hydrophobized surfaces [118].

Finally, it should be mentioned that these results are of relevance for the development of gene delivery vehicles in the sense that they provide an insight into the interactions of DNA–cationic amphiphile complexes with negatively charged surfaces that could be seen as a simple model for cell membranes.

11.3.2
Dual Polarization Interferometry (DPI)

DPI is a relatively new technique that so far has been used mainly for studying biomolecular interactions at interfaces. The technique is fully described in the literature [120–122]. Briefly, a laser beam is passed through a polarizer switch and directed into a dual waveguide. The interference of the light emitted from the other side of the waveguide is registered by a photodiode array. It should be noted that the device utilizes two different polarization directions of the light simultaneously and independently. Thus, the refractive index and the layer thickness can be determined by combining two separate measurements, which is an important difference from ellipsometry. Adsorption events are registered by pumping a reference solution through one of the waveguides at the same time as the same solution with the added adsorbate is pumped through the other waveguide. The adsorption of the adsorbate changes the refractive index on the solution side of the sample waveguide surfaces. The interaction between the evanescent wave and the adsorbed layer affects the light reflected and hence the position of the interference fringes registered by the photodiode array. Since the interference fringes for two polarization directions are registered, it is possible to calculate the thickness and the refractive index of the adsorbed layer with very high accuracy. The adsorbed mass can then be calculated using the same procedure as for ellipsometric measurements. We note that just as in ellipsometric measurements the response of the instrument is evaluated using an optical model of the surface layer, typically a homogeneous smooth layer characterized by its refractive index and thickness. It is therefore the refractive index and the thickness of such an idealized layer that are returned by the analysis. To facilitate the measurements, a waveguide material with high refractive index is needed and in the DPI silicon oxynitride (SiO_xN_y) is used. The surface of the waveguide can be modified to change the surface chemistry. One important difference compared with ellipsometry is that the light is traveling through the waveguide, not through the solution, which is an advantage when investigating adsorption events from turbid solutions.

Some results that we obtained with DPI for a Class (Ip) system are illustrated in Fig. 11.18. The data show the change in refractive index, layer thickness and adsorbed amount when a preadsorbed layer of chitosan, a cationic polysaccharide, in 30 mM $NaNO_3$ is exposed to SDS solution at a concentration corresponding to the c.m.c. The preadsorbed chitosan layer is very thin (0.7 nm) and the adsorbed amount is 0.32 mg m^{-2}. Addition of SDS increases the layer thickness and the adsorbed amount. The refractive index change is small but significant. Rinsing with surfactant-free 30 mM $NaNO_3$, results in a decrease in layer thickness and adsorbed mass, but not entirely back to the original values. Clearly, not all SDS is removed by rinsing, which is consistent with the strong association between chitosan and SDS. Similar conclusions were drawn by Dedinaite et al. when investigating the effect of SDS on another preadsorbed layer of a highly charged cationic polyelectrolyte using surface force techniques [123].

Interfacial association between chitosan and SDS has also been studied with the surface force apparatus [124]. In this case, it was found that chitosan adsorbed on mica from 30 mM NaCl solution formed a significantly thicker layer than that formed on the silicon oxynitride surface used in the DPI. However, at lower ionic strength, the chitosan layer thickness on mica was found to be the same as that one obtained with DPI. Further, the association between SDS and the thin chitosan layer on mica is very similar to the SDS association with the thin chitosan layer on the silicon oxynitride surface. This comparison indicates that it is the initial state of the preadsorbed layer, rather than the ionic strength

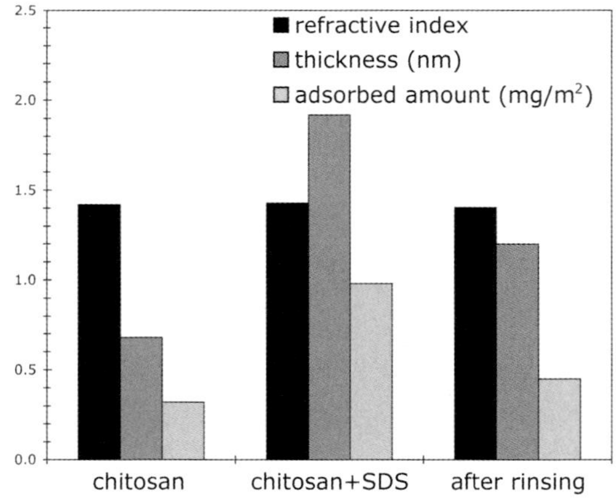

Fig. 11.18 Refractive index (black), layer thickness (dark gray) and adsorbed amount (light gray) determined for a layer of chitosan adsorbed on silicon oxynitride from 30 mM $NaNO_3$ solution (first group). The values of these quantities after exposure to a 1 c.m.c. SDS solution in 30 M $NaNO_3$ are shown by the second group of bars and the values after rinsing with surfactant-free 30 M $NaNO_3$ by the third group of bars.

of the solution, that determines the interfacial association between chitosan and SDS, at least up to an ionic strength of 30 mM.

11.3.3
Quartz Crystal Microbalance with Dissipation (QCM-D)

A QCM-D measurement allows the simultaneous determination of changes in the resonance frequency (Δf) and energy dissipation (ΔD) that occur as a result of an adsorption process. The resonance frequency of the crystal, f_0, decreases to a lower value, f. When the adsorbed mass is evenly distributed, rigidly attached and small compared with the mass of the crystal, $\Delta f = (f - f_0)$ can be related to the sensed mass per unit area (Δm) by the Sauerbrey equation [125]:

$$\Delta m = \frac{C \Delta f}{n} \tag{30}$$

where n is the vibrational resonance peak number (1, 3, 5, 7) and C is a constant that describes the sensitivity of the device to changes in mass. The Sauerbrey equation is derived for uniform ultra-thin rigid films in vacuum with material properties indistinguishable from those of the crystal resonator. It has been shown, however, that Eq. (30) is applicable also for measurements in liquids, provided that the adsorbed layer is homogeneous, thin and rigid [126]. One should note, however, that the mass sensed contains contributions from all material that oscillates with the crystal, i.e. both from the adsorbed species and from solvent associated with this layer. Hence the sensed mass should be expected to be larger than the adsorbed mass determined by, e.g., ellipsometry or DPI.

Energy losses occur in the crystal and due to the adsorbed material, which, when the driving voltage is turned off, leads to damping of the oscillation with an amplitude decay rate that is affected by all energy dissipative mechanisms in the system. The dissipation factor D is defined as

$$D = \frac{E_{\text{dis}}}{2\pi E_{\text{st}}} \tag{31}$$

where E_{dis} is the dissipated energy and E_{st} is the stored energy during one oscillation cycle. The dissipation change is defined as $\Delta D = D - D_0$, where D_0 denotes the dissipation when the sensor is immersed in solution prior to the start of the adsorption process and D is the dissipation at any given time during experiment. A large value of ΔD signifies a large energy dissipative power of the adsorbed layer and this is most often observed for thick and non-rigid layers.

For layers formed by mixtures of polyelectrolytes and surfactants, it is far from obvious that the simple Sauerbrey relation will hold. In such a case, the data analysis model of Johannsmann et al. [127] may be useful. They derived the following equation, which allows the true sensed mass to be calculated for viscoelastic layers:

$$\delta\hat{f} \approx -f_0 \frac{1}{\pi\sqrt{\rho_q \mu_q}} \left[f\rho d + \hat{J}(f) \frac{f^3 \rho^2 d^3}{3} \right] \tag{32}$$

where δf is the shift in the complex frequency, f_0 is the fundamental resonance frequency of the quartz crystal in air, f is the resonance frequency of the crystal in contact with solution, d is the thickness of the film and $\hat{J}(f)$ is the complex shear compliance. The quantities ρ_q and μ_q are the specific density and the elastic shear modulus of quartz, respectively, and ρ is the density of the fluid.

Equation (32) can be transformed into a more convenient form by using the equivalent mass [127], defined by

$$\hat{m}^* = -\frac{\sqrt{\rho_q \mu_q}}{2f_{\text{unl},1}} \frac{\delta\hat{f}}{f} \tag{33}$$

where $f_{\text{unl},1}$ is the fundamental resonance frequency of the unloaded quartz oscillator. One thus obtains

$$\hat{m}^* = m^0 \left[1 + \hat{J}(f) \frac{\rho f^2 d^2}{3} \right] \tag{34}$$

The true sensed mass, m^0, can then be calculated under the assumption that $\hat{J}(f)$ is independent of the frequency in the accessible frequency range where a plot of the equivalent mass against the square of the resonance frequency f^2 gives the true sensed mass as the intercept. Finally, we note that if the density and viscosity of the solution containing the adsorbing species differ from those of the reference solution without the adsorbate, then corrections have to be made for the changes in these values [128].

11.3.3.1 **Preadsorbed Polyelectrolyte Layers**

The adsorption of a 10% charged polyelectrolyte, AM-MAPTAC-10, on gold surfaces and the subsequent association between this preadsorbed layer and SDS has been monitored by QCM [129]. It was found that addition of a small amount of SDS had an insignificant effect, but once the SDS concentration had reached about 2 mM (~20% of the c.m.c. of SDS), the resonant frequency decreased and the dissipation increased (Fig. 11.19). This is a sign that SDS is incorporated in the adsorbed layer and the recharging of the layer results in a swelling. A further increase in SDS concentration to 5 mM caused a massive swelling that resulted in a large decrease in frequency and increase in dissipation. Once the SDS solution had been replaced by pure water, the layer once again collapsed on the surface, as evidenced by the changes in dissipation and frequency.

By comparison with other studies [130], it was found that the SDS concentration needed to induce swelling of the AM-MAPTAC-10 layer on gold was significantly higher than that needed to cause the same effect on a preadsorbed layer

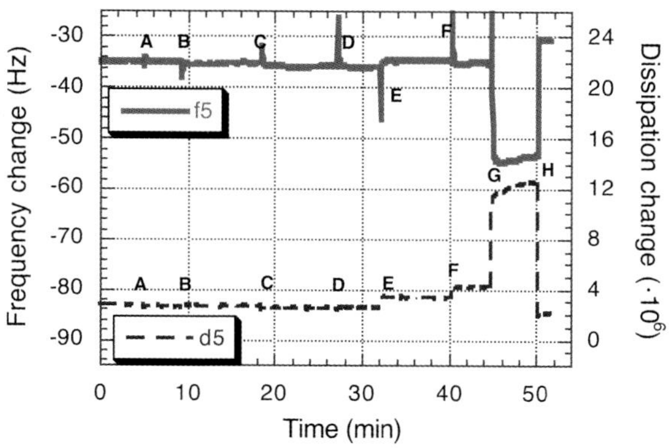

Fig. 11.19 Plot illustrating the swelling of a preadsorbed layer of a cationic polyelectrolyte with 10% of the segments charged due to the presence of anionic surfactant (SDS). The solid line represents the frequency change and the dashed line represents the dissipation change. Both curves are for the fundamental frequency (~5 MHz). The polyelectrolyte-coated gold surface was rinsed twice with water (at points A and B) and then exposed to SDS solutions with concentrations of 0.17 mM (point C), 0.34 mM (point D), 1.7 mM (point E), 5 mM (point F) and 16.6 mM (point G), and finally rinsed with pure water at point H. (Adapted with permission from [129]; Copyright 2002, American Chemical Society).

of the same polyelectrolyte on mica. This indicated that the polyelectrolyte was more strongly bound to gold than mica, which subsequently was attributed to image charge effects [131].

11.3.3.2 **Adsorbed Layers Formed by Preformed Aggregates**

The adsorption of preformed aggregates of polyvinylamine (PVAm) and SDS on polystyrene has been investigated as a function of the polyelectrolyte to surfactant ratio [132]. Since the QCM technique does not rely on optical evaluation, it works equally well in highly turbid and in transparent solutions. This is an important advantage when considering mixtures of polyelectrolytes and oppositely charged surfactants. It was found that both PVAm and SDS had an affinity for the surface and this mixture is therefore, according to the Tilton classification system, a Class (II) system.

At low surfactant concentrations, a mixed adsorbed layer was formed. In the colloidally stable region, both in excess polyelectrolyte and in excess surfactant, the adsorption was limited and an equilibrium value for the sensed mass was obtained within 1 h. Apparently, the electrostatic repulsion between adsorbed aggregates and aggregates present in solution limits the adsorption. However, in the colloidally unstable region, a slow deposition of large aggregates continued for a prolonged time (Fig. 11.20). In a large excess of surfactant, the mass sensed was the same as for the corresponding PVAm-free SDS solution, demonstrating a total dominance of the surfactant at the interface.

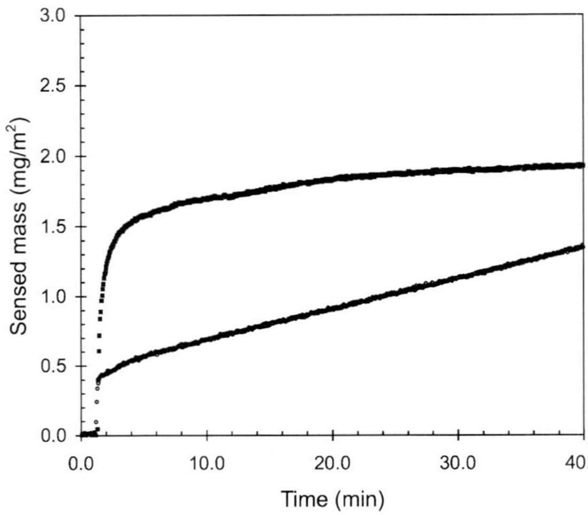

Fig. 11.20 Sensed mass as function of time for PVAm mixed with SDS. The PVAm concentration was 20 ppm in both cases. The upper curve corresponds a sample with an SDS concentration of 0.05 mM, where positively charged and stable complexes form. Here adsorption equilibrium is established after about 30 min. The lower curve corresponds to an SDS concentration of 0.3 mM, where uncharged and unstable aggregates form in solution. The sensed mass continues to increase linearly with time due to precipitation.

11.3.4
Neutron Reflectivity

Neutron reflectivity measurements are probably the experiments that provide the most detailed structural information on adsorbed layers. A very important advantage is that the contrast can be tuned by varying the solvent composition in terms of the H_2O to D_2O ratio and by deuterating one component in a mixture or a specific region of a component. This allows individual classes of molecules to be probed selectively, e.g. the surfactant in a polyelectrolyte–surfactant mixture or even separate parts of a molecule. It is normally assumed that the properties of a system are not affected by the H_2O to D_2O ratio, but some care needs to be taken when considering, for instance, weak polyelectrolytes and mixtures between such polyelectrolytes and surfactants [133] and also for compounds containing sugar groups [134], where significant solvent isotope effects have been reported. This is presumably related to the exchange of hydrogen for deuterium within the structure of such compounds.

Neutron reflectivity studies have been reported for polyelectrolyte–oppositely charged surfactants adsorbed to the air/water interface [80, 135–140] and also more recently for such systems adsorbed on silica [141, 142] and polystyrene [143].

Penfold et al. showed that the highly charged cationic polyelectrolyte used in their experiments adsorbed in a very flat conformation on the silica surface

[141, 142], in agreement with results for similar systems investigated by force measurements, DPI and ellipsometry. The presence of the cationic polyelectrolyte on the surface promoted adsorption of anionic and counteracted adsorption of cationic surfactants, a result that is consistent with a predominantly electrostatic driving force for adsorption. The SDS layer adsorbed on the thin polyelectrolyte layer was found to be 4 nm thick [141] and adsorbed on top of the polyelectrolyte. Similar structures have been found for SDS adsorbing on mica surfaces precoated with PEI [110] and with chitosan [124]. The presence of the polyelectrolyte on the surface was also found to promote SDS adsorption from mixtures of SDS and non-ionic surfactant [141]. Penfold et al. also noted that a thicker adsorbed polyelectrolyte layer was more efficient in promoting SDS adsorption than a thin layer, particularly at low surfactant concentration [142]. Dedinaite et al. drew similar conclusions for the chitosan–SDS system on mica [124]. These findings are related to the easier access to the charged groups of the polyelectrolyte segments for the surfactant in thicker layers. Finally, we note that at high SDS concentrations the data of Penfold et al. demonstrated that a simple bilayer adsorption of SDS on the polyelectrolyte coated surface no longer described the structure, but a considerably thicker layer was obtained [80]. It is tempting to associate this structural change with the formation of large interfacial aggregates with an ordered internal structure, as reported by Dedinaite et al. for another polyelectrolyte–SDS system on mica [123].

11.3.5 Surface Forces

Polyelectrolyte–surfactant mixtures adsorbed on solid surfaces are bound to affect strongly the interactions between the surfaces and between such surfaces and other components present in solution. Considering the different structures that can form in bulk solutions containing polyelectrolytes and oppositely charged surfactants, it seems plausible that the forces induced by the presence of adsorbed mixtures of polyelectrolytes and surfactants also should display a rich behavior. This is indeed the case and some data available in the literature have been reviewed fairly recently [1], so here we limit the description to illustrate some typical effects that can be encountered.

11.3.5.1 Preadsorbed Polyelectrolyte Layers, Class (Ip) Systems

The case where the polyelectrolyte is preadsorbed and then the layer is exposed to a surfactant solution containing no polyelectrolyte in the bulk is the most commonly investigated situation. This is also by far the simplest case since no deposition of polyelectrolyte–surfactant aggregates from solution can take place. Nevertheless, a range of different processes can occur and these will be described below. All systems that will be discussed are Class (Ip) systems, i.e. the polyelectrolyte and surfactant associate and the surface is selective for the polyelectrolyte.

Desorption In cleaning applications, surfactants are added to promote the removal of unwanted surface species, "dirt", that has attached to the surface. It is therefore no surprise that in some cases, but far from all, addition of surfactant to a preadsorbed layer of polyelectrolyte induces desorption. An example where desorption is the dominant effect induced by surfactant exposure is illustrated in Fig. 11.21.

The force data shown in Fig. 11.21 were obtained for a low charge density polyelectrolyte [144], AM-MAPTAC-1, where the digit "1" signifies that 1% of the segments are the cationically charged MAPTAC unit whereas the other 99% of the segments are uncharged acrylamide (AM), adsorbed on negatively charged mica surfaces. The forces were probed in the absence and presence of the anionic surfactant SDS. In the absence of surfactant, a long-range repulsive force dominates. The decay length of this force is not consistent with a double-layer force at the ionic strength of the solution, demonstrating that a steric force is the dominant repulsive force. As the surfactant concentration is increased, a progressive reduction in the range of the steric repulsion is observed and the layer thickness obtained under high compression is also reduced. Complementary experiments show that the amount of AM-MAPTAC-1 remaining on the surface decreases with SDS concentration [145] and the changes in surface interactions are thus predominantly due to desorption of the polyelectrolyte layer. The removal must be due to formation of AM-MAPTAC-1–SDS complexes that have a lower affinity for the surface than the polyelectrolyte itself and the electrostatic repulsion between the complex and the surface is sufficient to cause detachment. This interpretation is supported by the observation that SDS desorbs the polyelectrolyte more readily than addition of simple salts such as NaCl

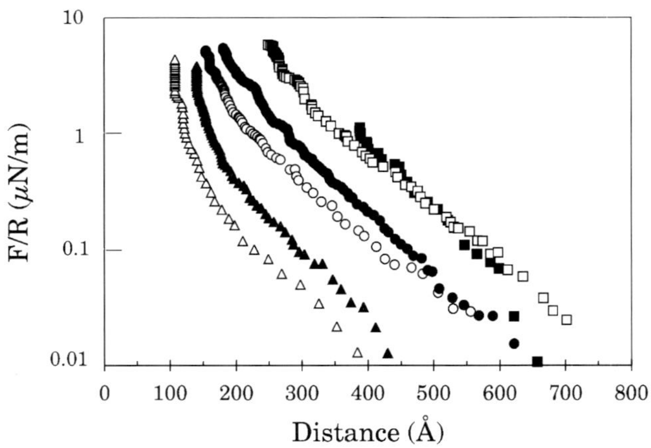

Fig. 11.21 Force normalized by radius as a function surface separation between mica surfaces precoated with AM-MAPTAC-1. The forces were measured across an aqueous solution containing SDS at the following concentrations: zero (■), 0.1 c.m.c. (□), 0.3 c.m.c. (●), 0.75 c.m.c. (○), 1 c.m.c. (▲) and 2 c.m.c. (△). (Adapted from [144]; Copyright 2001, with permission from Elsevier).

[145]. It should also be observed that desorption of the polyelectrolyte by SDS addition occurs less readily when the polyelectrolyte charge density is increased [145], which is due to the increased polyelectrolyte–surface affinity.

Swelling The polyelectrolyte–surface affinity, and also the polyelectrolyte–surfactant affinity, increase when the charge density of the polyelectrolyte is increased. Thus, AM-MAPTAC-10, with 10% of the segments carrying a permanent positive charge, adsorbs in a thinner layer on negatively charged mica than AM-MAPTAC-1. The forces acting between the AM-MAPTAC-10-coated surfaces display a weak attraction from a separation of about 20 nm, followed at smaller separations by a steep steric repulsion (Fig. 11.22).

Addition of a small amount of SDS results in dramatic swelling of the adsorbed layer [130] as the surfactant self-assembles around the adsorbed polyelectrolyte chains. The force data, displaying a sudden jump into contact from a separation of 50-nm, indicate that the surfactants are cooperatively desorbed upon compression (at $F/R = 15$ mN m^{-1}) when the electrostatic free energy penalty of confining the charges becomes larger than the free energy gain of association with the polyelectrolyte. Similarly on separation, a cooperative re-adsorption appears to occur when the compressive load is sufficiently released

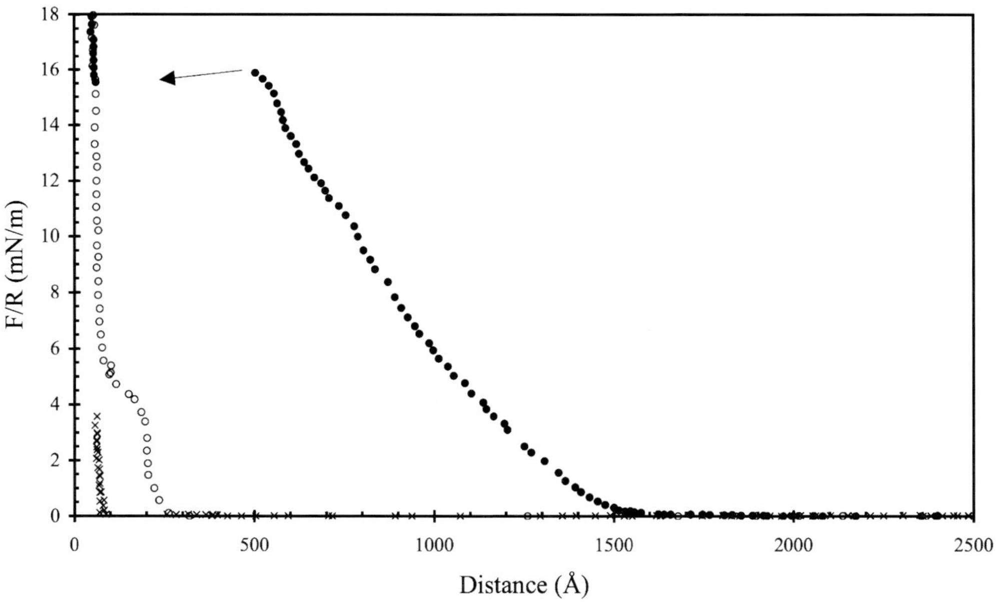

Fig. 11.22 Force normalized by radius as a function surface separation between mica surfaces precoated with AM-MAPTAC-10 in the presence of 0.8 mM SDS. Filled symbols represent data measured on approach and unfilled symbols forces measured on separation. The crosses represent the forces measured before addition of SDS on both approach and separation. The arrow indicates an inward jump. (Adapted from [130]; Copyright 1997, with permission from Elsevier).

($F/R = 4$–5 mN m^{-1}), which facilitates a further increase in surface separation. The higher polyelectrolyte-surface affinity clearly means that a larger amount of surfactant can be incorporated within the layer before desorption occurs, and it is the electrostatic repulsion between the excess surfactants that results in the very pronounced swelling of the layer. The forces in the presence of higher amounts of surfactants have not been probed, but adsorption studies have shown that desorption will eventually occur [145]. It should be noted that in bulk solution AM-MAPTAC-10 mixed with SDS forms extended connected structures [76], presumably due to SDS micelles forming cross-links between different polyelectrolyte chains.

Formation of Interfacial Complexes (PCMA) A further increase in polyelectrolyte charge density, as in MAPTAC and the structurally similar PCMA, results in a further increase in polyelectrolyte–surface affinity and, under low ionic strength conditions, very thin adsorbed layers are obtained. The forces are characterized by a long-range attraction due to bridging, followed by a hard wall repulsion at a separation of about 1 nm [146–148]. The polyelectrolyte–SDS affinity is also very high and in these particular cases compact polyelectrolyte–surfactant aggregates form in bulk with an internal hexagonal organization of rod-like micelles wrapped by the polyelectrolyte [79]. Once formed, these aggregates cannot be dissolved by adding water or excess surfactant, at least not over a period of days. Thus, since the complexes formed are insoluble, it is also difficult to remove

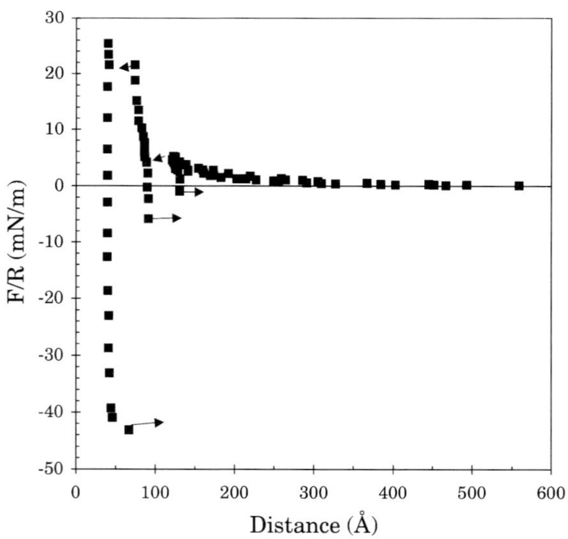

Fig. 11.23 Force normalized by radius as a function surface separation between mica surfaces precoated with PCMA, a cationic polyelectrolyte containing one permanent charge on each segment. The forces were measured across an aqueous 10^{-4} M KBr solution containing 1.6 mM SDS. The arrows illustrate inward and outward jumps. (Adapted from [148] with permission).

the preadsorbed layer by addition of SDS [145]. Nevertheless, the affinity between the adsorbed polyelectrolyte chain and the surfactants is high and SDS is readily incorporated within the layer. The resulting forces are intriguing and show oscillations between attraction and repulsion as the surfaces come together (Fig. 11.23).

The periodicity is very similar to the characteristic distance within the bulk aggregates, which leads to the conclusion that the force measurements reflect the internal structure of the layer [3, 123] and that the hexagonal units are oriented parallel to the surface. To facilitate the formation of the interfacial aggregates, some parts of the polyelectrolyte chain have to leave the surface and this is clearly observed by AFM imaging, which visualizes the transformation of a smooth and featureless polyelectrolyte layer before surfactant addition to a layer consisting of large aggregate structures [123]. This phenomenon has similarities to a dewetting process.

Adsorption in Layers Not all highly charged cationic polyelectrolyte layers transform to large interfacial polyelectrolyte–surfactant complexes when exposed to SDS solutions. Instead, the surfactant may simply adsorb on top of the polyelectrolyte, similarly to their adsorption on an oppositely charged surface. This has been observed to be the case for preadsorbed layers of chitosan [124], a cationic polysaccharide, and for preadsorbed layers of the branched PEI [110]. The case of chitosan is illustrated in Fig. 11.24.

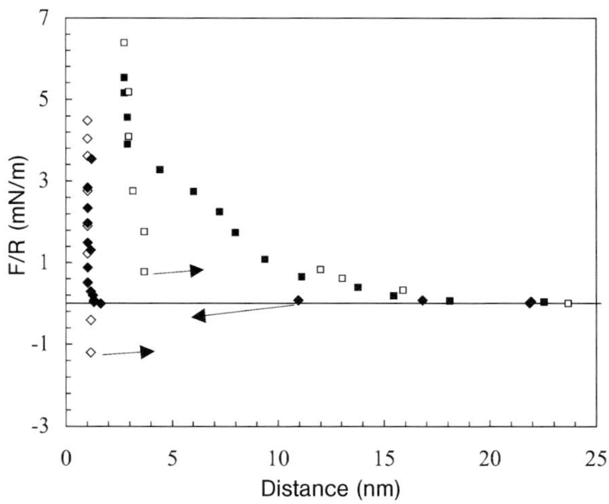

Fig. 11.24 Force normalized by radius as a function surface separation between mica surfaces precoated with Chitosan across a 0.1 mM $NaNO_3$ solution. Diamonds represent the forces measured in the absence of SDS and squares forces measured in the presence of 8 mM SDS. Filled and unfilled symbols represent forces measured on approach and separation, respectively. The arrows show inward and outward jumps. (Adapted with permission from [124]; Copyright 2003, American Chemical Society).

The adsorbed chitosan layer is slightly thicker than that of PCMA, but a long-range bridging attraction is also observed between chitosan layers. However, the magnitude of the adhesion force, measured by separating the layers from contact, is lower by about a factor of 50 in the case of chitosan. This is suggested to be due to the lower flexibility of chitosan and to the fact that only about 50% of the segments are charged within the layer [124]. Addition of SDS results in the formation of a charged bilayer on top of the chitosan layer. As a result of the SDS adsorption, a strong, repulsive, double-layer force develops (Fig. 11.24). There is no adhesion between the bilayers of SDS adsorbed on top of chitosan. The data that are available today indicate that increased flexibility of the polyelectrolyte chain favors the formation of surface aggregates as observed for PCMA, whereas stiffer and branched chains induce surfactant adsorption on top of the polyelectrolyte. However, too few systems have been investigated yet to allow a firm conclusion to be drawn.

De-Swelling All the data discussed above, except for the results with PEI, have been obtained with low background electrolyte concentration (0.1 mM or less in addition to the surfactant). This aspect is important since at higher ionic strengths the polyelectrolytes adsorb to form a thicker layer and, as a result, the steric repulsion between polyelectrolyte-coated surfaces increases in proportion [124, 146, 149]. The more open structure of the adsorbed layer also promotes the association with surfactants and it has even been observed that the c.a.c. at the surfaces can decrease with increasing salt concentration, in contrast to the increase observed in bulk [124]. When the preadsorbed polyelectrolyte layer ini-

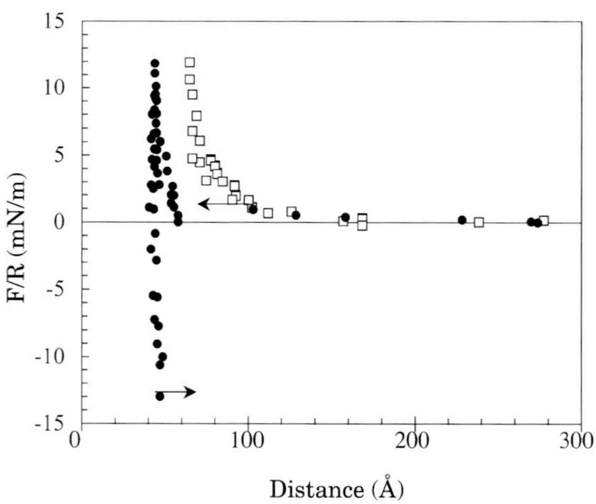

Fig. 11.25 Force normalized by radius as a function surface separation between mica surfaces precoated with Chitosan across a 30 mM $NaNO_3$ solution. Unfilled squares represent the forces measured in the absence of SDS and filled circles forces measured in the presence of 1.2 mM SDS. The arrows show inward and outward jumps. (Adapted with permission from [124]; Copyright 2003, American Chemical Society).

tially is extended for a highly charged polyelectrolyte, the incorporation of oppositely charged surfactants can give rise to compaction of the layer. This situation is illustrated in Fig. 11.25. This compaction is due to a reduction of the net charge within the layer and to hydrophobic interactions between surfactant tails within the layer.

11.3.5.2 **Adsorbed Layers Formed by Preformed Aggregates**

The forces induced by adsorption of preformed polyelectrolyte–surfactant aggregates have only been considered in a few cases [65, 150, 151]. Also in this case the results obtained differ significantly depending on the system investigated. Not surprisingly, aggregates with opposite sign of charge to that of the surface adsorb. It has also been observed, however, that negatively charged polyelectrolyte–surfactant aggregates adsorb on negatively charged surfaces to some extent. This process is presumably facilitated by expulsion of some negatively charged surfactants from the aggregate during the adsorption event [65]. Close to the composition of the bulk solution that results in the formation of uncharged polyelectrolyte–surfactant aggregates, extensive deposition is observed, resulting in the formation of very thick layers that readily adhere to each other when brought into contact.

11.4 Conclusions

The progress in understanding the bulk association of polyelectrolytes and oppositely charged surfactants has been considerable during the past 10 years. It has seen improvements in the interpretation of binding isotherms, new theoretical approaches and much new information has been gained through simulation methods. It is encouraging that the predictions made by the different approaches largely agree and that they are consistent with the main body of experimental results. The internal structure of polyelectrolyte–surfactant aggregates has been shown to exhibit large variation in response to changes in polyelectrolyte architecture and surfactant structure. A novel approach to represent the complex phase behavior has also been proposed. A remaining important issue is the presence of trapped, long-lived non-equilibrium states. It raises questions such as: how important is really the equilibrium situation in technical applications?; how should non-equilibrium states be studied experimentally?; how can we control the non-equilibrium state by mixing conditions?; how should simulations and theoretical approaches be developed for investigating such states?

The understanding of polyelectrolyte–surfactant association at solid/liquid interfaces has also been improved. Here the situation is such that the body of experimental data has increased significantly due to the use of several different techniques, and general patterns are beginning to be unraveled. A better understanding requires the development of theoretical approaches and simulation methods. In this way, the properties of the constituents can be systematically varied and trends in how, e.g., the surface-polyelectrolyte affinity and the polyelectrolyte architecture influence interfacial association can be deduced in the same manner as has been done for the case of bulk association. The presence of long-lived, trapped non-equilibrium states at surfaces is an even more severe difficulty than in bulk. This raises the same questions as above also for association phenomena at interfaces.

Acknowledgments

This work was supported by the European Commission under the 6th Framework Program, contract No. MRTN-CT-2004-512331-Project SOCON. P. M. C. and A. D. acknowledge financial support from the Swedish Research Council, VR. R. M. and I. V. also acknowledge support from the Hungarian Ministry of Economics under contract GVOP-3.2.1.-2004-04-0344.

References

1 Claesson, P. M., Dedinaite, A., Poptoshev, E. In *Physical Chemistry of Polyelectrolytes*, Radeva, T. (Ed.), *Surfactant Science Series*, Vol. 99, Marcel Dekker, New York, **2001**, p. 447.

2 Claesson, P. M., Poptoshev, E., Blomberg, E., Dedinaite, A. *Adv. Colloid Interface Sci.* **2005**, *114/115*, 173.

3 Bergeron, V., Claesson, P. M. *Adv. Colloid Interface Sci.* **2002**, *96*, 1.

4 Zana, R. In *Polymer–Surfactant Systems*, Kwak, J. C. T. (Ed.), *Surfactant Science Series*, Vol. 77, Marcel Dekker, New York, **1998**, p. 409.

5 Linse, P., Piculell, L., Hansson, P. In *Polymer-Surfactant Systems*, Kwak, J. C. T. (Ed.), *Surfactant Science Series*, Vol. 77, Marcel Dekker, New York, **1998**, p. 193.

6 Wei, Y.-C., Hudson, S. M. *Rev. Macromol. Chem. Phys.* **1995**, *C35*, 15.

7 Lindman, B., Thalberg, K. In *Interactions of Surfactants with Polymers and Proteins*, Goddard, E. D., Ananthapadmanabhan, K. P. (Eds.), CRC Press, Boca Raton, FL, **1993**, p. 203.

8 Piculell, L., Lindman, B., Karlström, G. In *Polymer–Surfactant Systems*, Kwak, J. C. T. (Ed.), *Surfactant Science Series*, Vol. 77, Marcel Dekker, New York, **1998**, p. 65.

9 Goddard, E. D. *Colloids Surf.* **1986**, *19*, 255.

10 Goddard, E. D. *Colloids Surf.* **1986**, *19*, 301.

11 Hayakawa, K., Kwak, J. C. T. In *Cationic Surfactants: Physical Chemistry*, Rubingh, D.N., Hollands, P.M. (Eds.), Marcel Dekker, New York, **1991**, p. 189.

12 Shirahama, K. In *Polymer–Surfactant Systems*, Kwak, J. C. T. (Ed.), *Surfactant Science Series*, Vol. 77, Marcel Dekker, New York, **1998**, 1998.

13 Gilanyi, T., Mészáros, R., Varga, I. *Colloid Polym. Sci.* **2001**, *117*, 136.

14 Gilanyi, T., Varga, I. *Langmuir* **1998**, *14*, 7397.

15 Varga, I., Gilanyi, T., Mészáros, R. *Prog. Colloid Polym. Sci.* **2001**, *117*, 136.

16 Tanford, C. *Physical Chemistry of Macromolecules*, Wiley, New York, **1961**.

17 Ising, E. *Z. Phys.* **1925**, *31*, 253.

18 Hill, T. L. *J. Phys. Chem.* **1953**, *57*, 324.

19 Kramers, H. A., Wannier, G. H. *Phys. Rev.* **1941**, *60*, 252.

20 Lifson, S. *J. Chem. Phys.* **1957**, *26*, 727.

21 Zimm, B.H., Bragg, J.K. *J. Chem. Phys.* **1959**, *31*, 526.

22 Schwarz, G. *Eur. J. Biochem.* **1970**, *12*, 442.

23 Satake, I., Yang, J.T. *Biopolymers* **1976**, *15*, 2263.

24 Shirahama, K., Yuasa, H., Sugimoto, S. *Bull. Chem. Soc. Jpn.* **1981**, *54*, 375.

25 Delville, A. *Chem. Phys. Lett.* **1985**, *118*, 617.

26 Witte, F.M., Engberts, J.B.F.N. *Colloids Surf.* **1989**, *36*, 417.

27 van Stam, J., Almgren, M., Lindblad, C. *Prog. Colloid Polym. Sci.* **1991**, *84*, 13.

28 Thalberg, K., van Stam, J., Lindblad, C., Almgren, M., Lindman, B. *J. Phys. Chem.* **1991**, *95*, 8975.

29 Almgren, M., Hansson, P., Mukhtar, E., van Stam, J. *Langmuir* **1992**, *8*, 2405.

30 Hansson, P., Almgren, M. *J. Phys. Chem.* **1995**, *99*, 16684.

31 Anthony, O., Zana, R. *Langmuir* **1996**, *12*, 1967.

32 Tanford, C. *The Hydrophobic Effect: Formation of Micelles and Biological Membranes*, Wiley-Interscience, New York, **1973**.

33 Nagarajan, R. *J. Chem. Phys.* **1989**, *90*, 1980.

34 Hill, A.V. *J. Physiol. (London)* **1910**, *40*, 190.

35 Gilanyi, T., Wolfram, E. *Colloids Surf.* **1981**, *3*, 181.

36 Nagarajan, R. *Colloids Surf.* **1985**, *13*, 1.

37 Nagarajan, R. *Adv. Colloid Interface Sci.* **1986**, *26*, 205.

38 Tanford, C. *Phys. Chem.* **1974**, *24*, 2469.

39 Ruckenstein, E., Nagarajan, R. *J. Phys. Chem.* **1975**, *79*, 2622.

40 Nagarajan, R., Ruckenstein, E. *J. Colloid Interface Sci.* **1979**, *71*, 580.

41 Ruckenstein, E., Huber, G., Hoffmann, H. *Langmuir* **1987**, *3*, 382.

42 Nikas, Y.J., Blankstein, D. *Langmuir* **1994**, *10*, 3512.

43 Gilanyi, T. *J. Phys. Chem. B* **1999**, *103*, 2085.

44 Zana, R., Lang, J., Lianos, P. In *Microdomains in Polymer Solutions*, Dubin, P.L. (Ed.), Plenum Press, New York, **1985**, p. 357.

45 Gilanyi, T., Varga, I., Mészáros, R. *Prog. Colloid Polym. Sci.* **2004**, *125*, 179.

46 Hansson, P. *Langmuir* **2001**, *17*, 4167.

47 Skerjanc, J., Kogei, K., Vesnaver, G. *J. Phys. Chem.* **1988**, *92*, 6382.

48 Löfroth, J.E., Johansson, L., Norman, A.C., Wettström, K. *Prog. Colloid Polym. Sci.* **1991**, *84*, 8.

49 Hansson, P., Almgren, M. *J. Phys. Chem.* **1996**, *100*, 9038.

50 Evans, D.F., Mitchell, D.J., Ninham, B.W. *J. Phys. Chem.* **1984**, *88*, 6344.

51 Mészáros, R., Thompson, L., Bos, M., Varga, I., Gilanyi, T. *Langmuir* **2003**, *19*, 609.

52 Wallin, T., Linse, P. *Langmuir* **1996**, *12*, 305.

53 Akinchina, A., Linse, P. *Macromolecules* **2002**, *35*, 508.

54 Wallin, T., Linse, P. *J. Phys. Chem.* **1996**, *100*, 17873.

55 Wallin, T., Linse, P. *J. Phys. Chem.* **1997**, *101*, 5506.

56 Akinchina, A., Linse, P. *J. Phys. Chem. B* **2003**, *107*, 8011.

57 Jonsson, M., Linse, P. *J. Chem. Phys.* **2001**, *115*, 3406.

58 Wallin, T., Linse, P. *Langmuir* **1998**, *14*, 2940.

59 Wallin, T., Linse, P. *J. Chem. Phys.* **1998**, *109*, 5089.

60 Jonsson, M., Linse, P. *J. Chem. Phys.* **2001**, *115*, 10975.

61 Nguyen, T.T., Shklovski, B.I. *J. Chem. Phys.* **2001**, *114*, 5905.

62 Schiessel, H., Bruinsma, R.F., Gelbart, W.M. *J. Chem. Phys.* **2001**, *115*, 7245.

63 Skepö, M., Linse, P. *Macromolecules* **2003**, *36*, 508.

64 Naderi, A., Claesson, P.M., Bergström, M., Dedinaite, A. *Colloids Surf. A* **2005**, *253*, 83.

65 Dedinaite, A., Claesson, P.M. *Langmuir* **2000**, *16*, 1951.

66 Dedinaite, A., Claesson, P.M., Nygren, J., Iliopoulos, I. *Prog. Colloid Polym. Sci.* **2000**, *116*, 84.

67 Guillot, S., Delsanti, M., Désert, S., Langevin, D. *Langmuir* **2003**, *19*, 230.

68 Guillot, S., McLoughlin, D., Jain, N., Delsanti, M., Langevin, D. *J. Phys.: Condens. Matter* **2003**, *15*, S219.

69 Naderi, A., Claesson, P.M. *J. Dispers. Sci. Technol.* **2005**, *26*, 329.

70 Antonietti, M., Burger, C., Effing, J. *Adv. Mater.* **1995**, *7*, 751.
71 Antonietti, M., Conrad, J., Thünemann, A. *Macromolecules* **1994**, *27*, 6007.
72 Antonietti, M., Kaul, A., Thünemann, A. *Langmuir* **1995**, *11*, 2633.
73 Antonietti, M., Maskos, M. *Macromolecules* **1996**, *29*, 4199.
74 Antonietti, M., Wenzel, A., Thünemann, A. *Langmuir* **1996**, *12*, 2111.
75 Claesson, P.M., Bergström, M., Dedinaite, A., Kjellin, M., Legrand, J.F., Grillo, I. *J. Phys. Chem. B* **2000**, *104*, 11689.
76 Bergström, L.-M., Kjellin, U.R.M., Claesson, P.M., Grillo, I. *J. Phys. Chem. B* **2004**, *108*, 1874.
77 Zhou, S., Hu, H., Burger, C., Chu, B. *Macromolecules* **2001**, *34*, 1772.
78 Bastardo, L., Garamus, V.M., Bergström, M., Claesson, P.M. *J. Phys. Chem. B* **2005**, *109*, 167.
79 Bergström, M., Kjellin, U.R.M., Claesson, P.M., Pedersen, J.S., Nielsen, M.M. *J. Phys. Chem. B* **2002**, *106*, 11412.
80 Penfold, J., Tucker, I., Thomas, R.K., Zhang, J. *Langmuir* **2005**, *21*, 10061.
81 Svensson, A., Piculell, L., Cabane, B., Ilekti, P. *J. Phys. Chem. B* **2002**, *106*, 1013.
82 Braem, A.D., Prieve, D.C., Tilton, R.D. *Langmuir* **2001**, *17*, 883.
83 Bain, C.D. *Curr. Opin. Colloid Interface Sci.* **1998**, *2*, 287.
84 Keddie, J.L. *Curr. Opin. Colloid Interface Sci.* **2001**, *6*, 102.
85 Zhmud, B., Tiberg, F. *Adv. Colloid Interface Sci.* **2005**, *113*, 21.
86 Alterovitz, S.A., Johs, B. *Thin Solid Films* **1998**, *313/314*, 124.
87 Barradas, N.P., Keddie, J.L., Sackin, R. *Phys. Rev. E* **1999**, *59*, 6138.
88 de Feijter, J.A., Benjamins, J., Veer, F.A. *Biopolymers* **1978**, *17*, 1759.
89 Tiberg, F., Landgren, M. *Langmuir* **1993**, *10*, 3714.
90 Varga, I., Mészáros, R., Gilanyi, T. *J. Phys. Chem. B*, in press.
91 Blaakmeer, J., Böhmer, M.R., Cohen Stuart, M.A., Fleer, G.J. *Macromolecules* **1990**, *23*, 2301.
92 Böhmer, M.R., Evers, O.A., Scheutjens, J.M.H.M. *Macromolecules* **1990**, *23*, 2288.
93 Linse, P. *Macromolecules* **1996**, *29*, 326.
94 Fleer, G.J., Cohen Stuart, M.A., Scheutjens, J.M.H.M., Cosgrove, T., Vincent, B. *Polymers at Interfaces*, Chapman and Hall, London, **1993**.
95 Samoshina, Y., Diaz, A., Becker, Y., Nylander, T., Lindman, B. *Colloids Surf. A* **2003**, *231*, 195.
96 van de Steeg, H.G.M., Cohen Stuart, M.A., de Keizer, A., Bijsterbosch, B.H. *Langmuir* **1992**, *8*, 2538.
97 Sennerfors, T., Solberg, D., Tiberg, F. *J. Colloid Interface Sci.* **2002**, *254*, 222.
98 Sierakowski, M.R., Freitas, R.A., Fujimoto, J., Petri, D.F.S. *Carbohydr. Polym.* **2002**, *49*, 167.
99 Le Berre, F., Malmsten, M., Blomberg, E. *Langmuir* **2001**, *17*, 699.
100 Mahltig, B., Gohy, J.-F., Antoun, S., Jerome, R., Stamm, M. *Colloid Polym. Sci.* **2002**, *280*, 495.
101 Mahltig, B., Muller-Bischbaum, P., Wolkenhauer, M., Wunnicke, O., Wiegand, S., Gohy, J.-F., Jerome, R., Stamm, M. *J. Colloid Interface Sci.* **2001**, *242*, 36.
102 Cohen Stuart, M.A., Hoogendam, C.W., de Keizer, A. *J. Phys.: Condens. Matter* **1997**, *9*, 7767.
103 Cohen Stuart, M.A., Fleer, G.J. *Annu. Rev. Mater. Sci.* **1996**, *26*, 463.
104 Mészáros, R., Varga, I., Gilanyi, T. *Langmuir* **2004**, *20*, 5026.
105 Decher, G. *Science* **1997**, *277*, 1232.
106 Decher, G., Hong, J.-D. *Ber. Bunsen-Ges. Phys. Chem.* **1991**, *95*, 1430.
107 Mészáros, R., Thompson, L., Bos, M., Varga, I., Gilanyi, T. *Langmuir* **2003**, *19*, 9977.
108 Mészáros, R., Thompson, L., Bos, M., de Groot, P. *Langmuir* **2002**, *18*, 6164.
109 Poptoshev, E., Claesson, P.M. *Langmuir* **2002**, *18*, 2590.
110 Dedinaite, A., Mészáros, R., Claesson, P.M. *J. Phys. Chem. B* **2004**, *108*, 11645.
111 El-Khouri, R.J., Johal, M.S. *Langmuir* **2003**, *19*, 4880.
112 Calculitan, N.G., Scudder, P.H., Rodriguez, A., Casson, J.L., Wang, H.-L., Robinson, J.M., Johal, M.S. *Langmuir* **2004**, *20*, 8735.

113 Samoshina, Y., Nylander, T., Lindman, B. *Langmuir* **2005**, *21*, 4490.
114 Samoshina, Y., Nylander, T., Claesson, P.M., Schillén, K., Iliopoulos, I., Lindman, B. *Langmuir* **2005**, *21*, 2855.
115 Terada, E., Samoshina, Y., Nylander, T., Lindman, B. *Langmuir* **2004**, *20*, 1753.
116 Shubin, V. *Langmuir* **1994**, *10*, 1093.
117 Terada, E., Samoshina, Y., Nylander, T., Lindman, B. *Langmuir* **2004**, *20*, 6692.
118 Cardenas, M., Braem, A.D., Nylander, T., Lindman, B. *Langmuir* **2003**, *19*, 7712.
119 Cardenas, M., Teran, J.C., Nylander, T., Lindman, B. *Langmuir* **2004**, *20*, 8597.
120 Cross, G.H., Reeves, A., Brand, S., Popplewell, J.F., Peel, L.L., Swann, M.J., Freeman, N.J. *Biosens. Bioelectron.* **2003**, *125*, 2173.
121 Lu, J.R., Swann, M.J., Peel, L.L., Freeman, N.J. *Langmuir* **2004**, *20*, 1827.
122 Swann, M.J., Peel, L.L., Carrington, S., Freeman, N.J. *Anal. Biochem.* **2004**, *329*, 190.
123 Dedinaite, A., Claesson, P.M., Bergström, M. *Langmuir* **2000**, *16*, 5257.
124 Dedinaite, A., Ernstsson, M. *J. Phys. Chem. B* **2003**, *107*, 8181.
125 Saurbrey, Z. *Z. Phys.* **1959**, *155*, 206.
126 Kanazawa, K.K., Gordon, J.G. *Anal. Chim. Acta* **1985**, *175*, 99.
127 Johannsmann, D., Mathauer, K., Wegner, G., Knoll, W. *Phys. Rev. B* **1992**, *46*, 7808.
128 Lundgren, S., Persson, K., Kronberg, B., Claesson, P.M. *Tribol. Lett.* **2006**, *22*, 15.
129 Plunkett, M.A., Claesson, P.M., Rutland, M.W. *Langmuir* **2002**, *18*, 1274.
130 Kjellin, U.R.M., Claesson, P.M., Audebert, R. *J. Colloid Interface Sci.* **1997**, *190*, 476.
131 Plunkett, M.A., Claesson, P.M., Ernstsson, M., Rutland, M.W. *Langmuir* **2003**, *19*, 4673.
132 Naderi, A., Claesson, P.M. *Langmuir* **2006**, *22*, 7639.
133 Bastardo, L., Mészáros, R., Varga, I., Gilanyi, T., Claesson, P.M. *J. Phys. Chem. B* **2005**, *109*, 16196.
134 Whiddon, C.R., Söderman, O. *Langmuir* **2001**, *17*, 1803.
135 Creeth, A., Staples, E., Thompson, L., Tucker, I., Penfold, J. *J. Chem. Soc., Faraday Trans.* **1996**, *92*, 589.
136 Penfold, J., Taylor, D.J.F., Thomas, R.K., Tucker, I., Thompson, L. *Langmuir* **2003**, *19*, 7740.
137 Taylor, D.J.F., Thomas, R.K., Hines, J.D., Humphreys, K., Penfold, J. *Langmuir* **2002**, *18*, 9783.
138 Taylor, D.J.F., Thomas, R.K., Penfold, J. *Langmuir* **2002**, *18*, 4748.
139 Staples, E., Tucker, I., Penfold, J., Warren, N., Thomas, R.K. *Langmuir* **2002**, *18*, 5139.
140 Staples, E., Tucker, I., Penfold, J., Warren, N., Thomas, R.K., Taylor, D.J.F. *Langmuir* **2002**, *18*, 5147.
141 Penfold, J., Tucker, I., Staples, E., Thomas, R.K. *Langmuir* **2004**, *20*, 7177.
142 Penfold, J., Tucker, I., Thomas, R.K. *Langmuir* **2005**, *21*, 11757.
143 Turner, S.F., Clarke, S.M., Rennie, A.R., Thirtle, P.N., Li, Z.X., Thomas, R.K., Langridge, S., Penfold, J. *Langmuir* **2005**, *21*, 10082.
144 Rojas, O.J., Claesson, P.M., Neuman, R.D. *J. Colloid Interface Sci.* **2001**, *237*, 104.
145 Rojas, O.R., Ernstsson, M., Neuman, R.D., Claesson, P.M. *Langmuir* **2002**, *18*, 1604.
146 Dahlgren, M.A.G., Waltermo, Å., Blomberg, E., Claesson, P.M., Sjöström, L., Åkesson, T., Jönsson, B. *J. Phys. Chem.* **1993**, *97*, 11769.
147 Dahlgren, M.A.G., Claesson, P.M., Audebert, R. *J. Colloid Interface Sci.* **1994**, *166*, 343.
148 Claesson, P.M., Dedinaite, A., Blomberg, E., Sergeyev, V.G. *Ber. Bunsen-Ges. Phys. Chem.* **1996**, *100*, 1008.
149 Dahlgren, M.A.G. *Langmuir* **1994**, *10*, 1580.
150 Claesson, P.M., Fielden, M., Dedinaite, A., Brown, W., Fundin, J. *J. Phys. Chem. B* **1998**, *102*, 1270.
151 Rojas, O.J., Claesson, P.M., Berglund, K.D., Tilton, R.D. *Langmuir* **2004**, *20*, 3221.

12
Non-ionic Micellar Films: Thinning and Stability

Alex D. Nikolov and Darsh T. Wasan

Abstract

Foams, emulsions and particle suspensions are often stabilized using non-ionic surfactant micellar solutions. In such colloidal dispersions, the bubbles or droplets in the dispersed phase interact via the intervening liquid films containing micelles. The stability of these micellar films and, thereby, dispersion stability are governed by the film size (i.e. drop or bubble size) and the capillary pressure. We used the capillary force balance technique in conjunction with the combined differential and common interferometric method to monitor simultaneously the film curvature, capillary pressure and film thickness. The effect of the film size on the lamella thickness transition was investigated. This chapter reports the experimental observations showing that the film stability depends more on the lamella or film size and micelle concentration and less on the capillary pressure. A novel feature of the film thinning phenomenon is also presented, showing that the lamella thickness transition is a reversible process for a non-ionic micellar system, that is, the film assumes its original thickness containing more micellar layers inside it when the film size is reduced. These new findings are of both theoretical and practical importance in controlling dispersion stability.

12.1
Introduction

The stability of dispersions is a central issue in colloid and interfacial science. In dispersions such as foams and emulsions, the bubbles or droplets are separated by liquid films. The stability of these films plays a crucial role in controlling the stability and texture of any foamed or emulsified products. In real dispersions, the bubbles or droplets are polydispersed in size, so the films are curved, as shown in the photograph in Fig. 12.1. This photograph was taken in common reflected white (polychromatic) light and it shows the "rainbow" colors normally seen in actual curved foam lamellae.

Colloids and Interface Science Series, Vol. 3
Colloid Stability and Application in Pharmacy. Edited by Tharwat F. Tadros

ISBN: 978-3-527-31463-8

Fig. 12.1 Photograph taken using common reflected white (polychromatic) light depicting the 'rainbow' colors normally seen in actual curved foam lamellae.

The films in real foam and emulsion systems are randomly distributed; some are nearly vertical and others are nearly horizontal. Several methods exist for studying the drainage behavior and stability of such films. Figure 12.2 shows films with plateau borders and the wire frame with three symmetrically arranged legs. Once again, one can see the different colors created by the variation in the reflected light. We used this technique in the past to investigate the interactions between vertical foam films, plateau borders and foaming and antifoaming agents. However, the most commonly investigated liquid films are horizontal films.

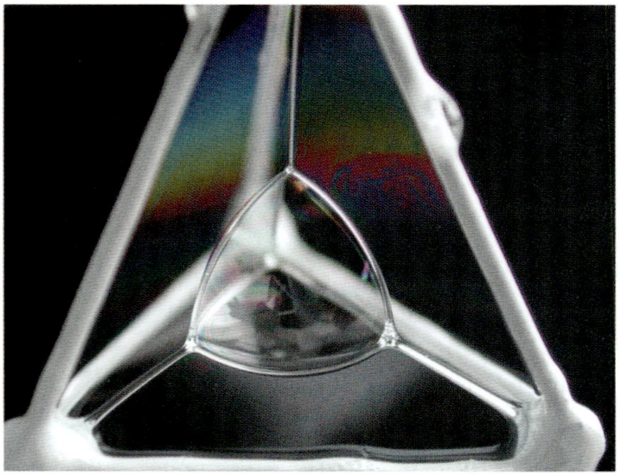

Fig. 12.2 Method for studying the foam cell with plateau borders by pulling the wire frame out of the surfactant solution. The foam lamella and plateau borders are visualized in reflected white light.

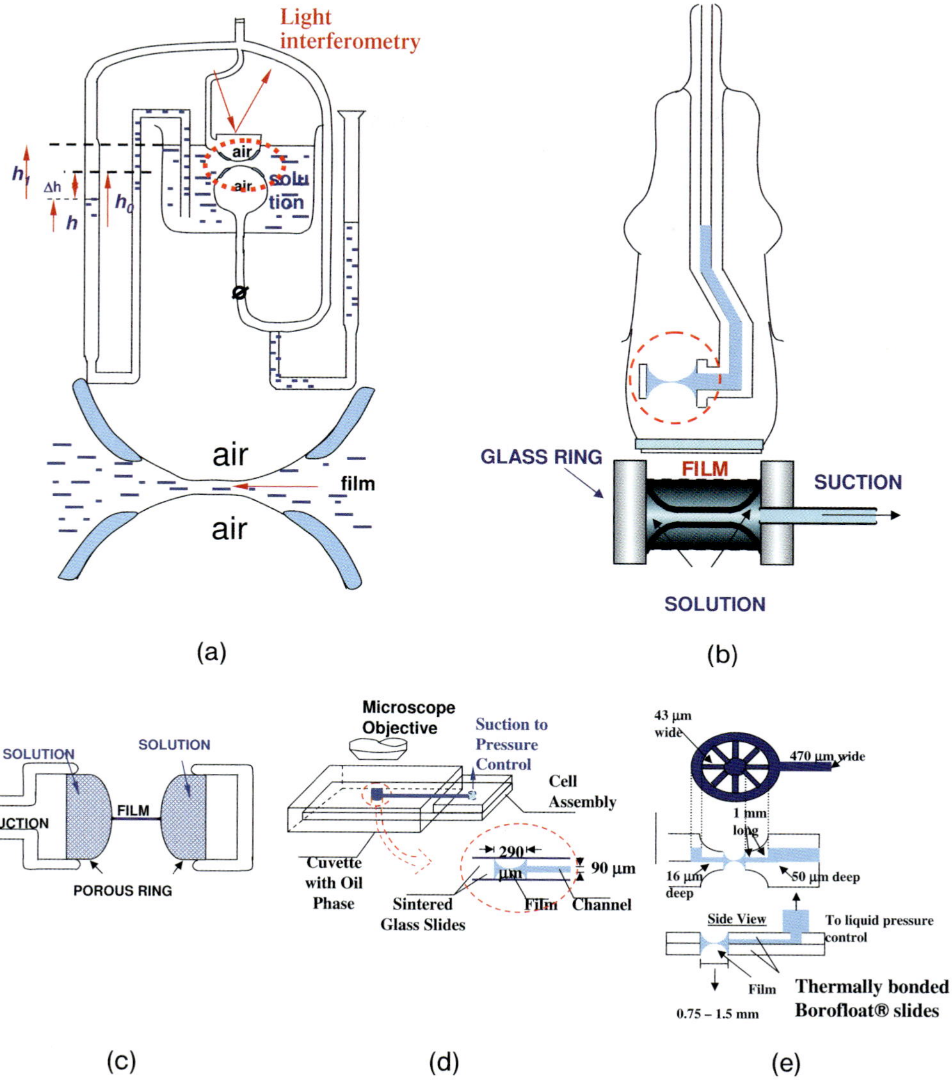

Fig. 12.3 Sketches of various cell designs which have evolved over the years: (a) Derjaguin and coworkers [1]; (b) Scheludko and Exerowa [3]; (c) Mysels and Jones [2]; (d) Velev et al. [4]; (e) Cascão Pereira et al. [5].

Several experimental cells for creating and observing the behavior of thin horizontal films using common reflected light interferometry are discussed in the literature. Derjaguin and coworkers [1], Mysels and Jones [2], Scheludko and Exerowa [3], Velev et al. [4], Cascão Pereira et al. [5] and Nikolov and coworkers [6] have all used different cell designs to investigate the thinning and stability of foams, emulsions and pseudoemulsion films. Figure 12.3 shows sketches of the various cell designs that have evolved over the years. Several review articles, including those by Scheludko [7] and Claesson et al. [8], give more detailed descriptions of these techniques, but most of the previous methods for investigating micellar film thinning and stability phenomena are concerned with only plane-parallel circular liquid films. None of the other investigators studied the dynamic behavior of curved films containing micelles or the effects of both the film size and capillary pressure on their stability. Therefore, we have recently designed and built an apparatus for investigating curved thin liquid surfactant films and used it in conjunction with the combined differential and common interferometric techniques to monitor the film size and capillary pressure. We report here, for the first time, some new and intriguing experimental observations on the effects of film size and capillary pressure on the lamella thickness stability for a non-ionic surfactant micellar system.

12.2 Micellar Film Thinning Probed by Capillary Force Balance

We used our force balance technique to monitor the film thinning and thickness stability of micellar films. The apparatus consists of two parts: a specially designed glass cell with a thick-walled capillary (having an inner diameter varying from 200 to 1000 μm) where curved or flat films of different sizes, types (e.g. foam, emulsion and pseudoemulsion) and capillary pressures are formed, and an optical bench equipped with both differential and common interferometry to monitor the film size, curvature and thickness as a function of time. A sketch of the cell and optical arrangement is shown in Fig. 12.4.

The main part of the apparatus is a glass cell, which consists of three vertically oriented concentric glass cylinders. A small bubble or droplet is slowly expelled from the inner cylinder (capillary) via a micro-syringe pump. The bubble or droplet approaches the interface and a curved film (lamella) is formed. The bottom part of the bubble or droplet forms a meniscus, which is held up by the capillarity. The positions of the top and bottom interfaces and the direction of the movement of the liquid in the capillary determine the configuration of the film. The curvature of the film is measured using a Epival Interfako differential interferometry microscope in combination with a fiber-optic probe (also used to study the film thinning). The microscope is equipped with a Max Zhender interferometer, which produces a differential interference pattern. A charge-coupled device (CCD) camera, a digital video recorder and an image analyzer were used to monitor the interference patterns as a function of time. The patterns are used to determine the lamella curvature and lamella meniscus. The details of this method are given elsewhere by Nikolov and coworkers [6].

The relationship between the capillary pressure, P_c (and therefore the disjoining pressure) and the lamella (film) curvature, $1/R_f$, is given by the Laplace equation; for a spherical lamella, it is

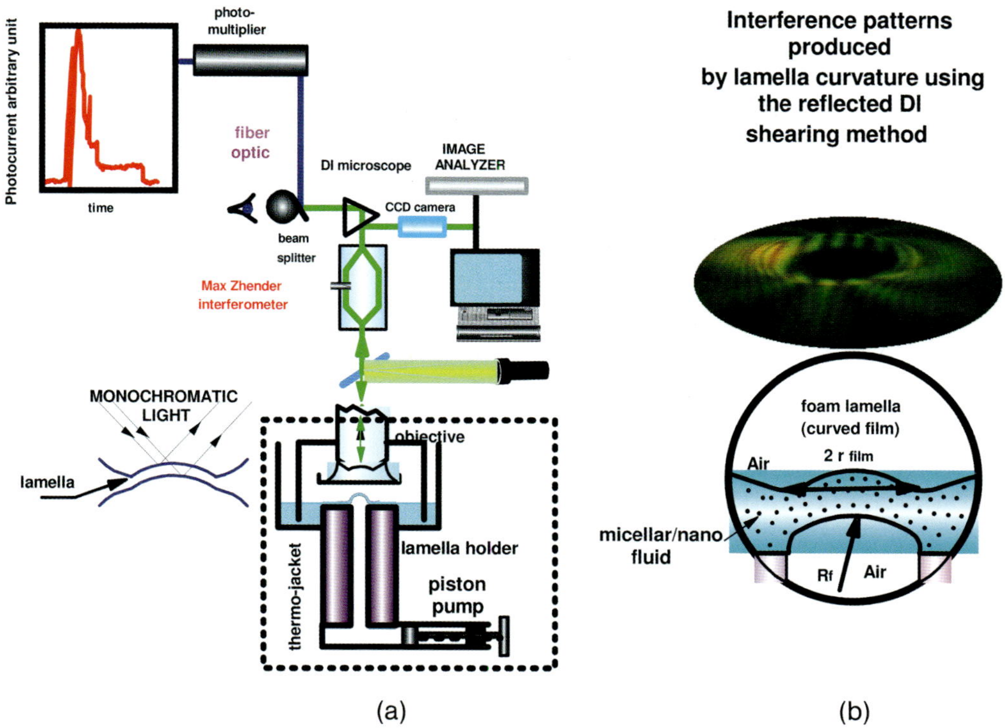

Fig. 12.4 Sketch of the optical and cell arrangement for monitoring the thinning of curved and flat films. From Nikolov and Wasan [6].

$$P_c = 4\sigma/R_f \tag{1}$$

where σ is the surface tension.

The relationship between the lamella area ($S_{\text{area}}^{\text{lamella}}$), lamella radius curvature (R_f) and the lamella three-phase contact radius (r_{film}) is

$$S_{\text{area}}^{\text{lamella}} = 2\pi R_f \left(R_f - \sqrt{R_f{}^2 - r_{\text{film}}{}^2} \right) \tag{2}$$

12.3 Results and Discussion

The capillary force balance technique was used to monitor the thinning behavior manner and stability of aqueous micellar films formed from a non-ionic surfactant, ethoxylated alcohol (NEODOL-R-56-13), containing a 15–16 *n*-carbon chain length and 13 ethylene oxide groups, a product of Shell Development Company. Three concentrations, 2, 4 and 8 wt%, corresponding to about 325, 650 and 1300 times the critical micelle concentration (c.m.c.), respectively were studied and all experiments were conducted at 20 °C. The surface tension was measured by the Wilhemy plate method. Aqueous surfactant solutions were prepared using millipore purified water. The c.m.c. of the NEODOL-R-56-13 solution was 5×10^{-6} mol L^{-1}.

The radius of curvature of the film (R_f), film radius (r_{film}) and capillary pressure (P_c) were determined using differential and common interference patterns. A typical photocurrent versus time plot (interferogram) for a micellar film made from a surfactant concentration of 325×c.m.c. is shown in Fig. 12.5. For a curved film with a diameter of 175 µm, an area of 2.6×10^{-4} cm^2 and a capillary pressure of 40.7 Pa, two stepwise thickness transitions are observed, one at a film thickness of 42.7 nm and the other at 33.2 nm (Fig. 12.5 a). The amplitude of each thickness transition is about 9 nm and this value is approximately equal to the effective micelle diameter. After the two thickness transitions, the film remains stable at a thickness of 25 nm with one micellar layer inside it. However, when the film diameter increases to 250 µm (with an area of 5.9×10^{-4} cm^2 and a capillary pressure of 38.5 Pa), one more stepwise thickness transition is observed (Fig. 12.5 b). The analysis of the film thinning interferogram reveals two essential features of the stepwise film thinning phenomenon: as the film thickness decreases in a stepwise manner, the time for the next thickness transition to occur increases, and the stepwise thickness transition occurs more rapidly for the larger film.

A novel feature of the stepwise thinning phenomenon occurs that is specific to non-ionic micellar films, and has never been observed for ionic micellar films. When the size of the thin film formed from the non-ionic micellar solution is reduced from 250 to 175 µm, the film slowly increases in thickness from 17 back to 25 nm (i.e. from one micellar layer to two micellar layers). This observation shows that the micellar lamella layer–layer thinning phenomenon is a reversible process. The micellar lamella with fewer layers inside increases in

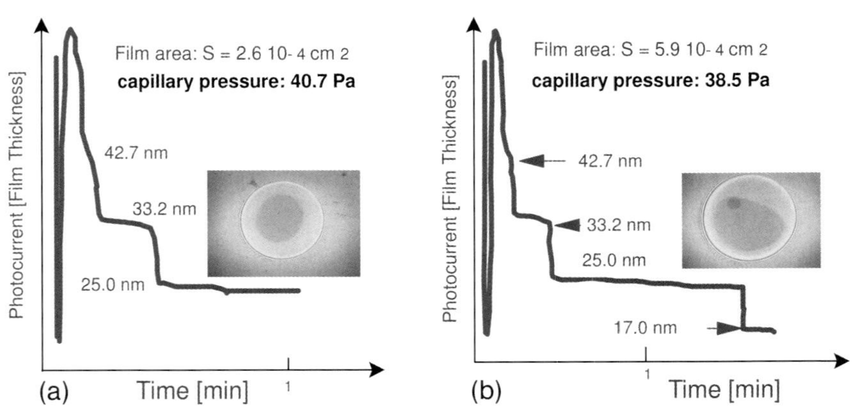

Fig. 12.5 Photocurrent versus time interferograms of the thinning of a curved foam film at a surfactant concentration of 2 wt% (325×c.m.c.) depicting the effect of film size on film stepwise thinning. Film area: (a) 2.6×10^{-4} and (b) 5.9×10^{-4} cm^2.

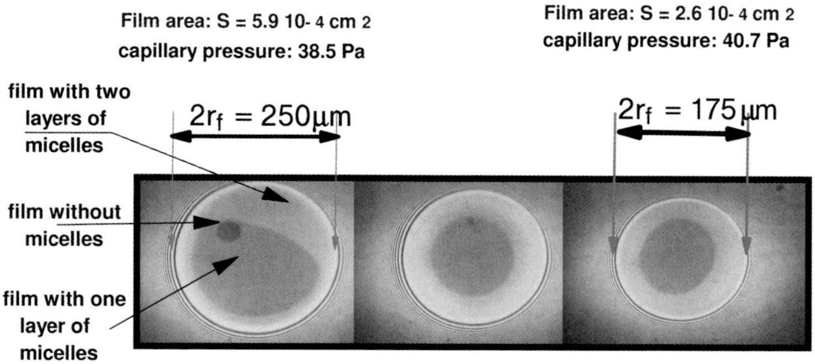

Fig. 12.6 Effect of micellar film size on film thickness.

thickness in a stepwise manner when its size is reduced. The photomicrographs in Fig. 12.6 depict the effect of the film size on the film thickness reversibility. This unusual thinning behavior of the non-ionic micellar film cannot be explained by the previously proposed mechanisms for the stepwise thinning process for either anionic or cationic micellar films [9].

A detailed analysis of the phenomenon depicted in Figs. 12.5 and 12.6 reveals that a short time after the dispersed phase (air) is expelled from the capillary against the liquid (i.e. the micellar solution), a thick, curved film (lamella) is formed on the tip of the capillary (its size is controlled by the syringe piston). Under the combined effects of capillary and gravitational forces, the liquid from the film begins to drain and, after about 10 s (Fig. 12.5 a), the film reaches its thickness of 104 nm (corresponding to the last light intensity maximum). Then the film begins to drain slowly and the first stepwise thickness transition occurs at a thickness of 42.7 nm. A dark dot is formed inside the film at the beginning of the film stepwise thickness transition. The area of the dark dot is thinner (i.e. one micellar layer less) than the surrounding film. The dark dot area reflects less light than the surrounding thicker film (Fig. 12.5 a) and the dot area appears darker when the film is observed in reflected light. If the film diameter increases slightly (as in Fig. 12.5 b), the dark dot area also increases. However, a critical film size exists beyond which the dark dot increases in size with increasing film size until the whole film area is covered. The film temporarily rests at a uniform thickness and then the film stepwise thinning process begins again; this process is repeated until the film reaches its equilibrium thickness. We previously reported the phenomenon of the formation of a dark spot in anionic micellar films based on the micellar vacancy diffusive-osmotic mechanism, which assumed that the film at each stepwise transition is in a metastable state [9].

According to the micellar vacancy diffusive-osmotic mechanism of the stepwise film thinning process proposed by Kralchevsky et al. [9], the thickness transition formation from h_n to h_{n-1} (where n is the number of micelle layers) occurs due to the decrease in the free energy (Fig. 12.7). This mechanism has been used successfully to explain the stepwise transition phenomenon observed in liquid films formed from both anionic (e.g. sodium dodecyl sulfate) and cationic [e.g. cetyltrimethylammonium bromide (CTAB)] micellar solutions [10] and also films formed from suspensions of charged nanoparticles (e.g. silica and latexes) [11]. The phenomenon of film thickness reversibility was not observed in any of these systems. However, for the films formed from non-ionic surfactant solutions, such as NEODOL-R2-56-13, we observed that a micellar film that is stable in one state (with one micellar layer, Fig. 12.5) can revert to its original stable state (with two micelle layers, Fig. 12.5 a) when it shrinks back to its original size. This is due to the fact that the energy of both stable states is nearly the same even if the capillary pressure (i.e. the disjoining pressure) for the smaller film (e.g. 175 μm) is somewhat larger than that for the larger (250 μm) film. Hence the vacancy diffusive–osmotic mechanism as put forth earlier fails to explain the phenomenon observed here. These observations clearly suggest that the stability of the film depends much more on the film size and less on the capillary pressure (or disjoining pressure).

We also observed the effect of the film size on the layer-by-layer thickness transition phenomenon for the thinning of the film formed from a cationic micellar solution of CTAB. For a CTAB concentration of 8.4×10^{-4} mol l^{-1} (corresponding to 100 times the c.m.c.), the film with an area of 2×10^{-4} cm^2 and a high capillary pressure of 180 Pa remained at a greater thickness (with one micellar layer inside it), whereas the film with the larger area of 6×10^{-4} cm^2 and a low capillary pressure of 60 Pa remained thin (without any micellar layers inside it).

The micellar concentration is also an important factor in film thickness stability and dispersion stability. For example, when we increased the micellar concentration for the non-ionic NEODOL R-56-13 sample from 2 to 4 to 8 wt% for a film of the same size (250 μm), the number of thickness transitions increased from 2 to 4 to 5, respectively. We concluded that the film thickness stability depends more on the micelle concentration and film size and less on the capillary pressure.

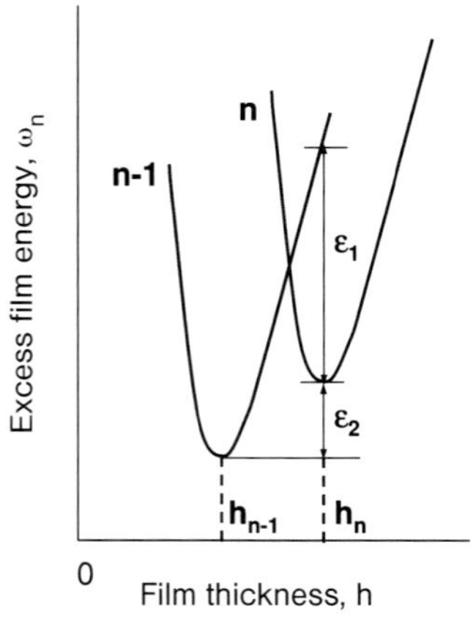

Fig. 12.7 Schematic presentation of the excess film energy of ionic micellar film with n and $n-1$ micellar layers.

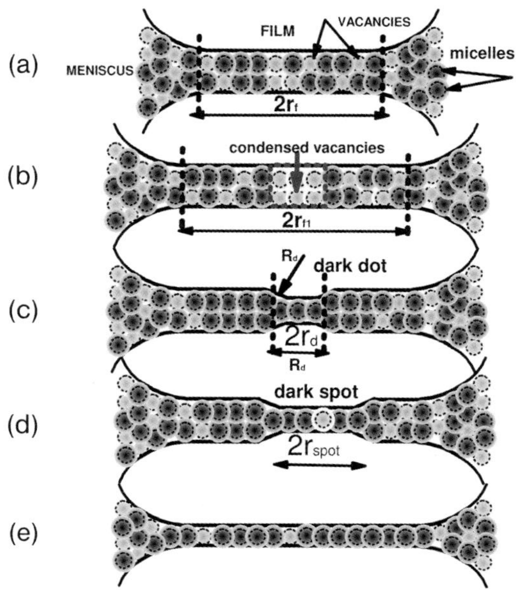

Fig. 12.8 Schematic representation of the mechanism of dark dot and spot formation inside the stratifying micellar film.

The layer-by-layer thinning of the micellar films is triggered by the formation of black spots inside the film. An analysis of the normalized micelle/nanoparticle density profile obtained by Monte Carlo computer simulations inside the film reveals that the average micelle/nanoparticle concentration inside the film is slightly higher than that in the bulk/meniscus region [12]. The micelles diffuse from the film area to the meniscus and vacancies (dislocations) appear inside the film in their place. The vacancies tend to condense at a sufficient concentration inside the film and a tiny dark dot appears (with a size of 4–10 μm). The dark dot moves randomly around the film area without increasing its size. At this stage, the capillary pressure supported by the curvature (which is promoted by the difference in the film thicknesses between the dark dot and film), the line tension (κ) and the structural disjoining pressure of the film are at equilibrium. However, if the film size increases and more vacancies are generated at the film periphery or inside the film, then to restore the equilibrium between the vacancies and the dark dot, more vacancies condense, causing the dark dot to grow. As a result, the dark dot meniscus curvature ($1/R_{\mathrm{dot}}$) increases and the capillary force decreases by a factor of $1/R_{\mathrm{dot}}$; the equilibrium between the capillary pressure line tension and structural disjoining pressure vanishes and the dark dot begins to expand (forming the dark spot). Figures 12.8 and 12.9 depict schematically the mechanisms of dark dot and spot formation inside the micellar/particle film. The dark dot is the nucleus of the dark spot. For thick films (several micellar layers), the dark spot is formed at the film periphery where the vacancy concentration is high for a film thicker than two micelle diameters. Films thinner than three micellar diameters have dark spots that form in the central part of the film. The films with two and one micellar layer

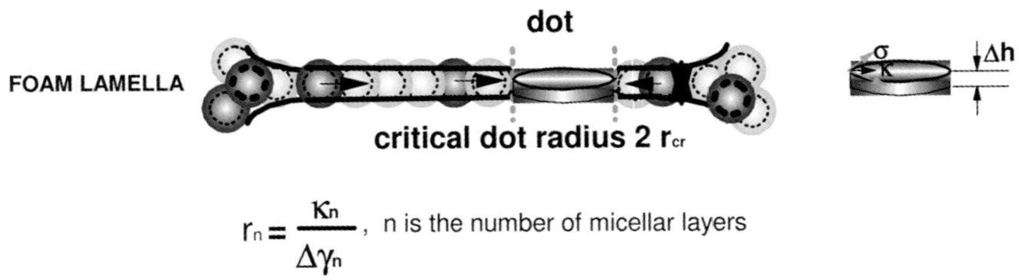

Fig. 12.9 Mechanism of dark dot formation.

thicknesses are more structured [12]; the concentration of vacancies inside the film is low and the vacancies diffuse inside the film. Condensation of the vacancies occurs, forming a dark dot in a larger film area. Once the dark dot is formed, it needs vacancies to condense in it in order to grow and form a dark spot. The concentration of vacancies inside the film depends on its size, thickness and micellar concentration. At a fixed micellar concentration and film thickness, the size of the film governs the vacancy concentration inside the film. The vacancy concentration inside the film decreases as the film size decreases. The dark spot begins to shrink and vacancies from the spot diffuse into the film area (the spot begins to dissolve) to restore the equilibrium between the spot and the vacancies inside the film. The dark spot is 18 μm for micellar films with 250 μm diameters that are 25 nm thick. The effect of the film size on the evolution of dark spots clearly demonstrates the proposed "vacancies condensation" mechanism [9].

According to the nucleation mechanism of dark spot formation, the formation of dark spots is given by the increment of the grand thermodynamic potential [9]:

$$\Delta\Omega_n = -\pi {r_{nc}}^2 \Delta\gamma_{\mathrm{r}} + 2\pi r_{nc} \kappa_n \tag{3}$$

where r_{nc} is the critical radius of the dark dot, γ is the film tension and κ is the line tension. The first term on the right-hand side is the work done by the film tension and the second term is the work for the formation of the dot. The critical radius of the dark dot is related to the film tension and the line tension as follows:

$$r_{nc} = \frac{\kappa_n}{\Delta\gamma_n} \tag{4}$$

where the subscript n refers to the film with n micelle layers inside and $\Delta\gamma_n$ refers to the film tension difference between the film and dot. Figure 12.8 illustrates the proposed model for the work of dark spot formation in the stepwise thinning micellar film. The film tension is given by

$$\gamma_n = 2\sigma_n^{\mathrm{f}} + P_{\mathrm{c}} h_n \tag{5}$$

with

$$\sigma_n^{\mathrm{f}} = \sigma \cos \theta_n \tag{6}$$

where σ and P_c are the surface tension and the capillary pressure of the meniscus, respectively, h_n and σ_n^{f} are the thickness and the surface tension of the film, respectively, and θ_n is the contact angle between the film and the meniscus. Both h_n and θ_n were measured via interferometry in reflected light using the experimental set-up in Fig. 12.4. The film tension difference, $\Delta\gamma_n$, is calculated from the relation

$$\Delta\gamma_n = 2\sigma(\cos\theta_n - \cos\theta_{n-1}) + P_\mathrm{c}(h_n - h_{n-1}) \tag{7}$$

The experimentally determined values of the dark spot radius, r_{nc}, at two film thicknesses corresponding to one and two micellar layers, the corresponding film tension difference, $\Delta\gamma_n$, the contact angle, θ_n, the capillary pressure, P_c, and the film thicknesses, h_n and h_{n-1}, are given in Table 12.1 for a 4 wt% micellar solution of the non-ionic NEODOL-R-56-13. Using these values, the line tension for two film thicknesses, h_n and h_{n-1}, were determined to be nearly identical and κ was 3×10^{-7} mN. The positive sign for the line tension indicates that the dark dot expansion is resisted by the line tension and the critical size of the dark dot exists and above that the dark dot usually forms.

Table 12.1 Measured values of the film thickness, h_n, contact angle, θ_n, capillary pressure, P_c, film tension difference, $\Delta\gamma$, and size of dark spot, r_{exp}, for a 4 wt% non-ionic micellar solution.

n	h_n (nm)	θ_n	P_c (Pa)	$\Delta\gamma$ (dyn cm^{-1})	r_{exp} (μm)	r_{cal} (μm)
0	17	1.41	38.5	1.44×10^{-2}	18	22
1	25	0.91	38.5	3.74×10^{-2}	77	86
2	33.2	0.74	38.5			

12.4 Conclusions

Thinning and stability of liquid films formed from non-ionic surfactant micellar solutions were investigated using a specially designed capillary glass cell for forming curved films having different sizes, types and different capillary or disjoining pressures. The capillary force balance, equipped with a new cell in conjunction with differential interference microscopy, was used to measure the film thickness transitions and disjoining (or capillary) pressure. The results for both non-ionic and cationic micellar solutions indicated that the film thickness stability depends more on the film size and micellar concentration and less on disjoining (or capillary) pressure; the concept of the disjoining pressure isotherm alone cannot predict film thickness stability. The film thickness transition and the black dot and spot formations appear to be due to the "vacancies condensation" mechanism controlled by the film size for a fixed micellar concentration. Our novel finding is that the film thickness transition is a reversible process for the non-ionic micellar film and the film can assume its original thickness containing more micellar layers when the film shrinks (containing fewer micellar layers) back to its original size. The theory based on the micelle–vacancy diffuse–osmotic mechanism, which assumes that the film thickness transition occurs due to the decrease in the film energy and that each stepwise transition is in a metastable state [9] cannot explain the phenomenon observed for the non-ionic system investigated in the present study.

Acknowledgments

This work was supported in part by a grant from the National Science Foundation and in part by the US Department of Energy Grant No. DE-FG02-05ER6400.

References

1 B.V. Derjaguin, A.S. Titijevskaia, *Colloid J. USSR, 15* (**1953**) 416; B.V. Derjaguin, A.S. Titijevskaia, I.I. Abricossova, A.D. Malkina, *Discuss. Faraday Soc., 18* (**1954**) 24; B.V. Derjaguin, A.S. Titijevskaia, *Discuss. Faraday Soc., 18* (**1957**) 27; B.V. Derjaguin, *Theory of Stability of Colloids and Thin Films*, Consultants Bureau, New York, **1989**.

2 K.J. Mysels, M.N. Jones, *Discuss. Faraday Soc., 44* (**1966**) 44.

3 A. Scheludko, D. Exerowa, *Kolloid-Z., 155* (**1957**) 39.

4 O.D. Velev, G.N. Constantinides, D.G. Avraam, A.C. Payatakes, R. P. Borwankar, *J. Colloid Interface Sci., 175* (**1995**) 68.

5 L.G. Cascão Pereira, C. Johansson, H.W. Blanch, C.J. Radke, *Colloid Surf. A, 186* (**2001**) 103.

6 A.D. Nikolov, D.T. Wasan, *Colloids Surf. A, 123* (**1997**) 375; K. Kumar, A.D. Nikolov, D.T. Wasan, *J. Colloid Interface Sci., 256* (**2002**) 194.

7 A. Scheludko, *Adv. Colloid Interface Sci., 1* (**1967**) 391.

8 P.M. Claesson, T. Ederth, V. Bergeron, M.W. Rutland, *Adv. Colloid Interface Sci., 67* (**1996**) 119.

9 P.A. Kralchevsky, A.D. Nikolov, D.T. Wasan, I.B. Ivanov, *Langmuir, 6* (**1990**) 1180.

10 A.D. Nikolov, D.T. Wasan, presented at the 11th International Symposium on Surfactant in Solutions, June 9–13, 1996, Jerusalem, Israel; *Colloids Surf. A, 128* (**1997**) 243.

11 D.T. Wasan, A.D. Nikolov, in *Particulate Two-phase Flow*, M.C. Roco (Ed.), Butterworth-Heinemann, Boston, **1993**, Chapter 11, pp. 325–354; D.T. Wasan, A.D. Nikolov, P. Kralchevsky, I.B. Ivanov, *J. Colloids Surf., 67* (**1992**) 139; A.D. Nikolov, D.T. Wasan, *Langmuir, 8* (**1992**) 2985; G. Sethumadhavan, A.D. Nikolov, D.T. Wasan, *J. Colloid Interface Sci., 240* (**2001**) 105.

12 X. Chu, A.D. Nikolov, D.T. Wasan, *Langmuir, 10* (**1994**) 4403; D.T. Wasan, A.D. Nikolov, *ACS Symp. Ser., 736* (**1999**) 40.

13
PEGylated Polymer-based Nanoparticles for Drug Delivery to the Brain*

Hyun Ryoung Kim, Karine Andrieux, and Patrick Couvreur

Abstract

Polyethylene glycol (PEG) coating is extensively applied in the design of drug delivery carriers. Avoiding the recognition by the mononuclear phagocyte system (MPS), the PEGylated nanocarriers have a prolonged circulation time in the bloodstream. They offer the advantage of an enhanced permeability and retention (EPR) effect in diseased sites. This chapter focuses on PEGylated and polymer-based nanoparticles and it describes their surface decoration by OX26 or transferrin for active brain targeting. In the first part, the methodologies for surface modification of nanoparticles, their preparation, their physicochemical characterization and their interaction with blood components are described. Application to brain targeting is discussed. The second part describes the recently developed PEGylated nanogels consisting of a cross-linked polymer network and their application to brain delivery. Finally, OX26 or transferrin conjugation with PEGylated nanoparticles for transferrin receptor targeting in the brain tissue is discussed.

13.1
Introduction

Various types of colloidal systems have been developed to achieve controlled drug delivery or targeting to a specific tissue. The aim in using colloidal carriers is generally to increase the specificity towards cells or tissues, to improve the bioavailability of drugs by increasing their diffusion through biological membranes and/or to protect them against enzyme inactivation. Colloidal carriers include liposomes and polymer-based nanoparticles but also polymer-based micelles, nanogels and metallic nanoparticles.

* A list of abbreviations can be found at the end of this chapter.

Colloids and Interface Science Series, Vol. 3
Colloid Stability and Application in Pharmacy. Edited by Tharwat F. Tadros

ISBN: 978-3-527-31463-8

After intravenous administration, colloidal particles interact with plasma proteins, especially with immunoglobulins, albumin, the elements of the complement, fibronectin, etc. This process, known as "opsonization", is crucial in driving the subsequent fate of the administered colloidal particles. Colloidal particles which present hydrophobic surface properties are efficiently coated with these opsonins and rapidly removed from the blood circulation through recognition by the mononuclear phagocyte system (MPS), which detects them as foreign products. Essentially, macrophages located in the reticuloendothelial system (RES) (of which the Kupffer cells in the liver contain 85–95% of the total intravascular phagocytic capacity [1]) play a crucial role in the clearance of the injected particles because these cells express the corresponding receptors for these opsonins. Thus, the first generation of colloidal carriers, also often designed as "conventional" colloidal carriers, has encountered the problem of rapid blood clearance due to the uptake by the MPS, resulting in liver and spleen accumulation.

Polyethylene glycol (PEG) polymers are the most popular materials to modify particulate surfaces in order to avoid recognition by MPS. As PEG provides a "cloud" of hydrophilic chains at the particle surface, plasma proteins are repelled and the corresponding particles are sterically stabilized and have a reduced clearance by MPS and a long circulation in the blood [2]. Jeon et al. [3] theoretically supported the hypothesis that PEGylation can confer opsonization-resistant properties on materials through the exploitation of the interaction between proteins and PEGylated surfaces. In the theory, a protein approaches the PEG surface by diffusion. The hydrophilic and flexible nature of the surface PEG chains allows them to take on an extended conformation. When opsonins and other proteins are attracted to the particle surface, a van der Waals force operates between the PEG surface and the proteins. Further approach of the protein initiates compression of PEG chains into a more condensed and higher energy conformation. This change creates an opposing repulsive force that, when great enough, can completely balance and/or overpower the attractive force between the opsonin and the particle surface. It is worth noting that effective repulsion of opsonins occurs when the thickness of the PEG layer at the surface of particles exceeds a threshold that depends on the PEG chain length (molecular weight), the surface chain density and the conformation [2]. This effect, so-called "Stealth", is a representative property of the second generation of colloidal carriers which are "invisible" to macrophages by using a surface coating of a hydrophilic carbohydrate or polymer, usually PEG.

On the other hand, in growing solid tumors, and also regions of infection and inflammation, the blood capillaries have increased permeability as a result of the disease process (e.g. tumor angiogenesis). Pore diameters in solid tumor vasculature vary from 100 to 800 nm [4, 5]. Drug-containing colloidal carriers that have diameters in the approximate range 50–60 nm are small enough to extravasate from the blood into the tumor interstitial space through these pores [6]. Hence the particle size and surface characteristics are key parameters for the biodistribution of colloidal carriers, because these factors can prevent their uptake by MPS macrophages. Hence a high curvature (size <100 nm) and/or a hydrophilic surface (as opposed to the hydrophobic surface of conventional nanoparticles) are needed to reduce opsonization and subsequent clearance by macrophages [7]. The Stealth colloidal carriers have been applied in cancer therapy because the solid tumors possess a well-documented feature of leaky vasculature and poor lymphatic drainage. Even if the mechanisms by which those particles diffuse into the tumors and release their drug content are not completely understood, it is believed that small size and prolonged circulating time permit selective extravasation through the small defects of the fenestrated, leaky vasculature [8] and concentration into the tumors owing to the lack of an effective lymphatic drainage system [9]. This leads to the so-called enhanced permeability and retention (EPR) effect, which results in a higher drug accumulation in tumor tissues than in plasma and in other tissues [10, 11].

This chapter focuses on PEGylated nanoparticles by describing in detail their preparation, their physicochemical characterization, their interaction with plasma proteins and their application to brain delivery. It also reviews the recently developed PEGylated nanogels. Finally, surface modifications of PEGylated nanoparticles for active brain targeting using antibody, especially OX26 and transferrin, are discussed.

13.2 Biodegradable Polymer-based PEGylated Nanoparticles

Nanoparticles are desirable for drug delivery because of a number of properties. They offer some advantages over liposomes such as better stability in blood and the possibility of better control of drug release. By definition, nanoparticles are submicron (<1 μm) colloidal systems made of polymers [12] or lipids [13, 14]. Only polymer-based PEGylated nanoparticles will be described in this chapter. According to the differences in the morphological and structural organization of nanoparticles, they are classified into two categories: nanospheres and nanocapsules. Whereas nanospheres are matrix systems constituted by a solid core with a dense polymeric network in which the drug is physically and uniformly dispersed, nanocapsules are vesicular systems in which the drug is confined into an aqueous or oily cavity surrounded by a thin polymeric membrane that is regarded as a "reservoir" system. This section focuses on the PEGylation of biodegradable polymer-based nanospheres composed of poly(lactic acid) (PLA), poly (lactic-*co*-glycolic acid) (PLGA), poly(ε-caprolactone) (PCL) and poly(alkyl cyanoacrylate) (PACA), which exhibit the Stealth effect mentioned above.

13.2.1 Methodology for Surface Modification of Nanoparticles by PEG

For the purpose of modifying the surface of nanoparticles by PEG, two options have been considered: (i) simple PEG adsorption and (ii) PEG covalent grafting on to the surface.

13.2.1.1 Adsorption of PEG Chains on the Nanoparticle Surface

Coating of particles by the adsorption of artificial surfactants such as polyoxamer and polyoxamine that contain blocks of hydrophilic PEG and blocks of hydrophobic poly(propylene glycol) (PPG) could be achieved. PPG segments adsorb on the particle surface and PEG blocks stick out from the surface [15]. The study of surfactant-coated polystyrene (PS) particles demonstrated an effective reduction in opsonization and a delay in particle clearance which depended on the surfactant (polyoxamer/polyoxamine) chain length [16, 17]. However, non-degradable PS particles are not therapeutically acceptable systems. On the other hand, the biodegradable PLGA particles were coated by adsorption of amphiphilic diblock PLA–PEG copolymer which was responsible for their stability due to the occurrence of steric repulsions and created a repulsive barrier to albumin adsorption [18]. Nevertheless, it was evidenced that the adsorbed polyoxamer 407 and polyoxamine 908 could be desorbed in significant amounts from the surface of PS or PLGA nanospheres after a long period of incubation in serum [19], leading to the loss of their Stealth properties.

13.2.1.2 Covalent Attachment of PEG Chains to the Nanoparticle Surface

In order to avoid the desorption of adsorbed surfactant, the chemical attachment of PEG chains on the surface of particles could be achieved by the self-assembly process using PEG block or branched copolymers [20, 21].

Synthesis of PEG Block Copolymer Covalently linked PEG chains are stable in comparison with simple adsorption and remain on the surface during the biodegradation process *in vivo*. Various types of PEGylated polymers are illustrated in Fig. 13.1. A lot of work using PLA, PLGA, PCL, polyanhydride and PACA has been dedicated to the manufacture of PEG grafted nanoparticles with biodegradable cores. Amphiphilic diblock or multiblock copolymers with PEG chain attachment have been used for the formation of Stealth nanoparticles [22, 23].

Diblock PEG–polyester (PLA, PLGA, PCL) copolymers (Fig. 13.1 a) were synthesized by ring-opening catalytic polymerization of monomers (glycolide and/or lactide, caprolactone, respectively) in the presence of monomethoxy-PEG (mPEG) [20, 24]. Triblock R–PEG–R copolymers, where R is a PLA or PCL, have been synthesized by anionic polymerization or by ring-opening polymerization of monomers initiated by the two hydroxyl groups of PEG ends [25, 26]. Multiblock PEG*n*–R*m* copolymers have also been synthesized. First, several mono-

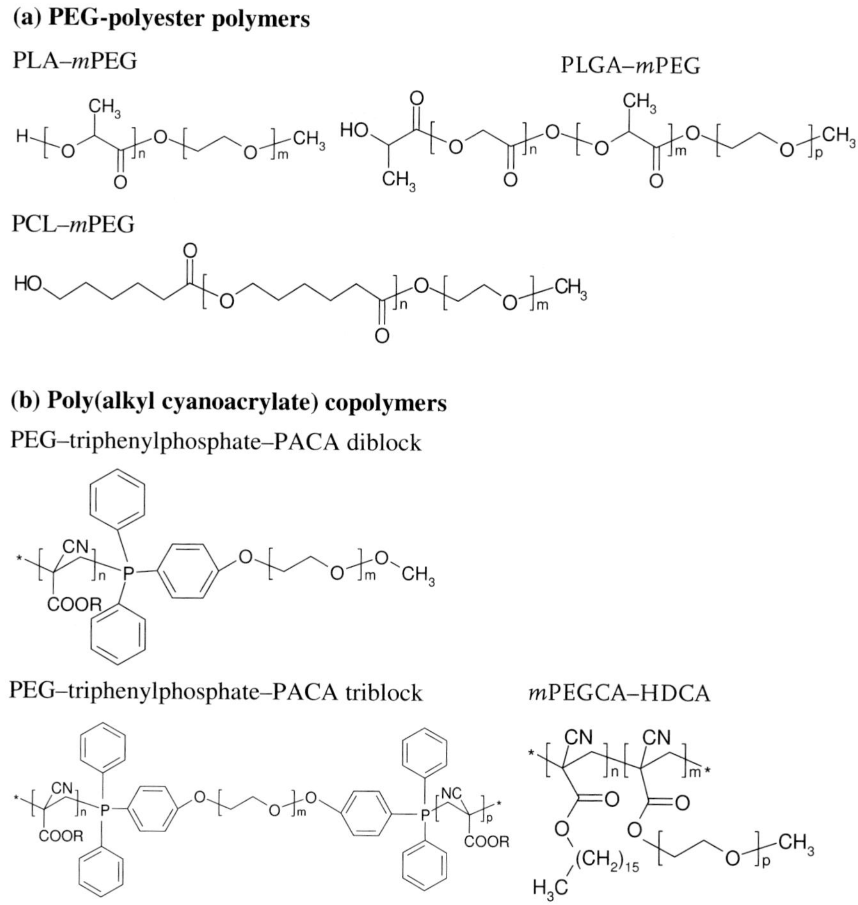

Fig. 13.1 Structures of various PEGylated polymers: (a) PEG–polyester polymers and (b) poly(alkyl cyanoacrylate) copolymers.

methoxy-monoamine-PEG (mPEG-NH_2) chains were attached together at one chain end by reaction with a polyfunctional molecule (e.g. citric, mucic or tartaric acid). The remaining hydroxyl groups were used to initiate the ring-opening polymerization of R. Branched copolymers were prepared by coupling one end-group of "star" PEG (several PEG chains coupled at one chain end) to a polyester chain [27].

Biodegradable PACA polymers were synthesized by emulsion polymerization that was initiated by the hydroxyl ions of water and elongation of the polymer chains occurred according to an anionic polymerization mechanism [28, 29]. To obtain Stealth nanoparticles, block PEG–PACA copolymers (Fig. 13.1 b) have been synthesized by initiating the polymerization of alkyl cyanoacrylates either by monomethoxy (MeO)–PEG–triphenylphosphine or by triphenylphosphine–PEG–triphenylphosphine, leading to diblock MeO–PEG–PACA or triblock

MeO–PACA–PEG–PACA–OMe linear copolymers, respectively [30]. Branched PEG–PACA copolymer is obtained by the Knoevenagel reaction of block copolymers, involving the condensation of a cyanoacetate derivative with formaldehyde. This monomethoxy or monoamino poly(PEG–cyanoacrylate-*co*-hexadecyl cyanoacrylate) (PEG–PHDCA) copolymer allows the production of very stable nanospheres and avoids the use of the triphenylphosphine group, which may be toxic [31].

Nanoparticle Formation Using PEG Block Copolymers With these copolymers, nanoparticles are usually prepared by nanoprecipitation or emulsification–solvent evaporation. Nanoprecipitation is based on the formation of colloidal polymer particles during phase separation induced by the addition of a non-solvent of the polymer to a dilute polymer solution. This method usually consists of dissolving the polymer in an organic solvent miscible with water (e.g. acetone) and mixing it with an aqueous solution, leading to the precipitation of the polymer and the formation of nanospheres spontaneously. The organic solvent is removed from the suspension by rotoevaporation. In the emulsification–solvent evaporation method, first the polymer is dissolved in a water-immiscible organic solvent such as methylene chloride or ethyl acetate and then the polymer solution is emulsified in an aqueous phase by help of a high-pressure homogenizer or a microfluidizer to produce emulsion droplets of very small size. Finally, the polymer solvent is evaporated, inducing polymer precipitation as nanospheres [32].

13.2.2 Physicochemical Characterization of Nanoparticles

The hydrodynamic diameter of nanoparticles and their size distribution are usually measured by quasi-elastic light scattering (QELS). The polymer concentration used in the organic phase is one parameter to optimize the particle size because the size decreases with decrease in polymer concentration [27]. For example, PEG–PHDCA nanoparticles displayed a mean diameter in the range 80–250 nm depending on the polymer concentration in the organic phase but also on the preparation procedure (nanoprecipitation or emulsification–solvent evaporation) and on the nature of the organic solvent used [33].

The particle size may also be reduced by increasing the concentration of surfactant or stabilizing agent used for their preparation, such as sodium cholate (NaCA), poly(vinyl alcohol) (PVA) or Pluronic F68 (polyoxyethylene–polyoxypropylene block). For example, the size of PEGylated PACA nanospheres varied from 50 to 300 nm as a function of the concentration of Pluronic F68 (1–0.5%) used as a stabilizing agent [21].

In the spite of washing the nanoparticles, traces of surfactants can remain associated with the surface of the nanospheres, hence influencing their physicochemical characteristics such as the zeta potential. For example, NaCA at 14 mM was added to reduce both the size and polydispersity index of PEG–PLA nanospheres and a small amount (<6 wt%) of this compound remained after the washing procedure.

The presence of PEG chains on the nanoparticle surface was found to change their zeta potential dramatically. Non-PEGylated PLA nanoparticles had a negative surface charge, ranging from –4 mV (when non-ionic PVA surfactants were used) to –12 mV (when anionic surfactants were employed). These negative surface charges were attributed to the presence of carboxyl groups located near the surface and of adsorbed anionic residual surfactants. On the contrary, PEGylated PLA nanoparticles had a zeta potential near zero due to the non ionic PEG chains on the particle surface, which completely masked the surface charge [27]. In the same way, PEG–PHDCA nanoparticles prepared from mPEG-*co*-HDCA copolymer also showed zeta potential values close to neutrality (–5/–10 mV), compared with non-PEGylated nanoparticles (–25/–30 mV) [33]. In general, the zeta potential increases (from negative charge to neutral charge) when the PEG density and length increase. In contrast, PMN (phagocytosis by polymorphonuclear cells) uptake decreases with increase in the PEG content of the nanoparticles [34]. Thus, the length and density of PEG chains is shown to play an important role in the surface charge and the PMN uptake of nanoparticles.

The hydrophilicity of nanoparticles, due to the PEGylation, could be estimated using hydrophobic interaction chromatography. In a column filled with propylamine–agarose gel, PEG–PHDCA nanoparticles can be eluted easily through the gel phase, whereas hydrophobic PHDCA nanoparticles are removed only by washing with TritonX-100 [33]. However, this method did not allow the separation of nanoparticles with different hydrophilic material coatings.

13.2.3
Interaction with Blood Components and Plasma Proteins

PLA nanoparticles strongly activate the complement system; however, this activation decreases with increasing PEG surface density. PEG–PLA nanoparticles possessing an estimated average density exceeding 0.2 PEG molecules per square nanometer were found to consume a very small amount of complement [35]. In the case of PEGylated poly(isobutyl cyanoacrylate) nanoparticles, it was reported that complement consumption was affected by the PEG surface density and also by the PEG configuration [36]. A PEG chain anchored by both ends to the surface ("loop" configuration, Fig. 13.2 b) is more effective in preventing opsonization than a PEG chain anchored by one end ('brush' configuration, Fig. 13.2 a).

Two-dimensional polyacrylamide gel electrophoresis (2D-PAGE) allowed the detection of the proteins competitively adsorbed from plasma to the polymeric surface. In general, the PEGylated (polyester and PACA) nanoparticles showed a reduced total amount of proteins adsorbed as compared with their conventional counterparts. The adsorption of opsonic proteins such as fibrinogen and immunoglobulin G (IgG) was also decreased because of the presence of PEG chains on the particle surface [27, 37]. Concerning apolipoproteins, the amounts

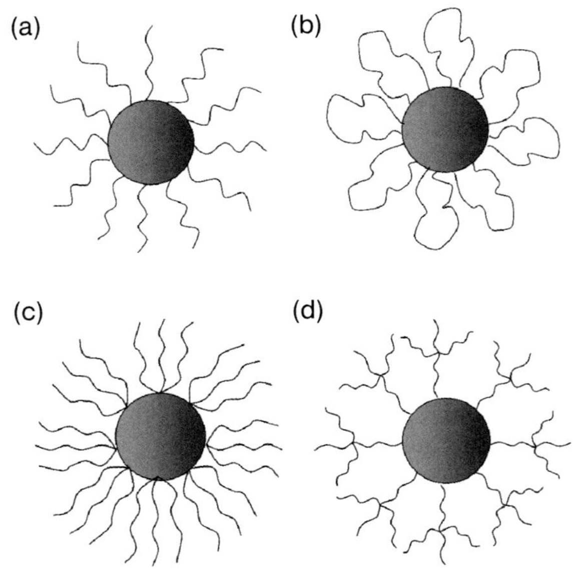

Fig. 13.2 Various configurations formed by attachment of PEG to the surface of nanoparticles. Depending on the structure of the copolymer used, the PEG chains would be at the nanoparticle surface in different configurations. The "brush" (a) would be obtained with diblock PEG–R copolymer and a "loop" (b) would be achieved with triblock R–PEG–R copolymers. R–PEG*n* copolymers can form a "bunch"-type coating (c) and a "star" PEG configuration (d) would be obtained when the star copolymers are used.

of ApoJ and C-III adsorbed were drastically reduced on all types of PEG–PLGA particles in comparison with PLGA nanoparticles. The authors hypothesized that these specific apolipoproteins might play a role in the recognition process of the particles by the MPS [38].

Recently, by means of 2D-PAGE, the influence of the length (or thickness) and density of PEG coating on PLA nanoparticles on plasma protein adsorption has been investigated to optimize the surface modification [34]. According to this study, there is a threshold in terms of PEG molecular weight between 2000 and 5000 g mol^{-1}, allowing the maximum reduction of plasma protein adsorption. When the distance (*D*) between two terminally attached PEG chains was 5.8 nm (corresponding to 0.5 wt% PEG in the blends), the overall amount of protein adsorbed was reduced by 50% compared with the PLA control particles and when *D* was decreased below 2 nm (5 wt% PEG), no further reduction in protein adsorption was observed.

According to Szleifer [39], each protein exhibits a maximum reduction in adsorption on the particle surface with a specific PEG molecular weight. For instance, in the case of ApoE on PLA nanoparticles, adsorption of ApoE was drastically reduced when the PEG molecular weight was increased from 2000 to 5000 g mol^{-1} and progressively reduced above 5000 g mol^{-1} (up to 20000 g mL^{-1}). Similar observations were made for IgG, fibrinogen and ApoJ. On the other hand, ApoA-IV and ApoC-

III were less adsorbed when PEG had a molecular weight above 5000 [34]. In conclusion, there are protein-dependent thresholds of the PEG chain length to obtain maximum reduction of protein adsorption.

PEG–PHDCA nanoparticles had not only a reduced total amount of proteins adsorbed compared with PHDCA but they also showed, in general, a lower adsorptive capacity with regard to each individual protein [37]. Although 2D-PAGE is considered as a reference tool for the study of plasma protein adsorption on nanoparticles, this methodology fails to be accurate in the case of plasma proteins which are represented by very small amounts in the blood or when the molecular weight is high. Complementary analytical methods such as capillary electrophoresis, protein Lab-on-a-Chip and Western blotting could be employed for studying protein adsorption in theses cases [40]. Thus, after incubation of nanoparticles with rat serum, Western blotting revealed that ApoE adsorbed more on PEG–PHDCA than on PHDCA nanoparticles. Also, capillary electrophoresis detected that ApoB-100 (550 kDa), a very high molecular weight protein, adsorbed on the surface of PEG–PHDCA nanoparticles. Moreover, protein Lab-on-a-Chip methodology allowed the adsorption kinetics of rat plasma proteins on nanoparticle surface to be followed [40].

In an attempt to assess the influence of the nature of the nanoparticle core on protein adsorption patterns, PEG–PLA, PEG–PLGA and PEG–PCL nanoparticles with the same high surface coating thickness were prepared by covalent coating of PEG of 5000 g mol^{-1} and analyzed by 2D-PAGE. The extent of protein adsorption differed distinctly among the three types of nanoparticles and especially concerning the adsorption of apolipoproteins and of the light chains of the immunoglobulins. The PEG–PLA particles were covered by albumin, fibrinogen, IgG, ApoA-I, A-IV, C-III and E, PEG–PLGA particles by fibrinogen, IgG, ApoA-I, A-IV and E and PEG–PCL nanoparticles by albumin, fibrinogen, IgG, ApoA-I, C-II and C-III. These results demonstrated that the nature of the core also plays a key role in the plasma protein adsorption pattern [27, 34].

13.2.4
Application of PEGylated Nanoparticles to Brain Targeting

Due to its low permeability, the blood–brain barrier (BBB) is an insuperable obstacle for a large number of drugs, such as antibiotics, antineoplastic agents, neuropeptides and other active drugs for the central nervous system. In addition to the pioneering research performed by Kreuter's group using polysorbate 80-coated poly(butyl cyanoacrylate) nanoparticles [41–43], PEGylated PACA nanoparticles have been also considered for brain delivery.

Recently, *in vivo* studies have evidenced that PEG–PHDCA nanoparticles were able to penetrate into the healthy brain in a significantly higher proportion than other colloids [44]. In addition, a biodistribution study of PEG–PHDCA nanoparticles after administration to rats bearing an experimental allergic encephalomyelitis demonstrated that the concentration of PEG–PHDCA nanoparticles in the central nervous system, especially in the white matter, was greatly increased in comparison with conventional nanoparticles. Nanoparticle translocation is, indeed, generally higher in pathological situations where the BBB permeability is augmented and/or macrophages have infiltrated the brain tissue. Passive diffusion through the permeabilized BBB and macrophage uptake in inflammatory lesions seem to be the mechanisms underlying the brain penetration of nanoparticles in these cases [45]. Further, these PEGylated nanoparticles also showed a higher uptake by the brain of scrapie-infected animals, which may be useful for targeting drugs for the treatment of prion disease [46].

However, doxorubicin (DOX)-loaded PEG–PHDCA nanospheres have shown negative preclinical results in a rat 9L gliosarcoma model. It is noteworthy that the zeta potential of nanospheres was increased by DOX loading from -24 ± 1 mV (DOX non-loaded) to $+15 \pm 2$ mV, resulting in nanosphere accumulation in the lungs. Moreover, the zeta potential of the DOX-loaded nanoparticles displayed $+15 \pm 5$ and -25 ± 6 mV before and after 30 min of incubation with rat plasma, respectively, demonstrating the electrostatic aggregation of positively charged DOX-loaded Stealth nanoparticles with negatively charged plasma proteins [47].

Drug-unloaded PEG–PHDCA nanoparticles have also been found in the brain of healthy animals [44, 48]. A pharmacokinetic study suggested that the accumulation of PEGylated nanoparticles in healthy brain occurred by diffusion/convection not only because of a prolonged blood circulation time but also because of a specific affinity of the surface of nanoparticles for the endothelial cell membrane of the BBB [48].

In order to validate the translocation of PEG–PHDCA nanoparticles in the intact BBB, an *in vitro* BBB model has been established using rat brain endothelial cells (RBEC) and astrocytes in a transwell system [49]. This *in vitro* rat BBB model retained the specific brain endothelial cell protein, P-glycoprotein and tight junction proteins such as occludin and ZO-1. Moreover, the model demonstrated high values of transendothelial electrical resistance and low permeability coefficients of marker molecules. Using this *in vitro* model, we observed an increased uptake of PEG–PHDCA nanoparticles by RBEC in comparison with PHDCA nanoparticles [49]. As the BBB consists primarily of endothelial cells forming the wall of brain microvessels and these endothelial cells specifically regulate and mainly contribute to the transport of substances into the brain [50], RBEC was further used for the study of the transport mechanism of nanoparticles across the BBB. Experiments on cell internalization and intracellular distribution of nanoparticles in RBEC have suggested that the cellular uptake of PEG–PHDCA nanoparticles was due to a specific endocytosis [51]. More recently, it has been shown that nanoparticle internalization by RBEC was inhibited by using anti-LDLR monoclonal antibody [52] and, in contrast, was enhanced by the preincubation of these nanoparticles into ApoE or ApoB-100 solutions [40]. These results suggested that ApoE and/or ApoB-100 were adsorbed on the surface of PEG–PHDCA nanoparticles after intravenous administration, which allowed the recognition by LDL receptor (LDLR) of endothelial cells and their mediated endocytosis. This opens up new perspectives to use these PEG–

PHDCA nanospheres for the treatment of brain diseases where the BBB remains intact, such as neurodegenerative diseases.

13.3
Nanogels

As biodegradable polyester and PACA nanoparticles have a hydrophobic core, the efficacy of hydrophilic drug loading is limited. Vinogradov and coworkers [53, 54] proposed a new family of carrier system, so-called "nanogels", which are hydrophilic nanosized particles consisting of a cross-linked cationic polymer network. In fact, PEG-*cl*-PEI nanogels are formed by a dispersed copolymer network in which cationic chains alternate with hydrophilic nonionic chains through cross-linking of branched PEI (25 kDa) with bis-activated PEG (8 kDa) [53]. To obtain a fine dispersed system, the cross-linking reaction was performed by an emulsification–solvent evaporation method. As PEG-*cl*-PEI nanogels exhibited combined properties of a swollen polyelectrolyte and a hydrophilic nonionic network, polyelectrolyte chains were able to be associated with oppositely charged counterions or macromolecules by electrostatic interactions, leading to the collapse of the gel, which resulted in a decreased volume and size of the particles. Because the steric stabilization of the PEG chains prevented precipitation, the collapsed nanogels formed an aqueous stable dispersion with a mean particle size of 80 nm.

Moreover, ligand-coupling strategies were performed both by direct covalent attachments at free amino groups of PEI fragments in the PEG-*cl*-PEI nanogels and by biotinylation allowing the attachment of ligands using standard biotin–avidin coupling chemistry [54, 55]. Drug loading was usually achieved through polyplex formation by self-assembly mechanisms involving noncovalent interactions (e.g. electrostatic interaction, hydrogen bond formation and hydrophobic interaction) or cleavable covalent bond (immobilization in the polymer gels) between the drug moiety and the polymer matrix [54].

The PEG-*cl*-PEI nanogels have been tested as carriers for brain delivery of oligodeoxyribonucleotides (ODN) by using polarized monolayers of bovine brain microvessel endothelial cells [56]. The studies performed in this *in vitro* model showed that ODN incorporated into the nanogels were effectively transported across the cell layer. Biodistribution experiments in a mouse model showed that the accumulation of ODN in the brain increased by over 15-fold whereas in the liver and spleen it decreased 2-fold compared with the ODN free case. No toxic effect was observed, as shown by the unmodified permeability of a paracellular transport marker [56]. This study suggested that this nanogel system could be a promising carrier candidate for the brain delivery of negative charged drugs such as ODN.

13.4
Active Targeting of PEGylated Nanoparticles

Active targeting using long-circulating PEGylated nanoparticles can still increase specific tissue accumulation thanks to molecular recognition processes. In this view, specific antibodies and ligands such as proteins which recognize receptors expressed on cancer cell membranes or endothelial cells of blood vessels generated by angiogenesis in tumors have been conjugated with long-circulating nanoparticles [57]. They might control the delivery of drugs both by prolonging drug circulation and by targeting the drug to the site of action in a specific manner. In this section, antibody- and ligand-coupled PEGylated nanoparticles will be discussed with regard to their ability to perform active targeting to the brain.

13.4.1
PEG Derivatives for Protein Attachment

The concept of antibody- or ligand-coupled particles was first applied in many liposomal-based constructions. These studies demonstrated that the antibody conjugation at the PEG terminus was more efficient in many cases of immunoliposomes for maintaining their long circulating properties. More recently, the terminus-modifying PEG derivatives of biodegradable PLA–PEG block copolymer have been synthesized to allow the antibody to be linked at the PEG terminus of nanoparticles [57–60]. Various derivatives of PEG-grafted polymer are illustrated in Fig. 13.3. Tessmar et al. [58] developed a synthesis strategy in which the methyl ether group of the polymer (MeO–PEG–PLA) is exchanged for a primary amine in order that thiol-reactive (maleinimido propionate) or amine-reactive (succinimidyl tartrate or succinimidyl succinate) linkers can be attached. Yamamoto et al. [59] reported the synthesis of amine-reactive PLA–PEG–aldehyde derivatives through acid treatment (pH 2) of PLA–PEG–acetal to constitute long-circulating polymeric immunomicelles. Olivier et al. [57] succeeded in the conjugation of thiolated OX26 Mab (transferrin receptor monoclonal antibody) with PEG–PLA by using thiol-reactive hydroxyl–PEG–maleimide bifunctional polymer.

On the other hand, indirect protein coupling between avidin-conjugated protein and nanoparticles composed of biotin-conjugated PEGylated copolymer was developed using the noncovalent biological interaction between avidin–biotin (association constant 10^{15} M^{-1}). Bond formation between biotin and avidin (or streptavidin) is very rapid and this binding is unaffected by pH, temperature, organic solvents or other denaturing agents. Moreover, this approach offers the advantage of easily functionalizing nanoparticles with a large variety of antibodies and ligands [61]. Biotinylated (Biot) PEG–PLA copolymers in which biotinylated ends were grafted at the terminus of PEG were synthesized and characterized [62] (Fig. 13.3 c). In spite of the vigorous conditions of synthesis, the binding capabilities of the biotin moieties were not altered [62]. Biotin–PEG–poly-

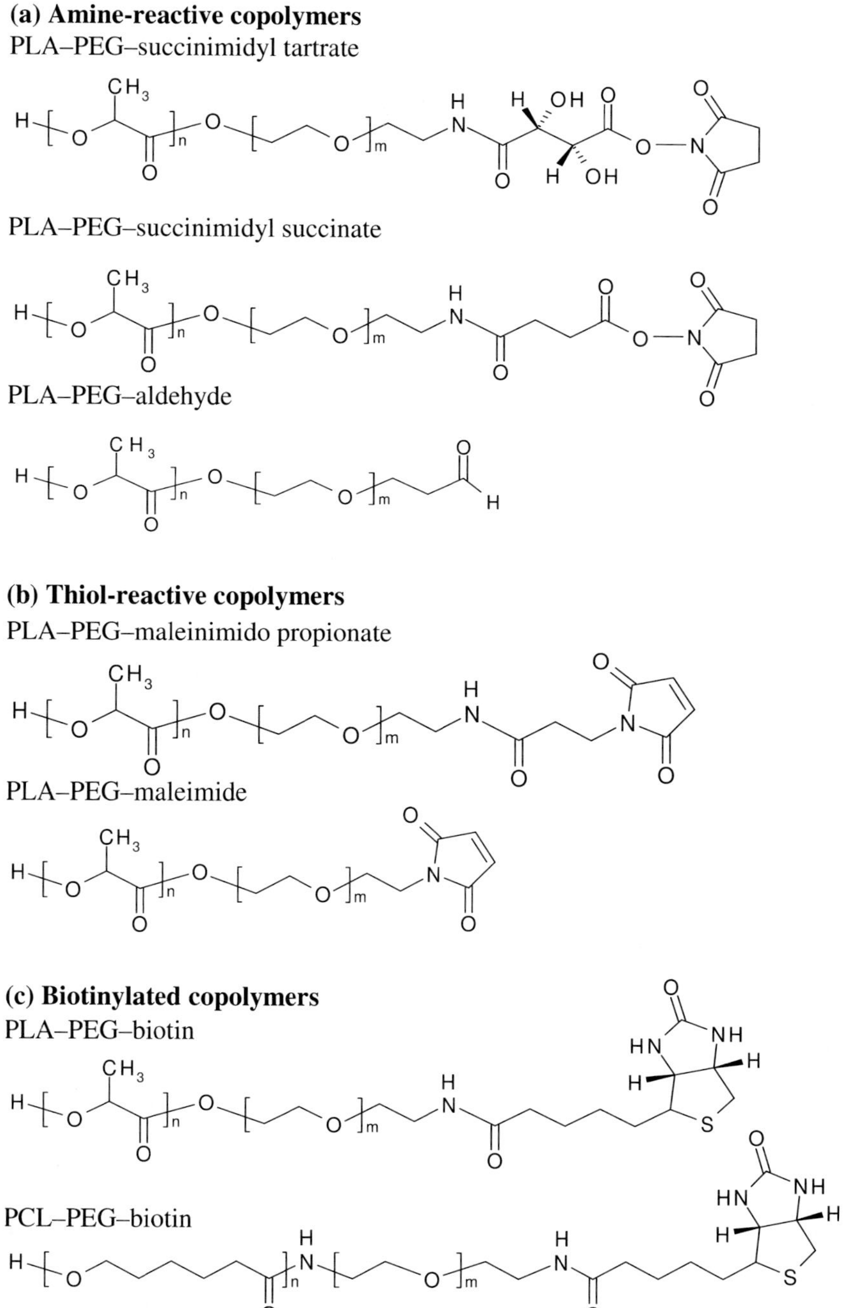

Fig. 13.3 Derivatives of PEG for protein conjugation. Direct conjugation of protein can be performed thanks to (a) amine- or (b) thiol-reactive linkers. Indirect conjugation of avidinylated protein was obtained with biotinylated PEG derivatives (c).

caprolactone (PCL) copolymers were also synthesized (Fig. 13.3c) and used to prepare nanoparticles of around 100 nm. Biotinylated wheat germ agglutinin (WGA), which can specifically recognize *N*-acetyl-D-glucosamine and *N*-acetylneuraminic acid, carbohydrates located on various cell surfaces, were used to prepare PCL–PEG–biotin–avidin–biotin–WGA nanoparticles. The PEG layer thickness and density, the biotin PEG spacer length and the core material need to be studied further to optimize this system [61].

13.4.2
Targeting of Transferrin Receptor for Brain Delivery

TfR is a 180-kDa dimeric transmembrane glycoprotein which is ubiquitously expressed in various tissues including the luminal membrane of the brain capillary endothelium and is overexpressed on tumor and leukemia cells [63, 64]. The mechanism of TfR-mediated endocytosis in cells is also well documented [65, 66]. An approach using an active targeting strategy for brain tissue is based on the use of either OX26, a TfR antibody, or transferrin itself.

13.4.2.1 OX26-conjugated PEGylated Nanoparticles

OX26 is directed against a functional group of TfR which is not the transferrin (Tf) binding site. Thanks to this characteristic, the efficacy of OX26 is not limited by the competition with endogenous Tf, which has high concentration in plasma [67, 68]. This strategy does not allow receptor-mediated endocytosis but increases the accumulation of nanoparticles at the surface of the endothelial cells.

OX26-conjugated PEG–PLA immunonanoparticles have been synthesized by using bifunctional PEG derivatives that bridged nanoparticles and OX26 [57]. The diameter of these nanoparticles measured by QELS was 125 ± 5 nm with a monodisperse size distribution. The average number of OX26 conjugated per nanoparticle was calculated to be 67 ± 4, which might be sufficient for brain targeting [57]. Most recently, chitosan (CS) nanoparticles conjugated with PEG bearing OX26 have been developed [69]. The streptavidin (SA)–OX26 and biotinylated (Biot) PEG–chitosan nanoparticles were conjugated through avidin–biotin technology and the peptide Z–DEVD–FMK, a caspase-3 inhibitor, was encapsulated in these nanoparticles. The conjugation of OX26 facilitated the controlled release of the peptide from the nanoparticles. Moreover, *in vivo* studies evidenced the efficacy of OX26 conjugation to reach the brain of healthy animals after intravenous administration [69].

Although antibody conjugation provides nanocarriers with high affinity for the corresponding antigen-expressed cells, in general, immunoparticulate (antibody-conjugated) systems have some limitations that hinder their clinical application: (1) the targeting ligand may be immunogenic, (2) the rigorous procedure of conjugation chemistry could lead to partial denaturation of the antibody and (3) rapid clearance in the blood could occur due to Fc receptor-mediated MPS uptake.

13.4.2.2 Transferrin Conjugation

As explained above, the alternative strategy is represented by coupling Tf at the surface of the nanoparticles. In this approach, which allows specific receptor-mediated endocytosis, the entrapped drug may accumulate more efficiently in the cells.

Mechanism of Transferrin-mediated Endocytosis The principal biological function of Tf is thought to be related to iron-binding properties. Serum Tf has the role of transporting ferric ion from the sites of intake into the systemic circulation to the cells and tissues [70]. Tf has been used as a target ligand to deliver anticancer drugs to primarily proliferating cancer cells that overexpress TfR [64, 70, 71].

After binding of Tf to the receptors on the cell surface, the TfR–Tf complexes are internalized to form endosomes through clathrin-coated vesicles by the process of receptor-mediated endocytosis. The iron-loaded Tf releases its iron at low endosomal pH and iron-free Tf remains bound to the receptor. Thanks to exocytic vesicles, these complexes are further released at the cell surface. The entire Tf cycle takes only 4–5 min with a mean transit time of about 10 min and avoids the degradation of TfR by the lysosomal enzymes.

Transferrin-conjugated PEGylated Nanoparticles The strategy of Tf-conjugated nanoparticles has been developed in order to mimic the Tf cycle [71–74]. The internalization mechanism by receptor-mediated endocytosis is also observed with Tf-conjugated nanoparticles. First, it was demonstrated that doxorubicin (DOX)-encapsulated Tf-conjugated liposomes have undergone receptor-mediated endocytosis in C6 glioma [72]. Another study evidenced the interference of the PEG chain with Tf recognition by TfR whereas modification of the DNA/Tf–PEI nanocomplex surface with moderate amounts of PEG did not block ligand-mediated internalization. However, a higher degree of PEGylation reduced the internalization of Tf complexes to a level similar to that of ligand-free complexes [75]. In order to avoid the interference of the PEG chain and to increase the exposure of Tf on the surface, Tf should be attached on the PEG chain terminus.

Another problem is control of the intracellular localization of the particles after receptor-mediated endocytosis has occurred. In the case of the small protein horseradish peroxidase (HRP)-loaded Tf–PEG liposomes, experiments on *in vitro* bovine brain endothelial cells suggested the intracellular degradation of endocytosed HRP in lysosome after their internalization [76].

The strategy of using Tf-coupled nanoparticles has been applied with success in tumor targeting [77, 78] and also in brain drug delivery. PEGylated albumin nanoparticles loaded with the water-soluble antiviral drug azidothymidine (AZT) were coupled with Tf by an ultra-emulsification method using chemical cross-linking by glutaraldehyde. The nanoparticle size produced was less than 150 nm and the zeta potential was $+8.52 \pm 0.65$ mV. A significant enhancement ($21.1 \pm 1.8\%$) of brain localization of AZT was observed 4 h after intravenous administration in rat compared with the non-targeted nanoparticles counterpart.

The authors interpreted the results by suggesting that multiple factors of biological events might be involved in this high accumulation, such as the shielding of MPS uptake by the PEG chain, effective receptor-mediated endocytosis and the contribution of the positively charged nanoparticle surface [79].

Nevertheless, the use of the transferrin approach for the intravenous administration of addressed nanoparticles has the main drawback that the high endogenous plasma concentration of transferrin may act as a competitor towards the targeted receptors. Additionally, the possibility of disruption of iron transport should not be overlooked. Also, after internalization by receptor-mediated endocytosis, the drug has to escape the endosomes/lysosomes compartment, a process that may depend on the method of drug encapsulation and nanoparticle composition [80].

13.5 Conclusions

In solid tumors or inflammatory sites, extensive angiogenesis and hence hypervasculature, defective vascular structure and impaired lymphatic drainage that are not observed in normal tissues or organs are generated. PEGylated drug delivery nanocarriers have been developed because the advantage of their prolonged circulation time provides the EPR effect, allowing their penetration into diseased tissues. PEGylated polymer-based nanoparticles as reviewed here have demonstrated interesting properties for the treatment of brain diseases when the BBB is intact or disrupted. Noteworthily, the translocation of PEG–PHDCA nanoparticles through the intact BBB by LDLR-mediated endocytosis opens up interesting perspectives for the treatment of neurodegenerative diseases.

Now, the concept of active targeting which is to address the diseased site by specific ligand recognition may still improve the control of biodistribution. Thus, if the Stealth effect and active targeting are combined, a synergistic effect could be expected. In this view, various PEG derivatives were synthesized for the covalent attachment of ligand protein at the terminus of PEG chains. Because of the abundant expression of TfR in endothelial cells in brain tissue, active targeting of this receptor has been investigated. OX26-conjugated PEGylated nanoparticles showed promising results in preclinical assays of brain delivery. As another approach, Tf-coupled PEGylated nanoparticles were also discussed because they were able to perform a specific receptor-mediated endocytosis. In spite of some specific drawbacks of each of these systems, the combined strategies of long-circulating and specific targeting appear promising for the design of new brain delivery carriers.

List of Abbreviations

Apo	apolipoprotein
AZT	azidothymidine
BBB	blood–brain barrier
DOX	doxorubicin
EPR	enhanced permeability and retention
FR	folate receptor
LDLR	low-density lipoprotein receptor
MPS	mononuclear phagocyte system
NaCA	sodium cholate
ODN	oligodeoxyribonucleotide
PACA	poly(alkyl cyanoacrylate)
PCL	poly(ε-caprolactone)
PEG	polyethylene glycol
PEG–PHDCA	poly(PEG–cyanoacrylate-*co*-hexadecyl cyanoacrylate)
PEI	polyethylenimine
PHDCA	poly(hexadecyl cyanoacrylate)
PLA	poly(lactic acid)
PLGA	poly(lactic-*co*-glycolic acid)
PPG	poly(propylene glycol)
PS	polystyrene
PVA	poly(vinyl alcohol)
QELS	quasi-elastic light scattering
RBEC	rat brain endothelial cell
RES	reticuloendothelial system
Tf	transferrin
TfR	transferrin receptor

References

1 Saba, T.M. (**1970**) *Arch. Intern. Med. 126,* 1031–1052.

2 Owens, D.E., III, Peppas, N.A. (**2006**) *Int. J. Pharm. 307,* 93–102.

3 Jeon, S.I., Lee, J.H., Andrade, J.D., De Gennes, P.G. (**1991**) *J. Colloid Interface Sci. 142,* 149–158.

4 Yuan, F., Dellian, M., Fukumura, D., Leunig, M., Berk, D.A., Torchilin, V.P., Jain, R.K. (**1995**) *Cancer Res. 55,* 3752–3756.

5 Hobbs, S.K., Monsky, W.L., Yuan, F., Roberts, W.G., Griffith, L., Torchilin, V.P., Jain, R.K. (**1998**) *Proc. Natl. Acad. Sci. USA 95,* 4607–4612.

6 Yuan, F., Leunig, M., Huang, S.K., Berk, D.A., Papahadjopoulos, D., Jain, R.K. (**1994**) *Cancer Res. 54,* 3352–3356.

7 Storm, G., Belliot, S.O., Daemen, T., Lasic, D.D. (**1995**) *Adv. Drug Deliv. Rev. 17,* 31–48.

8 Moghimi, S.M., Hunter, A.C., Murray, J.C. (**2001**) *Pharmacol. Rev. 53,* 283–318.

9 Seymour, L.W. (**1992**) *Crit. Rev. Ther. Drug Carrier Syst. 9,* 135–187.

10 Maeda, H., Wu, J., Sawa, T., Matsumura, Y., Hori, K. (**2000**) *J. Control. Release 65,* 271–284.

11 Noguchi, Y., Wu, J., Duncan, R., Strohalm, J., Ulbrich, K., Akaike, T., Maeda,

H. (**1998**) *Jpn. J. Cancer Res. 89*, 307–314.

12 Soppimath, K. S., Aminabhavi, T. M., Kulkarni, A. R., Rudzinski, W. E. (**2001**) *J. Control. Release 70*, 1–20.

13 Wissing, S. A., Kayser, O., Muller, R. H. (**2004**) *Adv. Drug Deliv. Rev. 56*, 1257–1272.

14 Manjunath, K., Reddy, J. S., Venkateswarlu, V. (**2005**) *Methods Find. Exp. Clin. Pharmacol. 27*, 127–144.

15 Illum, S. L., Davis, S. S. (**1983**) *J. Pharm. Sci. 72*, 1086–1089.

16 Illum, L., Jacobsen, L. O., Muller, R. H., Mak, E., Davis, S. S. (**1987**) *Biomaterials 8*, 113–117.

17 Moghimi, S. M., Muir, I. S., Illum, L., Davis, S. S., Kolb-Bachofen, V. (**1993**) *Biochim. Biophys. Acta 1179*, 157–165.

18 Stolnik, S., Dunn, S. E., Garnett, M. C., Davies, M. C., Coombes, A. G., Taylor, D. C., Irving, M. P., Purkiss, S. C., Tadros, T. F., Davis, S. S., Illum, L. (**1994**) *Pharm. Res. 11*, 1800–1808.

19 Neal, J. C., Stolnik, S., Schacht, E., Kenawy, E. R., Garnett, M. C., Davis, S. S., Illum, L. (**1998**) *J. Pharm. Sci. 87*, 1242–1248.

20 Gref, R., Minamitake, Y., Peracchia, M. T., Trubetskoy, V., Torchilin, V., Langer, R. (**1994**) *Science 263*, 1600–1603.

21 Peracchia, M. T., Fattal, E., Desmaele, D., Besnard, M., Noel, J. P., Gomis, J. M., Appel, M., d'Angelo, J., Couvreur, P. (**1999**) *J. Control. Release 60*, 121–128.

22 Layre, A., Couvreur, P., Chacun, H., Richard, J., Passirani, C., Requier, D., Benoit, J. P., Gref, R. (**2006**) *J. Control. Release 111*, 271–280.

23 Gref, R., Minamitake, Y., Peracchia, M. T., Domb, A., Trubetskoy, V., Torchilin, V., Langer, R. (**1997**) *Pharm. Biotechnol. 10*, 167–198.

24 Bazile, D., Prud'homme, C., Bassoullet, M. T., Marlard, M., Spenlehauer, G., Veillard, M. (**1995**) *J. Pharm. Sci. 84*, 493–498.

25 Cohn, D., Younes, H. (**1988**) *J. Biomed. Mater. Res. 22*, 993–1009.

26 Li, Y., Kissel, T. (**1993**) *J. Control. Release 27*, 247–257.

27 Gref, R., Domb, A., Quellec, P., Blunk, T., Muller, R. H., Verbavatz, J. M., Langer, R. (**1995**) *Adv. Drug Deliv. Rev. 16*, 215–233.

28 Couvreur, P., Kante, B., Roland, M., Guiot, P., Bauduin, P., Speiser, P. (**1979**) *J. Pharm. Pharmacol. 31*, 331–332.

29 Vauthier, C., Dubernet, C., Fattal, E., Pinto-Alphandary, H., Couvreur, P. (**2003**) *Adv. Drug Deliv. Rev. 55*, 519–548.

30 Choi, Y. K., Kim, Y. H., Kim, S. W. (**1995**) *Macromolecules 28*, 8419–8421.

31 Peracchia, M. T., Desmael, D., Couvreur, P., D'Angelo, J. (**1997**) *Macromolecules 30*, 846–851.

32 Baraton, M. I. (Ed.) (**2003**) Synthesis, functionalization and surface treatment of nanoparticles. In: *Advances in Nanoscale Materials and Nanotechnology, Book Series*, American Scientific Publishers, 302 pp.

33 Peracchia, M. T., Vauthier, C., Desmaele, D., Gulik, A., Dedieu, J. C., Demoy, M., d'Angelo, J., Couvreur, P. (**1998**) *Pharm. Res. 15*, 550–556.

34 Gref, R., Luck, M., Quellec, P., Marchand, M., Dellacherie, E., Harnisch, S., Blunk, T., Muller, R. H. (**2000**) *Collids Surf. B 18*, 301–313.

35 Vittaz, M., Bazile, D., Spenlehauer, G., Verrecchia, T., Veillard, M., Puisieux, F., Labarre, D. (**1996**) *Biomaterials 17*, 1575–1581.

36 Peracchia, M. T., Vauthier, C., Passirani, C., Couvreur, P., Labarre, D. (**1997**) *Life Sci. 61*, 749–761.

37 Peracchia, M. T., Harnisch, S., Pinto-Alphandary, H., Gulik, A., Dedieu, J. C., Desmale, D., d'Angelo, J., Muller, R. H., Couvreur, P. (**1999**) *Biomaterials 20*, 1269–1275.

38 Leroux, J. C., Gravel, P., Balant, L., Volet, B., Anner, M., Allemann, E., Doelker, E., Gurny, R. (**1994**) *J. Biomed. Mater. Res. 28*, 471–481.

39 Szleifer, I. (**1997**) *Biophys. J. 72*, 592–612.

40 (a) Kim, H. R., Andrieux, K., Delomenie, C., Chacun, H., Appel, M., Desmaële, D., Taran, F., Georgin, D., Couvreur, P., Taverna, M. (**2007**) *Electrophoresis*, in press. (b) Andrieux, K., Gil, S., Taverna, M., Chacun, H., Desmaële, D., Taran, F., Georgin, D., Couvreur, P. (**2007**) *Biomacromolecules*, in press.

41 Kreuter, J. (**2001**) *Adv. Drug Deliv. Rev. 47*, 65–81.

42 Kreuter, J., Shamenkov, D., Petrov, V., Ramge, P., Cychutek, K., Koch-Brandt, C., Alyautdin, R. (**2002**) *J. Drug Target. 10*, 317–325.

43 Ambruosi, A., Khalansky, A. S., Yamamoto, H., Gelperina, S. E., Begley, D. J., Kreuter, J. (**2006**) *J. Drug Target. 14*, 97–105.

44 Calvo, P., Gouritin, B., Chacun, H., Desmaële, D., d'Angelo, J., Noel, J. P., Georgin, D., Fattal, E., Andreux, J. P., Couvreur, P. (**2001**) *Pharm. Res. 18*, 1157–1166.

45 Calvo, P., Gouritin, B., Villarroya, H., Eclancher, F., Giannavola, C., Klein, C., Andreux, J. P., Couvreur, P. (**2002**) *Eur. J. Neurosci. 15*, 1317–1326.

46 Calvo, P., Gouritin, B., Brigger, I., Lasmezas, C., Deslys, J., Williams, A., Andreux, J. P., Dormont, D., Couvreur, P. (**2001**) *J. Neurosci. Methods 111*, 151–155.

47 Brigger, I., et al. (**2004**) *J. Control. Release 100*, 29–40.

48 Brigger, I., Morizet, J., Laudani, L., Aubert, G., Appel, M., Velasco, V., Terrier-Lacombe, M. J., Desmaële, D., d'Angelo, J., Couvreur, P., Vassal, G. (**2002**) *J. Pharmacol. Exp. Ther. 303*, 928–936.

49 Garcia-Garcia, E., Gil, S. Andrieux, K., Desmaële, D., Nicolas, V., Taran, F., Georgin, D., Andreux, J. P., Roux, F., Couvreur, P. (**2005**) *Cell. Mol. Life Sci. 62*, 1400–1408.

50 Abbott, N. J., Romero, I. A. (**1996**) *Mol. Med. Today 2*, 106–113.

51 Garcia-Garcia, E., Andrieux, K., Gil, S., Kim, H. R., Le Doan, T., Desmaële, D., d'Angelo, J., Taran, F., Georgin, D., Couvreur, P. (**2005**) *Int. J. Pharm. 298*, 310–314.

52 Kim, H. R., Gil, S., Andrieux, K., Nicolas, V., Appel, M., Chacun, H., Desmaële, D., Taran, F., Georgin, D., Couvreur, P. (**2007**) *Cell. Mol. Life Sci. 64*, 356–364.

53 Vinogradov, S., Batrakova, E., Kabanov, A. (**1999**) *Colloids Surf. B 16*, 291–304.

54 Vinogradov, S. V., Bronich, T. K., Kabanov, A. V. (**2002**) *Adv. Drug Deliv. Rev. 54*, 135–147.

55 Vinogradov, S. V., Zeman, A. D., Batrakova, E. V., Kabanov, A. V. (**2005**) *J. Control. Release 107*, 143–157.

56 Vinogradov, S. V., Batrakova, E. V., Kabanov, A. V. (**2004**) *Bioconjug. Chem. 15*, 50–60.

57 Olivier, J. C., Huertas, R., Lee, H. J., Calon, F., Pardridge, W. M. (**2002**) *Pharm. Res. 19*, 1137–1143.

58 Tessmar, J. K., Mikos, A. G., Gopferich, A. (**2002**) *Biomacromolecules 3*, 194–200.

59 Yamamoto, Y., Nagasaki, Y., Kato, Y., Sugiyama, Y., Kataoka, K. (**2001**) *J. Control. Release 77*, 27–38.

60 Olivier, J. C. (**2005**) *NeuroRx 2*, 108–119.

61 Gref, R., Couvreur, P., Barratt, G., Mysiakine, E. (**2003**) *Biomaterials 24*, 4529–4537.

62 Salem, A. K., Cannizzaro, S. M., Davies, M. C., Tendler, S. J., Roberts, C. J., Williams, P. M., Shakesheff, K. M. (**2001**) *Biomacromolecules 2*, 575–580.

63 Staber, P. B., Linkesch, W., Zauner, D., Beham-Schmid, C., Guelly, C., Schauer, S., Sill, H., Hoefler, G. (**2004**) *Oncogene 23*, 894–904.

64 Qian, Z. M., Li, H., Sun, H., Ho, K. (**2002**) *Pharmacol. Rev. 54*, 561–587.

65 Fishman, J. B., Rubin, J. B., Handrahan, J. V., Connor, J. R., Fine, R. E. (**1987**) *J. Neurosci. Res. 18*, 299–304.

66 Pardridge, W. M., Eisenberg, J., Yang, J. (**1987**) *Metabolism 36*, 892–895.

67 Seligman, P. A. (**1983**) *Prog. Hematol. 13*, 131–147.

68 Jefferies, W. A., Brandon, M. R., Hunt, S. V., Williams, A. F., Gatter, K. C., Mason, D. Y. (**1984**) *Nature 312*, 162–163.

69 Aktas, Y., Andrieux, K., Alonso, M. J., Calvo, P., Gursoy, R. N., Couvreur, P., Capan, Y. (**2005**) *Bioconjug. Chem. 16*, 1503–1511.

70 Li, H., Qian, Z. M. (**2002**) *Med. Res. Rev. 22*, 225–250.

71 Sarti, P., Ginobbi, P., D'Agostino, I., Arancia, G., Lendaro, E., Molinari, A., Ippoliti, R., Citro, G. (**1996**) *Biotechnol. Appl. Biochem. 24* (Pt. 3), 269–276.

72 Eavarone, D. A., Yu, X., Bellamkonda, R. V. (**2000**) *J. Biomed. Mater. Res. 51*, 10–14.

73 Maruyama, K., Ishida, O., Kasaoka, S., Takizawa, T., Utoguchi, N., Shinohara, A., Chiba, M., Kobayashi, H., Eriguchi, M., Yanagie, H. (**2004**) *J. Control. Release 98*, 195–207.

74 Chiu, S. J., Liu, S., Perrotti, D., Marcucci, G., Lee, R. J. (**2006**) *J. Control. Release 112*, 199–207.

75 Ogris, M., Steinlein, P., Carotta, S., Brunner, S., Wagner, E. (**2001**) *AAPS PharmSci 3*, E21.

76 Visser, C. C., Stevanovic, S., Voorwinden, L. H., van Bloois, L., Gaillard, P. J., Danhof, M., Crommelin, D. J., de Boer, A. G. (**2005**) *Eur. J. Pharm. Sci. 25*, 299–305.

77 Xu, Z., Gu, W., Huang, J., Sui, H., Zhou, Z., Yang, Y., Yan, Z., Li, Y. (**2005**) *Int. J. Pharm. 288*, 361–368.

78 Li, Y., Ogris, M., Wagner, E., Pelisek, J., Ruffer, M. (**2003**) *Int. J. Pharm. 259*, 93–101.

79 Mishra, V., Mahor, S., Rawat, A., Gupta, P. N., Dubey, P., Khatri, K., Vyas, S. P. (**2006**) *J. Drug Target. 14*, 45–53.

80 Andresen, T. L., Jensen, S. S., Jorgensen, K. (**2005**) *Prog. Lipid Res. 44*, 68–97.

Subject Index

Colloids and Interface Science Series, Vol. 3
Colloid Stability and Application in Pharmacy. Edited by Tharwat F. Tadros

ISBN: 978-3-527-31463-8

q

r

Colloids and Interface Science Series

Colloid Stability

The Role of Surface Forces, Part I

Volume 1

2007

ISBN 978-3-527-31462-1

Colloid Stability

The Role of Surface Forces, Part II

Volume 2

2007

ISBN 978-3-527-31503-1

Colloid Stability and Applications in Pharmacy

Volume 3

2007

ISBN 978-3-527-31463-8

Colloids in Cosmetics and Personal Care

Volume 4

2008

ISBN 978-3-527-31464-5

Colloids in Agrochemicals

Volume 5

2008

ISBN 978-3-527-31465-2

Colloids in Paints

Volume 6

2009

ISBN 978-3-527-31466-9